IFT | Press

The *IFT Press* series reflects the mission of the Institute of Food Technologists — to advance the science of food contributing to healthier people everywhere. Developed in partnership with Wiley, *IFT Press* books serve as leading-edge handbooks for industrial application and reference and as essential texts for academic programs. Crafted through rigorous peer review and meticulous research, *IFT Press* publications represent the latest, most significant resources available to food scientists and related agriculture professionals worldwide. Founded in 1939, the Institute of Food Technologists is a nonprofit scientific society with 18,000 individual members working in food science, food technology, and related professions in industry, academia, and government. IFT serves as a conduit for multidisciplinary science thought leadership, championing the use of sound science across the food value chain through knowledge sharing, education, and advocacy.

WILEY Blackwell

Edited by

Cristina M. Sabliov
Louisiana State University and LSU Agricultural Center, Baton Rouge, LA, USA

Hongda Chen
United States Department of Agriculture, Washington, DC, USA

Rickey Y. Yada
University of British Columbia, Vancouver, British Columbia, Canada

IFT Press

WILEY Blackwell

Registered Office
John Wiley & Sons, Ltd, The Atrium, Southern Gate, Chichester, West Sussex, PO19 8SQ, UK

Editorial Offices
9600 Garsington Road, Oxford, OX4 2DQ, UK
The Atrium, Southern Gate, Chichester, West Sussex, PO19 8SQ, UK
111 River Street, Hoboken, NJ 07030-5774, USA

For details of our global editorial offices, for customer services and for information about how to apply for permission to reuse the copyright material in this book please see our website at www.wiley.com/wiley-blackwell.

Library of Congress Cataloging-in-Publication Data

Nanotechnology and functional foods : effective delivery of bioactive ingredients / edited by Cristina M. Sabliov, Hongda Chen, Rickey Y. Yada.
pages cm. – (Institute of food technologists series)
Includes bibliographical references and index.
ISBN 978-1-118-46220-1 (hardback)
1. Food–Biotechnology. 2. Bioactive compounds–Biotechnology. 3. Functional foods. I. Sabliov, Cristina M., editor. II. Chen, Hongda, editor. III. Yada, R. Y. (Rickey Yoshio), 1954- editor.
TP248.65.F66N35 2015
664–dc23

2015000039

A catalogue record for this book is available from the British Library.

Wiley also publishes its books in a variety of electronic formats. Some content that appears in print may not be available in electronic books.

Cover image: © Thanida Chuacharoen, PhD candidate, LSU

Set in 9.5/11.5pt Times by SPi Global, Pondicherry, India
Printed and bound in Singapore by Markono Print Media Pte Ltd

1 2015

Titles in the *IFT Press* series

- *Accelerating New Food Product Design and Development* (Jacqueline H. Beckley, Elizabeth J. Topp, M. Michele Foley, J.C. Huang, and Witoon Prinyawiwatkul)
- *Advances in Dairy Ingredients* (Geoffrey W. Smithers and Mary Ann Augustin)
- *Anti-Ageing Nutrients: Evidence-based Prevention of Age-Related Diseases* (Delminda Neves)
- *Bioactive Compounds from Marine Foods: Plant and Animal Sources* (Blanca Hernández-Ledesma and Miguel Herrero)
- *Bioactive Proteins and Peptides as Functional Foods and Nutraceuticals* (Yoshinori Mine, Eunice Li-Chan, and Bo Jiang)
- *Biofilms in the Food Environment* (Hans P. Blaschek, Hua H. Wang, and Meredith E. Agle)
- *Calorimetry in Food Processing: Analysis and Design of Food Systems* (Gönül Kaletunç)
- *Coffee: Emerging Health Effects and Disease Prevention* (YiFang Chu)
- *Food Carbohydrate Chemistry* (Ronald E. Wrolstad)
- *Food Industry Design, Technology and Innovation* (Helmut Traitler, Birgit Coleman, and Karen Hofmann)
- *Food Ingredients for the Global Market* (Yao-Wen Huang and Claire L. Kruger)
- *Food Irradiation Research and Technology, second edition* (Christoper H. Sommers and Xuetong Fan)
- *Foodborne Pathogens in the Food Processing Environment: Sources, Detection and Control* (Sadhana Ravishankar, Vijay K. Juneja, and Divya Jaroni)
- *Food Oligosaccharides: Production, Analysis and Bioactivity* (F. Javier Moreno and Maria Luz Sanz)
- *Food Texture Design and Optimization* (Yadunandan Lal Dar and Joseph M. Light)
- *High Pressure Processing of Foods* (Christopher J. Doona and Florence E. Feeherry)
- *Hydrocolloids in Food Processing* (Thomas R. Laaman)
- *Improving Import Food Safety* (Wayne C. Ellefson, Lorna Zach, and Darryl Sullivan)
- *Innovative Food Processing Technologies: Advances in Multiphysics Simulation* (Kai Knoerzer, Pablo Juliano, Peter Roupas, and Cornelis Versteeg)
- *Mathematical and Statistical Methods in Food Science and Technology* (Daniel Granato and Gastón Ares)
- *Membrane Processes for Dairy Ingredient Separation* (Kang Hu, James Dickson)
- *Microbial Safety of Fresh Produce* (Xuetong Fan, Brendan A. Niemira, Christopher J. Doona, Florence E. Feeherry, and Robert B. Gravani)
- *Microbiology and Technology of Fermented Foods* (Robert W. Hutkins)
- *Multiphysics Simulation of Emerging Food Processing Technologies* (Kai Knoerzer, Pablo Juliano, Peter Roupas and Cornelis Versteeg)
- *Multivariate and Probabilistic Analyses of Sensory Science Problems* (Jean-François Meullenet, Rui Xiong, and Christopher J. Findlay)
- *Nanoscience and Nanotechnology in Food Systems* (Hongda Chen)
- *Nanotechnology and Functional Foods: Effective Delivery of Bioactive Ingredients* (Cristina M. Sabliov, Hongda Chen, and Rickey Y. Yada)
- *Natural Food Flavors and Colorants* (Mathew Attokaran)
- *Nondestructive Testing of Food Quality* (Joseph Irudayaraj and Christoph Reh)
- *Nondigestible Carbohydrates and Digestive Health* (Teresa M. Paeschke and William R. Aimutis)
- *Nonthermal Processing Technologies for Food* (Howard Q. Zhang, Gustavo V. Barbosa-Cánovas, V.M. Balasubramaniam, C. Patrick Dunne, Daniel F. Farkas, and James T.C. Yuan)
- *Nutraceuticals, Glycemic Health and Type 2 Diabetes* (Vijai K. Pasupuleti and James W. Anderson)
- *Organic Meat Production and Processing* (Steven C. Ricke, Ellen J. Van Loo, Michael G. Johnson, and Corliss A. O'Bryan)
- *Packaging for Nonthermal Processing of Food* (Jung H. Han)

- *Practical Ethics for the Food Professional: Ethics in Research, Education and the Workplace* (J. Peter Clark and Christopher Ritson)
- *Preharvest and Postharvest Food Safety: Contemporary Issues and Future Directions* (Ross C. Beier, Suresh D. Pillai, and Timothy D. Phillips, Editors; Richard L. Ziprin, Associate Editor)
- *Processing and Nutrition of Fats and Oils* (Ernesto M. Hernandez and Afaf Kamal-Eldin)
- *Processing Organic Foods for the Global Market* (Gwendolyn V. Wyard, Anne Plotto, Jessica Walden, and Kathryn Schuett)
- *Regulation of Functional Foods and Nutraceuticals: A Global Perspective* (Clare M. Hasler)
- *Resistant Starch: Sources, Applications and Health Benefits* (Yong-Cheng Shi and Clodualdo Maningat)
- *Sensory and Consumer Research in Food Product Design and Development* (Howard R. Moskowitz, Jacqueline H. Beckley, and Anna V.A. Resurreccion)
- *Sustainability in the Food Industry* (Cheryl J. Baldwin)
- *Thermal Processing of Foods: Control and Automation* (K.P. Sandeep)
- *Trait-Modified Oils in Foods* (Frank T. Orthoefer and Gary R. List)
- *Water Activity in Foods: Fundamentals and Applications* (Gustavo V. Barbosa-Cánovas, Anthony J. Fontana Jr., Shelly J. Schmidt, and Theodore P. Labuza)
- *Whey Processing, Functionality and Health Benefits* (Charles I. Onwulata and Peter J. Huth)

WILEY Blackwell

Contents

Contributors

Carlos E. Astete
Department of Biological and Agricultural Engineering
Louisiana State University and LSU Agricultural Center
Baton Rouge, Louisiana
USA

Deepak Bhopatkar
Whistler Center for Carbohydrate Research
Department of Food Science
Purdue University
West Lafayette, Indiana
USA

Jeffrey B. Blumberg
Antioxidants Research Laboratory
Jean Mayer USDA Human Nutrition Research Center on Aging
Tufts University
Boston, Massachusetts
USA

Toni Borel
Department of Biological and Agricultural Engineering
Louisiana State University and LSU Agricultural Center
Baton Rouge, Louisiana
USA

Diana Bowman
Department of Health Management and Policy and Risk Science Center
University of Michigan
Ann Arbor, Michigan
USA

Osvaldo H. Campanella
Whistler Center for Carbohydrate Research
Department of Food Science
Purdue University
West Lafayette, Indiana
USA

Laurence Castle
The Food and Environment Research Agency
Sand Hutton
York, UK

Qasim Chaudhry
The Food and Environment Research Agency
Sand Hutton
York, UK

C-Y. Oliver Chen
Antioxidants Research Laboratory
Jean Mayer USDA Human Nutrition Research Center on Aging
Tufts University
Boston, Massachusetts
USA

Hongda Chen
USDA-National Institute of Food and Agriculture
Washington, District of Columbia
USA

Huaiqiong Chen
Department of Food Science and Technology
University of Tennessee
Knoxville, Tennessee
USA

Oana Craciunescu
Department of Cellular Biology
National Institute R and D for Biological Sciences
Bucharest
Romania

Eric Andrew Decker
Department of Food Science
University of Massachusetts
Amherst, Massachusetts
USA

Stephanie R. Dungan
Department of Food Science and Technology
Department of Chemical Engineering and Materials Science
University of California
Davis, California
USA

Anna Gergely
Steptoe & Johnson LLP
Brussels
Belgium

Maria Fernanda San Martin-Gonzalez
Food Science Department
Purdue University
West Lafayette, Indiana
USA

Renuka Gupta
Department of Chemistry and Biology
Ryerson University
Toronto, Ontario
Canada

William K. Hallman
Department of Human Ecology
Rutgers University
New Brunswick, New Jersey
USA

Bruce R. Hamaker
Whistler Center for Carbohydrate Research
Department of Food Science
Purdue University
West Lafayette, Indiana
USA

Qingrong Huang
Department of Food Science
Rutgers University
New Brunswick, New Jersey
USA

Elizabeth J. Johnson
Antioxidants Research Laboratory
Jean Mayer USDA Human Nutrition Research Center on Aging
Tufts University
Boston, Massachusetts
USA

Alison Kamil
Antioxidants Research Laboratory
Jean Mayer USDA Human Nutrition Research Center on Aging
Tufts University
Boston, Massachusetts
USA

Ketinun Kittipongpittaya
Department of Food Science
University of Massachusetts
Amherst, Massachusetts
USA

David D. Kitts
Food Science, Food, Nutrition and Health
The University of British Columbia
Vancouver, British Columbia
Canada

Isao Kobayashi
Alliance for Research on North Africa (ARENA)
University of Tsukuba
and
Food Engineering Division
National Food Research Institute
NARO
Tsukuba, Ibaraki
Japan

Yunqi Li
Department of Food Science
Rutgers University
New Brunswick, New Jersey
USA

Loong-Tak Lim
Department of Food Science
University of Guelph
Guelph, Ontario
Canada

Yazheng Liu
Food Science, Food, Nutrition and Health
The University of British Columbia
Vancouver, British Columbia
Canada

David Julian McClements
Department of Food Science
University of Massachusetts
Amherst, Massachusetts
USA

Emily S. Mohn
Antioxidants Research Laboratory
Jean Mayer USDA Human Nutrition Research Center on Aging
Tufts University
Boston, Massachusetts
USA

Dorel Moldovan
Department of Mechanical and Industrial Engineering, and
Center for Computation and Technology
Louisiana State University
Baton Rouge, Louisiana
USA

Mitsutoshi Nakajima
Alliance for Research on North Africa (ARENA)
and
Faculty of Life and Environmental Sciences
University of Tsukuba
and
Food Engineering Division, National Food Research Institute, NARO
Tsukuba, Ibaraki
Japan

Marcos A. Neves
Alliance for Research on North Africa (ARENA)
and
Faculty of Life and Environmental Sciences
University of Tsukuba
and
Food Engineering Division, National Food Research Institute, NARO
Tsukuba, Ibaraki
Japan

Brian Novak
Department of Mechanical and Industrial Engineering, and
Center for Computation and Technology
Louisiana State University
Baton Rouge, Louisiana
USA

Mary L. Nucci
Department of Human Ecology
Rutgers University
New Brunswick, New Jersey
USA

Graciela Padua
Department of Food Science and Human Nutrition
University of Illinois
Urbana, Illinois
USA

Kang Pan
Department of Food Science and Technology
University of Tennessee
Knoxville, Tennessee
USA

Moumita Ray
Department of Chemistry and Biology
Ryerson University
Toronto, Ontario
Canada

Dérick Rousseau
Department of Chemistry and Biology
Ryerson University
Toronto, Ontario
Canada

Cristina M. Sabliov
Department of Biological and Agricultural Engineering
Louisiana State University and LSU Agricultural Center
Baton Rouge, Louisiana
USA

Lorena Salcedo
Department of Food Science
University of Massachusetts
Amherst, Massachusetts, USA

Paul B. Thompson
Departments of Philosophy, Community Sustainability and Agricultural, Food and Resource Economics
Michigan State University
East Lansing, Michigan
USA

Rohan V. Tikekar
Department of Nutrition and Food Science
University of Maryland
College Park, Maryland
USA

Mihaela Trif
Institute of Biochemistry of the Romanian Academy
Bucharest
Romania

Wan Wang
Department of Food Science and Technology
University of Tennessee
Knoxville, Tennessee
USA

Zheng Wang
Alliance for Research on North Africa (ARENA)
University of Tsukuba
and
Food Engineering Division
National Food Research Institute, NARO
Tsukuba, Ibaraki
Japan

Meocha Whaley
Department of Biological and Agricultural Engineering
Louisiana State University and LSU Agricultural Center
Baton Rouge, Louisiana
USA

Rickey Y. Yada
Faculty of Land and Food Systems
Department of Food Science
University of British Columbia
Vancouver, British Columbia
Canada

Genyi Zhang
Whistler Center for Carbohydrate Research
Department of Food Science
Purdue University
West Lafayette, Indiana
USA

Yue Zhang
Department of Food Science and Technology
University of Tennessee
Knoxville, Tennessee
USA

Qixin Zhong
Department of Food Science and Technology
University of Tennessee
Knoxville, Tennessee
USA

1 Introduction

Cristina M. Sabliov,[1] Hongda Chen,[2] and Rickey Y. Yada[3]

[1] *Department of Biological and Agricultural Engineering, Louisiana State University and LSU Agricultural Center, Baton Rouge, Louisiana, USA*
[2] *USDA-National Institute of Food and Agriculture, Washington, District of Columbia, USA*
[3] *Department of Food Science, University of British Columbia, Vancouver, British Columbia, Canada*

Antioxidants, polyunsaturated fatty acids, and proteins are common bioactives that can be added to food to improve its nutritional value and to prevent diseases such as cancer and heart disease for an improved overall health of the consumer. Bioactive stability, poor solubility in water, and low bioavailability are some of the challenges faced by the functional food industry interested in achieving optimum activity of the bioactives. It is generally accepted that nanoparticles offer distinct advantages for delivery of bioactives over traditional methods of delivery, such as improved stability, controlled release kinetics, and targeting of the bioactive for enhanced uptake and functionality of the bioactive. Nanodelivery systems, emulsions, solid lipid nanoparticles, polymeric nanoparticles, nanocomplexes, etc., are unique; their individual physical, chemical, and biological properties make them suitable for some specific food applications. No delivery system is superior above all others across the board. While the advantages of nanodelivery systems for food applications are supported by a wealth of data, the interaction of nanoparticles with the human body is complex and not fully understood. Due to their small size, nanoparticles have the potential to translocate to various parts of the body, raising concerns about their safety. The multitude of types of delivery systems and associated properties make safety assessment a challenging task for the researchers and regulatory agencies. Without compelling scientific data supporting safety of nanodelivery systems, their application in functional foods has no future, regardless of their proved beneficial impact on the functionality of the bioactive.

This book attempts to gather and present the latest data on all aspects of nanodelivery of bioactives ingredients to functional foods. It starts by describing the gastrointestinal (GI) tract and its function, with emphasis on uptake of macro- and micronutrients (Chapter 2). Nutrients can be effectively delivered by nanoparticles through two mechanisms: (i) the load is released from the delivery systems in the GI tract and absorbed by established bioactive-specific mechanisms; (ii) particles are absorbed intact and the load carried to the blood stream and cells, where the bioactive is released. Nanoparticle properties, composition, morphology, size, and surface properties among others play a key role in their interaction with biological systems. The effect of nanoparticle–cell interaction on bioactive uptake in the GI tract can be thoroughly understood by performing experimental studies accompanied by molecular dynamic simulations, as highlighted in Chapter 3.

Nanotechnology and Functional Foods: Effective Delivery of Bioactive Ingredients, First Edition.
Edited by Cristina M. Sabliov, Hongda Chen, and Rickey Y. Yada.

Several methods are available to synthesize nanoparticles of controlled properties out of biocompatible and biodegradable food-grade materials. Interfacial science is at the basis of nanoparticle formation, nanoparticle stability profiles, and release kinetics of the bioactive (Chapter 4). The process of emulsification is a key component of most nanoparticle synthesis methods, hence a thorough understanding of emulsion formations and ways to control emulsion size is provided in Chapter 5.

Various loadings, release properties, and nanoparticle stability profiles can be achieved by carefully selecting a synthesis method and associated parameters from the multitude of available processes (Chapter 6). More often than not, these properties are reported in the literature for newly synthesized nanoparticles. It is now understood that when particles are incorporated into the food or en-route through the GI tract, these properties are changed as a result of nanoparticle interaction with the food components or the media to which it is exposed. In general, methods for the detection of soft, nonmetallic nanoparticles incorporated into complex food matrixes are not readily available. Methods that are available for characterization of the nanoparticle itself or when suspended in a simple food medium include spectroscopic and microscopic technique as described in Chapter 7.

The most significant improvements that can be offered by nano-entrapment include enhanced stability and improved bioavailability of the bioactives. Chapter 8 provides an overview on the stability of bioactives entrapped in emulsions and stabilized emulsions, while Chapter 9 covers the stability of a particular bioactive folic acid, delivered with various polymeric encapsulants. Improved bioavailability of polyphenols delivered with polymeric nanoparticles is discussed in Chapter 10.

Organic, soft, nonmetallic nanodelivery systems designed for food applications are classified into two main groups: liquid (nanoemulsions, nanoliposomes, and nanopolymersomes) and solid (solid lipid nanoparticles, polymeric nanoparticles, nanocrystals, and complexes). A significant portion of the book (Chapters 11–18) is dedicated to different types of particles, emulsions, liposomes, solid lipid nanoparticles, polymeric nanoparticles, nanocomplexes, bi-continuous systems, and nanofibers, with an emphasis on synthesis methods, properties, and applications.

The type of nanoparticle, as well as physical and biological nanoparticle properties determine the route of clearance from the gastrointestinal system and possible toxic effects. Safety concerns stem from the potential of the nanoparticle to translocate to tissues due to their small size and the higher than physiological normal concentrations of the nanodelivered bioactive in this tissue. Involvement of scientists, risk assessors, and the broader public is necessary in addressing possible risks from nanotechnology for bioactive ingredient delivery (Chapter 19). If consumer attitude toward nanodelivery systems in foods is not addressed early, the technology has the risk of failing before reaching its potential. Consumer attitude must therefore be addressed to see the full potential of nano-enabled applications in foods (Chapter 20). In addition, safety assessment is needed to label a certain nanodelivery system safe under conditions of use. It is not surprising that with the wide-variety of nanodelivery systems and application significant roadblocks exist in assessing safety in a broad sense (Chapter 21). Regulatory agencies throughout the world are challenged to effectively regulate the risk of nano-enabled materials to be used as delivery systems for bioactives in functional foods (Chapter 22). The approaches are different in different countries and harmonization of regulations might be attempted in the future. It is hoped that with the evolving science, increasing consumer awareness, and recent developments in the regulatory field, nanotechnology can make a true and significant impact on the functional foods industry in the area of delivery of bioactives for improved consumer health.

2 Nutrient absorption in the human gastrointestinal tract

Emily S. Mohn and Elizabeth J. Johnson

Antioxidants Research Laboratory, Jean Mayer USDA Human Nutrition Research Center on Aging, Tufts University, Boston, Massachusetts, USA

2.1 INTRODUCTION

The human body possesses an intricate, but highly organized system for the digestion and absorption of nutrients. Research spanning hundreds of years has shed light on how exactly this process works, with new pieces of information still being discovered. This chapter focuses on the mechanisms by which our bodies isolate and obtain the various nutrients required for optimal health.

2.2 OVERVIEW OF THE GASTROINTESTINAL TRACT

The gastrointestinal (GI) tract is divided into different sections: the mouth, esophagus, stomach, small intestine, large intestine, rectum, and anus. The small intestine is further divided into three sections called the duodenum, jejunum, and ileum. The large intestine is made up of several parts called the ascending, transverse, descending, and sigmoid colon. Some of the sections of the GI tract are separated from one another by rings of muscles that act as valves by contracting and relaxing to control the movement of food between each part of the tract. These rings of muscle are known as sphincters and there are several located in various spots along the GI tract (Byrd-Brenner *et al.*, 2009). The upper esophageal sphincter separates the mouth and esophagus, the lower esophageal sphincter separates the esophagus and stomach, the pyloric sphincter separates the stomach and small intestine, the ileocecal valve separates the small and large intestine, and the internal and external anal sphincters control the defecation reflex of feces from the anus. All of these sphincters are involuntary muscles except for two, the upper esophageal and external anal sphincters, which are under voluntary control. In addition to each section, there are several accessory organs that work in cooperation with the GI tract to aid the digestion and absorption of food. These accessory organs include the liver, gallbladder, and pancreas (Byrd-Brenner *et al.*, 2009).

Nanotechnology and Functional Foods: Effective Delivery of Bioactive Ingredients, First Edition.
Edited by Cristina M. Sabliov, Hongda Chen, and Rickey Y. Yada.

2.3 THE GASTROINTESTINAL TRACT

The oral cavity is where food first enters the GI tract and it is considered to be the gateway to the digestive tract. The mouth consists of several different parts, including the tongue, teeth, and salivary glands. Each plays a role in either the lubrication or breakdown of food, both mechanically and chemically. Teeth begin to mechanically breakdown food into smaller pieces, which increases its surface area. This increased surface area allows for a greater amount of contact between the food and saliva (Salles *et al.*, 2011). Saliva, produced from the salivary glands, consists of mucus, lysozyme, and salivary amylase. Infants also contain an additional component in saliva called lingual lipase. The amount of saliva produced per day varies among individuals but, on average, the salivary glands can produce about 1 L of saliva in a day (Schipper *et al.*, 2007; Byrd-Brenner *et al.*, 2009). Saliva mixes with particles of food produced from chewing with help from the tongue. As food and saliva mix, lysozyme kills any bacteria and pathogens in the food, while the mucus lubricates food and holds it together. Salivary amylase begins the chemical breakdown of starches by hydrolyzing α 1–4 glycosidic bonds. Due to the limited amount of time food actually spends in the mouth, however, the salivary amylase provides minimal digestion (5%) of these carbohydrates. In infants, lingual lipase begins to chemically digest fats in the mouth. The presence of this enzyme in babies helps them digest fat found in breast milk. However, once more foods are introduced into the diet, the need to digest fat in the mouth lessens and the presence of this enzyme in saliva gradually declines (Gropper *et al.*, 2005; Byrd-Brenner *et al.*, 2009). Saliva is also essential for taste perception. That is, when eating, the food that gets dissolved in saliva is tasted because it is able to dissolve the taste-forming compounds found in foods. As the tongue mixes food and saliva, the food becomes known as bolus (Salles *et al.*, 2011).

The next step in the digestive process is moving the bolus out of the mouth and into the esophagus. This is known as swallowing. Since there are two openings in the back of the throat, the trachea and esophagus, the process requires particular coordination of the mouth and throat to prevent choking. The bolus must be able to enter the esophagus without getting into the trachea, which is the airway to the lungs. The organization of structures in the mouth and throat allows for this process to occur quite easily. When the bolus is ready to be swallowed, the tongue retracts back in the mouth toward the throat and pushes against the epiglottis, a flap of tissue, which then closes over the top of the trachea (larynx), causing breathing to stop (Salles *et al.*, 2011). At the same time, the upper esophageal sphincter relaxes, opening the esophagus and allowing the bolus to enter. Once the bolus has entered the esophagus, the upper esophageal sphincter contracts and the tongue moves back toward the mouth, releasing the epiglottis from the top of the trachea and allowing breathing to resume (Salles *et al.*, 2011).

The esophagus is a 10-inch (~25 cm) muscular tube which moves the bolus from the mouth to the stomach (Byrd-Brenner *et al.*, 2009). This is accomplished by peristalsis. Peristalsis is the coordinated movement of voluntary and involuntary muscle contractions and relaxations that push the bolus down the esophagus. As it is propelled forward, bolus is further lubricated by more mucus secreted from the esophagus. It takes approximately 10 s for the bolus to move from the top to the bottom of this section (Gropper *et al.*, 2005). Once it reaches the end, the lower esophageal sphincter relaxes and the bolus enters the stomach. The sphincter contracts after the bolus passes in order to block acidic gastric secretions from flowing into the esophagus and causing damage (Hershcovici *et al.*, 2011).

The stomach section of the GI tract serves as the main site of storage for partially digested food as well as the beginning of fat and protein digestion. No carbohydrate digestion occurs in the stomach (Whitney and Rolfes, 2011). Bolus from the esophagus enters the stomach corpus (or body), which is the holding area. This space is lined with a series of cells specific to the stomach. Each cell secretes a different substance to aid in the digestion process. Secretions from these cells are stimulated by the hormone gastrin, which is released when the bolus first enters the stomach (Schubert, 2008). The first of the stomach-specific cells are the parietal cells. These cells secrete hydrochloric acid (HCl), which serves several purposes. First, it destroys any remaining pathogens in food that cannot survive in an acidic environment. Second, it destroys the activity of proteins in the bolus and denatures them. Third, it dissolves any dietary minerals that may be present, and last, it activates the stomach enzyme pepsinogen, to its active form, pepsin (Schubert, 2009). Pepsin is a zymogen, which is an enzyme that is synthesized and stored in an inactive form in order to protect the surrounding areas of the body. These zymogens only become activated under certain conditions or are activated by other enzymes. Pepsinogen is secreted by peptic chief cells. Once activated, pepsin digests denatured proteins into smaller peptides by hydrolyzing peptide bonds. In addition to pepsinogen, these cells also secrete gastric lipase, which functions to breakdown dietary fat. Like pepsin, gastric lipase is active in the acidic environment (Gropper *et al.*, 2005). Another important cell type in the stomach is the mucus neck cell. These cells, as their name implies, secrete mucus. Again, this mucus works to lubricate the food; however, it also plays an important role in protecting the cells lining the stomach from the acidic environment created by the hydrochloric acid (Ensign *et al.*, 2012). Another important substance that is secreted in the stomach is intrinsic factor (IF). This protein is secreted by the parietal cells and is very important for the absorption of vitamin B12 (Byrd-Brenner *et al.*, 2009). Upon secretion, it binds to the vitamin and forms a complex that will be described later in the chapter.

Around the stomach body there is a complex network of muscles. This network consists of oblique, circular, and longitudinal muscles that wrap around the stomach in every direction. The coordinated contraction and relaxation of these muscles squeezes and relaxes the stomach body, which causes all of the secretions to mix well with the bolus (Kong and Singh, 2010). This provides enzymes with adequate exposure to the appropriate nutrients to cleave them. This is especially important for gastric lipase, since dietary fat separates out from the rest of the fluids because it is too hydrophobic to dissolve in the acid, and forms a layer that rests on top of the aqueous HCl layer. The contractions of the muscles around the stomach allow for the emulsification of the fat so that gastric lipase can make contact with the fat and hydrolyze bonds (Gropper *et al.*, 2005). Upon further digestion in the stomach, the food is now referred to as chyme.

Flow of chyme from the stomach to the small intestine is controlled by stomach contractions and the pyloric sphincter (Janssen *et al.*, 2011). The alternating relaxation and contraction of the stomach body and sphincter causes chyme to be ejected out of the stomach in small doses. The release of chyme in small doses is carried out in order to allow the small intestine to adequately neutralize the highly acidic chyme so that the small intestine is not damaged (Byrd-Brenner *et al.*, 2009). The speed at which gastric emptying occurs depends on the composition of the meal consumed and caloric content. For example, a high caloric meal containing large amounts of fat will empty out of the stomach more slowly, while less energy dense meals leave the stomach more quickly (Janssen *et al.*, 2011).

As the chyme leaves the stomach, it enters the first section of the small intestine known as the duodenum. At the beginning of the duodenum, the liver, pancreas, and gallbladder begin to aid in the digestion process. These accessory organs are connected to the duodenum through the common bile duct and the pancreatic duct. These ducts converge at the sphincter of Oddi, which controls the release of secretions from the accessory organs into the intestine. The main function of the liver is the production of bile. Bile, synthesized from cholesterol, is essential for the digestion of fat and the absorption of fat and fat-soluble vitamins. It works to emulsify fat and form micelles in order to increase their surface area so that the appropriate enzymes, which are soluble in aqueous solution, can make proper contact with the dietary fat. The liver synthesizes between 0.4 and 0.8 L of fresh bile everyday (Maldonado-Valderrama *et al.*, 2011). Bile from the liver travels through the common bile duct and is released into the duodenum through the sphincter of Oddi. Once bile has delivered fat and fat-soluble vitamins to intestinal absorptive cells, some of it is reabsorbed in the ileum and eventually returns to the liver for reuse. This process, called enterohepatic circulation, can occur several more times for the same batch of bile. Bile that is synthesized in the liver, but not used, is stored in the gallbladder (Hofmann, 2009). Bile stored in the gallbladder tends to be more concentrated since it sits in the body for longer. Bile from this organ is also secreted through the common bile duct to the sphincter of Oddi for release.

The main function of the pancreas is the release of bicarbonate and pancreatic enzymes, which is collectively known as the pancreatic juice. Bicarbonate is released into the duodenum in order to neutralize chyme from the stomach, which has a pH ranging from 1 to 2 (Byrd-Brenner *et al.*, 2009; Whitney and Rolfes, 2011). Not only does this protect the small intestine from damage, but it also inactivates gastric enzymes and creates a pH environment suitable for the pancreatic and intestinal enzymes. The amount of bicarbonate released is dependent on the quantity and acidity of the chyme. The pancreas also secretes pancreatic amylase and lipase in order to begin the digestion of carbohydrates and continue the digestion of fats. Proteases, which cleave peptides, are secreted from this organ as well (Pezzilli, 2009). Similar to pepsin, these enzymes are zymogens, and are stored and secreted in their inactive form. Examples of these proteases include trypsinogen, chymotrypsinogen, and procarboxypeptidase. Once in the duodenum, trypsinogen is converted to trypsin, its active form, by an enzyme called enteropeptidase. Free trypsin can then cleave trypsinogen to the active form as well. Chymotrypsin and procarboxypeptidase are also converted to their active forms, chymotrypsin and carboxypeptidase, by trypsin (Goodman, 2010). The targets of these enzymes will be discussed in the protein section.

Secretions from the liver, gallbladder, and pancreas into the duodenum are regulated by the hormones secretin and cholecystokinin (CCK). These hormones are released upon chyme leaving the stomach and entering the small intestine. Both stimulate pancreatic enzyme and bile secretion. The CCK stimulates contraction of the gallbladder to release stored bile while secretin stimulates the secretion of bicarbonate from the pancreas as well as the synthesis of bile in the liver. The amount of secretin and CCK that is secreted depends on the fat content and acidity of the chyme. If the chyme is highly acidic and high in fat there will be an increased amount of both secretin and CCK secreted. If the chyme is acidic but not high in fat, more secretin will be present than CCK and vice versa. After the chyme has left the stomach and digestion progresses, gastric inhibitory peptide is secreted. This hormone causes a decrease in the release of gastric juice and signals the end of digestion of the meal (Table 2.1) (Byrd-Brenner *et al.*, 2009).

The small intestine is made up of small finger-like structures, known as villi and microvilli, which significantly increase the surface area of the intestine. Microvilli are located on each villus, creating what is known as the "brush border." This forces the chyme to move slowly through the intestine, allowing it to mix with enzymes, bile, and bicarbonate, and giving it optimal contact for absorption (Byrd-Brenner *et al.*, 2009). A process called segmentation or, segmental mixing, occurs at this stage. Segmentation is the process in which circular muscles in the small intestine alternate contracting and relaxing in order to divide and mix the chyme (Whitney and Rolfes, 2011). This process increases the surface area of the chyme and facilitates its contact with the intestinal wall for absorption. The base of each villus, known as the crypt, is the location where the absorptive cells, or enterocytes, of the intestine originate from. These cells constantly undergo mitosis and slowly migrate toward the top of the villus. Once they reach the very top, they are usually less efficient at absorbing nutrients due to damage from digestive enzymes, and are sloughed off and excreted in the feces. The turnover rate of these cells is normally between 4 and 5 days. Goblet cells, which secrete mucus for lubrication, and endocrine cells that secrete hormones undergo this same turnover process (Vereecke *et al.*, 2011).

As mentioned earlier, the small intestine is divided into three separate parts: the duodenum, jejunum, and ileum. The duodenum and jejunum, which are the primary sites of macronutrient digestion, secrete many enzymes and hormones, including secretin and CCK. Nutrient and water absorption also occur at this stage, as well as in the ileum (Gropper *et al.*, 2005). Nutrients are mainly absorbed into enterocytes via four different mechanisms: passive diffusion, facilitated diffusion, active transport, and endocytosis. Passive diffusion occurs when the concentration of a nutrient is higher in the lumen than in the enterocytes. This creates a concentration gradient and nutrients can pass down the gradient into cells. Facilitated diffusion is necessary when nutrients cannot pass through cells on their own, even if a concentration gradient is present. Nutrients like this need a carrier protein to aid in the transport of the nutrient across the membrane. Active transport is necessary

Table 2.1 Major digestive secretions in the gastrointestinal tract.

Location	Secretion	Role in digestion	Result
Mouth	Saliva	Contains: mucus, lysozyme, amylase. Lubricates food, kills bacteria, begins starch breakdown	Food becomes bolus
Esophagus	Mucus	Lubricates food	Bolus moves down esophagus via peristalsis
Stomach	Gastrin	Hormone released when bolus enters stomach. Stimulates secretion of HCl and pepsinogen	Stomach secretions act on bolus Digestion in stomach begins
	Gastric Juice	Contains: HCl (denature proteins, kills pathogens, dissolves minerals, activates pepsinogen), pepsin (protein digestion) and gastric lipase (lipid digestion)	Bolus become chyme
	Mucus	Protects stomach cells from highly acidic environment. Lubricates food	No damage to stomach lining
	Intrinsic Factor	Binds to vitamin B_{12} to aid absorption in small intestine	B_{12}-IF complex ready for absorption in ileum
Liver and gallbladder	Bile	Emulsifies fat to aid in digestion and absorption of fats and fat-soluble vitamins. Bile produced in liver and stored in gallbladder	Micelles
Pancreas	Enzymes	Proteases (trypsin, chymotrypsin, carboxypeptidase) breakdown proteins into short peptides and amino acids, pancreatic lipase breaks down fats, pancreatic amylase breaks down starches.	Macronutrients are ready to be absorbed through luminal wall of intestine
	Bicarbonate	Neutralizes acidic chyme from stomach. Activates pancreatic and intestinal enzymes	Increased pH protects small intestine
Small intestine	Secretin and Cholecystokinin	Hormones stimulated by chyme entering duodenum. Lead to increased release of bile, enzymes, and bicarbonate	Digestion of macronutrients in small intestine begins
	Brush Border Enzymes	Example: Lactase, sucrase, maltase. Breakdown disaccharides into monosaccharides	Monosaccharides ready for absorption
	Mucus	Lubrication	
	Gastric Inhibitory Peptide	Hormone released as digestion occurs. Stimulates decrease in secretion of gastric juice and gastric motility	Digestion for meal slows down

IF, intrinsic factor.

when nutrients must be transported against their concentration gradient. That is, when the concentration of the nutrient is higher in enterocytes than it is in the lumen, energy in the form of adenosine triphosphate (ATP) is required to pump nutrients into cells. Endocytosis occurs when absorptive cells engulf nutrients by forming an invagination in the membrane that circles around the compound, forming a vesicle. This vesicle is then pinched off from the membrane and is transported into the cell (Byrd-Brenner *et al.*, 2009).

Nutrients and foods not absorbed in the small intestine move through the ileocecal sphincter into the large intestine, which is divided up into the cecum, ascending, transverse, descending, and sigmoid colon, and rectum. Most nutrients have been absorbed by the time the food contents reach the large intestine. Therefore, the major components that are absorbed at this stage are water and electrolytes, and they are absorbed within the first half of the colon. The large intestine usually absorbs the last 10% of water and remaining electrolytes that were not absorbed in the small intestine. The rest of the contents not absorbed consist of indigestible carbohydrates, bile, bacteria, and dead cells sloughed off from the small intestine (Byrd-Brenner *et al.*, 2009). As water is continually removed, the remaining contents harden and form feces. Peristalsis moves this through the remaining third of the large intestine and stores it in the rectum. The build up of feces in the rectum stimulates muscle contractions known as the defecation reflex. The involuntary internal anal sphincter and voluntary external anal sphincter relax, which leads to the expulsion of feces from the body (Bajwa *et al.*, 2009).

While absorption is an important role of the large intestine, it is not the only function. It also serves as the location for bacterial flora, which is important in the digestion and absorption process. Some of its functions include breaking down complex carbohydrates that cannot be digested in the small intestine. This process, known as fermentation, leads to the production of short chain fatty acids that are used as energy in the large intestine (O'Keefe, 2008). The digestion of fiber can also lead to the production of gas. In addition to aiding the digestive process, it has also been shown that microflora can synthesize vitamin K and biotin (O'Keefe, 2008). The extent of the bioavailability of these nutrients is still unclear, however, and more research is being undertaken in order to learn more about the benefits of this function. Due to the current interest in intestinal microflora, there is a rising interest in the study of prebiotics, which are nondigestible carbohydrates that promote the growth of good bacteria in the large intestine (De Preter *et al.*, 2011). These carbohydrates, such as inulin, can be found in foods including asparagus and bananas (Moshfegh *et al.*, 1999). Resistant starches, as their name implies, resist digestion in the small intestine, and are also considered to be prebiotics. These resistant starches undergo fermentation in the large intestine and lead to the production of small chain fatty acids. More research is being carried out in order to look at the possible benefits of these prebiotics (De Preter *et al.*, 2011).

2.4 MACRONUTRIENTS

The human body uses three main nutrients for energy: carbohydrates, fat, and protein. Together, they are known as macronutrients.

2.4.1 Carbohydrates

Carbohydrates are made up of carbon, hydrogen, and oxygen and have the general formula: $(CH_2O)_n$. The smallest unit of a carbohydrate is a monosaccharide, which consists of three to seven carbons atoms, hydroxyl groups, and a carbonyl group (Gropper *et al.*, 2005). Monosaccharides do not need to be digested any further for absorption into the body. Examples of monosaccharides include glucose, fructose, and galactose. Disaccharides are two monosaccharides linked together by a glycosidic bond, which forms between hydroxyl groups. Examples of disaccharides include sucrose (glucose and fructose), lactose (two galactose), and maltose (two glucose) (Goodman, 2010). Together, mono- and disaccharides are considered to be simple sugars. Carbohydrates consisting of more than two monosaccharides, on the other hand, are known as oligosaccharides (3–10) or polysaccharides (>10) and are classified as complex carbohydrates. Examples of complex carbohydrates include starch, glycogen, and fiber (Byrd-Brenner *et al.*, 2009). Carbohydrates can be classified further into digestible or indigestible carbohydrates. Simple sugars and starches are considered to be digestible carbohydrates because the body possesses the necessary enzymes to break down these sugars into monosaccharides. Specifically, these carbohydrates are linked together by α 1–4 glycosidic bonds, which can be cleaved by amylase in the body. Fiber, on the other hand, an indigestible carbohydrate, is made up of monosaccharides linked together by β 1–4 glycosidic bonds. The stereochemistry of this bond prevents amylase from cleaving it (Gropper *et al.*, 2005). Since it cannot be broken down any further in the

Table 2.2 Food sources of carbohydrates.

Carbohydrate	Example	Food sources
Digestible		
Monosaccharides	Glucose,[1] fructose, galactose[1]	Fruits, honey, high fructose corn syrup
Disaccharides	Maltose, sucrose, lactose	Maltose: alcoholic beverages, sprouted seeds. Sucrose: table sugar, sweets. Lactose: dairy
Complex	Starch	Legumes, bread, cereal, pasta
Indigestible		
Complex	Fiber	Legumes, celery, whole grains, vegetables, fruits, oat bran

[1]Rarely eaten as monosaccharides. Obtained from breakdown of disaccharides and polysaccharides.

intestine, fiber is not absorbable there either. Digestible carbohydrates function as a major energy source, and allow the body to conserve protein, which has numerous essential functions. Indigestible carbohydrates, such as fiber, enhance intestinal health by providing fuel for bacteria. Fiber has also been shown to lower blood glucose and reduce cholesterol absorption and bile reabsorption, which may help lower the risk for type II diabetes and cardiovascular disease (Ye *et al.*, 2012).

Monosaccharides are not usually obtained directly from food, but rather, are obtained from polysaccharides that are broken down into monosaccharides in the body. An exception to this is fructose, which can be obtained directly from foods such as fruit and honey (Gropper *et al.*, 2005). Sucrose is obtained from table sugar and sweets, while lactose is obtained from milk and dairy products (United States Department of Agriculture, 2011). Maltose is mostly obtained from digestion of polysaccharides but can be found in alcoholic beverages. The complex carbohydrate, starch, can be obtained from eating breads, pasta, and legumes. Fiber can be found in whole grains, vegetables, legumes, fruits, and oat bran (United States Department of Agriculture, 2011) (Table 2.2).

Digestion for carbohydrates begins in the mouth where salivary amylase begins breaking down starch. Food is in the mouth for such a short period of time, however, that very little digestion actually occurs. Starches that were not digested in the mouth are cleaved by pancreatic amylase in the small intestine to yield disaccharides. These disaccharides are then further digested by brush border enzymes, which break glycosidic bonds to produce monosaccharides. These brush border enzymes include maltase, sucrase, and lactase, which cleave maltose, sucrose, and lactose, respectively (Goodman, 2010). Once carbohydrates are broken down into monosaccharides, they are ready for absorption. Glucose and galactose are absorbed into cells via carrier-dependent active transport requiring sodium. The carrier protein is known as sodium-glucose transporter 1 (Goodman, 2010). Neither glucose nor galactose can bind to the carrier protein until sodium has bound. This transporter is dependent on the sodium–potassium pump in order to provide ATP energy for the process. Unlike glucose and galactose, fructose can enter intestinal cells by facilitated diffusion through the GLUT5 transporter (Goodman, 2010). Once inside the cell, fructose can be converted to glucose. Glucose, galactose, and any remaining fructose then exit the cell via facilitated diffusion and enter the bloodstream. Sodium is also actively pumped out of the cell via the Na^+/K^+ pump in order to maintain a concentration gradient (Goodman, 2010).

2.4.2 Fats

Another macronutrient utilized for energy in the body is fat. Similar to carbohydrates, fats contain the elements carbon, hydrogen, and oxygen in their structure. Unlike carbohydrates, however, fats are very hydrophobic and not soluble in water. Fats, also known as lipids, all consist of hydrocarbon chains of various lengths called fatty acids, as well as structural groups that classify lipids into different groups. The most common lipids found in the diet and the human body are triglycerides. Triglycerides contain three fatty acid chains that are bonded to a glycerol through its hydroxyl groups.

Table 2.3 Food sources of fatty acids.

Fat	Example	Food Sources
Saturated fatty acids		
Long chain	Stearic acid	Fat from meats, lard
Short chain	Decanoic acid	Coconut oil, palm oil
Unsaturated fatty acids		
Monounsaturated	Oleic acid	Olive oil, canola oil
Polyunsaturated	Omega 3 (alpha-linoleic acid)	Canola oil, fish, walnuts
	Omega 6 (linoleic acid)	Beef, poultry, sunflower oil
Trans	Elaidic acid	Partially hydrogenated products, margarine, fried foods

This glycerol is known as the backbone of the molecule (Innis, 2011). Triglycerides can be broken down into diglycerides, which have two fatty acid chains, or monoglycerides, which have one fatty acid chain (Ramirez *et al.*, 2001). Another form of lipid commonly found in the body and the diet are phospholipids. These lipids contain two fatty acid chains and a phosphate group attached to a glycerol. Because of the phosphate group, phospholipids are amphipathic, meaning they have both hydrophobic and hydrophilic regions in their structure (Küllenberg, 2012). A third kind of lipid seen in the body are sterols, the most common being cholesterol. Sterols form complex multiple ring structures consisting of four rings and an alcohol attached to the five-member ring (Gropper *et al.*, 2005).

The fatty acid component of each lipid described can vary in both chain length and degree of saturation (number of double bonds). Fatty acids that contain no double bonds are saturated, while those with double bonds are unsaturated. Unsaturated fatty acids can be further classified as either mono- or polyunsaturated depending on whether there are multiple double bonds in the chain or not. Double bonds are also classified as either *cis* or *trans* depending on whether the hydrogen atoms attached to the double-bonded carbons are on the same side (*cis*) or opposite side (*trans*). *Trans* fatty acids have straight chains similar to that of saturated fatty acids, while unsaturated fatty acids have bent chains (Byrd-Brenner *et al.*, 2009). The human body is able to synthesize most fatty acids except for those with double bonds before the ninth carbon. These types of fatty acids are known as essential fatty acids because the only way we can obtain them is from our diet. Examples of these essential fatty acids are alpha linolenic, an omega-3 fatty acid, and linoleic acid, an omega-6 fatty acid (Byrd-Brenner *et al.*, 2009).

In addition to their use as an energy source, lipids have many functions in the body. They protect and insulate the body as well as aid in the absorption of fat-soluble vitamins. Furthermore, lipids, especially phospholipids, are important for the composition of cell membranes by keeping them fluid. Phospholipids also play an important role as emulsifiers in the digestive process (Küllenberg, 2012). Cholesterol is used in the synthesis of steroid hormones and is also a component of bile (Byrd-Brenner *et al.*, 2009). Most people obtain saturated fatty acids from animal fat and lard. Saturated fatty acids with shorter chains are also found in some oils, such as palm oil. Unsaturated fatty acids are generally obtained from oils, since they have lower melting points than saturated fatty acids. They are found most commonly in olive oil and canola oil (United States Department of Agriculture, 2011). Essential fatty acids can also be found in olive oil as well as fatty fish, such as salmon, and walnuts. *Trans* fats are found most often in processed foods, where unsaturated fatty acids undergo partial hydrogenation, yielding *trans* fats. *Trans* fats are also present in fried foods and some margarines, although most companies now use nonhydrogenated fats in margarine production, causing it to have very little or no *trans* fat content (Table 2.3) (Hunter, 2005; Byrd-Brenner *et al.*, 2009).

By the time fats have reached the small intestine, they have undergone minimal digestion in the mouth (lingual lipase) and more substantial digestion in the stomach, with gastric lipase hydrolyzing triglycerides into mono- and diglycerides as well as free fatty acids, however, the majority of fat digestion occurs in the small intestine (Goodman, 2010). As mentioned earlier, however, due to the hydrophobic nature of lipids, it is difficult for water-soluble pancreatic lipase to make contact with bonds to cleave them. Secretions of bile from the liver and gallbladder fix this problem by emulsifying fat and forming micelles, allowing pancreatic lipase appropriate contact to break down triglycerides into free fatty acids and monoglycerides. Pancreatic lipase cannot function efficiently, however, without colipase, which is activated by trypsin. Colipase functions to help pancreatic lipase establish optimal contact with micelles so that it can adequately cleave triglycerides. Phospholipids are broken down by phospholipase, yielding free fatty acids, choline, glycerol, and phosphoric acid (Goodman, 2010). Once fats are broken down into their main components, they are ready for absorption into enterocytes in the duodenum and jejunum. Fatty acids, glycerol, monoglycerides, and other components are absorbed into cells from micelles via passive diffusion, although recent evidence indicates that protein-dependent mechanisms may also be involved (Mansbach II and Gorelick, 2007). Once inside the enterocytes, short and medium chain fatty acids diffuse into the portal vein for transport to the liver. Other fatty acids and monoglycerides are rebuilt into trigylcerides, packaged with phospholipids and cholesterol to form chylomicrons, and released into the lymphatic system (Mansbach and Gorelick, 2007; Goodman, 2010).

2.4.3 Proteins

The final main macronutrient used for energy is protein. Just like carbohydrates and fats, proteins are made up of carbon, hydrogen, and oxygen. Unlike the other two macronutrients, however, nitrogen is also a main component of protein. The building block structure of protein is the amino acid, which has the general structure: $NH_2RCHCOOH$, with R representing different side chains that distinguish one amino acid from another. Amino acids form peptide bonds with one another to form proteins, which are then folded to their appropriate shape (Byrd-Brenner *et al.*, 2009). All together, there are 20 amino acids that are necessary to build different proteins essential for functions in the body. Some of the amino acids, called nonessential amino acids, can be synthesized in the body, while others, named essential amino acids, must be obtained from the diet (Table 2.4). This means that protein synthesis in

Table 2.4 List of amino acids.

Name	Symbol	Classification
Alanine	Ala	Nonessential
Arginine	Arg	Nonessential
Asparagine	Asn	Nonessential
Aspartic acid	Asp	Nonessential
Cysteine	Cys	Nonessential
Glutamic acid	Glu	Nonessential
Glutamine	Gln	Nonessential
Glycine	Gly	Nonessential
Histidine	His	Essential
Isoleucine	Ile	Essential
Leucine	Leu	Essential
Lysine	Lys	Essential
Methionine	Met	Essential
Phenylalanine	Phe	Essential
Proline	Pro	Nonessential
Serine	Ser	Nonessential
Threonine	Thr	Essential
Tryptophan	Trp	Essential
Tyrosine	Tyr	Nonessential
Valine	Val	Essential

Table 2.5 Food sources of proteins.

Protein	Food sources	Amino acids
Complete	Meat, poultry, fish, eggs, milk, quinoa	His, Ile, Leu, Lys, Met, Phe, Thr, Trp, Val
Incomplete	Beans	His, Ile, Leu, Lys, Phe, Thr, Val **Limiting: Met, Trp*
	Grains and nuts	His, Ile, Leu, Phe, Thr, Val, Trp, Met **Limiting: Lys*

the body is limited by our ability to obtain the nine essential amino acids from the foods we eat. Therefore, protein consumed in food is classified as either complete or incomplete. Complete proteins contain all of the essential amino acids, while incomplete proteins are missing one or more essential amino acid (Byrd-Brenner *et al.*, 2009). Generally, complete proteins are obtained from meat, poultry, and animal products, while incomplete proteins are obtained from plant sources, such as legumes and grains (United States Department of Agriculture, 2011). However, there are some exceptions to this rule. Spirulina and quinoa are plant sources that have complete proteins (Table 2.5) (Minsky, 2006).

The function of protein as an energy source is only one of the many roles this macronutrient plays in the body. Proteins, such as collagen, provide structural support for body cells, while the transport proteins such as albumin aid in maintaining fluid balance in cells, while also transporting nutrients into cells from the bloodstream. Furthermore, numerous proteins are required for the synthesis of various hormones, enzymes, and neurotransmitters (Byrd-Brenner *et al.*, 2009).

Digestion of protein begins in the stomach when HCl and pepsin are secreted. As mentioned earlier in the chapter, HCl denatures proteins and pepsin begins cleaving peptide bonds to form smaller polypeptide chains. When the polypeptides reach the small intestine, the pancreatic enzymes trypsin, chymotrypsin, and carboxypeptidase are activated and continue cleaving peptide bonds to break down polypeptides into shorter chains and free amino acids. Trypsin cleaves after the basic amino acids, lysine and arginine, chymotrypsin cleaves after the aromatic residues, phenylalanine, tyrosine, and tryptophan, while carboxypeptidase cleaves the carboxy terminal amino acids (Goodman, 2010). Once proteins are broken down into amino acids, they are ready to be absorbed. The absorption of most amino acids into enterocytes in the small intestine requires sodium-dependent active transport. In this process, sodium first binds to a carrier protein, increasing the affinity of the carrier protein to the amino acid. Binding of the amino acid to the carrier induces a conformational change, allowing the amino acid and sodium to enter the cell. Eventually, the sodium ion is pumped back out of the cell via the Na^+/K^+ pump, and amino acids are released into the portal vein usually via facilitated diffusion at the basolateral membrane (Goodman, 2010).

2.5 ALCOHOL

While alcohol is not essential for humans, it is considered to be a major dietary component, since it is consumed frequently. Alcohol, also known as ethanol, is made up of carbon, hydrogen, and oxygen, similar to carbohydrates and fats, with the structural formula: CH_3CH_2OH. Alcohol is also similar to these nutrients in that the body uses it as an energy source. Humans obtain alcohol through the consumption of alcoholic beverages, such as beer, wine, and liquor. Alcohol is produced in these beverages through the fermentation of carbohydrates by yeast. For example, rum is made from sugar, beer made from wheat, and vodka from potatoes (Byrd-Brenner *et al.*, 2009).

Alcohol requires no digestion and is readily absorbed throughout the GI tract, particularly in both the stomach and small intestine, via simple diffusion (Seitz and Mueller, 2012). The speed by which alcohol is absorbed depends on food content in the stomach. Once absorbed, alcohol enters the bloodstream and disperses into cells, where it can damage cell membranes. Alcohol must be broken down and used for energy immediately because it cannot be stored. The main site of alcohol metabolism is the liver, though other tissues can break it down, as well. Here, alcohol is metabolized through the alcohol dehydrogenase (ADH) pathway. The enzyme, ADH, oxidizes alcohol to form acetaldehyde, which is then converted to acetyl CoA by aldehyde dehydrogenase (Seitz and Mueller, 2012). This is the main pathway of metabolism, however, if excess alcohol is consumed, the enzymes become saturated and the pathway is not sufficient to break it down. In this case, the body has a second pathway known as the microsomal ethanol oxidizing system (MEOS). The MEOS also produces acetaldehyde by oxidizing alcohol. This pathway requires energy and produces reactive oxygen species, which can lead to oxidative damage of various tissues. The activation of this pathway is dependent on the amount of ethanol in the body (Seitz and Mueller, 2012).

The consumption of alcohol has been shown to have profound effects on the digestion and absorption of other nutrients. Specifically, alcohol has been shown to decrease the absorption of vitamins B6, B12, thiamin, and folate as well as many minerals such as calcium, magnesium, zinc, and iron. Furthermore, excess consumption of alcohol can cause serious damage to the liver and pancreas, which secrete bile and enzymes essential for the digestion of macronutrients and absorption of fat-soluble vitamins. All of these problems can lead to vitamin deficiencies and maldigestion of macronutrients (DiCecco and Francisco-Ziller, 2006).

2.6 MICRONUTRIENTS

The following groups of nutrients, collectively called micronutrients, do not provide energy for the body, but rather they function to aid the growth, development, and maintenance of the body. They are called micronutrients because they are needed in only small quantities. However, they are essential because the body either does not make these nutrients or it does not synthesize them in sufficient quantities. These nutrients are better known as vitamins and minerals. Vitamins can further be classified as either fat-soluble or wate- soluble, while minerals are categorized as either major or trace minerals.

2.6.1 Fat-soluble vitamins

Fat-soluble vitamins, as their name implies, are hydrophobic and are soluble only in fat. There are four fat-soluble vitamins: A, D, E, and K. When consumed in the correct doses, humans can absorb half to almost the entire amount consumed (Byrd-Brenner *et al.*, 2009).

2.6.1.1 Vitamin A

The chemical name for the active form of vitamin A is retinoid. Retinol, retinal, and retinoic acid are three forms of retinoids (Table 2.6). Sources of vitamin A include liver, meat, fish, eggs, and fortified milk (United States Department of Agriculture, 2011). These foods contain what is known as preformed vitamin A. Vitamin A can also be made in the body by converting provitamin A carotenoids, such as β-carotene, into vitamin A. Carotenoids will be discussed later in this chapter. Vitamin A has been shown to be vital for embryonic development and cell differentiation. However, the most well known function of vitamin A is its role in vision, namely, it functions in the retina to help convert light into a nerve signal in the brain (D'Ambrosio *et al.*, 2011).

When vitamin A is consumed from food, it is often bound to other compounds, such as esters and proteins. Therefore, some digestion does occur in order to isolate vitamin A, usually in the form of

Table 2.6 Fat-soluble vitamins.

Vitamin	Structure	Sources	Function
A (retinol)	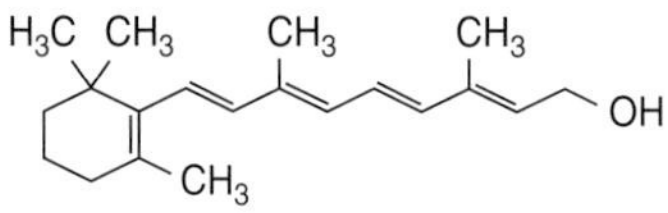 Wikipedia 2008b. Retinol [image online]	Liver, fish, eggs, fortified milk, dark green, yellow orange vegetables	Growth and development, vision
D (calcitriol)	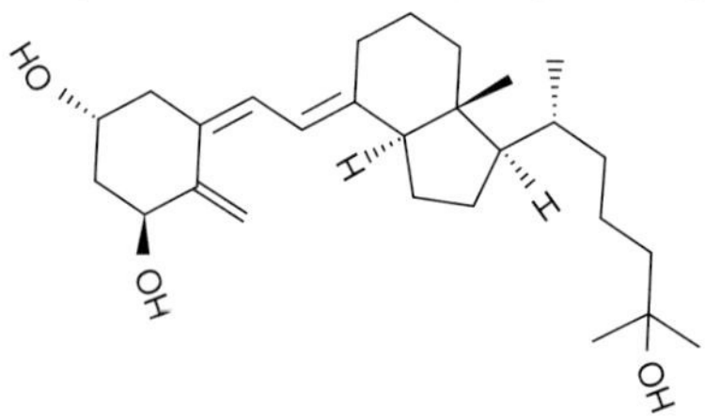 Wikipedia 2008a. Calcitriol [image online]	Sunlight, fatty fish, fortified milk	Regulate Ca and P levels, bone health
E (alpha tocopherol)	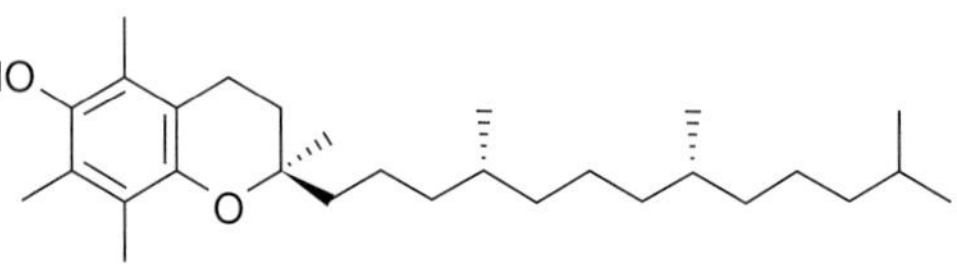 Wikipedia 2007a. Alpha Tocopherol [image online]	Asparagus, almonds, peanuts, salad dressing	Antioxidant, prevent lipid peroxidation
K (phylloquinone)	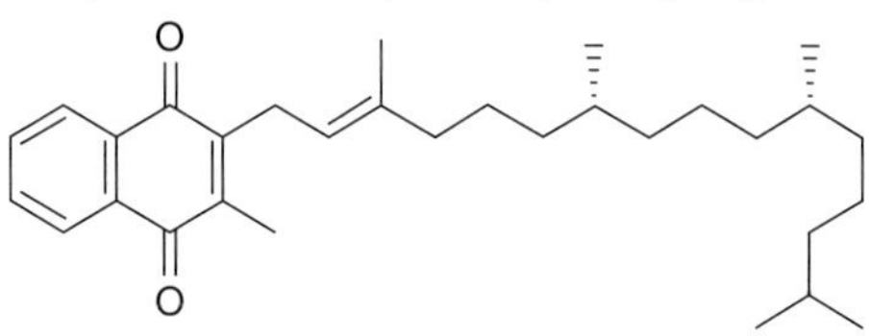 Wikipedia 2010. Phylloquinone. [image online]	Green leafy vegetables, broccoli, peas, green beans, bacteria in colon	Blood clotting

retinol. Retinol that is attached to protein is initially digested in the stomach by pepsin, this process continues in the small intestine with the help of pancreatic enzymes. Just like triglycerides, vitamin A is hydrophobic and, therefore, must be emulsified by bile so that enzymes can reach the vitamin in the micelle. Retinyl esters can be cleaved by pancreatic lipase and esterase enzymes in the brush border of the small intestine (D'Ambrosio *et al*., 2011), and once free, retinol is transported from micelles into enterocytes, by either carrier-mediated facilitated diffusion or simple diffusion. Inside cells, retinol is reattached to a free fatty acid to form a new retinyl ester, packed into a chylomicron and released into the lymphatic system (Harrison, 2012). Vitamin A has high bioavailability when eaten with the appropriate amount of dietary fat.

2.6.1.2 Vitamin D

Vitamin D is unique in that it is obtained not only from the diet, but can also be synthesized in the skin. Skin cells, in the presence of sunlight, can adequately synthesize vitamin D from a cholesterol derivative known as 7-dehydrocholesterol. This process produces a form of vitamin D known as D_3, or cholecalciferol. This product then enters the bloodstream for transport to the liver or kidneys, where hydroxyl groups are added, creating the active form of vitamin D: 1,25-dihydroxy D_3 (Table 2.6). The amount of vitamin D synthesized in the skin depends on many factors, including age, skin color, geographical location, and season of the year. Therefore, some people obtain enough from sunlight while others must obtain significant quantities from their diet (Kulie *et al.*, 2009). The main sources of vitamin D, which is usually in an inactive form known as D_2 (ergocalciferol), in the diet include fatty fish, such as salmon, swordfish, and halibut, as well as fortified milk (United States Department of Agriculture, 2011). The major function of vitamin D is to maintain blood levels of calcium and phosphorous. If levels of these minerals are low, vitamin D works to enhance their absorption in the small intestine, while also removing calcium and phosphorous that is stored in the bone so that cells can receive what they need. Vitamin D has also been shown to play a role in immune function and cell growth (Byrd-Brenner *et al.*, 2009; Kulie *et al.*, 2009).

2.6.1.3 Vitamin E

Vitamin E has many known forms that consist of two different types of compounds: tocopherols and tocotrienols. Each form has a certain amount of biological activity, with α-tocopherol being the most active (Table 2.6). γ-Tocopherol, however, is the most common form found in foods. Dietary sources of vitamin E include plant oils, such as canola and olive oil, nuts and seeds, as well as salad dressings. Wheat germ is also a high source of this vitamin (United States Department of Agriculture, 2011). Vitamin E is a potent antioxidant that functions to protect lipids in cell membranes and other lipid-rich regions from lipid peroxidation and oxidative damage (Traber and Stevens, 2011).

2.6.1.4 Vitamin K

Vitamin K, like vitamin E, has several forms depending on the source. Phylloquinones are found in plant sources and are the more active form of the vitamin (Table 2.6), while menaquinones are obtained from animal sources. Specifically, phylloquinones can be found in kale, broccoli, spinach, asparagus, and peas, while menaquinones are found in dairy products (Suttie and Booth, 2011; United States Department of Agriculture, 2011). In addition to dietary sources, menaquinone can also be synthesized in the body by microflora in the large intestine (O'Keefe, 2008). The most well known function of vitamin K is its role in blood clotting, where it is required for the carboxylation of blood-clotting proteins containing glutamic acid residues. Vitamin K may also play a role in bone health, inflammation, and energy metabolism regulation (Booth, 2009; Suttie and Booth, 2011).

Vitamins D, E, and K all undergo the same mechanism for absorption in the small intestine, with small differences. All three vitamins require emulsification and bile and are incorporated into micelles in the duodenum. They are then absorbed from these micelles via passive diffusion, however, recent evidence indicates that these vitamins may also be absorbed via transport proteins (Reboul and Borel, 2011). More work is currently being done to try and identify specific transporters and elucidate the mechanism of absorption. Once in cells, these vitamins are packaged into chylomicrons and released into the lymphatic system. Approximately 60–90% of dietary vitamin D is absorbed depending on fat content in the meal. It is absorbed throughout the small intestine, with the most rapid absorption occurring in the duodenum, but with significant amounts absorbed in that last two-thirds (Reboul and Borel, 2011). In the case of vitamin E, tocopherols are found in their free form in foods, but tocotrienols are esterified. Therefore, tocotrienols must be hydrolyzed by pancreatic and duodenal mucosal esterases before they can be absorbed. The majority of vitamin E absorption occurs in the jejunum, with its bioavailability ranging from 10 to 95% depending on dietary fat content in the meal. Phylloquinone and dietary menaquinone are absorbed mainly in the jejunum at a rate of about 10%, but this number increases when it is given as a supplement (80%). Menaquinone synthesized by microflora are absorbed via passive diffusion in the colon, however, more research must be done to better understand the bioavailability and use of this form of vitamin K to better understand its role in human health (Reboul and Borel, 2011).

2.6.2 Water-soluble vitamins

Water-soluble vitamins (Table 2.7) consist of the eight B vitamins: thiamin, riboflavin, niacin, pantothenic acid, biotin, B_6, folate, and B_{12}, as well as vitamin C. In general, the B vitamins all function as important coenzymes for reactions in pathways essential for energy metabolism (Byrd-Brenner *et al.*, 2009). Also, with the exception of vitamin C and niacin, the water-soluble vitamins are synthesized in some capacity by microflora in the large intestine: the physiological role of vitamins synthesized there is still being determined (Said and Mohammed, 2006; Said, 2011).

2.6.2.1 Thiamin (vitamin B_1)

Vitamin B_1 is better known as thiamin. Thiamin forms the coenzyme thiamin pyrophosphate (TPP), which is essential for amino acid and carbohydrate metabolism (Said and Mohammed, 2006). Thiamin is found in a wide variety of foods, including pork, seeds, nuts, and fortified cereal (United States Department of Agriculture, 2011). In plant sources, thiamin is present in its free form, but animal products contain a phosphorylated form of the vitamin (Said and Mohammed, 2006). Only free thiamin can be absorbed into enterocytes, therefore phosphorylated thiamin must be hydrolyzed by intestinal phosphatases prior to transport into cells. Once it is in the free form, thiamin is transported into intestinal cells via a pH-dependent, but sodium-independent, carrier-mediated mechanism (Said and Mohammed, 2006; Said, 2011). Other studies have shown that when thiamin is ingested in significantly higher concentrations, it can enter cells via passive diffusion (Gropper *et al.*, 2005). Inside cells, thiamin becomes phosphorylated again and is transported into the bloodstream in a sodium and energy-dependent manner. The bioavailability of free thiamin is quite high, but transport into and out of intestinal cells can be inhibited by alcohol consumption (Gropper *et al.*, 2005; Said and Mohammed, 2006; Said, 2011).

2.6.2.2 Riboflavin (vitamin B_2)

Vitamin B_2 (riboflavin) is a component of two major coenzymes, flavin mononucleotide (FMN) and flavin adenine dinucleotide (FAD), involved in oxidation/reduction reactions involving carbohydrates, amino acids, and lipids (Said, 2011). The main sources of riboflavin in the diet are dairy, fortified breads and cereals, meat, and green vegetables (United States Department of Agriculture, 2011). Riboflavin is often bound to other proteins in food sources and undergoes digestion prior to absorption. HCl, pepsin, and pancreatic proteases denature and hydrolyze protein in order to obtain the free vitamin. Additionally, riboflavin can also be found in foods in its coenzyme forms, FAD and FMN (Gropper *et al.*, 2005; Said and Mohammed, 2006; Said, 2011). These molecules are cleaved by intestinal phosphatases, producing free riboflavin. Once the vitamin is in its free form, riboflavin is absorbed via a sodium-independent carrier-mediated mechanism. Almost all of the riboflavin consumed in the diet is absorbed (95%). Transport out of intestinal cells is also accomplished via carrier-mediated transport (Gropper *et al.*, 2005; Said and Mohammed, 2006; Said, 2011).

Table 2.7 Water-soluble vitamins.

Vitamin	Structure	Source	Function
B_1 (Thiamin)	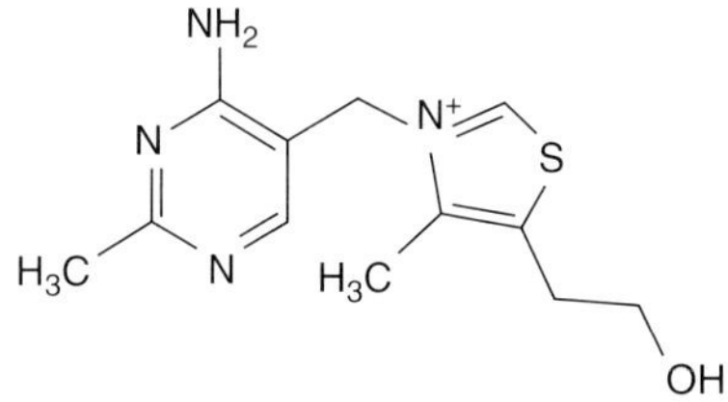 Wikipedia 2007f. Thiamin [image online]	Pork, sunflower seeds, legumes, fortified cereals	Forms coenzyme thiamin triphosphate (TPP) necessary for metabolism of carbohydrates and amino acids.
B_2 (Riboflavin)	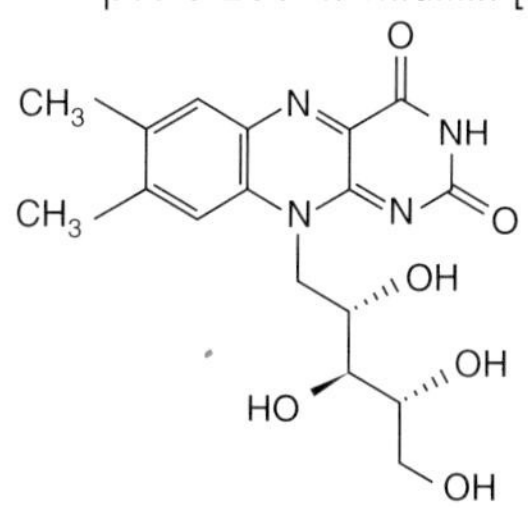 Wikipedia 2007e. Riboflavin [image online]	Milk products, fortified bread and cereals, eggs, meat. Enriched in green leafy vegetables, broccoli, asparagus	Componentflavin mononucleotide (FMN) and flavin adenine dinucleotide (FAD). These coenzymes are needed for redox functions in the body.
B_3 (Niacin)	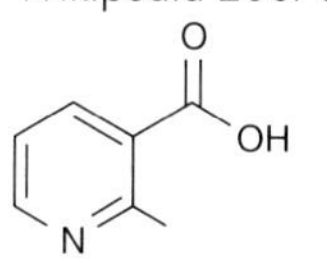 Wikipedia 2007c. Niacin [image online]	Poultry, meat, fish, fortified breads and cereals, foods high in protein	Form nicotinamide adenine dinucleotide (NAD) and nicotinamide adenine dinucleotide phosphate (NADP), used for redox reactions. Functions in reactions that produce adenosine triphosphate (ATP)
B_5 (Pantothenic acid)	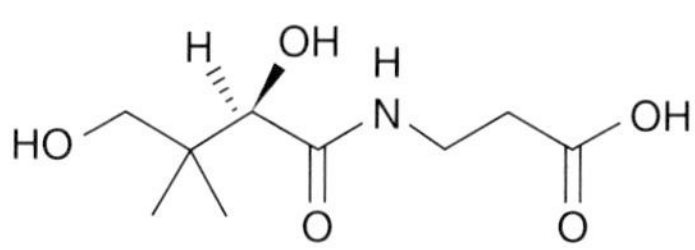 Wikipedia 2006b. Pantothenic Acid [image online]	Meat, milk, avocado, sunflower seeds, mushrooms, broccoli	Part of CoA, necessary for formation of acetyl CoA. Also forms part of the acyl carrier protein
B_6 (Pyridoxal)	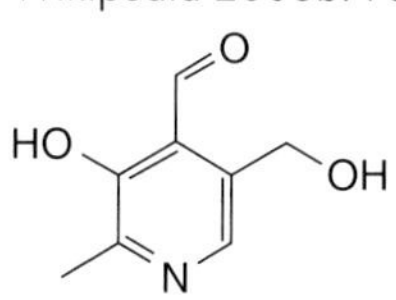 Wikipedia 2007d. Pyridoxal [image online]	Meat, fish, poultry, bananas, potatoes, fortified cereals	Pyridoxal 5′phosphate (PLP) needed for amino acid metabolism and release of glucose from glycogen
B_7 (Biotin)	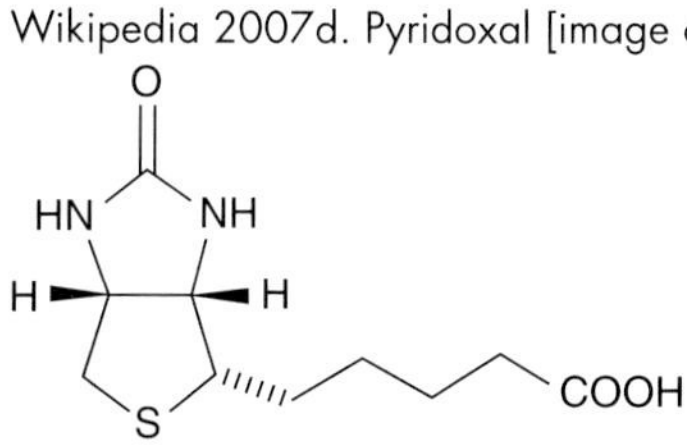 Wikipedia 2007b. Biotin [image online]	Whole grains, eggs, nuts, and legumes	Coenzyme for many carboxylase enzymes

(*continued*)

Table 2.7 (continued) Water-soluble vitamins.

Vitamin	Structure	Source	Function
B9 (Folate)	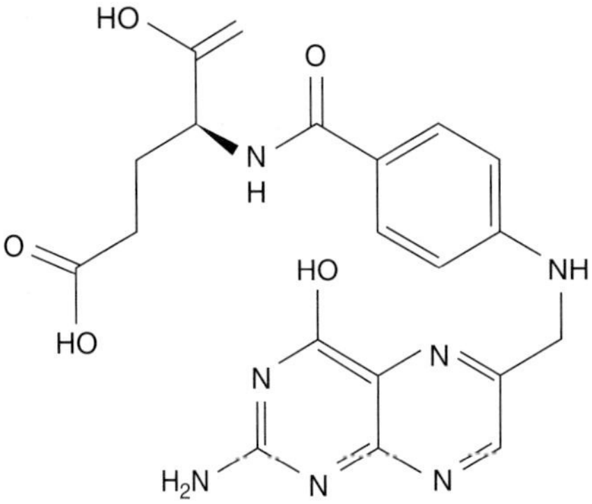 Wikipedia 2006a. Folic Acid [image online]	Liver, legumes, leafy green vegetables, avocados, oranges, fortified breads and cereals	Coenzymes required for synthesis and maintenance of cells. Essential for one-carbon metabolism
B_{12} (Cobalamin)	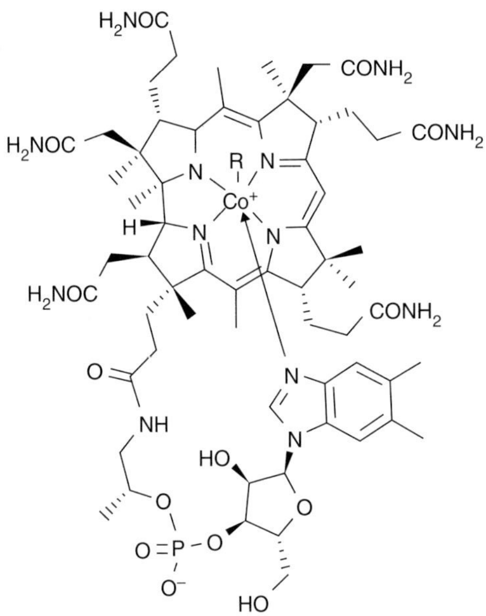 R = 5'-deoxyadenosyl, Me, OH, CN Wikipedia 2012. Cobalamin [image online]	Meat, poultry, seafood, eggs, and dairy products	Formation of methionine from homocysteine. Metabolism of fatty acids with an odd number of carbon molecules
C (Ascorbic acid)	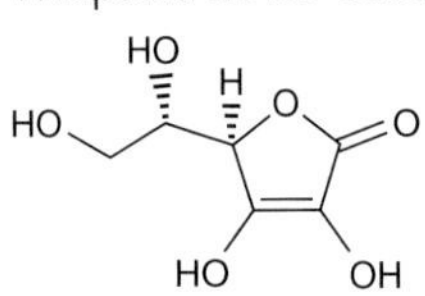 Wikipedia 2009. Ascorbic Acid [image online]	Citrus fruits, peppers, and green vegetables	Donates electrons in redox reactions, antioxidant, collagen synthesis, iron absorption, immune function

2.6.2.3 Niacin (vitamin B_3)

Vitamin B_3, or niacin, forms the coenzymes nicotinamide adenine dinucleotide hydrogen (NADH) and nicotinamide adenine dinucleotide phosphate hydrogen (NADPH), which are central to an enormous amount of oxidation/reduction reactions important for energy metabolism. Unlike most of its water-soluble vitamin counterparts, there is not significant synthesis of niacin by microflora in the colon (Said and Mohammed, 2006; Said, 2011). Niacin can be synthesized in the body, however, from the amino acid tryptophan, with approximately 60 mg of the amino acid needed to make 1 mg of niacin (Food and Nutrition Board, Institute of Medicine, 1998; Said, 2006). Therefore, foods containing tryptophan are good sources of non-preformed niacin. Food sources high in preformed niacin include meat, poultry, and fish (United States Department of Agriculture, 2011). Any NAD^+ and $NADP^+$ consumed in the diet can be hydrolyzed in the small intestine by glycohydrolase, releasing free niacin for absorption (Gropper *et al.*, 2005). The exact mechanism of niacin absorption is not completely understood. Current research indicates that niacin is absorbed via a pH-dependent but sodium-independent carrier-mediated mechanism (Said and Mohammed, 2006; Said, 2011). Other evidence, however, has indicated that a carrier-mediated mechanism dependent on sodium transports niacin, though with a much lower affinity. Furthermore, evidence indicates that at extremely high concentrations, niacin can enter intestinal absorptive cells via passive diffusion (Gropper *et al.*, 2005). Nothing is currently known about the transport mechanism of niacin out of intestinal cells.

2.6.2.4 Pantothenic acid (vitamin B_5)

Vitamin B_5, better known as pantothenic acid, makes up a portion of coenzyme A, which is central to the formation of acetyl CoA, making it essential for macronutrient metabolism. Pantothenic acid is also necessary for the formation of acyl carrier protein, which is needed for fatty acid biosynthesis (Said and Mohammed, 2006; Said, 2011). This vitamin is found mostly in meat, milk, fish, avocados, and broccoli (United States Department of Agriculture, 2011). Pantothenic acid occurs in these foods in both a free form and coenzyme A form. Coenzyme A undergoes hydrolysis several times by phosphatases and pyrophosphatases in order to release the vitamin from its bound form (Gropper *et al.*, 2005; Said and Mohammed, 2006; Said, 2011). Pantothenic acid is absorbed into enterocytes in the small intestine, primarily the jejunum, via a sodium-dependent, carrier-mediated mechanism. The carrier is known as the sodium-dependent multivitamin transporter because it has been shown to also transport the water-soluble vitamin, biotin, and lipoate, an antioxidant (Said and Mohammed, 2006; Said, 2011). At extremely high concentrations, this vitamin can be absorbed via passive diffusion. About half the pantothenic acid consumed in the diet is absorbed (Gropper *et al.*, 2005). Similar to niacin, there is little information on the mechanism of pantothenic acid transport out of intestinal cells.

2.6.2.5 Pyridoxine (vitamin B_6)

Vitamin B_6 is converted to its coenzyme form through phosphorylation. There are three different forms of this vitamin that become phosphorylated, the most common form being pyridoxal (Table 2.7). Pyridoxal phosphate, also known as PLP, is essential for amino acid metabolism, in particular, nonessential amino acid synthesis. Without PLP, every amino acid becomes essential because the body is not able to synthesize any. Pyridoxal phosphate is also important for the release of glucose from glycogen (Byrd-Brenner *et al.*, 2009). The other two forms of vitamin B_6, pyridoxine and pyridoxamine, form the coenzymes PNP and PMP, respectively. These coenzymes are also important for transamination reactions in amino acid metabolism (Gropper *et al.*, 2005; Byrd-Brenner *et al.*, 2009). Main sources of vitamin B_6 in the diet include meat, fish, poultry, bananas, chickpeas, and fortified cereals (United States Department of Agriculture, 2011). Vitamin B_6 can be found in both its phosphorylated and unphosphorylated forms, but only the unphosphorylated form can be absorbed. Brush border phosphatases, dependent on zinc, can hydrolyze any of the three forms of vitamin B_6 (Gropper *et al.*, 2005). Once in its free form, vitamin B_6 is absorbed into intestinal cells in the jejunum via a pH-dependent but sodium-independent, carrier-mediated mechanism (Said and Mohammed, 2006; Said, 2011). Like most of the other water-soluble vitamins, it can be absorbed via passive diffusion if consumed in very large doses. Bioavailability of vitamin B_6 varies between food sources and can be lost through processing (Gropper *et al.*, 2005).

2.6.2.6 Biotin (vitamin B_7)

Vitamin B_7 (biotin) is a coenzyme for many carboxylase reactions essential for macronutrient metabolism (Said and Mohammed, 2006; Said, 2011). The main sources of biotin in the diet are whole grains, meat, fish, eggs, and avocados (Staggs *et al.*, 2004). Most biotin is found in a bound form in foods. Specifically, biotin is either bound to protein or the amino acid, lysine, which is sometimes called biocytin. Protein-bound biotin is digested by pancreatic proteases to yield free biotin, biocytin, or biotinyl peptidases, which can be further hydrolyzed by brush border proteases or, in the case of biocytin, by biotinidase (Said, 2011). Once in the free form, biotin is absorbed via the sodium-dependent multivitamin transporter mechanism used for pantothenic acid absorption in the jejunum and ileum. Like the other water-soluble vitamins, biotin can be absorbed via passive diffusion at very high concentrations. Bioavailability of biotin is high, but can be hindered by alcohol consumption. The mechanism of biotin transport out of intestinal cells occurs via a sodium-independent carrier-mediated mechanism (Gropper *et al.*, 2005; Said and Mohammed, 2006; Said, 2011).

2.6.2.7 Folate (vitamin B_9)

Vitamin B_9, better known as folate, makes up different coenzymes necessary for cell synthesis and maintenance, DNA synthesis, and amino acid metabolism. Specifically, folate plays an important role in one-carbon metabolism (Said and Mohammed, 2006; Said, 2011). The main sources of folate in the diet include fortified cereals, liver, legumes, avocados, leafy green vegetables, and oranges (United States Department of Agriculture, 2011). The main form of folate in foods is the polyglutamate form, however, only monoglutamate forms of folate can be absorbed. Therefore, hydrolases and glutamate carboxypeptidases from the pancreas and small intestine hydrolyze polyglutamates before absorption occurs. Once the vitamin is in monoglutamate form, it is absorbed via a pH-dependent but sodium-independent, carrier-mediated mechanism in the jejunum (Said and Mohammed, 2006; Said, 2011). Absorption via passive diffusion can occur at very high concentrations. Alcohol consumption can decrease absorption of folate and bioavailability varies depending on the food source, since some food sources contain enzyme inhibitors that prevent hydrolysis of polyglutamate forms (Gropper *et al.*, 2005).

2.6.2.8 Cobalamin (vitamin B_{12})

Vitamin B_{12}, also known as cobalamin, forms two different coenzymes. The first, methylcobalamin, is required for the conversion of homocysteine to methionine, which later forms S-adenosyl methionine, a methyl donor important for the regulation of DNA and RNA. The second coenzyme, 5-deoxyadenosylcobalamine, converts methylamlonyl CoA to succinyl CoA, which is required for the metabolism of odd-chain fatty acids. The main sources of B_{12} come from animal products (Said and Mohammed, 2006; Byrd-Brenner *et al.*, 2009). Specifically, meat, seafood, poultry, fortified cereals, and dairy products are all good sources of this vitamin (United States Department of Agriculture, 2011). The digestion and absorption of B_{12} is slightly more complex than the other water-soluble vitamins and begins in the stomach. Vitamin B_{12} is bound to polypeptides in food sources, and must be released from them by pepsin and HCl in the stomach. A protein found in saliva and gastric juice, haptocorrin or R protein, can bind to the protein prior its release from polypeptides as well. Cobalamin, bound to R protein, enters the intestine along with an intrinsic factor that is secreted from parietal cells. In the duodenum, R protein is hydrolyzed by pancreatic proteases, releasing the vitamin. The free form is quickly bound by the intrinsic factor and the complex travels to the ileum, where the receptor, cubilin, is located on intestinal cells (Said and Mohammed, 2006). The mechanism of absorption is not well understood but it is thought that the complex is endocytosed (Said and Mohammed, 2006). The amount of B_{12} absorbed depends largely on the amount of the intrinsic factor produced in the stomach, which tends to decrease with age. About half the B_{12} consumed is absorbed (Gropper *et al.*, 2005; Byrd-Brenner *et al.*, 2009). When it is ingested in very large concentrations, which can occur with supplement use, some of the B_{12} can be absorbed via passive diffusion. Since supplemental B_{12} is in the free form, it can bypass the mouth and stomach digestive process, however, absorption may still be limited by the amount of the intrinsic factor present (Gropper *et al.*, 2005; Office of Dietary Supplements, National Institute of Health, 2011).

2.6.2.9 Ascorbic acid (vitamin C)

Ascorbic acid plays an important role in numerous oxidation/reduction reactions and is essential for collagen synthesis and immune function. It also possesses antioxidant abilities (Traber and Stevens, 2011). The best sources of vitamin C are citrus fruits, asparagus, and broccoli (United States Department of Agriculture, 2011). Unlike the other water-soluble vitamins, microflora do not appear to synthesize any vitamin C. The vitamin exists in both its oxidized and reduced form in foods. The reduced form is either oxidized to its other form or it can be absorbed via a sodium-dependent, carrier-mediated mechanism. The oxidized form is absorbed via a sodium-independent, carrier-mediated mechanism (Said and Mohammed, 2006; Said, 2011). Research indicates that the oxidized form is absorbed more readily. The bioavailability of vitamin C increases with increased need and vice versa. Therefore, its absorption can range from approximately 15–100% of the amount consumed (Gropper *et al.*, 2005).

2.7 WATER AND MINERALS

2.7.1 Water

Water is necessary in some capacity for every biological process in the body, which is why it makes up approximately 60% of it, depending on the amount of lean body mass a person has. Major functions of water include carrying nutrients and serving as a solvent for these nutrients so that they may participate in various metabolic reactions, maintaining blood volume, regulating body temperature, carrying waste, lubricating joints, and, participating in metabolic reactions (Byrd-Brenner *et al.*, 2009; Whitney and Rolfes, 2011). Water is distributed into two major areas of the body: the intracellular and extracellular compartments, with about two-thirds of the total water located in cells and one-third outside cells (Gropper *et al.*, 2005). The main dietary source of water is water and other beverages but humans can obtain water through other dietary sources, such as fruits and vegetables, which are approximately 90% water by weight, and meats and cheeses, which can be up to 50% water by weight (Byrd-Brenner *et al.*, 2009; Whitney and Rolfes, 2011). Water does not undergo any digestion and is mainly absorbed in the small and large intestine. It is absorbed into intestinal cells via osmosis through water pores or channels known as aquaporins (Kunzelmann and Mall, 2002). Osmosis of water is influenced by the concentration of solutes, particularly electrolytes, on either side of the cell membrane. Therefore, the absorption and removal of water depends on the concentrations of electrolytes. As electrolytes are absorbed into intestinal cells, a concentration gradient builds

and Rolfes, 2011). Chloride ions are absorbed mostly in the small intestine but can also be absorbed at the beginning of the large intestine. Chloride is generally absorbed from the intestine by either transcellular or paracellular transport. The transcellular mechanism has already been discussed in the sodium section, and is called the electroneutral cotransport system. In this process, chloride is transported into cells via the Cl^-/HCO_3^- exchange while sodium is simultaneously absorbed via the Na^+/H^+ exchange. Chloride can also be absorbed paracellularly through tight junctions into intracellular space between cells. This process is driven by the active absorption of sodium to maintain electroneutrality. Chloride ions absorbed into intestinal cells enter the bloodstream via passive diffusion through channels or by the KCl cotransporter (Kunzelmann and Mall, 2002; Kato and Romero, 2011).

2.7.6 Calcium

Dietary sources of calcium include dairy products such as milk and cheese, as well as leafy greens (United States Department of Agriculture, 2011). However, calcium found in leafy green vegetables, such as spinach, usually has low bioavailability due to the presence of substances like oxalic acid that bind to minerals and prevent their transport across the intestinal membrane (Byrd-Brenner *et al.*, 2009). Almost all of the calcium in the body is stored in the bones, where it functions to promote bone development and maintenance. The calcium that is not found in the bones is located in the extracellular and intracellular fluid, where it plays a role in nerve impulses along with sodium and potassium, as well as maintaining blood pressure. Calcium has also been shown to play a role in blood clotting (Whitney and Rolfes, 2011). Calcium ions that are attached to salts are dissociated in the acidic environment of the stomach. Once it reaches the small intestine, calcium, like chloride, can be absorbed either transcellularly or paracellularly. Calcium is mainly absorbed via the transcellular mechanism in the duodenum. In this process, calcium binds to the carrier protein, calbindin, and is transported across the plasma membrane into intestinal cells via calcium channels (TRPV6). The mechanism is dependent not only on energy, but also on vitamin D (Bronner, 1998, 2009). The release of vitamin D promotes the synthesis of calbindin protein so that there are plenty of carriers for calcium to bind to for transport into cells. Once inside cells, the calbindin-bound calcium diffuses to the basolateral membrane, where it is released and pumped into the bloodstream via a Ca^{2+}/Mg^{2+} ATPase (Bronner, 1998, 2009). The paracellular absorption mechanism occurs throughout the entire small intestine and is driven by a chemical gradient. Just like chloride, calcium is absorbed through tight junctions and then through the basolateral region into the bloodstream. The amount of calcium absorbed via the paracellular mechanism depends on how much calcium was consumed (Bronner, 1998; Gropper *et al.*, 2005).

2.7.7 Magnesium

Dietary sources of magnesium include green leafy vegetables, broccoli, beans, nuts, and whole grains (United States Department of Agriculture, 2011). Like its other electrolyte counterparts, magnesium is important for nerve impulses and muscle contractions. Additionally, it also is important for DNA and RNA synthesis, as well as bone metabolism (Byrd-Brenner *et al.*, 2009). Absorption of magnesium occurs in the second half of the small intestine and can be absorbed via either trans- or paracellular transport. At low concentrations, magnesium can be absorbed via a carrier-mediated active transporter. At higher concentrations, magnesium can be absorbed paracellularly via simple diffusion. Transport of magnesium into the bloodstream may occur via Na^+/Mg^{2+} antiport systems (Vormann, 2003; Konrad *et al.*, 2004). Since magnesium is often obtained from plant sources its bioavailability may be affected by substances mentioned in the section on calcium, including phosphate and phytate (Vormann, 2003).

2.7.8 Phosphorus

Phosphorous is one of the most abundant minerals in the body. It is often found with calcium in the bone, but is also the major intracellular anion. It is a component of DNA and RNA, ATP, and phospholipids found in cell membranes (Uribarri, 2007). Dietary sources of phosphorus include dairy products, meat, whole grains, and cereals (United States Department of Agriculture, 2011). Absorption of phosphorous must occur in its inorganic form, therefore, ingested phosphorous that is bound to lipids or other organic compounds must be hydrolyzed prior to transport. Absorption of phosphorous is similar to that of magnesium. It can either be absorbed via a carrier-mediated sodium/phosphate cotransporter transcellularly, or by paracellular diffusion at higher concentrations; the absorption of this mineral via the transporter is stimulated by vitamin D. Just like the other minerals, compounds such as phytate can decrease the bioavailability of phosphorous, since the intestine does not produce the phytase enzyme that would release phosphorous from the dietary compound (Uribarri, 2007). Other minerals can also impair absorption, including magnesium and calcium (Gropper *et al.*, 2005).

2.7.9 Sulfur

Almost all dietary sulfur comes from the amino acids methionine and cysteine. Therefore, protein-rich foods are good dietary sources. Sulfur is important for acid-base balance and is a component of proteins and vitamins. The absorption of sulfur occurs via the same mechanisms as amino acid absorption (Byrd-Brenner *et al.*, 2009).

2.8 TRACE MINERALS

Microminerals are so named because the body only needs small amounts to maintain good health. Trace minerals include iron, zinc, copper, manganese, selenium, chromium, iodine, and fluoride. Although they are present only in small quantities, trace minerals play important roles in numerous biological processes in the body (Table 2.9).

2.8.1 Iron

Iron is an essential part of multiple proteins including hemoglobin and myoglobin, which are vital in the transport and storage of oxygen, transferrin (a transport protein), ferritin (a storage protein), and cytochromes in the electron transport chain. It is very valuable for its ability to participate in oxidation/reduction reactions by interchanging between ferrous (Fe^{2+}) and ferric (Fe^{3+}) iron (Gropper *et al.*, 2005; Byrd-Brenner *et al.*, 2009). There are two main dietary forms of iron, heme and non-heme iron (Morgan and Oates, 2002). Heme iron is obtained from animal products such as beef, poultry, and seafood and is in the form of hemoglobin or myoglobin; non-heme iron is obtained from vegetables, such as spinach (United States Department of Agriculture, 2011). These two different forms of iron are absorbed via two different mechanisms. Heme iron is more bioavailable than non-heme iron and is much easier to absorb. When consumed, hemoglobin and myoglobin are digested by proteases in the stomach and small intestine in order to release heme iron, which is still attached to the porphyrin ring. Together, this is known as the iron-protoporphyrin complex. Once in this form, it is readily absorbed into intestinal cells throughout the small intestine, but particularly in the duodenum. This process is thought to occur via endocytosis, where it is then released from the porphyrin ring by an enzyme known as heme oxygenase (Morgan and Oates, 2002; Gropper *et al.*, 2005). Similar to heme iron, non-heme iron must first be released from its bound state to food components. This is achieved

Table 2.9 Trace minerals.

Mineral	Source	Function
Iron (Fe)	Heme-iron: beef, pork, seafood, and poultry. Non-heme iron: spinach, beans, grains, supplements	Redox reactions, essential part of hemoglobin and myoglobin, role in electron transport chain (ETC)
Zinc (Zn)	Seafood, beef, lamb, pork, nuts, beans, whole grains	DNA and RNA synthesis, alcohol metabolism, heme synthesis, bone formation, immune function, reproduction, growth and development, antioxidant, stabilizes structures of cell membrane proteins
Copper (Cu)	Liver, grain products, nuts, seeds, mushrooms, dark chocolate	Ceruloplasmin involved in transport of Fe in blood, functions as part of superoxide dismutase enzymes, cytochrome c oxidase involved in the electron transport chain
Manganese (Mn)	Whole grain cereals, nuts, legumes, tea	Nitrogen and carbohydrate metabolism, cholesterol and urea synthesis, cartilage formation, antioxidant
Selenium (Se)	Seafood, meats, cereals, grains, nuts	Part of glutathione peroxidase enzymes, thioredoxin reductase enzymes, lipid peroxidation defense, thyroid metabolism
Chromium (Cr)	Processed meats, eggs, whole grain products, broccoli, mushrooms, nuts, dark chocolate	May enhance insulin action, promote glucose uptake
Fluorine (F)	Fluoridated water, tea, fluoridated toothpaste, seafood	Deposits Ca and P in teeth and bones
Iodine (I)	Saltwater seafood, iodized salt, molasses	Component of thyroid hormone

through the secretion of HCl and pepsin in the stomach, as well as different proteases in the small intestine. In its free form, non-heme iron is present as ferric (Fe^{3+}) iron, but is much better absorbed in the ferrous (Fe^{2+}) state since Fe^{3+} is more susceptible to form complexes in the alkaline environment of the intestine (Gropper *et al.*, 2005). Therefore, most ferric iron is reduced prior to absorption into cells by ferric reductases. The absorption process occurs via a carrier protein known as the divalent metal transporter 1(DMT1). Research has suggested that there may also be other carriers involved in iron absorption as well (Morgan and Oates, 2002). Once inside cells, it is thought that iron from both heme and non-heme forms undergo the same processes. In enterocytes, free iron has the ability to attach to amino acids or protein for transport across cells to the basolateral membrane, where it is transported out of cells via the carrier ferroportin. Not all iron is transported out of cells, however; some is stored in cells, where it is either used by the cell or lost when the cell is sloughed off for excretion (Anderson *et al.*, 2009). Bioavailability of iron varies depending on its source. An acidic environment increases absorption due to its ability to reduce iron to the ferrous form, while dietary factors such as polyphenols, phytates, copper and zinc can also limit iron absorption (Gropper *et al.*, 2005; Arredondo *et al.*, 2006).

2.8.2 Zinc

Zinc is involved, either directly or indirectly, with almost every biological process or structure in the body. It has been shown to be involved in cell growth, metabolism, immune function, and antioxidant defense (Byrd-Brenner *et al.*, 2009). The main sources of zinc in the diet are animal products, including meat and seafood, however, whole grains, vegetables, and seeds can also contribute zinc to the diet (United States Department of Agriculture, 2011). Although zinc absorption is not completely understood, it has been shown that zinc is mainly absorbed in the proximal end of the small intestine via a carrier-mediated process. It is thought that free zinc forms complexes with different compounds that vary in their ability to be absorbed (Krebs, 2000). Many transporters have been identified that are thought to play a role in absorption but more research must be done to better understand this process. Similar to iron, bioavailability of zinc varies depending on the substances around it as well as the body's need for it. Phytate, iron, and calcium can all inhibit its absorption (Krebs, 2000).

2.8.3 Copper

Copper is an important component of many enzymes due to its ability to undergo oxidation and reduction. Copper is a component of proteins involved in the oxidation of iron (ceruloplasmin) as well as the elimination of free radicals (superoxide dismutase) (Byrd-Brenner *et al.*, 2009). Dietary sources of copper include shellfish, meat, nuts, seeds, beans, and dark chocolate (United States Department of Agriculture, 2011). Although it is known that copper is absorbed in the small intestine, the precise mechanism by which this occurs is not well understood. While previous studies have indicated that copper is absorbed via an energy-dependent, carrier-mediated transporter, recent research has shown that copper absorption may be dependent on chloride, indicating that anion exchangers may play an important role in absorption into intestinal cells. Once inside cells, copper is transported into the bloodstream via the hCTR1 transporter located on the basolateral membrane (Zimnicka *et al.*, 2011). The same factors that influence iron and zinc absorption also affect copper absorption, with iron also inhibiting its absorption (Gropper *et al.*, 2005; Arredondo, 2006).

2.8.4 Manganese

Similar to the other trace minerals, manganese is an important cofactor for enzymes involved in metabolism and antioxidant defense (Byrd-Brenner *et al.*, 2009). The best dietary sources of manganese are whole grains, nuts, and legumes (United States Department of Agriculture, 2011). Although it is known that manganese is absorbed throughout the small intestine, the mechanism by which it is absorbed is not well understood, though the amount absorbed appears to depend on age and gender. Research also indicates that iron inhibits its absorption, possibly via competition for the divalent metal transporter (DMT-1). Phytates, copper, and vitamin C also decrease the absorption of manganese. (Aschner and Aschner, 2005).

2.8.5 Selenium

Selenium is a component of various different enzymes essential for the antioxidant defense network, as well as the body's immune system. The best sources of selenium are seafood, meats, nuts, and grains (United States Department of Agriculture, 2011). The selenium content of foods depends on the amount present in the soil in which plants are grown and where animals feed. Selenium is consumed mostly bound to the amino acids methionine or cysteine, though some inorganic forms of the mineral, such as selenate, are found in vegetables (Fairweather-Tait *et al.*, 2011). These forms all appear to be absorbed efficiently, though amino-acid-bound selenium is absorbed slightly better. While it is known that selenium bound to either methionine or cysteine is transported into intestinal cells via amino acid transporters, the mechanism of absorption for selenate is not well understood, with transporters yet to be identified. Preliminary data indicate that anion exchangers may play a role in their absorption, however, more work must be carried out to elaborate on this finding (Fairweather-Tait *et al.*, 2011). Vitamin C, A, and E all appear to increase selenium absorption, while phytates are thought to inhibit it (Gropper *et al.*, 2005). Unlike many other trace minerals, absorption does not appear to be affected by the body's need for it (Byrd-Brenner *et al.*, 2009).

2.8.6 Chromium

Chromium is thought to be important for insulin action, therefore it is important for glucose homeostasis (Krejpcio, 2001). Chromium content in foods is still being determined and can be highly variable in certain foods, but currently known sources of the mineral include processed meats, eggs, whole-grain products, broccoli, and some fruits (Food and Nutrition Board, Institute of Medicine, 2001). The mechanism of chromium absorption is currently unknown, but is suspected to occur via a carrier-mediated transporter or by diffusion (Gropper, 2005). The overall bioavailability of chromium is extremely low, with absorption increasing with need. It is estimated that only about 0.4–3% of chromium intake is absorbed into the body, although it is thought that vitamin C and certain amino acids may enhance absorption. Just like the other trace minerals, phytate is thought to decrease absorption, along with zinc and iron (Krejpcio, 2001).

2.8.7 Iodine

Iodine is essential to the thyroid. Specifically, it is a component of the thyroid hormones T_3 and T_4. Due to this, iodine is important for metabolism, growth and development, and regulation of energy expenditure. Iodine content in most foods is relatively low. The best sources of iodine are saltwater seafood, seaweed, and iodized salt (Food and Nutrition Board, Institute of Medicine, 2001). The form of iodine in foods is either in its free form, in an iodate form (IO^{3-}), or bound to amino acids. Iodate can be reduced to iodine by glutathione but it is not necessary for absorption. The specific mechanism of absorption is not known, but it is known that all forms of iodine are absorbed in the small intestine and can even be absorbed in the stomach, with free iodide being the most efficiently absorbed form of the mineral (Gropper *et al.*, 2005). Current research indicates that the Na^+/I^- symporter used to transport iodine into the thyroid gland may also play a role in transporting iodine into enterocytes. However, more research looking into the result is needed (Nicola *et al.*, 2009). Free iodide is transported out of the cells into the blood, but the mechanism for this process is not known. The bioavailability of iodine is very high, with almost 100% absorption of the ingested mineral (Food and Nutrition Board, Institute of Medicine, 2001).

2.8.8 Fluoride

Fluoride, is very important for dental health. It functions to aid in the deposition of calcium and phosphorous into teeth and bones, which helps to prevent the formation of dental caries. The main source of fluoride is fluoridated water, however, it can also be obtained from tea and seafood. Although not a food, fluoridated toothpaste can also be used to obtain this mineral (Rocha *et al.*, 2012). Fluoride is thought to be absorbed rapidly throughout the GI tract, particularly in the stomach and small intestine when consumed as fluoridinated water. The bioavailability of this fluoride is very high, with almost 100% absorption. Some evidence indicates that absorption occurs via passive diffusion, however, some studies provide evidence for an active component in transport that may or may not be pH-dependent. The bioavailability of fluoride drops significantly when consumed in foods and absorption does not appear to be affected by the body's need for it (Rocha *et al.*, 2012).

2.9 PHYTOCHEMICALS

2.9.1 Carotenoids

Another group of compounds important for human health are phytochemicals. Although they are not considered to be essential, they have been shown to provide many health benefits (Byrd-Brenner *et al.*, 2009). Two types of phytochemicals that are known to provide health benefits are carotenoids and flavonoids. Carotenoids, which are pigments found in fruits, vegetables, and animal products, can be classified into two groups: provitamin A and non-provitamin A (D'Ambrosio *et al.*, 2011; Harrison, 2012). As their name implies, provitamin A carotenoids, such as β-carotene, are converted to vitamin A in the body, but this is not their only function. These carotenoids, along with those classified as non-provitamin A, have been shown to have antioxidant properties that protect against oxidative damage in lipid-rich regions of the body (Whitney and Rolfes, 2011). The main sources of carotenoids in the diet are carrots, tomatoes, eggs, peppers, avocados, kale, papayas, and spinach (Kotake-Nara *et al.*, 2011; United States Department of Agriculture, 2011). Despite the fact that there are approximately 50 known carotenoids found in the diet, humans can absorb only 20–30 of them. Similar to vitamin A, carotenoids are usually bound to other food components when ingested and are released from this bound form via the same processes. Due to their hydrophobic nature, carotenoids must be solubilized into micelles, along with other fat-soluble components, by bile prior to absorption. Once in micelles, intestinal absorption is thought to occur via facilitated and passive diffusion (Gropper *et al.*, 2005; Kotake-Nara *et al.*, 2011; Harrison, 2012). The efficiency of carotenoid absorption is less than that of vitamin A, with bioavailability ranging from absorption of 5%–60% of carotenoids consumed depending on the food source, whether the food was raw or cooked, and the fat content of the meal. Fiber and vitamin E intake may also decrease the absorption of carotenoids (Gropper *et al.*, 2005; Kotake-Nara *et al.*, 2011).

2.9.2 Flavonoids

Similar to carotenoids, flavonoids can also be classified into different classes, including flavonols, flavanones, isoflavones, and flavones. All of these forms of flavonoids possess antioxidant activity and, therefore, protect lipid bilayers from oxidative damage due to free radicals. The main sources of flavonoids in the diet are blueberries, strawberries, tea, kale, and apples (United States Department of Agriculture, 2003). Dietary flavonoids are consumed in their glycoside form, which are thought to be cleaved prior to absorption in the small intestine into their aglycone form. Although the exact mechanism of absorption is not fully known, current research indicates that a number of intestinal membrane transporters, including sodium-dependent glucose transporter 1, as well as an anion exchanger, may play a role in flavonoid absorption (Walle, 2004). More research needs to be carried out in order to shed light on the exact mechanism of absorption. Like carotenoids, flavonoid bioavailability can vary substantially depending on the food source.

2.10 IMPLICATIONS IN HEALTH AND DISEASE

When parts of the digestion and absorption system are not working properly, the body is not able to obtain the nutrients it needs to function properly. This can lead to a wide variety of health problems. A list of digestive and absorptive disorders, their symptoms, and effects are listed in Table 2.10. Most of these disorders prevent macro- and micronutrients from entering the body and performing their respective functions, causing energy deprivation as well as serious problems in metabolism and growth.

Table 2.10 Health issues affecting digestion and absorption.

Disease type	Name	Description	Symptoms	Effects
Digestive disorders	Gastro esophageal reflux disease	Loosening of lower esophageal sphincter, allows stomach contents to enter esophagus	Heartburn, ulcers, bleeding in esophagus	Injury to esophagus
	Ulcer	Erosion of cells in stomach or intestine	Pain after eating, nausea, vomiting	Chyme escapes GI tract, causing infection
	Gallstones	Cholesterol forms crystal-like structure in bile. Stones can lodge in ducts, preventing release into duodenum	Abdominal pain, nausea, gas	Maldigestion and malabsorption of fat and fat-soluble vitamins
	Lactose intolerance	Deficiency of lactase enzyme in the brush border of the small intestine	Bloating, abdominal pain, diarrhea	Cannot digest or absorb lactose, leading to its bacterial fermentation in colon
	Constipation	Slow movement of fecal matter.	Difficult and infrequent bowel movements	Water absorbed out of large intestine causing feces to dry and harden.
	Diarrhea	Intestine secretes fluid into small intestine, inefficient absorption of liquid	Loose, watery stool expelled multiple times per day	Indigestible foods cause secretion of water into intestine. Dehydration can occur over long periods
Malabsorption disorders	Inflammatory bowel diseases (ex. Crohn's)	Inflammation of the intestine. Brush border enzymes cannot make contact with nutrients. Transporters for minerals are blocked	Diarrhea, dehydration, blood in feces	Malabsorption of carbohydrates, fats, and protein, as well as iron, B12, calcium, and magnesium
	Pancreatitis	Inflammation of pancreas. Decrease in release of pancreatic enzymes and HCO_3^-	Pain while eating, nausea, diarrhea	Maldigestion and malabsorption of macronutrients and fat-soluble vitamins

2.11 USE OF NANOPARTICLES TO ENHANCE ABSORPTION OF NUTRIENTS

The study of nanoparticles to increase the absorption of nutrients and nutraceuticals has grown immensely over the last few years. Current research looks toward designing nanoparticles that protect nutrients and aid in their delivery into intestinal cells. In doing so, bioactive compounds, such as phytochemicals, vitamins, nutrients, or minerals can be preserved against destructive conditions that occur when consumed orally and absorbed through the GI tract (Kaya-Celiker and Mallikarjunan, 2012). An example of this would be the process of micellation around active substances, which has been shown to increase penetration and intestinal resorption, as well as create thermal, mechanical, and pH stability (Kaya-Celiker and Mallikarjunan, 2012). For more information on improved bioavailability of phytochemicals delivered with nanoparticles, the reader is directed to Chapter 10.

Nanoparticles can also be absorbed directly to enhance the absorption of bioactive compounds. There are two proposed pathways for the absorption of nanoparticles. The first is paracellular absorption, which tends to be rare since most nanoparticles are too large to fit into the spaces between cells. The second mechanism of action is through transcellular uptake through enterocytes, either passively or through binding to receptors that show chemical specificity (Acosta, 2009).

At this time, research on the effect of nanoparticle properties on GI uptake is inconsistent; most data, however, show that particle size plays an important role in the ability of the nanoparticle to increase bioavailability of a nutrient through its direct absorption or through protecting it from the varying environments of the GI tract. Decreased particle or droplet size increases the solubility and bioavailability of the nutrient (Acosta, 2009; Huang *et al.*, 2010). While research in this area is ongoing, current results indicate that nanoparticles, liposomes and surfactants improve the solubility and bioavailability of phytochemicals, such as curcumin (Huang *et al.*, 2010; Yu and Huang, 2012) and β-carotene (Nik *et al.*, 2012). For the absorption of minerals, however, the results are less clear. While a study from 2010 demonstrated that iron absorption was increased with the use of iron nanoparticles (Hilty, 2010), another study looking at polystyrene nanoparticles discovered a decrease in iron absorption in the GI tract of chickens, which has been shown to be an accurate model for iron absorption in humans (Mahler *et al.*, 2012). Preliminary work looking into increased copper bioavailability indicates that silica nanoparticles loaded with copper may increase its solubility (Maniprasad and Santra, 2012), however, more research must be carried out in order to confirm and expand on this result. Continued research examining nanoparticles and nutrient delivery is needed to determine the full capabilities of this technology. The nanoparticle GI uptake mechanisms of nanoparticle uptake and the effect of nanoparticle properties on intestinal uptake are discussed in more detail in Chapter 3.

References

Acosta, E. (2009) Bioavailability of nanoparticles in nutrient and nutraceutical delivery. *Current Opinion in Colloid and Interface Science,* 14, 3–15.

Anderson, G.J., Frazer D.M. and McLaren, G.D. (2009) Iron absorption and metabolism. *Current Opinion in Gastroenterology*, 25, 129–135.

Arredondo, M., Martínez, R., Núñez, M.T., Ruz, M. and Olivares, M. (2006) Inhibition of iron and copper uptake by iron, copper, and zinc. *Biological Research,* 39(1), 95–102.

Aschner, J.L. and Aschner, M. (2005) Nutritional aspects of manganese homeostasis. *Molecular Aspects of Medicine*, 26, 353–362.

Bajwa, A. and Emmanuel, A. (2009) The physiology of continence and evacuation. *Best Practice and Research In Clinical Gastroenterology*, 23(4), 477–485.

Booth, S.L. (2009) Roles for vitamin K beyond coagulation. *Annual Review of Nutrition,* 29, 89–110.

Bronner, F. (1998) Calcium absorption – a paradigm for mineral absorption. *Journal of Nutrition*, 128, 917–920.

Bronner, F. (2009) Recent developments in intestinal calcium absorption. *Nutrition Reviews,* 67(2), 109–13.

Brown, D., Bouley, R., Păunescu, T.G., Breton, S. and Lu, H.A. (2012) New insights into the dynamic regulation of water and acid-base balance by renal epithelial cells. *American Journal of Physiology – Cell Physiology*, 302(10), C1421–C1433.

Byrd-Brenner, C., Moe, G., Beshgetoor, D. and Berning, J. (2009) *Wardlaw's Perspectives in Nutrition*, 8th edn. McGraw-Hill, New York.

D'Ambrosio, D.N., Clugston, R.D. and Blaner, W.S. (2011) Vitamin A metabolism: An update. *Nutrients*, 3, 63–103.

De Preter, V., Hamer H.M., Windey, K. and Verbeke, K. (2011) The impact of pre- and/or probiotics on human colonic metabolism: Does it affect human health? *Molecular Nutrition and Food Research,* 55, 46–57.

DiCecco, S.R. and Francisco-Ziller, N. (2006) Nutrition in alcoholic liver disease. *Nutrition in Clinical Practice,* 21(3), 245–254.

Ensign, L.M., Cone, R. and Hanes, J. (2012) Oral drug delivery with polymeric nanoparticles: The gastrointestinal mucus barriers. *Advanced Drug Delivery Reviews*, 64, 557–570.

Fairweather-Tait, S.J., Bao, Y., Broadley, M.R. *et al.* (2011) Selenium in human health and disease. *Antioxidants and Redox Signaling,* 14(7), 1337–1383.

Food and Nutrition Board, Institute of Medicine. (1998) *Dietary Reference Intakes for Thiamin, Riboflavin, Niacin, Vitamin B-12, Pantothenic Acid, Biotin, and Choline.* National Academy Press, Washington, DC.

Food and Nutrition Board, Institute of Medicine. (2001) *Dietary Reference Intakes for Vitamin A, Vitamin K, Boron, Chromium, Copper, Iodine, Iron, Manganese, Molybedenum, Nickel, Silicon, Vanadium, and Zinc.* National Academy Press, Washington, DC.

Goodman, B.E. (2010) Insights into digestion and absorption of major nutrients in humans. *Advances in Physiology Education,* 34, 44–53.

Gropper, S.S., Smith, J.L. and Groff, J.L. (2005) *Advanced Nutrition and Human Metabolism*, 4th edn. Thomson Wadsworth, Belmont, CA.

Harrison, E.H. (2012) Mechanisms involved in the intestinal absorption of dietary vitamin A and provitamin A carotenoids. *Biochimica et Biophysica Acta*, 1821(1), 70–77.

Hershcovici, T., Mashimo, H. and Fass, R. (2011) The lower esophageal sphincter. *Neurogastroenterology and Motility*, 23, 819–830.

Hilty, F.M., Arnold, M., Hilbe, M. *et al.* (2010) Iron from nanocompounds containing iron and zinc is highly bioavailable in rats without tissue accumulation. *Nature Nanotechnology*, 5, 374–380.

Hofmann, A.F. (2009) The enterohepatic circulation of bile acids in mammals: Form and functions. *Frontiers in Bioscience*, 14, 2584–2598.

Huang, Q., Yu, H. and Ru, O. (2010) Bioavailability and delivery of nutraceuticals using nanotechnology. *Journal of Food Science*, 75(1), R50–R57.

Hunter, E.J. (2005) Dietary levels of trans-fatty acids: basis for health concerns and industry efforts to limit use. *Nutrition Research*, 25, 499–513.

Innis, S.M. (2011) Dietary triacylglycerol structure and its role in infant nutrition. *Advances in Nutrition*, 2(3), 275–283.

Janssen, P., Vanden Berghe, P., Verschueren, S., Lehmann, A., Depoortere, I. and Tack, J. (2011) Review article: the role of gastric motility in the control of food intake. *Alimentary Pharmacology and Therapeutics*, 33, 880–894.

Kato, A. and Romero, M.F. (2011) Regulation of electroneutral NaCl absorption by the small intestine. *Annu Rev Physiol.* 73, 261–281.

Kaya-Celiker, H.K. and Mallikarjunan, K. (2012) Better nutrients and therapeutics delivery in food through nanotechnology. *Food Engineering Reviews*, 4, 114–123.

Kong, F. and Singh, R.P. (2010) A human gastric simulator (HGS) to study food digestion in the human stomach. *Journal of Food Science*, 75(9), 627–635.

Konrad, M., Schlingmann, K.P. and Gudermann, T. (2004) Insights into the molecular nature of magnesium homeostasis. *American Journal of Physiology*, 286, F599–F605.

Kotake-Nara, E. and Nagao, A. (2011) Absorption and metabolism of xanthophylls. *Marine Drugs,* 9, 1024–1037.

Krebs, N.F. (2000) Overview of zinc absorption and excretion in the human gastrointestinal tract. *Journal of Nutrition*, 130(Supplement 5), 1374S–1377S.

Krejpcio, Z. (2001) Essentiality of chromium for human nutrition and health. *Polish Journal of Environmental Studies*, 10(6), 399–404.

Kulie, T., Groff, A., Redmer, J., Hounshell, J. and Schrager, S. (2009) Vitamin D: An evidence-based review. *Journal of the American Board of Family Medicine*, 22(6), 698–706.

Küllenberg, D., Taylor, L.A., Schneider, M. and Massing, U. (2012) Health effects of dietary phospholipids. *Lipids in Health and Disease*, 11(3). doi: 10.1186/1476-511X-11-3.

Kunzelmann, K. and Mall, M. (2002) Electrolyte transport in the mammalian colon: mechanisms and implications for disease. *Physiology Reviews*, 82, 245–289.

Mahler, G.J., Esch, M.B., Tako, E. *et al.* (2012) Oral exposure to polystyrene nanoparticles affects iron absorption. *Nature Nanotechnology*, 7, 264–271.

Maldonado-Valderrama, J., Wilde, P., Macierzanka, A. and Mackie, A. (2011) The role of bile salts in digestion. *Advances in Colloid Interface Science*, 165, 36–46.

Maniprasad, P. and Santra, S. (2012) Novel copper (Cu) loaded core-shell silica nanoparticles with improved Cu bioavailability: synthesis, characterization, and study of antibacterial properties. *Journal of Biomedicine and Nanotechnology*, 8(4), 558–566.

Mansbach II, C.M. and Gorelick, F. (2007) Development and physiological regulation of intestinal lipid absorption. II. Dietary lipid absorption, complex lipid synthesis, and the intracellular packaging and secretion of chylomicrons. Journal of Physiology – Gastrointestinal and Liver *Physiology*, 293, G645–G650.

Minsky, B. (2006) *Super Foods*, article, Nutritional Concepts. [Online] Available: http://nutritionalconcepts.com/Articles/Super%20Foods%204-24-06.pdf (accessed 10 July 2012).

Morgan, E. and Oates, P.S. (2002) Mechanisms and regulation of intestinal iron absorption. *Blood Cells, Molecules, and Diseases*, 29(3), 384–399.

Moshfegh, A.J., Friday, J.E., Goldman, J.P. and Chug Ahuja, J.K. (1999) Presence of inulin and oligofructose in the diets of Americans. *Journal of Nutrition*, 129(Supplement 7), 1407S–1411S.

Nicola, J.P., Basquin, C., Portulano, C., Reyna-Neyra, A., Paroder, M. and Carrasco, N. (2009) The Na+/I- symporter mediates active iodide uptake in the intestine. *American Journal of Physiology* – Cell *Physiology*, 296, C654–C662.

Nik, A.M., Langmaid, S. and Wright, A.J. (2012) Digestibility and β-carotene release from lipid nanodispersions depend on dispersed phase crystallinity and interfacial properties. *Food and Function*, 3, 234–245.

Office of Dietary Supplements, National Institute of Health. (2011) *Dietary Supplement Fact Sheet: B12*. [Online] Available: http://ods.od.nih.gov/factsheets/VitaminB12-HealthProfessional/ (accessed 5 September 2012).

O'Keefe, S.J.D. (2008) Nutrition and colonic health: the critical role of the microbiota. *Current Opinion in Gastroenterology*, 24, 51–58.

Pezzilli, R. (2009) Chronic pancreatitis: maldigestion, intestinal ecology, and intestinal inflammation. *World Journal of Gastroenterology*, 15(14), 1673–1676.

Ramírez, M., Amate, L. and Gil, A. (2001) Absorption and distribution of dietary fatty acids from different sources. *Nutrition and Infant Pathology*, 65(2), S95–S101.

Reboul, E. and Borel, P. (2011) Proteins involved in uptake, intracellular transport, and basolateral secretion of fat-soluble vitamins and carotenoids by mammalian enterocytes. *Progress in Lipid Research*, 50(4), 388–402.

Rocha, R.A., Vélez, D. and Devesa, V. (2012) *In vitro* evaluation of intestinal fluoride absorption using different cell models. *Toxicology Letters*, 210(3), 311–317.

Said, H.M. (2011) Intestinal absorption of water-soluble vitamins in health and disease. *Biochemical Journal*, 437, 357–372.

Said, H.M. and Mohammed, Z.M. (2006) Intestinal absorption of water-soluble vitamins: an update. *Current Opinion in Gastroenterology*, 22(2), 140–146.

Salles, C., Chagnon, M.C., Feron, G. *et al*. (2011) In-mouth mechanisms leading to flavor release and perception. *Critical Reviews in Food Science and Nutrition*, 51, 67–90.

Schipper, R.G., Silletti, E. and Vingerhoeds, M.H. (2007) Saliva as research material: biochemical, physicochemical, and practical aspects. *Archives of Oral Biology*, 52, 1114–1135.

Schubert, M.L. (2008) Gastric secretion. *Current Opinion in Gastroenterology*, 24, 659–664.

Schubert, M.L. (2009) Gastric exocrine and endocrine secretion. *Current Opinion in Gastroenterology*, 25(6), 529–536.

Seitz, H.K. and Mueller, S. (2012) Metabolism of alcohol and its consequences. In: *Metabolism of Drugs and Other Xenobiotics*, Ed. Anzenbacher, P. and Zanger, U.M. Wiley-VCH, Weinheim, Germany; pp. 493–516.

Staggs, C.G., Sealey, W.M., McCabe, B.J., Teague, A.M. and Mock, D.M. (2004) Determination of the biotin content of select foods using accurate and sensitive HPLC/avidin binding. *Journal of Food Composition and Analysis*, 17(6), 767–776.

Suttie, J.W. and Booth, S.L. (2011) Vitamin K. *Advances in Nutrition*, 2(5), 440–441.

Traber, M.G. and Stevens, J.F. (2011) Vitamins C and E: beneficial effects from a mechanistic perspective. *Free Radical Biology and Medicine*, 51, 1000–1013.

United States Department of Agriculture. (2003) *USDA Database for the Flavonoid Content of Selected Foods*. [Online] Available: http://www.nal.usda.gov/fnic/foodcomp/Data/Flav/flav.pdf (accessed 5 September 2012).

United States Department of Agriculture. (2011) *USDA National Nutrient Database for Standard Reference, Release 24*. [Online] Available: http://www.ars.usda.gov/Services/docs.htm?docid=22114 (accessed 5 September 2012).

Uribarri, J. (2007) Phosphorous homeostasis in normal health and in chronic kidney disease patients with special emphasis on dietary phosphorous intake. *Seminars in* Dialysis, 20(4), 295–301.

Vereecke, L., Beyaert, R. and van Loo, G. (2011) Enterocyte death and intestinal barrier maintenance in homeostasis and disease. *Trends in Molecular Medicine*, 17(10), 584–593

Vormann, J. (2003) Magnesium: nutrition and metabolism. *Molecular Aspects of Medicine*, 24(1–3), 27–37.

Walle, T. (2004) Absorption and metabolism of flavonoids. *Free Radical Biology and Medicine*, 36(7), 829–837.

Whitney, E. and Rolfes, S.R. (2011) *Understanding Nutrition*, 12th edn. Wadsworth Cengage Learning, Belmont, CA.

Wikipedia. (2006a) Folic acid. [Image online] Available: http://en.wikipedia.org/wiki/File:Folic_acid.svg (accessed 27 June 2012).

Wikipedia. (2006b) Pantothenic acid. [Image online] Available: http://en.wikipedia.org/wiki/File:Pantothenic_acid.svg (accessed 27 June 2012).

Wikipedia. (2007a) Alpha tocopherol. [Image online] Available: http://en.wikipedia.org/wiki/File:Tocopherol,_alpha-.svg (accessed 27 June 2012).

Wikipedia. (2007b) Biotin. [Image online] Available: http://en.wikipedia.org/wiki/File:Biotin_structure_JA.png (accessed 27 June 2012).

Wikipedia. (2007d) Pyridoxal. [Image online] Available: http://en.wikipedia.org/wiki/File:Pyridoxal.png (accessed 27 June 2012).

Wikipedia. (2007c) Niacin. [Image online] Available: http://en.wikipedia.org/wiki/File:Niacin_structure.svg (accessed 27 June 2012).

Wikipedia. (2007e) Riboflavin. [Image online] Available: http://en.wikipedia.org/wiki/File:Riboflavin.svg (accessed 27 June 2012).

Wikipedia. (2007f) Thiamin. [Image online] Available: http://en.wikipedia.org/wiki/File:Thiamin.svg (accessed 27 June 2012).

Wikipedia. (2008a) Calcitriol. [Image online] Available: http://en.wikipedia.org/wiki/File:Calcitriol.svg (accessed 27 June 2012).

Wikipedia. (2008b) Retinol. [Image online] Available: http://en.wikipedia.org/wiki/File:All-trans-Retinol2.svg (accessed 27 June 2012).

Wikipedia. (2009) Ascorbic acid. [Image online] Available: http://en.wikipedia.org/wiki/File:L-Ascorbic_acid.svg (accessed 27 June 2012).

Wikipedia. (2010) Phylloquinone. [Image online] Available: http://en.wikipedia.org/wiki/File:Vitamin_K1.png (accessed 27 June 2012).

Wikipedia. (2012) Cobalamin. [Image online] Available: http://en.wikipedia.org/wiki/File:Cobalamin.png (accessed 27 June 2012).

Ye, E.Q., Chacko, S.A., Chou, E.L., Kugizaki, M. and Liu, S. (2012) Greater whole-grain intake is associated with lower risk of type 2 diabetes, cardiovascular disease, and weight gain. *Journal of Nutrition*, 142(7), 1304–1313.

Yu, H. and Huang, Q. (2012). Improving the oral bioavailability of curcumin using novel organogel-based nanoemulsions. *Journal of Agriculture and Food Chemistry*, 60(21), 5373–5379.

Zimnicka, A.M., Ivy, K. and Kaplan, J.H. (2011) Acquisition of dietary copper: a role for anion transporters in intestinal apical copper uptake. *American Journal of Physiology Cell Physiology*, 300, C588–C599.

3 Cellular fate of delivery systems and entrapped bioactives

Cristina M. Sabliov,[1] Dorel Moldovan,[2] Brian Novak,[2] Toni Borel,[1] and Meocha Whaley[1]

[1]*Department of Biological and Agricultural Engineering, Louisiana State University and LSU Agricultural Center, Baton Rouge, Louisiana, USA*
[2]*Department of Mechanical and Industrial Engineering, and Center for Computation and Technology, Louisiana State University, Baton Rouge, Louisiana, USA*

3.1 CELLULAR FATE OF NANOPARTICLES – AN EXPERIMENTAL PERSPECTIVE

Nanoparticles (NPs) are a potentially beneficial delivery system for improving nutrient delivery. To fully take advantage of the nanodelivery systems for delivery of nutrients, it is important to understand the fate of the NPs and the entrapped bioactive, which will dictate the efficacy of the bioactive and the safety of the NP construct. The properties of the biological membranes and the physical–chemical characteristics of the nanodelivery systems will dictate the interaction of NPs with biological systems, and the fate of the NPs and entrapped bioactive.

Studies on cellular uptake of NPs reported in the literature (Table 3.1) have been performed with particles of different properties (size, morphology, zeta potential), in various cells. HeLa cells are derived from cervical cancer and are used in many different cell studies; these cells contain an active telomerase enzyme that enables an extended period of proliferation and makes them a good model for cancer research. Caco-2 are human epithethial colorectal adenocarcinoma cells suitable models for intestinal cellular uptake, as endocytosis can be evaluated, specific modulators or mechanisms can be identified, and colocalization studies can be easily performed in these cells. HepG2 are human liver hepatocellular carcinoma cells used for various studies, including studies of intracellular reactive oxygen species (ROS) generation. HT29 cells are human colon carcinoma cells that were subcloned to produce HT29-MTX cells; these cells serve as excellent models for intestinal uptake because of their mucous secreting ability. Uptake of NPs in these cells and others has been demonstrated and the extent of the uptake has been correlated with specific NP properties such as size, charge, and morphology. Uptake was quantified most commonly by measuring fluorescence following incubation of different particles with cells for various amounts of times (1, 2, 3, 4, 12, and 24 h), depending on the cell type and NP properties (Sahoo *et al.*, 2002; Win and Feng, 2005; Harush-Frenkel *et al.*, 2007; Gratton *et al.*, 2008; Zhang *et al.*, 2008; Hamdy *et al.*, 2009; Gaumet and Gurny, 2010; Li *et al.*, 2010; Yoo and Mitragotri, 2010; Swarnakar *et al.*, 2011; Hu *et al.*, 2012; Wang *et al.*, 2012).

Nanotechnology and Functional Foods: Effective Delivery of Bioactive Ingredients, First Edition.
Edited by Cristina M. Sabliov, Hongda Chen, and Rickey Y. Yada.

Table 3.1 Physicochemical properties of nanoparticles and cell uptake.

Cell	Polymer	Size (nm)	Charge/wall material	Shape/ morphology	Analysis methods	Observations	Reference
Caco-2	PLGA, Chitosan/ PLGA	100, 300 (diameter)	−8. ± 5.5 to −12.1 ± 0.1 mV zeta	Spherical	Spectroscopy, confocal micrsocopy	Smaller particles readily internalized, NPs > 300 nm remained associated with the apical membrane, higher affinity of chitosan/PLGA for Caco-2 cells compared to PLGA nanoparticles at sizes < 1000 nm	Gaumet and Gurny, 2010
HeLa	Poly(ethylene glycol) hydrogels	100,150, 200 (diameter)		Cylindrical	Flow cytometry	Highest aspect ratio internalized fastest, positively charged NPs were internalized in 84% of cells while same-shaped negative counterparts less than 5%	Gratton *et al.*, 2008
CHO	Diblock and triblock copolymer combinations	11, 20, 30 (cross-sectional diameter) N.A., 180 ± 120, 970 ± 900 (length)		Spherical, cylindrical	Fluorescent microscopy	Smaller, spherical particles had a higher cell uptake than larger cylindrical nanoparticles, CPP showed a noticeable increase in NP uptake where highest CPP load 2% NPs produced approximately fivefold increase in cell uptake	Zhang *et al.*, 2008
Dendritic	PLGA			Spherical		Size and uptake efficiency by DCs were reported as having an inverse relationship, DCs preferred particles of a viral size and macrophages preferred larger bacterial size NPs	Hamdy *et al.*, 2011

Human arterial smooth muscle cells (HASMCs)	PLGA	380, 421, 468, 522	0.5, 1.0, 2.0, 5.0% w/v PVA −8.0 ± 2.3, −12.5 ± 0.4, −11.3 ± 0.8, −15.4 ± 0.8	Spherical	HPLC (detection of entrapped dye)	Smaller size, uptake for NPs synthesized with 5% PVA was one-third of NPs synthesized with 2% PVA, and was associated with surface hydrophilicity	Sahoo *et al.*, 2002
Human colon adenocarcinoma cell Caco-2 cells	Poly(lacti-go-glycolic acid) coumarin-6, vitamin E-TPGS, polystrene	50, 100, 200, 500, 1000	2% PVA , 0.03% VitE TPGS	Spherical	Fluorescent microscopy confocal, TEM, cryo-SEM	Optimum size of 100–200 nm, vitamin E-TPGS coated NPs had an uptake 1.4-fold higher than PVA coated NPs	Win *et al.*, 2005
HeLa	PLA, mPEG-PLA	89.8 ± 4, 96.36 ± 6	32.8 ± 8.19, −26.1 ± 1 mv	Spherical	Confocal microscopy, FACS (flow cytometry)	Show a two- to three-fold higher uptake of positive particles, positively charge NPS uptaken through dynamic- and clathrin-dependent endocytic pathways	Harush-Frenkel *et al.*, 2007
Macrophage	PLGA, PLGA-ester	160 (diameter) 5.5 (aspect ratio)		Elliptical disc, spherical	SEM, time-lapse video microscopy	Internalization was not seen for the PLGA-ester elliptical disc micro particles until the shape shifted to spherical	Yoo *et al.*, 2010

3.1.1 Nanoparticle detection and quantification

Particle detection methods can be a limiting factor when developing NP cell interaction studies (Win and Feng, 2005). Methodology must consider the characteristics of NPs and the biological nature of cells. Earlier methods were typically confined to qualitative analysis, but advances in the field have produced more refined methods of detection, including quantification. Each experimental method is not without its advantages and disadvantages.

Of the more common methods confocal microscopy is frequently used in conjunction with florescent or radioactive labels (Win and Feng, 2005). Cultured cells are typically treated with the labeled NPs then incubated for a period of time. Cells are then rinsed with a buffer to remove excess non-attached NPs. Cells are fixed and an optional nuclei stain is applied before being mounted for examination. Imaging software is used to process images. This method provides visualization of particle localization, but is limited for quantitative analysis (Win and Feng, 2005; Cartiera *et al.*, 2009). Initial *in vitro* fluorescent marker release studies are recommended to distinguish between NP uptake or fluorescent dye leaching and dissociation in the medium (Win and Feng, 2005). Due to cell fixation, continuous same-cell observations for NP uptake are not possible.

Fluorescence microscopy is an alternative, qualitative method for NP uptake. Cells are seeded in well plates and incubated for monolayer cell-sheet growth. A NP suspension is introduced to the seeded cells via the medium. The NP suspension is removed and the cells are washed with phosphate buffer solution (PBS) to remove excess or non-internalized particles. Nanoparticle cellular uptake is detected by a fluorescence microscope at some pre-established time point(s). Untreated cells are used as a positive control and cells treated with C-6 as a negative control (Pamujula *et al.*, 2012).

Transmission electron microscopy (TEM) consists of visualizing cells following incubation of cultured cells with NPs after they are washed, pre-fixed, post-fixed, and then dehydrated by alcohols and resin permeation. Thinly sectioned samples are placed on a coated copper grid, stained, and blotted before observation under TEM (Win and Feng, 2005). Similar to confocal microscopy, TEM primarily provides qualitative analysis. Analysis is limited by imaging software capabilities. Complex sample preparation may limit the scope of these studies.

Cryo scanning electron microscopy (SEM) is traditionally used for unstructured biomaterial studies. Cells are incubated with NPs for cellular internalization for a set time, followed by washing to remove excess or non-internalized particles. Cells are fixed using glutaraldehyde then frozen by submersion in a nitrogen slurry, and fractured, etched, and sputter coated for observation by SEM. While limited to qualitative studies, this method provides examination of bulk biomaterials. Simple sample preparation decreases incidences of damage to cells and allows cells to be observed directly in the existing state (Win and Feng, 2005). Limited background and information on this method has been reported in the literature (Win and Feng, 2005).

Unlike the previously discussed methods, flow cytometry can quantitatively measure cellular uptake (Win and Feng, 2005; Pamujula *et al.*, 2012). Cells are grown in tissue culture plates for 24 h. The cell sheet is exposed to a NP suspension and incubated for an additional 24 h. Cells are then washed and trypsinized to remove cells from the culture plate. Removed cells are centrifuged to a pellet and resuspended in PBS to be counted for analysis by a flow cytometer. Increase of mean fluorescence intensity indicates intracellular delivery (Pamujula *et al.*, 2012).

3.1.2 Effect of NP properties on cell uptake

To take full advantage of opportunities offered by polymeric NPs there is an established need for a thorough understanding of the NP–cell interaction and the effect of this interaction on the entrapped bioactive function and localization within the cell. Nanoparticle size, charge, and surface characteristics among other parameters, affect NP uptake (Win and Feng, 2005; Gaumet and Gurny, 2010; Semete *et al.*, 2010). The drug solubility, whether hydrophobic or hydrophilic (Shea *et al.*, 2005; Asati

Table 3.2 Description of endocytic pathway comparable with particle size. (Source: Verma and Stellacci, 2010. Reproduced with permission from Wiley.)

Type of endocytosis	Description	Size of vesicle formed
Phagocytosis	Internalization of solid particles such as bacteria and yeast by specialized cells	Dependent on particle being engulfed
Pinocytosis	Fluid-phase uptake of extracellular molecules. Multiple pinocytic pathways are possible	See below
Macropinocytosis	Trapping of large fluid pockets by formation and enclosure of membrane protrusions	>1 μm
Clathrin-mediated endocytosis	Concentration of transmembrane receptors and bound ligands in "coated pits" on the plasma membrane formed by the assembly of cytosolic proteins, the main assembly unit being clathrin	≈120 nm
Caveolae-mediated endocytosis	Flask-shaped invagination in the plasma membrane that mediate uptake of extracellular molecules into the cell by specific receptor binding	≈50–60 nm

et al., 2010), will determine the localization of the bioactive within the cell and will dictate the effect of entrapment on its activity. With a better understanding of how NP physicochemical properties affect cellular uptake and localization of polymeric NPs, the benefits of polymeric NPs as cellular delivery systems can be fully achieved.

In eukaryotic cells and organelles the selectively permeable membrane acts as a considerable barrier to drug delivery and delivery systems (Zorko and Langel, 2005; Verma *et al.*, 2008). Necessary ions, polar molecules, and nanomaterials must be assisted across the lipid bilayer. Assistance is in the form of specialized membrane-transport protein channels or major endocytoic pathways where molecules are enclosed in membrane vesicles for transport into the cells. Five predominant cell-uptake mechanisms (Table 3.2) are hypothesized for NP cellular uptake, including phagocytosis, macropinocytosis, caveolar endocytosis, clathrin/caveole-independent endocytosis, and clathrin-mediated endocytosis (Sahay *et al.*, 2010; Verma and Stellacci, 2010; Lawlor *et al.*, 2011). When looking at the present literature it is necessary to asses proposed uptake mechanisms not only in the context of cell type and structure but also NP characteristics.

Upon entrance, NPs may be trafficked through endocytic pathways by vesicles in the cells to endo-lysosomal compartments, exocytosed, or trafficked by other subcellular compartments (Verma and Stellacci, 2010; Lawlor *et al.*, 2011). The final locations of these molecules are relevant to polymeric NP delivery systems, as the acidic pH of lysosomes may degrade NPs and entrapped bioactives, and critical applications of NPs and entrapped bioactives require access to cytosol and cytosolic components (Verma and Stellacci, 2010; Lawlor *et al.*, 2011).

Cellular uptake of NPs is dictated by the NP properties. Size is an important property of NPs because it affects both mucosal penetration and cellular uptake. Mucoadhesion occurs when NPs adhere to the mucosal lining of the intestine. Due to negative charges on the surface of the mucus, electrostatic interactions may occur with positively charged particles, causing them to adhere to the mucus. The mucosal lining is also hydrophobic, so hydrophobic interactions between NPs and mucus may also occur. Cellular uptake occurs via diffusion and endocytosis (Nam *et al.*, 2012). Two categories of endocytosis are phagocytosis and pinocytosis, both are depicted in Figure 3.1. Macrophages and neutrophils engulf large particles via phagocytosis. Pinocytosis occurs in four ways and in all cells. Macropinocytosis is one type of pinocytosis that occurs via actin. Clathrin mediated endocytosis is another type of pinocytosis and a coated pit formed by clathrin absorbs the particle. Caveolea-mediated

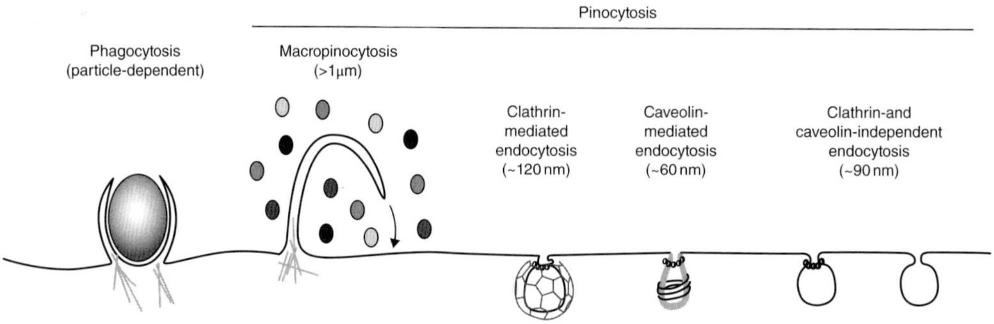

Figure 3.1 Diagram of the major forms of cellular uptake via endocytosis. (Nam *et al.*, 2012. Reproduced with permission from Elsevier.)

endocytosis is the absorption through the cell's plasma membrane. Independent endocytosis, while not well understood, is another means of cellular uptake via pinocytosis, and it does not involve clathrin or caveolea, but it is not well understood.

Nanoparticles of sizes > 500 nm may be taken up only by specific cells (e.g., macrophages) through phagocytosis; NPs < 500 nm are taken up by endocyotic pathways (Verma and Stellacci, 2010; Yoo and Mitragotri, 2010). Smaller particles are readily internalized while particles > 300 nm remain associated with the apical membrane (Gaumet and Gurny, 2010). In general, the size of NPs is inversely proportional to uptake (He *et al.*, 2010). The optimal size range for cellular uptake appears to be 100–200 nm, as confirmed by Win and Feng (2005) when comparing polystyrene particles and vitamin E-TPGS (D-α tocopheryl polyethylene glycol 1000 succinate) PLGA (poly(lactic-co-glycolic acid) particles of 50, 100, 200, 500, 1000 nm.

The very nature of cell interaction with foreign matter is dictated by the characteristics of the material's surface (Yoo and Mitragotri, 2010). Along with size, charge is claimed to be the most important factor influencing NP uptake in cells (Harush-Frenkel *et al.*, 2007). Cell membranes are largely dense with negative charges due to phospholipid heads and polysulfated proteoglycans. Therefore, positively charged NPs enter the cell more easily due to electrostatic interactions with the positive charges. Negatively charged NPs enter the cell more readily than particles with neutral charges due to small areas of positive charge on the cell membrane. Neutral NPs have the lowest percentage of cellular uptake (He *et al.*, 2010).

Charge of the NP wall is significant, but material composition of the wall material may also significantly impact NP uptake. For example, residual amounts of surfactants such as poly(vinyl alcohol) (PVA) on the NP surface, affecting NP hydrophilicity, is also a factor (Sahoo *et al.*, 2002). Hydrophobicity and hydrophilicity are properties that describe a molecule's or NP's affinity for water. Hydrophobic molecules do not have affinity for, and are insoluble in, water. Hydrophilic molecules are soluble in water. Hydrophobic and hydrophilic molecules interact differently with the phospholipid bilayer of the cell membrane. Hydrophilic molecules or NPs have multiple ways of entering the cell through active transport; cellular mechanisms use the cell's energy and specific membrane channels for the entry of hydrophilic molecules. Small hydrophilic molecules such as doxorubicin enter the cell by binding to the polar-head groups on the outside of the bilayer, and when the polar-head groups flip, they diffuse into the cytoplasm from the inner membrane (Wang *et al.*, 2012). Other pathways of hydrophilic molecules include diffusion through specific membrane channels or endocytosis. Hydrophobic molecules permeate cells more easily via passive diffusion; their insolubility creates an active driving force that allows them to cross into the membrane (Acosta, 2009). It is therefore expected that hydrophobic NPs may more likely be able to penetrate the cellular membrane and more efficiently transport the entrapped bioactive into the intracellular compartments of cells.

While size and charge were well investigated early on as aspects to control cell interaction and uptake, more recently functionalization of NPs has seen an increased interest. Nanoparticles can be functionalized to specifically target cells and ensure cell penetration and increased uptake (Verma *et al.*, 2008). The presence of ligands on the NP surface was determined to increase uptake through receptor-mediated endocytosis and NP specificity (Hamdy *et al.*, 2009). Successful targeting was highly dependent on cell type and targeting moieties.

As another viable aspect of NP functionality to improve uptake, cell-penetrating peptides can be attached to the NP surface. Cell-penetrating peptides are classified as amphiphilic peptides capable of minimal invasive penetration of cell membranes for enhanced NP internalization in cells by non-exerting energy mechanisms (Zorko and Langel, 2005; Verma *et al.*, 2008). Uptake of shell-cross-linked NPs of different morphologies was improved fivefold by the presence of cell-penetrating peptides (Zhang *et al.*, 2008).

Morphology describes shape, structure, and internal composition of NPs. Internal composition describes the molecular composition and whether it is homogeneous or heterogeneous. Homogeneous particles are composed of a single type of polymer or material. Heterogeneous particles are composed of multiple polymer or material types. Advances in NP synthesis methodologies have resulted in the ability to produce nanostructrures of various shapes and morphologies (Yoo and Mitragotri, 2010; Venkataraman *et al.*, 2011). Nanoparticles are commonly spherical, disc-shaped, or tubular, but they can also be complex nonspherical particles (Bouwmeester *et al.*, 2009). The shape of NPs plays a critical role in cellular uptake (Huang *et al.*, 2010). Rod-shaped or tubular particles made of mesoporous silica have shown a higher uptake in HeLa cells and A347 cells when compared to spherical particles (Huang *et al.*, 2010; Hao *et al.*, 2012). Gratton *et al.* (2008) found that rod-shaped particles have an advantage in the rate of cellular uptake over cubic particles of poly(ethylene glycol) (PEG) hydrogels. However, in other studies, spherical particles are shown to interact more efficiently with cells, for improved uptake. Data can be found for gold NPs in HeLa cells, where spherical gold NPs showed a higher percentage of uptake than rod-shaped gold NPs, with aspect ratios of 1:3 and 1:5 (Chithrani *et al.*, 2006). Yoo and Mitragotri (2010) concluded the ability of particles to be phagocytized was dependent on particle shape, geometry, and orientation. Elongated surfaces provided for higher contact surface area that correlated to higher cell targeting when compared to spherical particles. Yet near-spherical shapes are preferred for macrophage uptake by phagocytosis. This suggests that other NP properties must also influence cellular uptake or internalization (Venkataraman *et al.*, 2011), thus there is no clear consensus on ideal morphological properties for a maximum cellular uptake.

In summary, spherical particles, < 300 nm, demonstrated increased uptake in Caco-2, CHO, and HASMCs cells (Sahoo *et al.*, 2002; Win and Feng, 2005; Zhang *et al.*, 2008; Gaumet and Gurny, 2010), whereas uptake of larger particles of "viral" and "bacterial" size was preferred by dendritic and macrophage cells respectively (Hamdy *et al.*, 2009; Yoo and Mitragotri, 2010). Gaumet and Gurny (2010) determined that NPs > 300 nm remained attached to the apical membrane of Caco-2 cells (Gaumet and Gurny, 2010). At smaller sizes uptake was influenced by NP charge, wall material, functionality, and morphology. Positively charged NPs have shown as much as a two- to threefold increase in uptake compared to negatively charged particles (Harush-Frenkel *et al.*, 2007). An increase in the presence of PVA during synthesis, or increased surface hydrophilicity, decreased NP uptake by one-third as compared to that of NP synthesized with a smaller amount of PVA (Sahoo *et al.*, 2002). Though spherical particles were favored over cylindrical particles in terms of uptake, high aspect ratios of particle morphologies were most associated with increased NP uptake (Gratton *et al.*, 2008; Zhang *et al.*, 2008; Yoo and Mitragotri, 2010) (Table 3.1). It is apparent that no one single NP characteristic is solely responsible for the NP–cell interaction. The mechanisms of interaction and uptake kinetics are functions of several combinations of NP properties, of which size, surface properties, morphology, and solubility are most important. Nanoparticle cellular uptake may be optimized by engineering NPs that are hydrophobic, positively charged, either rod-shaped or spherical, and 100–200 nm in size due to their mucoadhesion properties, ability to penetrate the mucus, and be efficiently endocytosed by various cells. Nanoparticle cellular fate is a critical determinant of the functionality of the entrapped bioactive, as will be described further.

3.1.3 Fate of loaded NPs in the cell with implications on bioactive functionality

After absorption into the cell, a NP will either enter the cytoplasmic vesicles, nucleus, mitochondria, or cytoplasm. Once in the cytoplasm, the particle may be secreted back out into the serosal spaces. The fate of loaded NPs can be determined by several methodologies including flow cytometry, laser scanning confocal microscopy, and fluorescence microscopy (Garinot *et al.*, 2007), as previously mentioned. The NP type, size, and surface charge should be optimal for the delivery of the nutrient or nutraceutical.

Progress is currently being made in understanding the factors that contribute to improved NP uptake and colocalization of the particles within the cellular environment. Win and Feng (2005) found that PLGA particles stabilized by vitamin E TPGS were found primarily in the endoplasm, and some portion surrounding the nucleus of Caco2 cells after a 1-h incubation of cells with NPs. Cartiera *et al.* (2009) observed by confocal microscopy that PLGA(Rhodamine B) particles were localized around the Golgi complex and endosomes of HBE and OK cells after NP incubation of 2 h. Chui *et al.* (2010) found using confocal microscopy that particles made of *N*-palmitoyl chitosan (NPCS) were in the endosomes and lysosomes of Caco2 cells after a 3-h incubation.

Release kinetics of the NP carrier should also be determined in order to optimize or improve functionality of the entrapped nutraceutical in the cell environment. Release profiles are of importance in the event that NPs are not endocytosed as a whole and the bioactive is released in the cell vicinity, or to understand the exposure of the cell environment to the contents of the particles once absorbed. Three different types of release into cells occur via NP delivery, such as: drugs enter cells after being released from NPs, NPs release drugs from the inner leaflet of the cell membrane into the cytoplasm, and lastly the NP may release the drugs from inside the cell (Wang *et al.*, 2012).

It is important for bioactive functionality to be preserved during entrapment. Bioactive functionality can be tested via various assays, both *in vitro* and intracellularly. For antioxidants, the effect of entrapment on antioxidant activity can be understood by comparing the activity of free and encapsulated antioxidants. Several studies have compared antioxidants such as quercetin or epigallocatechin gallate (EGCG). Kumari *et al.* (2010) found that free quercetin and quercetin loaded in poly(lactic acid) (PLA) NPs had similar antioxidant activity using the 1,1-diphenyl-2-picrylhydrazyl (DPPH) assay as the free components. Peres *et al.* (2011) tested free EGCG and EGCG/gum arabic maltodextrin particles using a DPPH radical scavenging assay. The free EGCG and EGCG NPs had equivalent radical quenching at all concentrations tested. In contrast, bioactive antioxidant activity was found to be improved intracellularly by Swarnakar *et al.* (2011). Cells incubated for 12 h with coenzyme Q10 entrapped in NPs showed the highest uptake of NPs; the reduction in lipopolysaccharide (LPS)-stimulated ROS generation was concentration dependent. The coenzyme Q10 NPs were found to increase the intracellular antioxidant activity of coenzyme Q10 at a dose of 100 μg/mL compared to free coenzyme Q10 at the same dose. It is understood that NP physical characteristics, size, surface properties, and morphology will determine where the particles will interact with the cell, and for various drugs and antioxidants entrapped within the polymeric matrix this could have either a negative or positive effect on functionality.

3.2 CELLULAR UPTAKE OF SMALL MOLECULES AND NPs BY MEMBRANE PENETRATION – A MOLECULAR SIMULATION PERSPECTIVE

In molecular simulations, submolecular-level details of a system can be observed in order to gain insight into the behavior of a system or mechanism of a process. In atomistic simulations, all of the atoms are represented explicitly. Molecular simulations are computationally expensive and are usually limited to nanometer-size scales and timescales that are on the order of a microsecond or less.

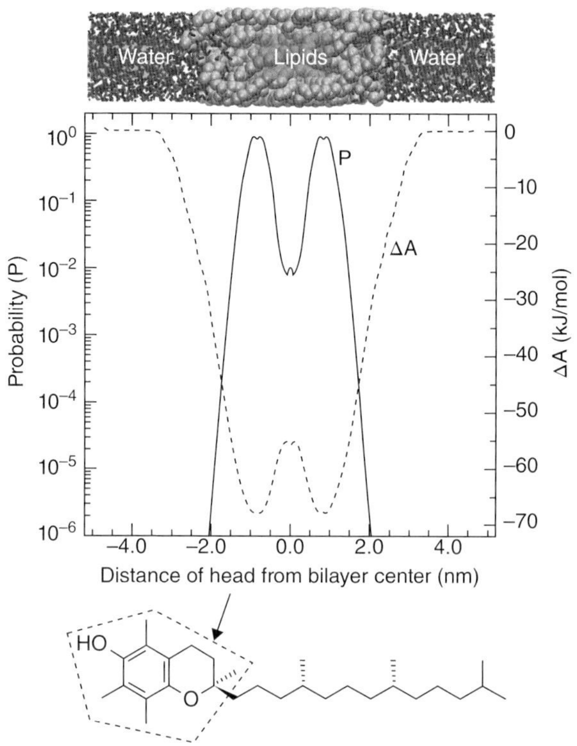

Figure 3.2 Plot of free-energy difference (ΔA) and probability (*P*) as a function of distance of the head group of α-tocopherol from the center of a dimyristoylphosphatidylcholine bilayer. The structure of α-tocopherol with the head group encircled with a dashed pentagon is shown below the plot. The picture above the plot is a slice of the simulation system with the lipids shown in light greyshade and water in dark greyshade.

Coarse-grained models in which several atoms are represented as a single site allow for longer timescales or larger physical dimensions to be simulated at the expense of losing atomistic-level details. Coarse-grained simulations can easily reach the microsecond to millisecond time range, which is necessary to observe many processes. Biases of various types can also be used in simulations to speed up processes or sampling. The effect of those biases can be removed after the simulations.

Due to the complexity of trying to model a cell membrane that contains many types of lipids, cholesterol, proteins, and other molecules, simulations usually focus on bilayers with one or a few types of lipids and perhaps cholesterol or a single membrane protein.

3.2.1 Small molecules and drugs interacting with lipid bilayers

Some small molecules can pass through bilayers without the assistance of pores, membrane proteins, or endocytosis. Due to the relative simplicity of this basal permeation process, it has been studied extensively with molecular simulations. Molecular simulations are often used to obtain insight into the most probable locations and orientations of molecules in a system consisting of a bilayer in aqueous solution or into their effect on the bilayer structure, or to calculate the permeation rates of the molecules through a bilayer. All of these can be accomplished by calculating a free energy as a function of a reaction coordinate(s). The free energy of a state is related to the probability of observing that state. A minimum in free energy corresponds to a maximum in probability. When considering the partitioning of a molecule between the bilayer and aqueous phase, the reaction coordinate is usually the distance between the molecule or part of the molecule and the bilayer center in the bilayer normal direction. Adequate sampling of all the relevant states from the aqueous phase to the center of the bilayer is usually not possible in unbiased simulations. Two common ways to accomplish this are multiple biased simulations in which the molecule is restrained (umbrella sampling) (Torrie and Valleau, 1974; Kastner, 2011) or constrained to different locations (constraint force method) (den Otter and Briels, 1998; Sprik and Ciccotti, 1998). These simulations are then combined to obtain the free-energy curve. Considering the orientation of the molecules and their effects on the bilayer as a function of the reaction coordinate is often useful for obtaining detailed information on how the molecules interact with a bilayer. In addition, the diffusion coefficient of the molecule as a function of the reaction coordinate can be estimated. Combined with the free-energy profile, this diffusion coefficient can be used to calculate basal permeabilities and mean first-passage times (Marrink and Berendsen, 1994). Permeabilities can be compared with experimental measurements.

Simulations of many types of small molecules have been performed. These include amino acids, pore-inducing molecules such as dimethyl sulfoxide (DMSO) and alcohols, drugs, hormones, fatty acids, vitamins, antioxidants, toxic substances, and fluorescent probes. The following is a sampling of the work published on the topic. The free energies of many small molecules have been calculated as a function of distance from a bilayer center. These include amino acid side chains (MacCallum *et al.*, 2008), DMSO (Hughes *et al.*, 2012; Lin *et al.*, 2012), the anticonvulsant valproic acid (Ulander and Haymet, 2003), β-blocker drugs and steroid hormones (Orsi and Essex, 2010), the anticancer drug doxorubicin with a bilayer containing various amounts of cholesterol (Yacoub *et al.*, 2011), and the antimicrobial compounds triclocarban and triclosan (Orsi *et al.*, 2011). Information about the most probable location and orientation of some molecules as well as their effect on the bilayer can be obtained without calculating a free-energy profile. Simulations of this type include studying the effect of relatively large methanol concentrations (11.3 mol%) on bilayer structure (Pinisetty *et al.*, 2006); the orientation, location, and lateral diffusivity of the fatty acids oleic and 2-hydroxyoleic acid in a bilayer as a function of fatty acid concentration (Cerezo *et al.*, 2011); the location and orientation of toxic sulfur mustard in a bilayer compared to heptane (Muller and Muller-Plathe, 2009); the location and orientation of the fluorescent probe *N,N*-dimethyl-6-propionyl-2-naphthylamine (PRODAN) in a bilayer as a function of polarization of its polar groups (Nitschke *et al.*, 2012); the location, orientation, and lateral diffusivity of α-tocopherol (vitamin E) as a function of the type of lipid in the bilayer and temperature (Qin *et al.*, 2009; Qin and Yu, 2011); and the location and orientation of the antioxidant quercetin and its metabolites in a bilayer (Kosinova *et al.*, 2012).

As an example of the information that can be obtained from molecular simulations of small molecules interacting with lipid bilayers, the free energy and orientation of α-tocopherol were calculated as a function of the distance of the α-tocopherol head group from a hydrated fluid phase (323 K) dimyristoylphosphatidylcholine (DMPC) bilayer center. α-Tocopherol is mostly hydrophobic, with a single hydrophilic hydroxyl group on the head group. Figure 3.3 shows the free-energy profile as well as the associated probability density. Notice that the probability density is very low for the head being outside the bilayer. The probability density for the head being in the center of the bilayer is only about two orders of magnitude lower than the maximum probability. This means that the α-tocopherol head can easily flip from one leaflet of the bilayer to the other. In fact, this happens with a frequency on the order of tens of nanoseconds and can be observed during simulations. Figure 3.3c shows the number densities of various types of DMPC and α-tocopherol atoms as a function of distance from the bilayer center averaged over a trajectory in which the α-tocopherol did not flip from one leaflet to another. These atoms are defined in Figure 3.3a and b. The α-tocopherol hydroxyl oxygens are found in the same region as the DMPC ester groups, and the end of the α-tocopherol tail can be found anywhere between about ±1.2 nm from the bilayer center, which means that it can move around freely in the tail region of the bilayer.

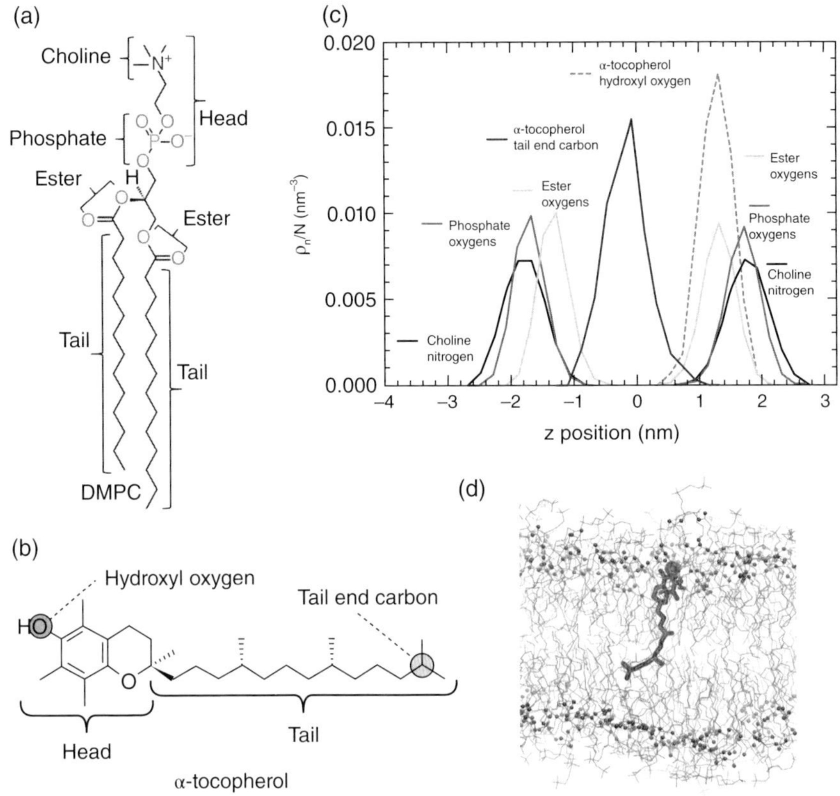

Figure 3.3 Orientation of α-tocopherol in a dimyristoylphosphatidylcholine (DMPC) bilayer. (a) Structure of DMPC. (b) Structure of α-tocopherol. (c) Number density over the number of atoms for various types of atoms as a function of distance from the bilayer center. (d) A snapshot showing α-tocopherol in a DMPC bilayer. Water is not shown. The α-tocopherol is shown as thick lines except for the hydroxyl oxygen which is a large sphere. DMPC is shown as thin lines except the ester oxygens which are shown as small spheres.

Several general conclusions can be drawn from the literature on simulations of small molecules interacting with lipid bilayers.

1. Uncharged, amphiphilic molecules tend to position themselves such that the hydrophilic part of the molecule is associated with the ester groups of the lipids, just below the head groups. The hydrophobic part points inward or aligns with the hydrophobic lipid tails. Amphiphilic molecules with hydrophobic tails extending to near or through the bilayer center tend to flip from one leaflet of the bilayer to the other. These molecules include α-tocopherol, valproic acid, and cholesterol.
2. Groups with a net charge tend not to penetrate past the head groups of the lipids.
3. Charged groups may change their protonation state and become neutral when entering a bilayer. In an amino acid study (MacCallum *et al.*, 2008) and the study of valproic acid (Ulander and Haymet, 2003), the protonation state was considered by calculating free energies of protonated and non-protonated species. The dominant species is that with the lowest free energy at a given distance from the bilayer center. It was found that molecules with negatively charged carboxylate groups (glutamate and aspartate side chains and valproate) become protonated in the head region of the bilayer, making the molecules neutral in the bilayer. This is a mechanism for carboxylates to cross

bilayers by basal permeation. The positively charged side chain of lysine loses a proton partway through the tail region of the bilayer to become neutral. The arginine side chain still has about a 50% chance of being charged even at the center of the bilayer. Both the positively charged side chains drag water into the bilayer, which helps to stabilize them. Despite this stabilization, the barrier for them to cross the bilayer is still large.

4. As pointed out by Orsi and Essex (2010), the difference between the free-energy maximum and the free-energy minimum is the major factor determining the rate of basal permeation of a molecule through bilayers. This means that to maximize basal permeability, molecules must not be either too hydrophilic or too hydrophobic. If they are too hydrophilic, there will be a large barrier for them to cross the hydrophobic tail region of a bilayer: this is the case for water. If they are too hydrophobic, they will go into the bilayer tail region and not easily come back out: this is the case for α-tocopherol.

3.2.2 Polymers and NPs interacting with lipid bilayers

Simulations of small molecules interacting with lipid bilayers are often simulated using atomistic simulations, but simulations of bilayers interacting with larger molecules or particles such as peptides and other linear polymers, carbon nanotubes and fullerenes, dendrimers, and other NPs are often simulated using coarse-grained simulations. The mode of permeation of bilayers may be through pore formation or endocytosis instead of basal permeation. In any case, the larger size of particles requires more rearrangements of the bilayer occurring over longer timescales, often making coarse-grained simulations necessary.

The interactions of large molecules and particles with bilayers can be divided into three types: (i) basal permeation or absorption into the bilayer; (ii) pore formation; (iii) endocytosis. The following paragraphs are a sample of the literature in the area of simulation of particle–bilayer interactions.

As in the case of small molecules, the hydrophobicity of a particle determines whether it will enter a lipid bilayer and whether it will pass through or remain absorbed in the hydrophobic tail region of the bilayer. Hydrophobic/hydrophilic patterning and the shape of particles also have an effect on permeation versus absorption. Nonfunctionalized carbon nanotubes (Hofinger *et al.*, 2011) and fullerenes (Jusufi *et al.*, 2011) are hydrophobic and tend to go into the tail region of the bilayer. Carbon nanotubes (Pogodin *et al.*, 2011) or fullerenes (Gkeka and Angelikopoulos, 2011; Li *et al.*, 2012) with bands of hydrophilic and hydrophobic surface modifications can more easily translocate bilayers. The hydrophobic nature of fullerenes and carbon nanotubes allows those functionalized with a moderate number of hydrophilic hydroxyl groups (Qiao *et al.*, 2007; D'Rozario *et al.*, 2009) or amino derivatives (Kraszewski *et al.*, 2011, 2012) to enter lipid bilayers and enhances their permeability relative to the nonfunctionalized particle. Hydrophilic particles with detachable hydrophobic ligands attached to their surface by noncovalent interactions can pass through bilayers (Ding *et al.*, 2012). Obviously, the shape of NPs also affects their ability to penetrate bilayers (Yang and Ma, 2010). For charged groups which can gain or lose protons, it is necessary to be aware that the protonation state will likely change when they enter a bilayer in order to neutralize them (Kraszewski *et al.*, 2011, 2012).

Some particles induce pores, which could enhance the permeation of the particles themselves or their cargoes, but also could potentially damage or kill cells. Nanotubes with hydrophilic tethers on the ends adopt a transmembrane configuration (Dutt *et al.*, 2011). The nanotube can then act as a pore. Cationic polyamidoamine (PAMAM) dendrimers (Lee and Larson, 2009), gold NPs (Lin *et al.*, 2010, 2011), polycations (Kepczynski *et al.*, 2012), and cyclic (Cirac *et al.*, 2011) and helical (Leontiadou *et al.*, 2006; Santo and Berkowitz, 2012) antimicrobial peptides induce pores. In the case of antimicrobial peptides, and perhaps polycations, the goal is to kill cells. In other cases, the goal is to increase permeation of the cell membrane with minimal damage to the cell.

If NPs do not penetrate the bilayer, they may still be taken up by endocytosis. The process of independent endocytosis (see Figure 3.1) of small NPs has been studied for spheres as a function of their interaction with the bilayer (Ruiz-Herrero *et al.*, 2012), and for spherical and elliptical particles (Yang and Ma, 2011).

As an example of the initial effect of a NP on a lipid bilayer, a 3-nm diameter PLGA NP in contact with a DMPC bilayer was simulated using molecular dynamics. The molecules were represented using a united atom model in which all hydrogens except the polar ones (those bonded to O, N, S, etc.) are combined with the atom they are bonded to, and all other atoms are represented explicitly. A PLGA chain in aqueous solution is negatively charged due to a carboxylate group on one end. Eight PLGA 20-mers were used to construct the particle. The simulation system is shown in Figure 3.4(a). The particle was found to interact strongly with the bilayer: adsorbing to the bilayer surface when placed very close, but moving away when placed slightly further away. This is indicative of a barrier for adsorption to the bilayer, and could be quantified using a free-energy calculation. Once adsorbed, the PLGA particle sank into the bilayer somewhat, caused a local thinning of the bilayer, and locally changed the orientation of the lipid head groups in both leaflets of the bilayer. Further penetration of the bilayer by the particle may have been prevented by the small size of the bilayer used in the simulation.

Figure 3.4b shows the bilayer width measured between the average positions of the phosphorus atoms in each leaflet as a function of distance in the bilayer plane (x–y) from the particle center. There

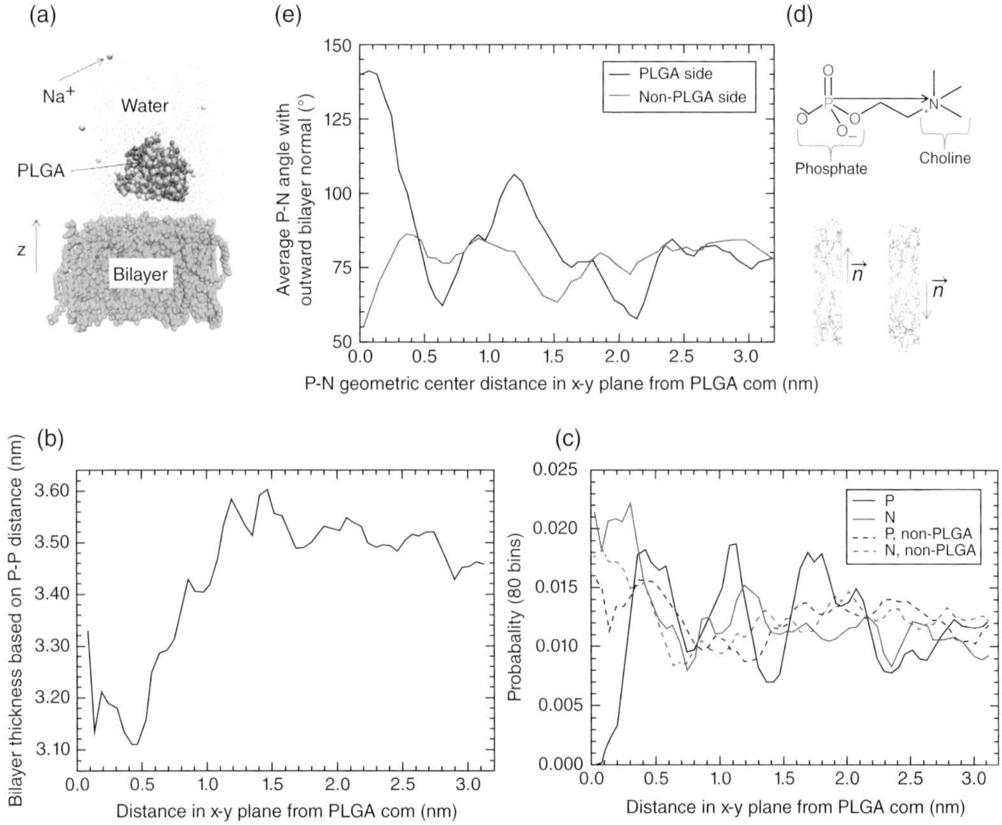

Figure 3.4 Result of the simulation of a 3-nm diameter poly(lactic-co-glycolic acid) (PLGA) nanoparticle in contact with a dimyristoylphosphatidylcholine bilayer. A 24-ns equilibration period was followed by an 8-ns production run from which the above plots were constructed. (a) Simulation system consisting of a lipid bilayer, water, sodium ions, and PLGA. (b) Bilayer thickness as a function of distance from the particle. (c) Probability of finding P and N atoms as a function of distance from the particle. (d) Definition of P–N vector and outward bilayer normal. (e) Angle between P–N vector and outward bilayer normal as a function of distance from the particle.

is a local thinning of the bilayer of about 10%. Figure 3.4c shows the probability based on 80 bins of observing phosphorus (P) or nitrogen (N) atoms as a function of distance in the bilayer plane (*x–y*) from the particle center. There was an excess of N and almost no P near the particle center in the leaflet the particle was adsorbed to. This is likely due to the fact that the choline groups containing the N atoms are positively charged, the phosphate groups containing the P atoms are negatively charged, and the PLGA is negatively charged. The positively charged choline groups are attracted to the negatively charged PLGA, and the negatively charged phosphate groups are repelled by the PLGA. Note that there is an ordering far from the particle center, especially in the P. There may be a corresponding ordering on the particle surface. In this simulation, the carboxylate groups were all considered to be unprotonated. It is likely that any carboxylate groups on the particle surface that are between the particle and the lipid heads are actually protonated. In that case, Figure Figure 3.4c could look very different since the PLGA in direct contact with the lipids would be neutral instead of negatively charged.

Figure 3.4e shows the angle of the P–N vector with the outward bilayer normal as a function of distance between the P–N geometric center and the center of the particle. This is a measure of the orientation of the lipid heads. The P–N vector and outward bilayer normal are defined in Figure 3.4d. In the leaflet that the PLGA is adsorbed to, the heads near the particle center have an average angle greater than 90°, meaning that they tend to point towards the bilayer center. In the other leaflet, the heads tend to point away from the bilayer center. As in the case of the phosphorus positioning, there is ordering far from the particle center in the leaflet that the PLGA is adsorbed to.

3.3 CONCLUSIONS

Nanodelivery systems have many proven advantages for drug and nutraceutical delivery. Cellular interactions play a large role in the efficacy of nanodelivery systems. Understanding how NPs interact with cells is important for designing future nanodelivery systems. Size, surface charge, morphology, and hydrophobicity are NP properties that will determine the mechanism of entry through the cell membrane. Nanoparticle localization within the cell environment will affect the functionality of the entrapped bioactive. For example, α-tocopherol, a small mostly hydrophobic molecule, will remain inside the hydrophobic region of the cell membrane, but if entrapped inside a particle it may be able to get to other organelles such as the nucleus or mitochondria and reduce oxidation of lipids in those areas. Optimization of NP properties will improve delivery of the bioactive by offering several advantages: protection against degradation, controlled-release at the target of interest, and delivery to the cell compartment of interest.

References

Acosta, E. (2009) Bioavailability of nanoparticles in nutrient and nutraceutical delivery. *Current Opinion in Colloid and Interface Science,* 14(1), 3–15.

Asati, A., Santra, S., Kaittanis, C., and Perez, M. (2010) Surface-charge-dependent cell localization and cytotoxicity of cerium oxide nanoparticles. *ACS Nano*, 4(9), 5321–5331.

Bouwmeester, H., Dekkers, S., Noordam, M.Y., *et al.* (2009) Review of health safety aspects of nanotechnologies in food production. *Regulatory Toxicology and Pharmacology*, 53(1), 52–62.

Cartiera, M.S., Johnson, K.M., Rajendran, V., Caplan, M.J. and Saltzman, W.M. (2009) The uptake and intracellular fate of PLGA nanoparticles in epithelial cells. *Biomaterials* 30(14), 2790–2798.

Cerezo, J., Zuniga, J., Bastida, A., Requena, A. and Ceron-Carrasco, J.P. (2011) Atomistic molecular dynamics simulations of the interactions of oleic and 2-hydroxyoleic acids with phosphatidylcholine bilayers. *Journal of Physical Chemistry B*, 115(40), 11727–11738.

Chithrani, B.D., Ghazani, A.A. and Chan W.C.W. (2006) Determining the size and shape dependence of gold nanoparticle uptake into mammalian cells. *Nano Letters*, 6(4), 662–668.

Chiu, Y.L., Ho, Y.C., Chen, Y. M., *et al.* (2010) The characteristics, cellular uptake and intracellular trafficking of nanoparticles made of hydrophobically-modified chitosan. *Journal of Controlled Release*, 146(1), 152–159.

Cirac, A.D., Moiset, G., Mika, J.T., *et al.* (2011) The molecular basis for antimicrobial activity of pore-forming cyclic peptides. *Biophysical Journal*, 100(10), 2422–2431.

D'Rozario, R.S.G., Wee, C.L., Wallace, E.J. and Sansom, M.S.P. (2009) The interaction of C–60 and its derivatives with a lipid bilayer via molecular dynamics simulations. *Nanotechnology*, 20(11), 7.

Den Otter, W.K. and Briels W.J. (1998) The calculation of free-energy differences by constrained molecular-dynamics simulations. *Journal of Chemical Physics*, 109(11), 4139–4146.

Ding, H.M., Tian, W.D. and Ma, Y.Q. (2012) Designing nanoparticle translocation through membranes by computer simulations. *ACS Nano*, 6(2), 1230–1238.

Dutt, M., Kuksenok, O., Little, S.R. and Balazs, A.C. (2011) Forming transmembrane channels using end-functionalized nanotubes. *Nanoscale*, 3(1), 240–250.

Garinot, M., Fiévez, V., Pourcelle, V., *et al.* (2007) PEGylated PLGA-based nanoparticles targeting M cells for oral vaccination. *Journal of Controlled Release*, 120(3), 195–204.

Gaumet, M. and Gurny, R. (2010) Interaction of biodegradable nanoparticles with intestinal cells: The effect of surface hydrophilicity. *International Journal of Pharmaceutics*, 390(1), 45–52.

Gkeka, P. and Angelikopoulos, P. (2011) The role of patterned hydrophilic domains in nanoparticle–membrane interactions. *Current Nanoscience*, 7(5), 690–698.

Gratton, S.E., Ropp, P.A., Pohlhaus, P.D., *et al.* (2008) The effect of particle design on cellular internalization pathways. *Proceedings of the National Academy of Sciences USA*, 105(33), 11613–11618.

Hamdy, N.M., Suwailem, S.M. and El-Mesallamy, H.O. (2009) Influence of vitamin E supplementation on endothelial complications in type 2 diabetes mellitus patients who underwent coronary artery bypass graft. *Journal of Diabetes and Its Complications*, 23(3), 167–173.

Hao, N., Li, L., Zhang, Q., *et al.* (2012) The shape effect of PEGylated mesoporous silica nanoparticles on cellular uptake pathway in Hela cells. *Microporous and Mesoporous Materials*, 162, 14–23.

Harush-Frenkel, O., Debotton, N., Benita, S. and Altschuler, Y. (2007) Targeting of nanoparticles to the clathrin-mediated endocytic pathway. *Biochemical and Biophysical Research Communications*, 353(1), 26–32.

He, C., Hu, Y., Yin, L., Tang, C. and Ying, C. (2010) Effects of particle size and surface charge on cellular uptake and biodistribution of polymeric nanoparticles. *Biomaterials*, 31(13), 3657–3666.

Hofinger, S., Melle-Franco, M., Gallo, T., *et al.* (2011) A computational analysis of the insertion of carbon nanotubes into cellular membranes. *Biomaterials*, 32(29), 7079–7085.

Hu, B., Ting, Y., Zeng, X., and Huang, Q. (2012) Cellular uptake and cytotoxicity of chitosan–caseinophosphopeptides nanocomplexes loaded with epigallocatechin gallate. *Carbohydrate Polymers*, 89(2), 362–370.

Huang, X., Teng, X., Chen, D., Tang, F. and He, J. (2010) The effect of the shape of mesoporous silica nanoparticles on cellular uptake and cell function. *Biomaterials*, 31(3), 438–448.

Hughes, Z.E., Mark, A.E. and Mancera, R.L. (2012) Molecular dynamics simulations of the interactions of DMSO with DPPC and DOPC phospholipid membranes. *Journal of Physical Chemistry B*, 116(39), 11911–11923.

Jusufi, A., DeVane, R.H., Shinoda, W. and Klein, M.L. (2011) Nanoscale carbon particles and the stability of lipid bilayers. *Soft Matter*, 7(3), 1139–1146.

Kastner, J. (2011) Umbrella sampling. *Wiley Interdisciplinary Reviews – Computational Molecular Science*, 1(6), 932–942.

Kepczynski, M., Jamroz, D., Wytrwal, M., Bednar, J., Rzad, E. and Nowakowska, M. (2012) Interactions of a hydrophobically modified polycation with zwitterionic lipid membranes. *Langmuir*, 28(1), 676–688.

Kosinova, P., Berka, K., Wykes, M., Otyepka, M. and Trouillas, P. (2012) Positioning of antioxidant quercetin and its metabolites in lipid bilayer membranes: implication for their lipid-peroxidation inhibition. *Journal of Physical Chemistry B*, 116(4), 1309–1318.

Kraszewski, S., Tarek, M. and Ramseyer, C. (2011) Uptake and translocation mechanisms of cationic amino derivatives functionalized on pristine C–60 by lipid membranes: a molecular dynamics simulation study. *ACS Nano*, 5(11), 8571–8578.

Kraszewski, S., Bianco, A., Tarek, M. and Ramseyer, C. (2012) Insertion of short amino-functionalized single-walled carbon nanotubes into phospholipid bilayer occurs by passive diffusion. *Plos One*, 7(7), e40703.

Kumari, A., Yadav, S.K., Pakade, Y.B., Singh, B. and Yadav, S.C. (2010) Development of biodegradable nanoparticles for delivery of quercetin. *Colloids Surfaces B Biointerfaces*, 80(2), 184–192.

Lawlor, C., Kelly, C., O'Leary, S., *et al.* (2011) Cellular targeting and trafficking of drug delivery systems for the prevention and treatment of MTb. *Tuberculosis*, 91(1), 93–97.

Lee, H. and Larson R.G. (2009) Multiscale modeling of dendrimers and their interactions with bilayers and polyelectrolytes. *Molecules*, 14(1), 423–438.

Leontiadou, H., Mark, A.E. and Marrink, S.J. (2006) Antimicrobial peptides in action. *Journal of the American Chemical Society*, 128(37), 12156–12161.

Li, P.Y., Lai, P.S., Hung, W.C. and Syu, W.J. (2010) Poly(L-lactide)-vitamin E TPGS nanoparticles enhanced the cytotoxicity of doxorubicin in drug-resistant MCF–7 breast cancer cells. *Biomacromolecules*, 11(10), 2576–2582.

Li, Y.F., Li, X.J., Li, Z.H. and Gao, H.J. (2012) Surface-structure-regulated penetration of nanoparticles across a cell membrane. *Nanoscale*, 4(12), 3768–3775.

Lin, J., Novak, B. and Moldovan, D. (2012) Molecular dynamics simulation study of the effect of DMSO on structural and permeation properties of DMPC lipid bilayers. *Journal of Physical Chemistry B*, 116(4), 1299–1308.

Lin, J.Q., Zhang, H.W., Chen, Z. and Zheng, Y.G. (2010) Penetration of lipid membranes by gold nanoparticles: insights into cellular uptake, cytotoxicity, and their relationship. *ACS Nano*, 4(9), 5421–5429.

Lin, J.Q., Zheng, Y.G., Zhang, H.W. and Chen, Z. (2011) A simulation study on nanoscale holes generated by gold nanoparticles on negative lipid bilayers. *Langmuir* 27(13), 8323–8332.

MacCallum, J.L., Bennett, W.F.D. and Tieleman, D.P. (2008) Distribution of amino acids in a lipid bilayer from computer simulations. *Biophysical Journal*, 94(9), 3393–3404.

Marrink, S.J. and Berendsen, H.J.C. (1994) Simulation of water transport through a lipid-membrane. *Journal of Physical Chemistry*, 98(15), 4155–4168.

Muller, T.J. and Muller-Plathe, F. (2009) A comparison of sulfur mustard and heptane penetrating a dipalmitoylphosphatidylcholine bilayer membrane. *Journal of Hazardous Materials*, 168(1), 13–24.

Nam, J., Won, N., Bang, J., *et al.* (2012) Surface engineering of inorganic nanoparticles for imaging and therapy. *Advanced Drug Delivery Reviews*, 65(5), 622–648.

Nitschke, W.K., Vequi-Suplicy, C.C., Coutinho, K. and Stassen, H. (2012) Molecular dynamics investigations of PRODAN in a DLPC bilayer. *Journal of Physical Chemistry B*, 116(9), 2713–2721.

Orsi, M. and Essex, J.W. (2010) Permeability of drugs and hormones through a lipid bilayer: insights from dual-resolution molecular dynamics. *Soft Matter*, 6(16), 3797–3808.

Orsi, M., Noro, M.G. and Essex, J.W. (2011) Dual-resolution molecular dynamics simulation of antimicrobials in biomembranes. *Journal of the Royal Society Interface*, 8(59), 826–841.

Pamujula, S., Hazari, S., Bolden, G. *et al.* (2012) Cellular delivery of PEGylated PLGA nanoparticles. *Journal of Pharmacy and Pharmacology*, 64(1), 61–67.

Peres, I., Rocha, S., Gomes, J., Morais, S., Pereira, M. C. and Coelho, M. (2011) Preservation of catechin antioxidant properties loaded in carbohydrate nanoparticles. *Carbohydrate Polymers*, 86(1), 147–153.

Pinisetty, D., Moldovan, D. and Devireddy, R. (2006) The effect of methanol on lipid bilayers: An atomistic investigation. *Annals of Biomedical Engineering*, 34(9), 1442–1451.

Pogodin, S., Slater, N.K.H. and Baulin, V.A. (2011) Surface patterning of carbon nanotubes can enhance their penetration through a phospholipid bilayer. *ACS Nano*, 5(2), 1141–1146.

Qiao, R., Roberts, A.P., Mount, A.S., Klaine, S.J. and Ke, P.C. (2007) Translocation of C–60 and its derivatives across a lipid bilayer. *Nano Letters*, 7(3), 614–619.

Qin, S.S. and Yu, Z.W. (2011) Molecular dynamics simulations of alpha-tocopherol in model biomembranes. *Acta Physico-Chimica Sinica*, 27(1), 213–227.

Qin, S.S., Yu, Z.W. and Yu, Y.X. (2009) Structural and kinetic properties of alpha-tocopherol in phospholipid bilayers, a molecular dynamics simulation study. *Journal of Physical Chemistry B*, 113(52), 16537–16546.

Ruiz-Herrero, T., Velasco, E. and Hagan, M.F. (2012) Mechanisms of budding of nanoscale particles through lipid bilayers. *Journal of Physical Chemistry B*, 116(32), 9595–9603.

Sahay, G., Alakhova, D.Y. and Kabanov, A. V. (2010) Endocytosis of nanomedicines. *Journal of Controlled Release*, 145(3), 182–195.

Sahoo, S.K., Panyam, J., Prabha, S. and Labhasetwar, V. (2002) Residual polyvinyl alcohol associated with poly (d,l-lactide-co-glycolide) nanoparticles affects their physical properties and cellular uptake. *Journal of Controlled Release*, 82(1), 105–114.

Santo, K.P. and Berkowitz, M.L. (2012) Difference between magainin–2 and melittin assemblies in phosphatidylcholine bilayers: results from coarse-grained simulations. *Journal of Physical Chemistry B*, 116(9), 3021–3030.

Semete, B., Booysen, L., Lemmer, Y., *et al.* (2010) In vivo evaluation of the biodistribution and safety of PLGA nanoparticles as drug delivery systems. *Nanomedicine: Nanotechnology, Biology and Medicine*, 6(5), 662–671.

Shea, T.B., Ortiz, D., Nicolosi, R.J., Kumar, R. and Watterson, A.C. (2005) Nanosphere-mediated delivery of vitamin E increases its efficacy against oxidative stress resulting from exposure to amyloid beta. *Journal of Alzheimer's Disease*, 7(4), 297–301.

Sprik, M. and Ciccotti, G. (1998) Free energy from constrained molecular dynamics. *Journal of Chemical Physics*, 109(18), 7737–7744.

Swarnakar, N.K., Jain, A.K., Singh, R. P., Godugu, C., Das, M. and Jain, S. (2011) Oral bioavailability, therapeutic efficacy and reactive oxygen species scavenging properties of coenzyme Q10-loaded polymeric nanoparticles. *Biomaterials*, 32(28), 6860–6874.

Torrie, G.M. and Valleau, J.P. (1974) Monte-Carlo free-energy estimates using non-Boltzmann sampling – Application to subcritical Lennard–Jones fluid. *Chemical Physics Letters*, 28(4), 578–581.

Ulander, J. and Haymet, A.D.J. (2003) Permeation across hydrated DPPC lipid bilayers: Simulation of the titrable amphiphilic drug valproic acid. *Biophysical Journal*, 85(6), 3475–3484.

Venkataraman, S., Hedrick, J.L., Ong, Z.Y., *et al.* (2011) The effects of polymeric nanostructure shape on drug delivery. *Advanced Drug Delivery Reviews*, 63(14–15), 1228–1246.

Verma, A. and Stellacci, F. (2010) Effect of surface properties on nanoparticle–cell interactions. *Small*, 6(1), 12–21.

Verma, A., Uzun, O., Hu, Y. *et al.* (2008) Surface-structure-regulated cell-membrane penetration by monolayer-protected nanoparticles. *Nature Materials*, 7(7), 588–595.

Wang, J., Wang, Y. and Liang, W. (2012) Delivery of drugs to cell membranes by encapsulation in PEG–PE micelles. *Journal of Controlled Release*, 160(3), 637–651.

Win, K.Y. and Feng, S.S. (2005) Effects of particle size and surface coating on cellular uptake of polymeric nanoparticles for oral delivery of anticancer drugs. *Biomaterials*, 26(15), 2713–2722.

Yacoub, T.J., Reddy, A.S. and Szleifer, I. (2011) Structural effects and translocation of doxorubicin in a DPPC/Chol bilayer: the role of cholesterol. *Biophysical Journal*, 101(2), 378–385.

Yang, K. and Ma, Y.Q. (2010) Computer simulation of the translocation of nanoparticles with different shapes across a lipid bilayer. *Nature Nanotechnology*, 5(8), 579–583.

Yang, K. and Ma, Y.Q. (2011) Wrapping and internalization of nanoparticles by lipid bilayers: a computer simulation study. *Australian Journal of Chemistry*, 64(7), 894–899.

Yoo, J.-W. and Mitragotri, S. (2010) Polymer particles that switch shape in response to a stimulus. *Proceedings of the National Academy of Sciences*, 107(25), 11205–11210.

Zhang, K., Fang, H., Che, Z., Taylor, J.S.A. and Wooley, K.L. (2008) Shape effects of nanoparticles conjugated with cell-penetrating peptides (HIV Tat PTD) on CHO cell uptake. *Bioconjugate Chemistry*, 19(9), 1880–1887.

Zorko, M. and Langel, Ü. (2005) Cell-penetrating peptides: mechanism and kinetics of cargo delivery. *Advanced Drug Delivery Reviews*, 57(4), 529–545.

4 Interfacial science and the creation of nanoparticles

Stephanie R. Dungan

Department of Food Science and Technology, Department of Chemical Engineering and Materials Science, University of California, Davis, California, USA

4.1 INTRODUCTION

According to the US National Science Foundation definition of nanotechnology, nanoparticles must be between 1 and 100 nm in size, with special properties associated with this very small characteristic dimension (National Science Foundation, 2000). In particular, for food applications, this nanoscale size gives nanoparticles the potential to act as tiny containers, hosting ingredients in a form that enhances their value through improved stability, reactivity, loading and/or delivery of the ingredient. In this book, several promising applications of nanoparticles for foods are discussed, including increasing oxidative stability of bioactives (Chapters 8 and 9) and delivery of phytochemicals of improved bioavailability (Chapter 10). Nanoparticle structures are already found naturally in our foods, where they often provide similar enhanced properties. Examples include casein micelles (de Kruif *et al.*, 2012), which influence the delivery of protein and calcium in milk, bile salt/phospholipid micelles and vesicles, which influence absorption of lipids within the gut, and lipoprotein particles, which are specialized biological aggregates for the distribution of water-insoluble lipids within an aqueous matrix.

The small size of nanoparticles gives them extremely large interfacial areas of contact with the surrounding medium. If we imagine for simplicity that all the particles in a mixture are spheres of equal radius a, then the specific area A_{sp} surrounding the particles per total volume V_T of the mixture is given by $A_{sp} = 4\pi a^2 N/V_T$, where N is the number of particles in the total system. The total volume can be rewritten as the ratio of the volume of the particle phase divided by their volume fraction ϕ: $V_T = 4\pi a^3 N/(3\phi)$. This yields for the specific area the relation

$$A_{sp} = 3\phi / a. \tag{4.1}$$

Thus, the smaller the size a of the particle, the greater its surface area per unit volume. The very high specific surface area of nanoparticles means that aspects of interfacial science play a major role in their formation, stability, and other attributes. This connection between fundamental interfacial properties and our ability to form and improve nanovehicles for use in foods is the focus of this chapter.

It is often the case that nanoparticles do not mix thermodynamically with the surrounding matrix, and therefore they represent a highly subdivided and dispersed second phase. In this case, the particle

Nanotechnology and Functional Foods: Effective Delivery of Bioactive Ingredients, First Edition.
Edited by Cristina M. Sabliov, Hongda Chen, and Rickey Y. Yada.

always pays an energetic penalty per unit area at its outer surface. These types of multiphase nanoparticle mixtures are described as lyophobic ("solvent-fearing") in the colloids literature (Hunter, 2001). To form lyophobic particles requires a series of steps involving addition of energy, because of the need to create additional unfavorable interface as the particle size is reduced. The characteristics of the nanoparticles formed will depend on the specific processing steps used, the efficiency with which energy is added, and the ability of the manufacturer or scientist to modify the interfacial properties of the surface as it is created. Many recent advances in the formation of new types of nanoparticles have come about because of an improved ability to control interfacial tension during particle formation or to deliver energy more efficiently to the particle surface.

As part of a dispersion of two or more phases, lyophobic nanoparticles are thermodynamically unstable, and tend to evolve over time towards larger particles, with a decreased interfacial area of contact. Thus, nanoparticles will coarsen, aggregate, and separate from the continuous phase over time. For similar reasons the particles will also be attracted and bind to other types of surfaces that are more favorable for the particles than the continuous phase. Controlling the rate of these processes is critically important for maintaining nanoparticle stability in foods, and for enhancing their performance as delivery vehicles. Interactions between surfaces play a key role in this control.

Nanoparticles can also form spontaneously to form a single-phase mixture within a continuous phase. In contrast to lyophobic systems, the interfacial interaction of these nanostructures is sufficiently low that the particles will preferentially disperse homogeneously throughout the mixture, and remain in this state indefinitely. Molecules exhibiting such behavior include amphiphilic compounds such as fatty acid esters, phospholipids, sucrose esters, and proteins such as casein. The spontaneous formation of these particles is termed self-assembly, with the resulting structures representing an energy minimum for the system. For these reasons, these self-assembled aggregates have predictable sizes and morphologies that are independent of the method of their formation, and these preferred structures will remain stable as long as the mixture temperature, pressure, and composition do not change.

4.2 FUNDAMENTALS OF INTERFACIAL SCIENCE

4.2.1 Equilibrium surface properties

Interfacial adsorption of molecules is an important feature of nanoparticle dispersions, due to their large surface areas and the need to reduce the energy penalty associated with that interface. At equilibrium and at constant temperature T, the adsorption of individual molecules to an interfacial region is governed thermodynamically by the Gibbs adsorption equation:

$$-\mathrm{d}\gamma = \sum_i \Gamma_i \mathrm{d}\mu_i. \tag{4.2}$$

In equation (4.2), γ is the interfacial tension, Γ_i is the surface excess concentration of component i, and μ_i is the chemical potential of that same component in the bulk solution. The summation may be taken over all components other than the solvent, if the interface is described as a region containing no surface excess of solvent, relative to its properties in the bulk (Hunter, 2001). For a typical surface-active component s, the concentration c_s in the bulk phase will be low, and the chemical potential of that component can be described assuming an ideal dilute solution, so that $\mathrm{d}\mu_s \approx kT\,\mathrm{d}\ln c_s$, where k is Boltzmann's constant. This assumption makes it possible to conduct experiments in which variations in bulk concentration of an amphiphile can be connected to changes in surface concentration and interfacial tension, through the expression

$$-\frac{1}{kT}\frac{\mathrm{d}\gamma}{\mathrm{d}\ln c_s} = \Gamma_s\left[1+\frac{1}{kT}\sum_{i\neq s}\frac{\Gamma_i}{\Gamma_s}\frac{\mathrm{d}\mu_i}{\mathrm{d}\ln c_s}\right]. \tag{4.3}$$

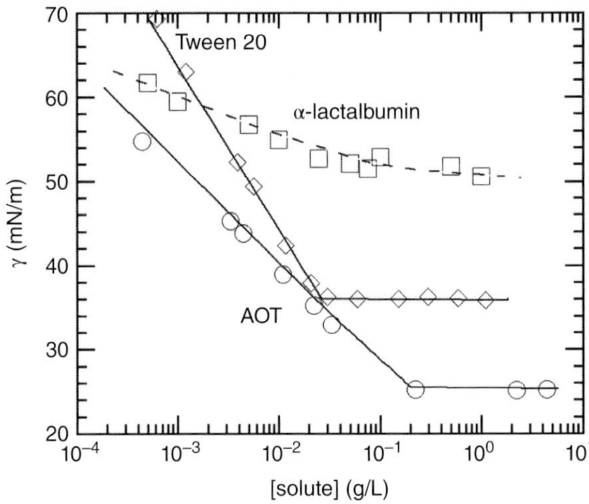

Figure 4.1 Equilibrium surface tension of aqueous solutions containing Tween 20 (Niño and Patino, 1998), Aerosol-OT (AOT), or α-lactalbumin. The AOT and α-lactalbumin solutions contained 0.125 M NaCl.

If s is the only surface-active component, or if the chemical potential of other components are unaffected by changes in concentration c_s, then equation (4.3) simplifies to the familiar form

$$-\frac{1}{kT}\frac{d\gamma}{d\ln c_s} = \Gamma_s. \tag{4.4}$$

For components whose surface concentration and chemical potential are linked to that of the solute s, however, the term in brackets in equation (4.3) will be important. For example, for charged amphiphiles in the absence of excess salt, there is a coupling between the amphiphile and its counterion(s), giving rise to an additional factor m in the result:

$$-\frac{1}{mkT}\frac{d\gamma}{d\ln c_s} = \Gamma_s. \tag{4.5}$$

$m = 2$ for a charged surfactant such as sodium dodecyl sulfate at low salt concentrations. In the presence of salt concentrations well in excess of the amphiphile, the surface behavior of charged solutes again reduces to equation (4.4).

Figure 4.1 presents surface tension data from our laboratory for the anionic, two-tailed Aerosol-OT (AOT) and the whey protein α-lactalbumin, as well as literature data for the nonionic surfactant Tween 20 (Niño and Patino, 1998). Measurements for the first two amphiphiles were conducted in the presence of 0.125 M NaCl, far in excess of the concentrations of surfactant or protein. The steeper slope of the data on the semi-log plot for AOT or Tween 20 than for the protein is a reflection of the significantly higher surface concentration Γ_s for the low molecular weight surfactants, as indicated by equation (4.4). Despite the more complex nature of the protein molecule and its potential for nonideal behavior, Krishnan and coworkers (2003) have recently found that globular proteins in the presence of excess salt exhibit behavior in reasonably good agreement with the predictions of equation (4.4). Figure 4.1 shows that the higher molecular packing of the low molecular weight surfactants at the interface causes them to lower the surface tension far more significantly over this range of concentrations than does the protein. The constant slope of lines through the data for AOT and Tween 20 also indicates that at these bulk concentrations, the interface is maximally packed or saturated with surfactant (i.e., Γ_s is constant).

From this review of the Gibbs adsorption equation, several practical consequences for nanodroplet formulations in food applications can be identified. First, this equation provides a pathway for using simple measurements of surface or interfacial tension to estimate the surface concentration Γ_s of a solute. Multiplying Γ_s by the specific area of the nanoparticle surfaces provides an estimate of the amount of amphiphile needed to coat the particles and lower their surface energy. As this calculation and equation (4.1) shows, the smaller the radius a of the nanoparticles, the larger their specific area, and the larger the amounts of amphiphile that are required at their surfaces. This represents one limitation of applications of nanoparticles in foods: to yield the very small particle sizes desired may require amounts of amphiphile that are costly, or may have adverse effects on the food palatability. For this reason, nanoparticles in foods are often most appropriately used in small quantities overall.

Second, knowing how densely the solute will pack at the particle surface is also important, because this determines the surface properties of the particle and how it interacts with its surroundings. The amount per unit area of surface charge, of bulky surface groups, or of specific chemical functional groups applied to the particle, are all related to the value of Γ_s. Such surface groups are very important in determining the tendency of particles to aggregate, to bind to surfaces, or to interact with enzymes at their surfaces.

Although the amount of amphiphile needed to saturate the droplet surfaces depends inversely on the particle size, the reverse is not also true: the particle size cannot be decreased without limit simply by increasing the amount of surfactant, as the Gibbs adsorption relation also shows. The surface or interfacial tension will be lowered only to a certain extent before the amphiphile in the bulk begins to self-assemble (at the critical micelle concentration) or to phase-separate from the solution. Reduction in droplet size requires the application of mechanical stress τ sufficient to overcome the Laplace pressure difference $\Delta\Pi$ across the droplet surface; for a spherical droplet of radius a this is given by

$$\Delta\Pi = 2\gamma / a. \tag{4.6}$$

For a given mechanical stress, there will therefore be an associated limiting size $a \sim \gamma/\tau$ below which droplet size cannot be further reduced, and this size is constrained by the minimum value of γ for the mixture. The nature of the stress – viscous, turbulent, or pressure-driven – depends on the nature of the processing device used. Thus, reducing the minimum surface or interfacial tension obtainable by choosing an appropriate amphiphile type, and knowing the amphiphile concentration necessary to attain that minimum, are two important aspects of developing the nanoparticle formation process. Often mixtures of surfactant will behave synergistically, so that in combination they lower surface energies more effectively than either component individually; nanoparticles developed for drug or nutraceutical delivery applications often employ surfactant combinations (Huang *et al.*, 2010; Mazzoferro *et al.*, 2013). Such synergistic effects occur typically in the presence of attractive interactions between amphiphiles at the interface (Siddiqui and Franses, 1996).

4.2.2 Dynamic surface properties

Because the process of deforming and breaking larger droplets into smaller involves the creation of a new interface, it is also important to consider whether sufficient surfactant can be delivered to the new droplet surface over the timescale of its formation, in order to attain the minimum possible tension. Consequently, mass transfer rates of amphiphile to the particle surface are also of great significance. Figure 4.2 shows data for the time-dependent surface tension of a 0.125 M NaCl solution in the presence of 10^{-5} M AOT or α-lactalbumin. The AOT surfactant exhibits an induction time of a few seconds where little change in tension is observed, followed by a rapid decrease in tension. The shape of this change is consistent with theoretical predictions for surfactant adsorption kinetics (Miller and Kretzschmar, 1991). No induction delay is observed with the protein solutions, in contrast, and the initial more rapid decrease in tension is followed by a later, slow decline in the tension. This slower

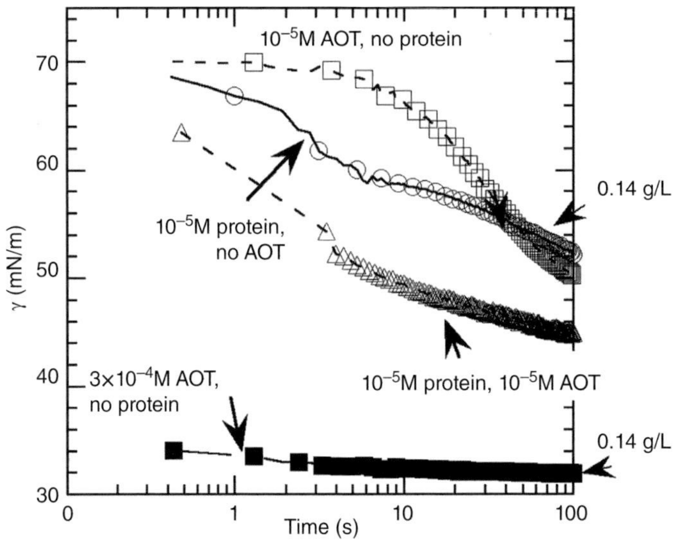

Figure 4.2 Dynamic surface tension of aqueous solutions containing Aerosol-OT (AOT) and/or α-lactalbumin and 0.125 mM NaCl. The α-lactalbumin solution concentration was 10^{-5} M (○, 0.14 g/L); AOT concentrations were 10^{-5} M (□, 0.003 g/L) or 3×10^{-4} M (■, 0.14 g/L). The AOT/α-lactalbumin mixtures contained 10^{-5} M of each component (Δ).

rate of decrease at longer times is likely a result of the larger size and slower diffusion coefficient of the protein; prediction of protein adsorption kinetics remains a complex challenge (Miller *et al.*, 2004). If instead, protein and surfactant solutions are compared at the same mass concentration (0.14 g/L), the faster adsorption of the low molecular weight surfactant is even more stark. At this higher concentration (3×10^{-4} M), the AOT is approaching its apparent critical micelle concentration (Figure 4.1), and the adsorption kinetics at this point become very rapid. Finally, we note that combining the protein and surfactant yields kinetic behavior very similar to the protein solution alone, but the mixture attains a lower surface tension, suggesting complex formation between α-lactalbumin and AOT. Such protein–surfactant complexes have been observed previously with other proteins (Miller *et al.*, 2004), and the potential for such complexes to improve interfacial properties in forming food nanoparticles should be explored more extensively.

4.2.3 Self-assembly and phase separation

Data for Tween 20 and AOT in Figure 4.1 show a break point, corresponding to a critical concentration, above which the surface tension shows little change. Above this concentration, not only is the interface saturated and no longer affected directly by additional surfactant (indeed, the interface is saturated over the entire constant slope region in Figure 4.1), but the activity of the surface in aqueous solution becomes only weakly dependent on the surfactant concentration. For the more hydrophilic surfactant Tween 20, this break corresponds to the critical micelle concentration (CMC), above which the amphiphile self-assembles to form spherical micelles containing approximately 80 surfactant molecules. The micelle radius is only a few nanometers in dimension, making these structures thermodynamically stable nanoparticles that form spontaneously without the use of high-energy equipment.

Micelle formation is driven by the hydrophobic interactions of amphiphile "tail" groups with water, and the resulting cluster therefore contains a hydrophobic region within its tail-rich core. For this reason, micelles can host or "solubilize" poorly water-soluble molecules, enhancing their solubility in the aqueous solution to a very large extent. As a greater number of solute molecules are incorporated within the micelle, they increasingly influence its energy and properties, ultimately swelling the structure sufficiently to form a microemulsion droplet. Microemulsion droplets are thermodynamically stable droplets – typically a few nanometers in dimension – that contain a core of liquid solute surrounded by surfactants. Like micelles, these droplets form spontaneously without the need for high-energy processing, and have predictable structures based on thermodynamics, rather than being dependent on their processing history during formation.

Micelles and microemulsions are amongst the smallest of nanostructures, and thus have very high specific areas at their surface. The formation of these aggregates, as a consequence, requires very low interfacial energy penalties at that surface. This low-energy penalty is achieved by the surfactant's ability to adopt an optimal molecular packing within the aggregate. The optimal packing depends heavily on the molecular architecture of the amphiphile and its associated intermolecular interactions, and largely determines the self-assembled size and shape of the micelle. Micelle shapes include spheres, ellipsoids, cylinders, and disks. Incorporation of significant solute in the core alters the interfacial packing. The amount of ingredient that can be solubilized within a microemulsion droplet depends on this interfacial packing and its relationship to the so-called "curvature energy" of the interface (Helfrich, 1978). The latter arises from the difference between the actual curvature around the droplet, related to its size and shape, and the preferred or "spontaneous" curvature, that corresponds to the optimal packing and the lowest energy conformation. Structures that self-assemble in water favor a "positive" spontaneous curvature, with the interface curved towards the hydrophobic core, whilst negatively curved structures form in organic phases, with their interfaces curved towards a water or polar core (Figure 4.3). Between these extremes are the bicontinuous phases, with nearly flat curvature at their interfaces. This latter preferred curvature creates channels of both oil and water that span the mixture, allowing the system to solubilize comparable amounts of aqueous and organic liquids (Figure 4.3).

Tween 20 is a nonionic surfactant with a bulky head group and a single, 12-carbon tail group, making it a hydrophilic surfactant that forms positively curved structures in water, such as micelles or oil-in-water microemulsion droplets. The AOT, on the other hand, has two eight-carbon tail groups, with a molecular packing that prefers to curve away from the tail groups. In the presence of salt, this more hydrophobic surfactant does not form spherical micelles at the break point shown in Figure 4.1, but instead forms a mixture of bilayers and vesicles, followed shortly by a phase separation into two phases. The second phase is a surfactant-rich lamellar liquid crystal phase (Ghosh and Miller, 1987; Grillo *et al.*, 2003). This type of behavior is quite common amongst the major natural food surfactants, including phospholipids, mono- and diglycerides, and fatty acid derivatives. Their long-chain single or

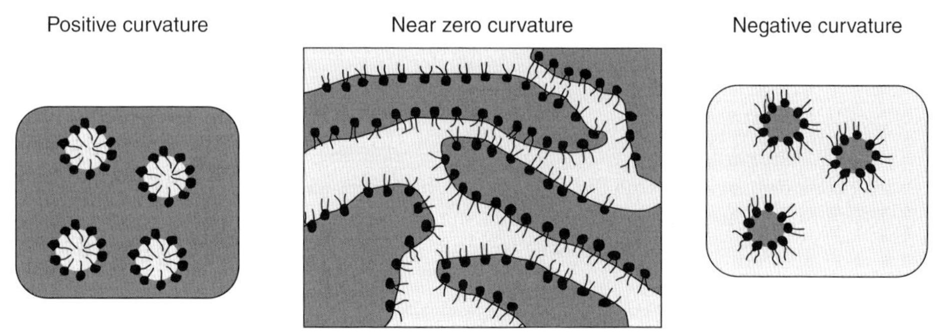

Figure 4.3 Cartoons showing self-assembled nanostructures formed at equilibrium in aqueous–organic–surfactant mixtures, and their associated interfacial curvatures (Adapted from Patel *et al.*, 2007).

double hydrophobic tails make these molecules very hydrophobic, and their overall structure favors separation from the water in the form of a liquid crystal phase. For example, bilayers may become stacked to yield lamellar structures with long-range order, with thin water regions separating the layers. The interfacial curvature in these lamellar phases is close to zero, as favored by the molecular packing properties of these hydrophobic lipids.

For example, soybean or egg lecithin (a phospholipid mixture dominated by phosphatidylcholine) separates from water at extremely low concentrations to form a lamellar liquid crystal phase (Rydhag and Wilton, 1981; Kabalnov *et al.*, 1996). Lecithin also has very low solubility in triglycerides, at less than 0.1 wt% (Rydhag and Wilton, 1981; Kabalnov *et al.*, 1996; von Corswant *et al.*, 1997). When mixed with water and oil, lecithin again forms a lamellar phase with very low internal amounts of water or triglycerides, in equilibrium with excess water and oil (a three-phase system). Because of the extremely limited ability of these amphiphiles to dissolve or to self-assemble in water or oil, their mass-transfer rates in multiphase mixtures are very slow (a very good property for a membrane structural component!). As a result, their adsorption rates to an interface from solution are also quite low, and they will not quickly reduce surface or interfacial tensions (Kabalnov *et al.*, 1995). On the other hand, once situated at equilibrium at the interface, lecithins are capable of considerable lowering of the tension, and their packing properties also help them to make conventional emulsions very stable (Kabalnov *et al.*, 1995; Mele *et al.*, 2004). Although the stiff and low curvature properties of phospholipids do not favor self-assembly with triglycerides, more compact or polar lipids/oils can penetrate the phospholipid tails, and allow formation of water-in-oil microemulsion structures to some extent (Kabalnov *et al.*, 1996; Ichikawa *et al.*, 2000).

4.2.4 Interactions at the interface

In adsorbing to the interface, amphiphilic molecules not only alter the interfacial tension, they introduce surface groups that affect the way the nanoparticle interacts with its surroundings. These surface groups play a large role in determining the stability of lyophobic nanoparticles to processes of aggregation and coalescence/fusion; they also influence the tendency of the particle to bind to other surfaces. Of particular importance are the introduction of charged groups at the surface of a particle in water, which creates the potential for relatively long-range electrostatic interactions with other particles (Eastman, 2010), and large bulky groups, which extend away from the particle surface and prevent the close approach of other surfaces or of macromolecules. The ability of the interface to deform and bend, temporarily adopting curvatures far from its preferred curvature, is a key determinant of how readily a fluid or deformable particle will coalesce or fuse with another (Kabalnov and Wennerström, 1996; Kabalnov *et al.*, 1996).

Interparticle interactions and breakdown mechanisms for particle dispersions are considerable topics in their own right, and will not be elaborated upon here. Binks (1998), Hunter (2001), and van Duijneveldt (2010) provide excellent discussions of these issues. Mechanisms of coarsening due to aggregation, coalescence and fusion work against the acknowledged goal of creating very small, nano-sized particles, and ultimately lead to a loss of dispersion stability. This is true not only during storage and utilization of the particles, but even during the formation process, when aggregation and coalescence can occur simultaneously with the process of droplet break-up. Two other dispersion destabilization mechanisms also need to be considered. One is gravitational separation (sedimentation or creaming). Caused by density differences between the particle and continuous phase, creaming and sedimentation rates scale approximately as the particle radius squared. Thus, a major advantage of forming nanoparticle dispersions is their excellent stability to gravitational separation. On the other hand, Ostwald ripening can be a particular concern for these systems. The Laplace pressure difference at the particle surface increases the chemical potential of molecules inside the particle, enhancing their tendency to partition into the phase outside the particle. According to equation (4.6), this effect becomes stronger for smaller particles, and leads to regions of higher concentration outside the smaller particles in the distribution of sizes, compared to that outside large particles. This concentration gradient drives molecular transport from small particles to large, causing coarsening ("ripening"). The Laplace pressure effect (equation 4.6) driving the process increases with increased interfacial tension and with decreased particle size, and it is the latter effect which makes it a particular concern in nanoparticle dispersions.

4.3 INTERFACIAL PROPERTIES IN NANOPARTICLE FORMATION

4.3.1 Lyotropic nanoparticles

The fundamentals of interfacial science discussed in the preceding section play a critical role in the formation of food nanoparticles. Most nanoparticles are lyotropic, representing a separate phase from water or other outside liquid. The formation of such nanoparticles therefore involves addition of energy through various processing steps, with emulsification of immiscible liquids often being central to this process. Table 4.1 provides several examples of methods for food nanoparticle production, along with the nature of the emulsion formed during the process and the key emulsification features. These examples are discussed in more detail below. For more information on various types of nanodelivery systems the reader is directed to Chapters 11–18 and to Chapter 5 for details on controlling properties of emulsions using various emulsification devices.

4.3.1.1 Nanoemulsions

Nanoemulsions are dispersions of one immiscible liquid within another (usually oil-in-water or water-in-oil), where the dispersed-phase particle size has been reduced to between 10 and 100 nm. As discussed above, the emulsion droplets are thermodynamically unstable, high-energy particles, due to the unfavorable energy penalty associated with their surfaces – a penalty which becomes more severe the

Table 4.1 Examples of the role of emulsification in nanoparticle formation, with key interfacial features.

Particle	Description	Liquids emulsified, surfactant	Features of emulsion formation
Nanoemulsions: high-energy processing	Emulsification using ultrahigh-pressure homogenizers, microfluidizers, or ultrasonicators	Oil and aqueous phases, various surfactants	High surfactant concentrations; rapid adsorption, low interfacial tensions
Nanoemulsions: phase inversion methods	Emulsification under conditions favoring ultralow tensions, followed by rapid change in temperature or composition	Oil and aqueous phases, typically nonionic surfactants, spontaneous curvature a strong function of temperature or composition	Emulsification at ultralow interfacial tension; then rapid change in preferred interfacial curvature to stabilize
Solid lipid nanoparticles	Emulsification at high temperature with liquid oil; cooling	High melting oil and aqueous phases, various surfactants	Dispersed phase at least partially solid at temperature of storage
Nanosized liposomes	Emulsification using ultrahigh-pressure homogenizers, microfluidizers, or ultrasonicators	Lamellar and aqueous phases, containing bilayer-forming amphiphiles	Deform and rupture large MLV into nanosized, unilamellar particles

MLV, multilamellar vesicle.

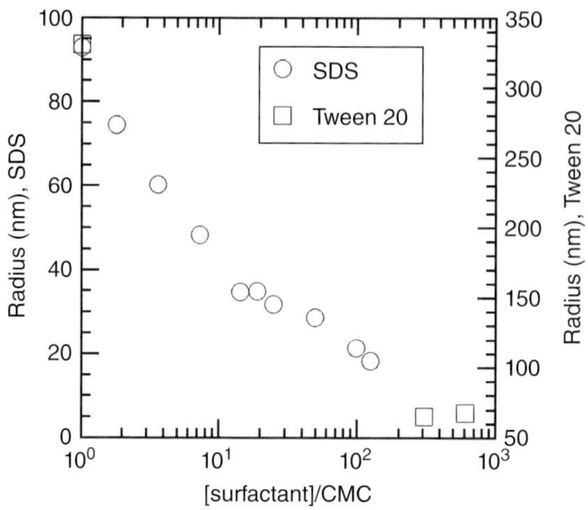

Figure 4.4 Effect of surfactant type and concentration on nanoemulsion droplet size. Left axis: volume-weighted droplet radius, measured by dynamic light scattering, as a function of sodium dodecyl sulfate (SDS) concentration. Right axis: Sauter mean radius measured by static light scattering, for emulsions formed with high-molecular-weight modified starch emulsifier and various concentrations of Tween 20. Data from Mason *et al.* (2006) and Jafari *et al.* (2006).

smaller the droplet sizes, according to equation (4.1). Compared to conventional emulsions, therefore, formation of nanosized emulsion droplets requires one to augment the energy transmitted to the droplet via fluid mechanical stresses, and/or reduce the interfacial tension of the droplet interface. Nanoemulsion formation via high-energy processing focuses more heavily on the former approach, taking advantage of recently developed high-pressure homogenizers, microfluidizers or ultrasonicators to overcome higher Laplace pressures (equation (4.6)) that are associated with smaller droplets (Meleson *et al.*, 2004; Mason *et al.*, 2006). Nevertheless, to be very effective, this high-energy processing requires a simultaneous lowering of the interfacial tension during droplet deformation and break-up. Meleson *et al.* (2004) found that droplet size decreased steeply with increasing concentrations of the hydrophilic surfactant sodium dodecyl sulfate (SDS), with concentrations chosen to be at or above the CMC (Figure 4.4). By increasing the surfactant concentration and thus the rate of surfactant adsorption during emulsification, the droplet size decreased from the upper end of the "nano" range (~100 nm), to less than 20 nm. Mason *et al.* (2006) further stated that: (i) surfactants used effectively to form nanoemulsions need to be able to adsorb to the interface rapidly from the adjacent liquid solvent – a condition achieved by adding the surfactant at levels well above its CMC; and (ii) the surfactant should not form liquid crystal phases in the solvents, since adsorption rates are very slow from these structures.

These two requirements would seemingly rule out many of the more hydrophobic food-grade surfactants, such as lecithin or long-chain mono- and diglycerides, as well as many proteins, because of their slow adsorption kinetics. Interestingly, reports of nanoemulsions formed using proteins, lecithin, or other liquid crystal-forming lipids generally find that high-energy processing yields droplets with sizes on the order of 100 nm or larger (Shukat and Relkin, 2011; Donsi *et al.*, 2012; Gupta and Ghosh, 2012), rather than the smaller droplets that might be attainable with smaller, more hydrophilic amphiphiles. This issue was demonstrated nicely in a paper by Jafari *et al.* (2007), in which nanoemulsions were formed using a microfluidizer from food oil, water, a biopolymer, and three different concentrations of Tween 20 (Figure 4.4). In the absence of Tween 20, droplets of radius close to 350 nm were formed; however, addition of 3 or 6 wt% Tween 20 reduced the droplet sizes to approximately 60 nm. Qian and McClements (2011) reached a similar conclusion by comparing low-molecular-weight surfactants to protein emulsifiers.

A clever approach for further reducing emulsion droplet sizes, even in the absence of optimal surfactant type and concentration, is by the solvent evaporation method (Horn and Rieger, 2001; Fryd and Mason, 2010; Lee and McClements, 2010). Here, a volatile organic solvent is incorporated into the dispersed phase during emulsification. After the droplets are formed, evaporation of this solvent is induced via low pressure, with a resulting shrinkage in the particle volume. Troncoso *et al.* (2011) used this approach to form liquid corn-oil-in-water nanodroplets, with sizes that were decreased by almost 70% from that obtained without the volatile solvent. The volatile solvent not only shrinks the size through evaporation, it lowers the viscosity of the drop during the initial emulsification step, and this too reduces the droplet radius.

It is also possible to form nanoemulsions with a low-energy process, by instead focusing on systems with ultralow interfacial energies. Solans (2005) provided a very useful review of this approach. Such low interfacial tensions are observed with surfactants whose self-assembly in oil–water mixtures is a strong function of temperature or composition. Nonionic surfactants with polyoxyethylene headgroup chains exhibit such behavior. Such surfactants have a specific "phase-inversion temperature" (PIT), where the spontaneous curvature of the surfactant monolayer switches from negative to positive. At this cross-over temperature, where a near-zero curvature interface is favored, the interfacial tension of the oil–water interface becomes very low, on the order of 10^{-2}–10^{-5} mN/m. Thus, even low-energy processing of these mixtures is able to achieve droplets of an extremely small size. However, coalescence between the droplets is unfortunately rapid at the PIT. Thus, the nanoemulsion is rapidly cooled upon emulsification, bringing the temperature into the region favoring positive curvature and stabilizing the newly formed emulsion. A similar concept can be employed using changes in oil–water ratios near the phase-inversion composition. Yang *et al.* (2012) compared the latter approach to high-energy formation of nanoemulsions, using nonionic, polyoxyethylene-containing food surfactants in the Tween series for both processes. It is important to note that, like in the high-energy nanoemulsification process, higher concentrations of surfactant in the PIT method will tend to reduce the droplet size (Solans, 2005; Yang *et al.*, 2012). Although in approaches utilizing phase-inversion behavior the energy requirements are much reduced, amphiphiles exhibiting a specific self-assembly behavior are required, and the processor must work within that known amphiphilic phase behavior.

4.3.1.2 Solid lipid nanoparticles

The small size of nanoparticles can have the disadvantage of speeding up the release of bioactive agents relative to that from larger particles (Mehnert and Mäder, 2001) (cf. section 4.4). Solid lipid nanoparticles have been proposed as a way to incorporate valuable ingredients within the solid or partially solid core of the nanoparticle, where it will be slow to release due to the solid matrix. The solid core may also retard migration into the particle of reactive molecules, such as oxygen, that can degrade the active agent over time (Helgason *et al.*, 2009a; Tikekar and Nitin, 2011). Solid lipid nanoparticles are often formed by high-energy emulsification at a temperature above the melting temperature of the solid core material. The resulting nanoemulsion is then quickly cooled, leading to crystallization of the internal phase. In addition to the central roles for the surfactant in influencing the particle-size distribution of the initial high-temperature nanoemulsion, interfacial interactions between the surfactant and the internal and external phases also appear to be important in affecting the crystal structure in the core and the tendency for particle aggregation (Helgason *et al.*, 2009a,b).

4.3.1.3 Liposomes

Formation of nanosized liposomes begins with a two-phase mixture of water and a liquid crystalline phase. The latter is typically a concentrated phase containing phospholipids in a lamellar arrangement, although other liquid crystalline morphologies are also used (Ulrich, 2002). It requires little energy input to disperse the phospholipid phase within the aqueous liquid, to form a dispersion of large (>0.1 μm) particles consisting of multiple bilayer shells – a "multilamellar vesicle" or MLV. As with emulsion droplets, these dispersed particles are not in their lowest energy state, but here the energy penalty arises less from the interfacial tension and more from the bending-energy penalty paid as the bilayer deviates from its preferred curvature or packing. Typically this preferred curvature for phospholipids is close to zero, and the large size of the MLV particles thus forces the bilayers to deviate from that curvature only to a small extent. Breaking the large MLV into smaller nanoparticles, each with a single bilayer ("unilamellar vesicles" or ULV) increases the bending-energy penalty at the interface considerably. It therefore requires the use of the same types of high-energy processing equipment as employed in forming nanoemulsions (Taylor *et al.*, 2005). Membrane-extrusion approaches, where the MLV are squeezed through small pores, have also been employed (Hwang *et al.*, 2011). In efforts to reduce and control the size of liposomes, knowledge of the influence of temperature, molecular architecture, and lipid composition on the curvature energies of the bilayers is key (Helfrich, 1978; Tayebi *et al.*, 2012).

4.3.2 Self-assembled nanoparticles

In contrast to lyophobic nanoparticles, self-assembled structures form spontaneously without the use of specific processing steps or high-energy input mechanisms. Incorporation of bioactive molecules in these self-assemblies must be driven by the lower energy that results when the agent is hosted within the surfactant aggregate. The simplest application of this approach is the solubilization of poorly water-soluble molecules within aqueous micelles. As an example, solid-phase microextraction headspace measurements were used to determine the partitioning of the flavor molecule limonene into aqueous SDS micelles (Lloyd *et al.*, 2011). The ratio of moles of limonene located in micelles, relative to those dissolved externally in the water, was determined to be ~2000 M^{-1} times the molar surfactant concentration. Thus, a 100 mM SDS solution would solubilize 200 moles of limonene in the micelles for every one in water. As a result, the aqueous micellar solution could "dissolve" far higher concentrations of the flavor than could water alone. This boost in aqueous concentrations, coupled with altered effects on chemical stability and release rates, makes micellar solubilization an attractive possibility for poorly water-soluble bioactives (Mitra and Dungan, 2001; Flanagan and Singh, 2006; Huang *et al.*, 2010; Rao and McClements, 2011; Hamed *et al.*, 2012; Ziani *et al.*, 2012). Surfactants used to solubilize the bioactive molecules should be water soluble with large or charged head groups and small tail groups; this molecular architecture favors molecular packing into positively curved self-assembled structures.

With higher concentrations of solute or a large solute molecular size, solubilization entails significant rearrangement of the surfactant away from its micellar packing. If this rearrangement is unfavorable, the number of solute molecules that can be accommodated in the structure will be limited. For example, long-chain triglycerides cannot be significantly solubilized within micelles, due to their large molecular structure. In other cases, the addition of solute may in fact allow the aggregate to pack more favorably, moving from micelles with highly positive curvature to larger spheres, to ellipsoids or rods with lower curvature. When such rearrangements are favored it facilitates the incorporation of significant amounts of solute, eventually creating a nanosized internal pseudo-phase. At this point the self-assembled structure is most clearly considered an oil-in-water microemulsion (Flanagan and Singh, 2006). Formation of oil-in-water microemulsions requires surfactants with somewhat positive spontaneous interfacial curvature, which also have interfacial "fluidity", meaning that there is only a low-energy cost for bending away from their optimal curvature (Lekkerkerker *et al.*, 1996; Solans and García-Celma, 1997). These features are often obtained with mixtures of surfactants, or surfactants mixed with smaller molecules residing at the interface known as cosurfactants (Flanagan *et al.*, 2006; Hu *et al.*, 2012). Introduction of surfactants with bulkier tail groups will shift phase behavior towards zero or even negative curvatures, enabling formation of bicontinuous microemulsions (Amar *et al.*, 2003; Garti *et al.*, 2005) or water-in-oil microemulsions (Patel *et al.*, 2006). Temperature and composition also are important in tuning the type of structures that form.

Most recently, self-assembled structures from amphiphilic biopolymers have attracted considerable interest (Huang *et al.*, 2010). Casein is well-known for displaying complex self-assembly behavior that is similar to micelle formation of conventional surfactants (Portnaya *et al.*, 2008), and several researchers have begun exploring the possibility of solubilizing valuable ingredients within casein nanostructures (Semo *et al.*, 2007; Yazdi and Corredig, 2012). Self-assembly of zein to form nanovehicles is also being investigated (Wang *et al.*, 2008; Wang and Padua, 2012). Food polysaccharides can also be modified to give them amphiphilic character, yielding a structure that self-assembles to form polymeric micelles (Zhang *et al.*, 2006; Yu and Huang, 2010; Li *et al.*, 2011).

4.4 INTERFACIAL EFFECTS IN DISTRIBUTION AND RELEASE

We close this chapter with a brief discussion of the role of nanoparticle interfaces in slowing down release of active ingredients from the particle. One of the notable features of nanovehicles is that their diffusion coefficient D_{np} increases with decreasing particle size, according to the Stokes–Einstein equation:

$$D_{np} = \frac{kT}{6\pi\eta a}. \tag{4.7}$$

Here η is the viscosity of the fluid surrounding the particle. Therefore the particle will transport its active ingredients (typically at a high loading relative to water) far more rapidly than a larger, conventional dispersed particle. As the particle circulates, however, a key question that must be considered is the rate at which the active compound will leave the particle to enter the continuous phase. At least for lyophobic particles, there will typically be a chemical potential driving force promoting this migration out of the particle.

Figure 4.5 shows a particle of radius a, which contains a solute at a concentration c_p. The nature of the nanoparticle will determine the relative affinity of the ingredient for the particle versus the external water phase, so that the solute will *partition* at the particle interface, giving at equilibrium an aqueous concentration near the surface of $c_w^{\,i} = c_p/H$, where H is the partition coefficient. If we assume for the moment that there is no resistance inside the particle to the mass transfer, the aqueous concentration near the particle will be higher than that far away, and this difference will drive release of the solute. Assuming for simplicity that the solution is unstirred and the concentration far from the particle is zero, the rate of mass transfer out of the particle is given by

$$\text{rate} = \frac{k}{H} A_{sp} c_p, \tag{4.8}$$

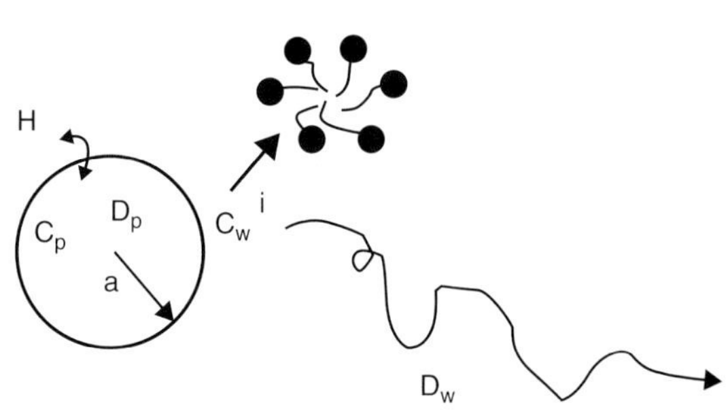

Figure 4.5 Mechanisms for solute release from a nanoparticle, in both the absence and presence of micelles.

where k is the mass transfer coefficient given by

$$k = 2D_w / a \tag{4.9}$$

and D_w is the solute diffusion coefficient in water. Hence the smaller the nanoparticle, the more rapid the release rate by simple molecular diffusion; however, a relatively high affinity of the particle for the solute relative to water (large H) will slow this rate considerably.

Solid lipid nanoparticles have been envisioned in part as a strategy for slowing solute release, by reducing the diffusion coefficient D_p of the solute *within* the particle (Mehnert and Mäder, 2001), and thereby introducing a mass transfer resistance internal to the particle. Mass transfer theory provides conditions for the diffusion coefficients, such that this internal resistance will substantially slow the release rate (Clift *et al.*, 1978):

$$H\sqrt{\frac{D_p}{D_w}} << 1 \text{ for short times;} \tag{4.10a}$$

$$H\frac{D_p}{D_w} << Sh \text{ for long times,} \tag{4.10b}$$

where Sh is the Sherwood number: $Sh = ka/D_w$, and $Sh = 2$ according to equation (4.9). When these conditions are satisfied, the slow internal mass-transfer rate will dominate the overall release. From equation (4.10) it can be concluded that if the partition coefficient H is very large, which it is for many nanoparticle–solute combinations, the diffusion within the particle must be *drastically* reduced to have an important effect on the release rate. On the other hand, the larger H is, the slower is the release rate according to equation (4.8), and an internally structured nanoparticle may not be needed.

Instead of modifying transport throughout the particle core, enhancing resistance just at the interface can be attempted. In the case of liposomes, for example, the partition coefficient may be nearly unity for hydrophilic solute distribution inside and outside the particle at equilibrium, because the liposome core consists of an aqueous phase similar to the external fluid. Retention of solute within the liposome therefore depends on the very high resistance to transport for a hydrophilic solute across the bilayer at the surface. This bilayer represents a lipid phase of nanometer thickness, unfavorable for accommodating the solute. Treating transport across this interfacial region with a simple membrane model gives us an interfacial mass transfer coefficient k_i across the bilayer of

$$k_i = H_i D_i / L \tag{4.11}$$

where H_i and D_i are the partition and diffusion coefficients for the bilayer, respectively (H_i is the ratio of concentrations inside and outside the bilayer), and L is the layer thickness. Choosing lipid structures that decrease H_i and D_i and increase L will reduce the release rate. A similar analysis may be applied to other strategies for augmenting the interfacial resistance – for example, by creating a dense polymeric coating around emulsion droplets using layer-by-layer technologies (Zeeb *et al.*, 2012). In this approach, resistances within three layers must be considered: the hydrophobic core of the droplet, the interfacial film, and the aqueous solution external to the particle. Partition coefficients for the first two regions and diffusion rates in all three zones must be evaluated, in order to determine whether the interfacial film contributes significantly to the overall transport rate.

It has been thus far assumed that the rate of release outside the particle is driven by simple diffusion. Convection in the nanoparticle dispersion will of course enhance the transport rate, because the convective boundary-layer thickness is significantly less than the particle radius. Less obviously, the presence of surfactant in the external liquid can also significantly speed up the release of hydrophobic solutes, if its concentration is above the CMC (Kabalnov and Weers, 1996; Dungan *et al.*, 2003). (Note that the concentration of surfactant external to nanoemulsions or other particles can be determined through a combination of equations (4.1) and (4.4).) There has been much debate on the role of micelles in mediating mass transfer in dispersed mixtures (Chen *et al.*, 1997, Todorov *et al.*, 2002; Sailaja *et al.*, 2003; Ariyaprakai and Dungan, 2008). However, the picture that has begun to emerge indicates that for solute molecules with some small aqueous solubility (e.g., alkanes with ten or fewer carbons), micelles speed up release by rapidly removing dissolved solute from the water, and therefore enhancing the aqueous concentration difference that drives migration external to the particle (Figure 4.5). On the other hand, when the solute is so hydrophobic that its water-solubility becomes negligible ($H >> 1$) and external transport through water is very slow, the micelles may interact with the particle interface, there picking up the solute directly in an interfacial step. Given that many nanoparticles are formed under conditions of high surfactant concentration, there is a strong likelihood that excess surfactant is present in many systems. For this reason and because this mechanism plays a central role in the bioabsorption of lipids in the gut, the role of self-assembled structures on solute release from particles deserves further attention.

Acknowledgments

I would like to acknowledge Josephine Christina, whose surface-tension measurements are presented in this chapter, and to thank Dominique Langevin for the opportunity to conduct dynamic surface-tension measurements in her laboratory.

References

Amar, I., Aserin, A. and Garti, N. (2003) Solubilization patterns of lutein and lutein esters in food grade nonionic microemulsions. *Journal of Agricultural and Food Chemistry*, 51, 4775–4781.

Ariyaprakai, S. and Dungan, S.R. (2008) Contribution of molecular pathways in the micellar solubilization of monodisperse emulsion droplets. *Langmuir*, 24, 3061–3069.

Binks, B. (1998) Emulsions – Recent advances in understanding. In: *Modern Aspects of Emulsion Science* (ed. B. Binks). Royal Society of Chemistry, Cambridge; 1–48.

Chen, B.H., Miller, C.A. and Garrett, P.R. (1997) Rates of solubilization of triolein into nonionic surfactant solutions. *Colloids and Surfaces A: Physicochemical Engineering Aspects*, 128, 129–143.

Clift, R., Grace, J.R. and Weber, M.E. (1978) *Bubbles, Drops and Particles*. Dover, Mineola, NY; 46–58.

De Kruif, C.G., Huppertz, T., Urban, V.S. and Petukhov, A.V. (2012) Casein micelles and their internal structure. *Advances in Colloid and Interface Science*, 171–172, 36–52.

Donsi, F., Annunziata, M., Vincensi, M. and Ferrari, G. (2012) Design of nanoemulsion-based delivery systems of natural antimicrobials: effect of the emulsifier. *Journal of Biotechnology*, 159, 342–350.

Dungan, S.R., Tai, B.H. and Gerhardt, N.I. (2003) Transport mechanisms in the micellar solubilization of alkanes in oil-in-water emulsions. *Colloids and Surfaces A: Physicochemical Engineering Aspects*, 216, 149–166.

Eastman, J. (2010) Stability of charge-stabilised colloids. In: *Colloid Science: Principles, Methods and Applications* (ed. T. Cosgrove). John Wiley & Sons, Ltd, Chichester, UK; 45–59.

Flanagan, J. and Singh, H. (2006) Microemulsions: a potential delivery system for bioactives in food. *Critical Reviews in Food Science and Nutrition*, 46, 221–237.

Flanagan, J., Kortegaard, K., Pinder, D.N., Rades, T. and Singh, H. (2006) Solubilisation of soybean oil in microemulsions using various surfactants. *Food Hydrocolloids*, 20, 253–260.

Fryd, M.M. and Mason, T.G. (2010) Time-dependent nanoemulsion droplet size reduction by evaporative ripening. *Journal of Physical Chemistry Letters*, 1, 3349–3353.

Garti, N., Spernath, A., Aserin A. and Lutz, R. (2005) Nano-sized self-assemblies of nonionic surfactants as solubilization reservoirs and microreactors for food systems. *Soft Matter*, 1, 206–218.

Ghosh, O. and Miller, C.A. (1987) Liquid-crystalline and microemulsion phase-behavior in alcohol-free aerosol-OT/oil/brine systems. *Journal of Physical Chemistry*, 91, 4528–4535.

Grillo, I., Kats, E.I. and Muratov, A.R. (2003) Formation and growth of anionic vesicles followed by small-angle neutron scattering. *Langmuir*, 19, 4573–4581.

Gupta, S.S. and Ghosh, M. (2012) *In vitro* study of anti-oxidative effects of β-carotene and α-lipoic acid for nano-capsulated lipids. *LWT – Food Science and Technology*, 49, 131–138.

Hamed, S.F., Sadek, Z. and Edris, A. (2012) Antioxidant and antimicrobial activities of clove bud essential oil and eugenol nanoparticles in alcohol-free microemulsion. *Journal of Oleo Science*, 61, 641–648.

Helfrich, W. (1978) Steric interaction of fluid membranes in multilayer systems. *Zeitschrift für Naturforschung*, 33, 305–315.

Helgason, T., Awad, T.S., Kristbergsson, K., Decker, E.A., McClements, D.J. and Weiss, J. (2009a) Impact of surfactant properties on oxidative stability of β-carotene encapsulated within solid lipid nanoparticles. *Journal of Agricultural and Food Chemistry*, 57, 8033–8040.

Helgason, T., Awad, T.S., Kristbergsson, K., McClements, D.J. and Weiss, J. (2009b) Effect of surfactant surface coverage on formation of solid lipid nanoparticles (SLN). *Journal of Colloid and Interface Science*, 334, 75–81.

Horn, D. and Rieger, J. (2001) Organic nanoparticles in the aqueous phase – theory, experiment, and use. *Angewandte Chemie*, 40, 4331–4361.

Hu, L.D., Jia, Y.H., Niu, F., Jia, Z., Yang, X. and Jiao, K.L. (2012) Preparation and enhancement of oral bioavailability of curcumin using microemulsions vehicle. *Journal of Agricultural and Food Chemistry*, 60, 7137–7141.

Huang, Q., Yu, H. and Ru, Q. (2010) Bioavailability and delivery of nutraceuticals using nanotechnology. *Journal of Food Science*, 75, R50-R57.

Hunter, R.J. (2001) *Foundations of Colloid Science*. Oxford University Press, Oxford, UK.

Hwang, T., Park, T.-J., Koh, W.-G., Cheong, I.W., Choi, S.-W. and Kim, J.H. (2011) Fabrication of nano-scale liposomes containing doxorubicin using Shirasu porous glass membrane. *Colloids and Surfaces A: Physicochemical Engineering Aspects*, 392, 250–255.

Ichikawa, S., Sugiura, S., Nakajima, M., Sano, Y., Seki, M. and Furusaki, S. (2000) Formation of biocompatible reversed micellar systems using phospholipids. *Biochemical Engineering Journal*, 6, 193–199.

Jafari, S.M., He, Y. and Bhandari, B. (2007) Optimization of nano-emulsions production by microfluidization. *European Food Research Technology*, 225, 733–741.

Kabalnov, A. and Weers, J. (1996) Kinetics of mass transfer in micellar systems: surfactant adsorption, solubilization kinetics, and ripening. *Langmuir*, 12, 3442–3448.

Kabalnov, A. and Wennerström, H. (1996) Macroemulsion stability: the oriented wedge theory revisited. *Langmuir*, 12, 276–292.

Kabalnov, A., Weers, J. Arlauskas, R. and Tarara, T. (1995) Phospholipids as emulsion stabilizers. 1. Interfacial tensions. *Langmuir*, 11, 2966–2974.

Kabalnov, A., Tarara, T., Arlauskas, R. and Weers, J. (1996) Phospholipids as emulsion stabilizers. 2. Phase behavior versus emulsion stability. *Journal of Colloid and Interface Science*, 184, 227–235.

Krishnan, A., Siedlecki, C.A. and Vogler, E.A. (2003) Traube–Rule interpretation of protein adsorption at the liquid-vapor interface. *Langmuir*, 19, 10342–10352.

Lee, S.J. and McClements, D.J. (2010) Fabrication of protein-stabilized nanoemulsions using a combined homogenization and amphiphilic solvent dissolution/evaporation approach. *Food Hydrocolloids*, 24, 560–569.

Lekkerkerker, H.N.W., Kegel, W.K. and Overbeek, J.T.G. (1996) Phase behavior of ionic microemulsions. *Berichte der Bunsen-Gesellschaft*, 100, 206–217.

Li, Q.A., Liu, C.G., Huang, Z.H. and Xue, F.F. (2011) Preparation and characterization of nanoparticles based on hydrophobic alginate derivative as carriers for sustained release of vitamin D-3. *Journal of Agricultural and Food Chemistry*, 59, 1962–1967.

Lloyd, N.W., Kardaras, E., Ebeler, S.E. and Dungan, S.R. (2011) Measuring local equilibrium flavor distributions in SDS solution using headspace solid-phase microextraction. *Journal of Physical Chemistry*, 115, 14484–14492.

Mason, T.G., Wilking, J.N., Meleson, K., Chang, C.B. and Graves, S.M. (2006) Nanoemulsions: formation, structure, and physical properties. *Journal of Physics: Condensed Matter*, 18, R635–R666.

Mazzoferro, S., Bouchemal, K. and Ponchel, G. (2013) Oral delivery of anticancer drugs III: formulation using drug delivery systems. *Drug Discovery Today*, 18, 99–104.

Mehnert, W. and Mäder, K. (2001) Solid lipid nanoparticles: production, characterization and applications. *Advanced Drug Delivery Reviews*, 47, 165–196.

Mele, S., Murgia, S., Caboi, F. and Monduzzi, M. (2004) Biocompatible lipidic formulations: phase behavior and microstructure. *Langmuir*, 20, 5241–5246.

Meleson, K., Graves, S. and Mason, T.G. (2004) Formation of concentrated nanoemulsions by extreme shear. *Soft Materials*, 2, 109–123.

Miller, R. and Kretzschmar, G. (1991) Adsorption kinetics of surfactants at fluid interfaces. *Advances in Colloid and Interface Science*, 37, 97–121.

Miller, R., Fainerman, V.B., Leser, M.E. and Michel, M. (2004) Kinetics of adsorption of proteins and surfactants. *Current Opinion in Colloid and Interface Science*, 9, 350–356.

Mitra, S. and Dungan, S.R. (2001) Cholesterol solubilization in aqueous micellar solutions of quillaja saponin, bile salts or nonionic surfactants. *Journal of Agricultural and Food Chemistry*, 49, 384–394.

National Science Foundation (2000) *Nanotechnology Definition*. [Online] Available: http://www.nsf.gov/crssprgm/nano/reports/omb_nifty50.jsp (accessed 13 February 2013).

Niño, R.R. and Patino, J.M.R. (1998) Surface tension of bovine serum albumin and Tween 20 at the air-aqueous interface. *Journal of the American Oil Chemists Society*, 75, 1241–1248.

Patel, N., Schmid, U. and Lawrence, M.J. (2006) Phospholipid-based microemulsions suitable for use in foods. *Journal of Agricultural and Food Chemistry*, 54, 7817–7824.

Portnaya, I., Ben-Shoshan, E., Cogan, U., *et al.* (2008) Self-assembly of bovine β-casein below the isoelectric pH. *Journal of Agricultural and Food Chemistry*, 56, 2192–2198.

Qian, C. and McClements, D.J. (2011) Formation of nanoemulsions stabilized by model food-grade emulsifiers using high-pressure homogenization: factors affecting particle size. *Food Hydrocolloids*, 25, 1000–1008.

Rao, J.J. and McClements, D.J. (2011) Formation of flavor oil microemulsions, nanoemulsions and emulsions: influence of composition and preparation method. *Journal of Agricultural and Food Chemistry*, 59, 5026–5035.

Rydhag, L. and Wilton, I. (1981) The function of phospholipids of soybean lecithin in emulsions. *Journal of the American Oil Chemists Society*, 58, 830–837.

Sailaja, D., Suhasini, K.L., Kumar, S. and Gandhi, K.S. (2003) Theory of rate of solubilization into surfactant solutions. *Langmuir*, 19, 4014–4026.

Semo, E., Kesselman, E., Danino, D. and Livney, Y.D. (2007) Casein micelle as a natural nano-capsular vehicle for nutraceuticals. *Food Hydrocolloids*, 21, 936–942.

Shukat, R. and Relkin, P. (2011) Lipid nanoparticles as vitamin matrix carriers in liquid food systems: on the role of high-pressure homogenisation, droplet size and adsorbed materials. *Colloids and Surfaces B: Biointerfaces*, 86, 119–124.

Siddiqui, F.A. and Franses, E.I. (1996) Surface tension and adsorption synergism for solutions of binary surfactants. *Industrial and Engineering Chemistry Research*, 35, 3223–3232.

Solans, C. (2005) Nano-emulsions. *Current Opinion in Colloid and Interface Science*, 10, 102–110.

Solans, C. and García-Celma, M.J. (1997) Surfactants for Microemulsions. *Current Opinion in Colloid and Interface Science*, 5, 464–471.

Tayebi, L., Vashaee, D. and Parikh, A.N. (2012) Stability of uni- and multilamellar spherical vesicles. *ChemPhysChem* 13, 314–322.

Taylor, T.M., Weiss, J., Davidson, P.M. and Bruce, B.D. (2005) Liposomal nanocapsules in food science and agriculture. *Critical Reviews in Food Science and Nutrition*, 45, 587–605.

Tikekar, R.V. and Nitin, N. (2011) Effect of physical state (solid vs. liquid) of lipid core on the rate of transport of oxygen and free radicals in solid lipid nanoparticles and emulsion. *Soft Matter*, 7, 8149–8157.

Todorov, P.D., Kralchevsky, P.A. Denkov, N.D., Broze, G. and Mehreteab, A. (2002) Kinetics of solubilization of *n*-decane and benzene by micellar solutions of sodium dodecyl sulfate. *Journal of Colloid and Interface Science*, 245, 371–382.

Troncoso, E., Aguilera, J.M. and McClements, D.J. (2011) Development of nanoemulsions by an emulsification-evaporation technique. In: *Food Process Engineering in a Changing World. Proceedings of the 11th International Congress on Engineering and Food, May 22–26, 2011, Athens, Greece* (eds P.S. Taoukis, N.G. Stoforos, V.T. Karathanos and G.D. Saravacos). Cosmosware, Athens; 929–930.

Ulrich, A.S. (2002) Biophysical aspects of using liposomes as delivery vehicles. *Bioscience Reports* 22, 129–150.

Van Duijneveldt, J. (2010) Effect of polymers on colloid stability. In: *Colloid Science: Principles, Methods and Applications* (ed. T. Cosgrove). John Wiley & Sons, Ltd, Chichester, UK; 181–194.

Von Corswant, C., Engström, S. and Söderman, O. (1997) Microemulsions based on soybean phosphatidylcholine and triglycerides. Phase behavior and microstructure. *Langmuir*, 13, 5061–5070.

Wang, Q., Lin, L. and Padua, G.W. (2008) Effect of hydrophilic and lipophilic compounds on zein microstructures. *Food Biophysics*, 3, 174–181.

Wang, Y. and Padua, G.W. (2012) Nanoscale characterization of zein self-assembly. *Langmuir*, 28, 2429–2435.

Yang, Y., Marshall-Breton, C., Leser, M.E., Sher, A.A. and McClements, D.J. (2012) Fabrication of ultrafine edible emulsions: comparison of high-energy and low-energy homogenization methods. *Food Hydrocolloids*, 29, 398–406.

Yazdi, S. and Corredig, M. (2012) Heating of milk alters the binding of curcumin to casein micelles. A fluorescence spectroscopy study. *Food Chemistry*, 132, 1143–1149.

Yu, H. and Huang, Q.R. (2010) Enhanced *in vitro* anti-cancer activity of curcumin encapsulated in hydrophobically modified starch. *Food Chemistry*, 119, 669–74.

Zeeb, B., Gibis, M., Fischer, L. and Weiss, J. (2012) Influence of interfacial properties on Ostwald ripening in crosslinked multilayered oil-in-water emulsions. *Journal of Colloid and Interface Science*, 387, 65–73.

Zhang, C., Ding, Y., Ping, Q.E. and Yu, L.L. (2006) Novel chitosan-derived nanomaterials and their micelle-forming properties. *Journal of Agricultural and Food Chemistry*, 54, 8409–8416.

Ziani, K., Fang, Y. and McClements, D.J. (2012) Encapsulation of functional lipophilic components in surfactant-based colloidal delivery systems: vitamin E, vitamin D, and lemon oil. *Food Chemistry*, 134, 1106–1112.

5 Controlling properties of micro- to nano-sized dispersions using emulsification devices

Zheng Wang,[1,3] Marcos A. Neves,[1,2,3] Isao Kobayashi,[1,3] and Mitsutoshi Nakajima[1,2,3]

[1]*Alliance for Research on North Africa (ARENA), University of Tsukuba, Tsukuba, Ibaraki, Japan*
[2]*Faculty of Life and Environmental Sciences, University of Tsukuba, Tsukuba, Ibaraki, Japan*
[3]*Food Engineering Division, National Food Research Institute, NARO, Tsukuba, Ibaraki, Japan*

5.1 INTRODUCTION

The properties of food depend largely on their micro-/nanoscale structure but the conventional processes in the preparation of foods is generally carried out at a large scale. As a consequence, the structure of the products cannot be precisely controlled. Recently, the formulation and characterization of foods has witnessed the continuous development micro-/nanotechnology (Kobayashi *et al.*, 2011). Food nanotechnology may incorporate emulsification, dispersion, mixing, and separation; that is, membrane emulsification, microdevice emulsification, micromixing, and the control of food rheology (Ribeiro *et al.*, 2010). Application of computational fluid dynamics (CFD) to food processing has also received considerable attention (Kobayashi *et al.*, 2007a, 2011).

Emulsions are dispersions of at least two immiscible liquids, one of which is dispersed as droplets in the other liquid, and stabilized by an emulsifier, for example oil and water. Emulsions have applications in many industries, such as food, pharmaceutical, cosmetics, agriculture, petrochemical, and other related industries (Chappat, 1994). According to their application, different types of emulsions can be formulated: among them, oil-in-water (O/W) or water-in-oil (W/O) emulsions are commonly used, aside from water-in-oil-in-water (W/O/W) and oil-in-water-in-oil (O/W/O) emulsions, also known as multiple or double emulsions. Also, depending on the emulsification process used, droplets of different size and size distribution may be formed, and these parameters can affect the physicochemical properties and overall stability of the resulting emulsions. For example, W/O food emulsions may contain flavoring agents in the dispersed phase, and different droplet sizes generally affect the flavor of the products. Hence, the droplet size and size distribution may determine the shelf-life stability, consistency, rheological properties, color, and possibly the taste of food emulsions.

Nanotechnology and Functional Foods: Effective Delivery of Bioactive Ingredients, First Edition.
Edited by Cristina M. Sabliov, Hongda Chen, and Rickey Y. Yada.

Emulsions are thermodynamically unstable, meaning that the dispersed droplets tend to coalesce, in order to minimize the interfacial area, and break into two separate phases in equilibrium, as time passes by. Physical instability may lead to changes in the spatial distribution or structural organization of the molecules, whereas chemical instability generally leads to the formation of different molecules. Basically, there are four types of physical instability in emulsions – flocculation, creaming, coalescence, and Ostwald ripening – whereas oxidation and hydrolysis are common examples of chemical instability (McClements, 2004; Tadros, 2009). Monodisperse emulsions with very narrow size distribution generally have improved stability against droplet coalescence, as well as making easier the control of major emulsion properties, as compared to polydisperse emulsions. Therefore, monodisperse emulsions are required for high quality and stability of the products (Kobayashi *et al.*, 2008a).

This chapter focuses on the monodispersity of emulsions, and discusses the conventional methods of producing emulsions, and the major factors that affect the efficiency of emulsification processes. The production of monodisperse emulsions using microdevice emulsification technology is reviewed in detail, as well as some applications of emulsions to different fields.

5.2 FUNDAMENTALS OF EMULSIFICATION PROCESSES

The fact that emulsions are thermodynamically unstable due to the excess free energy at the interface of the droplets dispersed is well known. Considering two immiscible liquids in contact with each other, the net forces at the interface, so-called interfacial tension, they tend to minimize the surface area of individual liquids. In emulsions, the process of dispersing one liquid into the other results in an increase in surface area between the dispersed droplets and dispersion medium, and surface free energy, which can be calculated using the Gibbs equation, as follows:

$$\Delta G = \gamma \Delta A \tag{5.1}$$

where ΔG is the interfacial free energy (N m/J), γ is the interfacial tension (N/m), and A is the total interfacial area of the dispersed phase (m^2) (Aserin, 2008). In order to reduce the total interfacial area and the free energy, the dispersed droplets tend to coalesce and separate, making emulsions thermodynamically unstable. Lowering of interfacial tension is one way in which the increased surface free energy can be reduced, leading to a more stable emulsion, due to the formation of smaller emulsion droplets with narrower droplets size distributions and greater kinetic stability (Burgess and Yoon, 1995). Emulsifying agents such as ionic, nonionic or zwitterionic emulsifiers, and proteins or amphiphilic polymers are employed to reduce the interfacial tension by forming a barrier between two immiscible liquids. Emulsifiers are molecules that consist of a nonpolar tail group and a polar head group capable of forming hydrogen bonds. Emulsifiers are classified as amphiphilic molecules because of their dual nature in that they are attracted to and repelled by water. The amphiphilic nature of emulsifiers can be expressed in terms of an empirical scale, the so-called hydrophile–lipophile balance (HLB), established by Griffin (1949). The HLB is described by a number that gives an indication of the relative affinity of an emulsifier molecule for the oil or aqueous phases (Davis, 1994). Each emulsifier is assigned an HLB number according to its chemical structure. More specifically, this value can be calculated from the number and type of hydrophilic and lipophilic groups present in the molecule of a certain emulsifier. A good emulsifying agent should have a limited solubility in both the oil and water phases of the system. Emulsifiers having HLB values from 3 to 6 are generally lipophilic and produce W/O emulsions, and those agents with HLB values from 8 to 18 produce O/W emulsions. An emulsifier with an intermediate HLB number (7–9) has no particular preference for either oil or water, and is considered a good "wetting agent." Molecules with HLB numbers below 3 (very hydrophobic) and above 18 (very hydrophilic) are not particularly surface-active, since they tend to accumulate preferentially either in bulk oil or water, rather than at an oil–water interface (Myers, 2006).

There are several theories that describe how emulsifying agents promote emulsification and maintain the stability of the resulting emulsion, the most prevalent of which are the oriented-wedge, surface-tension, and the interfacial film theories. In-depth discussions of these theories are beyond the scope of this chapter and can be found elsewhere in the literature. However, a general way in which emulsions are produced and stabilized will be discussed in this chapter. Emulsions do not form spontaneously when liquids are mixed and hence an input of energy is required to break up the liquids into small droplets. As the energy is applied, the interface between the oil and water phases is deformed, resulting in the formation of droplets. Emulsifier molecules are rapidly adsorbed at the interface formed between the droplets, thereby reducing the interfacial tension. After the formation of emulsions, emulsifiers prevent coalescence of newly formed droplets by providing a strong short-ranged interfacial repulsion (Myers, 2006). By lowering the interfacial tension, emulsifiers also reduce the energy needed to break-up the large droplets into smaller ones.

5.3 CONVENTIONAL MECHANICAL EMULSIFICATION

Conventional emulsification techniques are based on disrupting larger droplets in coarse premix emulsions into smaller ones, requiring high-energy input. This energy must override Laplace pressure, which increases when the droplets become smaller, using intense force fields (Dickinson, 1992). Generally, emulsifiers form part of the raw materials that go into the additives used in the processing of foods, but the range of emulsifiers that can be used is limited. Therefore, the use of mechanical power in addition to emulsifiers has played a major role in making emulsions with good stability (Takagi, 2013). Traditionally, emulsions have been prepared by mechanical disruption of droplets into smaller ones, which generally involves strong shear stress and requires high-energy input, leading to the formation of polydisperse emulsions, due to the difficulty in controlling droplet size. In such processes, different parameters, such as the volume ratio between dispersed and continuous phases, the type of emulsifying agent used, among others, are also known to play major roles during emulsification.

Different processes are available for emulsification. The most frequent ones are the rotor–stator systems (colloid mill, toothed disc dispersing machine), ultrasound homogenizers, and high-pressure homogenizers, and generally each one has advantages and disadvantages. The choice of homogenizer depends on various parameters, such as the production environment and scale, the equipment available, the volume of material to be homogenized, the desired throughput, the nature of the starting materials, the desired droplet size distribution, the required physicochemical properties of the final product, and the cost of purchasing and running the equipment. The general characteristics of these different homogenizers are presented in Table 5.1.

Emulsification processes may involve a single step or a number of consecutive steps, depending on the nature of the bulk materials and the method used. The size of the droplets produced by a homogenizer

Table 5.1 Comparison of the different types of emulsification methods. (Source: data from Karbstein and Schubert, 1995.)

Method	Energy input per volume (J/m^3)	Droplet size (μm)	Throughput scale (L/h)	
			Laboratory	Industry
High-speed mixer	10^6–10^7	2–10	0.005–10	10–5000
Colloid mill	10^6–10^7	1–10	2–10	10–10000
Microfluidizer	10^6–10^8	0.03–3	5–100	10–10000
Ultrasonic homogenizer	10^7–10^8	1–30	0.1–2	1–1000

depends on a balance between droplet disruption and droplet coalescence. A better understanding of the factors that influence these processes may lead food manufacturers to select the most appropriate ingredients and emulsification parameters required to produce particular food products.

Therefore, conventional emulsification devices may produce emulsions with a broad range of droplet sizes, depending on the application they are intended for. The most popular mechanical emulsification devices used are described below.

5.3.1 High-speed mixer

When fluid is mixed by a rotor in a small vessel that has a narrow clearance between its wall and the rotor, a high-speed thin-film spin flow is created in the fluid being processed. This flow possibly allows the peripheral speed to be over 30 m/s, which was unable to be obtained by a conventional high-speed mixer. As a result, the processing fluid obtains more homogeneous energy. In an experiment, we compared a high-speed thin-film spin mixer with a conventional high-speed mixer in an emulsification experiment using soybean oil (Shundo, 2009). High-speed mixers are the most commonly used method for directly homogenizing oil and aqueous phases in the food industry. In a batch process, the oil, water, and other ingredients are placed in a suitable vessel (Figure 5.1), which may contain as little as a few milliliters for laboratory use, or as large as several thousand liters for industrial applications. The design of the mixing head may determine the efficiency of the homogenization process, and a number of different types are available for different situations, for example, blades, propellers, and turbines (McClements, 2004). Typical throughputs for in-line devices range from a few tenths to 1000 L/h. Blending generally leads to an increase in the temperature of an emulsion because most of the mechanical energy is converted into heat. If any of the ingredients in the emulsion are sensitive to heat, it may be necessary to control the temperature of the vessel during homogenization. In general, droplets produced by a high-speed mixer range between 2 and 10 μm in diameter.

5.3.2 Colloid mill

Colloid mills are widely used in the food industry to homogenize medium- and high-viscosity liquids (McClements, 2004). A variety of different designs of colloid mills are commercially available, however, they all operate on similar physical principles (Figure 5.2). The mixing mechanisms

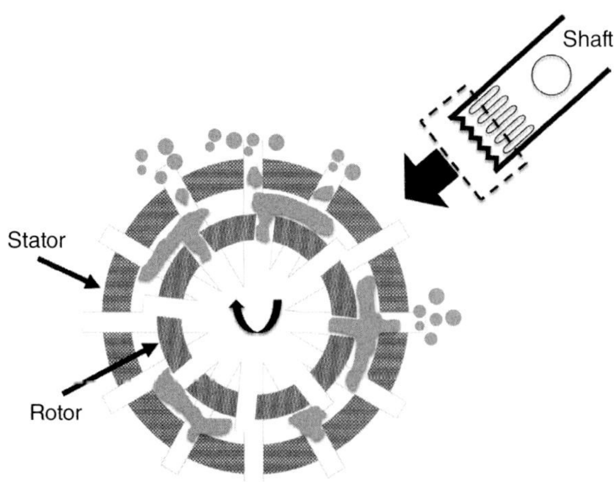

Figure 5.1 Schematic representation of a typical high-speed homogenizer.

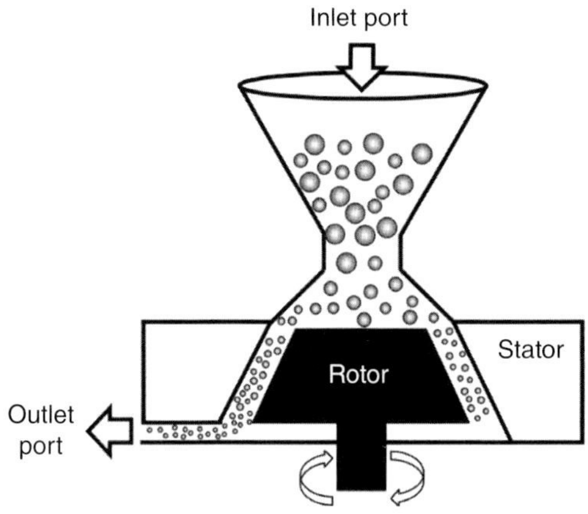

Figure 5.2 Schematic representation of colloid mills.

involved can be described as follows: initially, a sharp stator is utilized in the mixing zone to mill the fluid. Next, a series or fine teeth accelerate the fluid between the hardened surfaces of the spinning rotor and stator gap. Finally, centrifugal forces impinge fluid on the stator wall for additional refinement (Perrier-Cornet *et al.*, 2005). The liquids to be homogenized are usually fed into the colloid mill in the form of a coarse emulsion. The coarse emulsion is usually prepared directly from the oil, water, and other ingredients, using high-speed mixers. The rapid rotation of the rotor generates shear stress in the gap, which causes the larger droplets to be broken down into smaller ones, and generates a centrifugal force that causes the fluid to move from the center to the periphery of the disks, where it is either collected or conducted through a pipe to another operation. Conventionally, the term fine milling is used for sizes below 100 μm and ultrafine (or very fine) defines milling of particles less than 10 μm. Moreover, the size reduction of solids is energy intensive, where most of the energy input is converted into heat (Rhodes, 1998). Colloid mills are more suitable for homogenizing high-viscosity fluids (e.g., butter, fish, or meat pastes) than high-pressure valve or ultrasonic homogenizers. Typically, they can be used to produce emulsions with droplet diameters between 1 and 10 μm, with throughputs of 2–10 L/h (laboratory scale) and 10–10000 L/h (industrial scale).

5.3.3 High-pressure homogenizer (microfluidizer)

In the past, food processing under high pressure was not common. For example, milk-fat globules of a few micrometers in diameter have been traditionally reduced to submicron size by using pressures of around 25 MPa. On the other hand, the increasing demand for special food products has required more energy-intensive processes, such as the production of low-fat ice cream, which involves pressures generally above 100 MPa (Takagi, 2013). During high-pressure homogenization, the fluid is subjected to several simultaneous force-induced phenomena such as cavitation, turbulence, shear, friction, heat, compression, acceleration, rapid pressure drop, and impact (Roach and Harte, 2008). All these forces are significantly increased as pressure increases. Homogenization technology has evolved from systems operating at 10 MPa to devices working at up to 200 MPa (Bouaouina *et al.*, 2006).

With a long time history, high-pressure homogenizers have been recognized as the most efficient system when compared to classic homogenizers. To date this technique is used widely in the food processing industry, such as for milk treatment in order to avoid creaming (Wilbey, 1992), in this

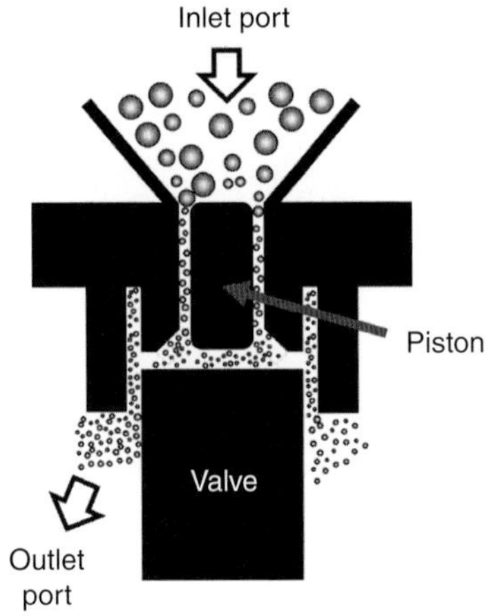

Figure 5.3 Schematic representation of high-pressure homogenizer.

case, the particle break-up is initiated by a combination of turbulence and laminar shear stress. Microfluidizers, one of the high-pressure homogenizers, can create emulsions with extremely small droplet sizes, using either primary or secondary homogenization (Strawbridge *et al.*, 1995). This equipment consists of a fluid inlet, a pumping device, and an interaction chamber containing two channels through which the fluids are forced, and then they interact with each other (Figure 5.3). Small emulsion droplets can be produced by passing an emulsion through a microfluidizer a number of times or by increasing the pressure (Strawbridge *et al.*, 1995). The mechanism for droplet size reduction is not well known, but different factors could contribute to this phenomenon: passing through a hole smaller than the droplets size (generally lower 100 μm), impact between droplets and/or the device wall, laminar and/or turbulent shear forces, cavitation and torsion (Perrier-Cornet *et al.*, 2005). Microfluidizers are available for laboratory throughput of 5–100 L/h and for large-scale production with 100–10000 L/h. Laboratory-scale microfluidizers can also produce small volumes, starting from around 10 mL, which is convenient for preparation of emulsions using ingredients that are costly or scarce. Maximum operating pressures up to 275 MPa and throughputs up to 12,000 L/h can be achieved using industrial microfluidizers. Similar operating pressures may be achieved using laboratory-scale models, which account for their ability to produce very fine droplets smaller than 0.1 μm.

5.3.4 Ultrasonic homogenizer

Ultrasonic homogenizers have been used for many years to produce emulsions. Ultrasonic homogenizers use high-intensity ultrasonic waves that generate intense shear and pressure gradients within a material, and disrupt the droplets mainly due to cavitation and turbulence (Canselier *et al.*, 2002). Laboratory-scale ultrasonic homogenizers (Figure 5.4) can be used for preparing emulsions with a volume of a few milliliters to 1 L. Industrial ultrasonic homogenizers are used for production of batches from 1 to 1000 L or a throughput of 0.1 to 10,000 L/h. Ultrasonic homogenizers can generate very small droplets, and they are usually more energy efficient than high-pressure homogenizers; that is to say less energy is required to produce droplets of the same size. Increased intensity of the ultrasonic radiation or homogenization time may lead to further decrease in droplet size (Abismail *et al.*, 1999). The resulting droplet sizes often range between 1 and 30 μm.

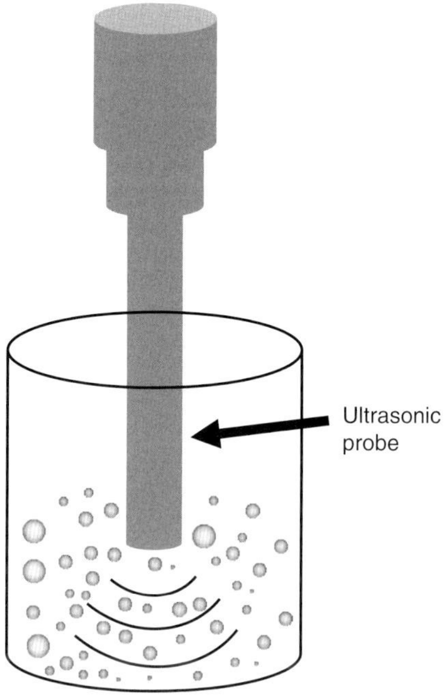

Figure 5.4 Schematic representation of ultrasonic homogenizer.

5.4 PREPARATION OF QUASI-MONODISPERSE EMULSIONS USING MEMBRANE EMULSIFICATION

Nakashima and Shimizu (1988) first proposed membrane emulsification, which uses microporous membranes with narrow pore-size distributions. In direct membrane emulsification (Figure 5.5a), the dispersed phase is pressurized so that it passes through the membrane pores, while the continuous phase flows along the membrane surface. Droplets grow at pore outlets until they detach, upon reaching a certain size. Shirasu porous glass (SPG) membranes developed by Nakashima and Kurki (1981) are the most commonly used membranes for membrane emulsification. The SPG membrane with relatively uniform interconnected pores is fabricated from Na_2O–CaO–MgO–Al_2O_3–B_2O_3–SiO_2 type glass, which is made from 'Shirasu', a Japanese volcanic ash. The SPG membranes have a wide spectrum of available mean pore sizes (0.05–30 μm) and high porosities (50–60%). Tubular SPG membranes typically with an outer diameter of 10 mm, a membrane thickness of 0.45–0.75 mm, and a length of 20–500 mm are generally used in membrane emulsification (Charcosset, 2009). Quasi-monodisperse emulsions with a mean droplet size of 0.3–60 μm and a coefficient of variation (CV) of approximately 10% can be produced by direct SPG membrane emulsification (Nakashima *et al.*, 1991; Vladisavljević and Schubert, 2002). The above-mentioned mean droplet size is basically proportional to the membrane pore size. Microsieve membranes and other microporous inorganic/organic membranes have also been used for membrane emulsification (Vladisavljević and Williams, 2005). Several research groups have proposed emulsification using dynamic membranes, such as rotating membranes, vibrating membranes, and oscillating membranes (Vladisavljević *et al.*, 2012a).

A key factor for achieving successful membrane emulsification is to prevent wetting of the dispersed phase covered by emulsifier molecules onto the membrane surface. For instance, hydrophilic

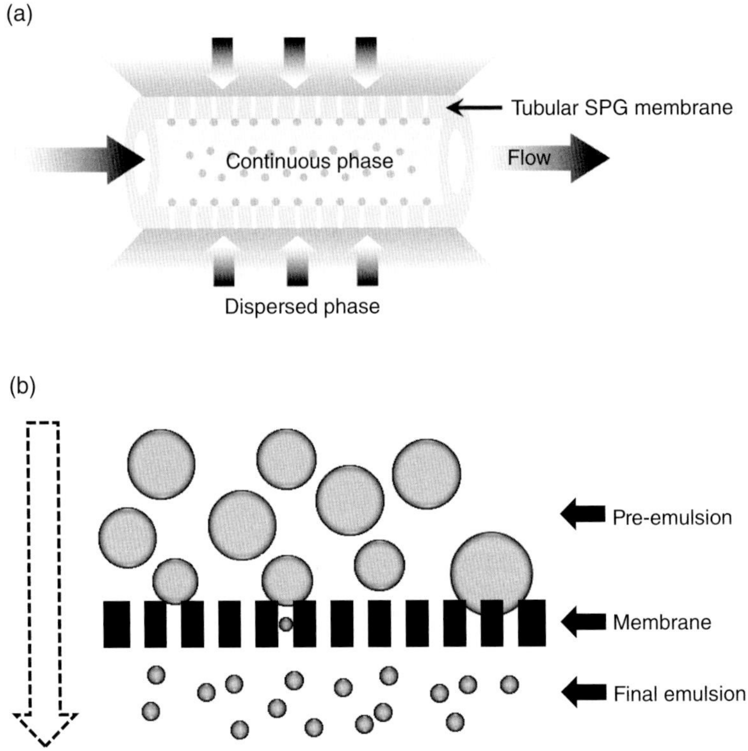

Figure 5.5 Schematic representation of membrane emulsification: (a) direct membrane emulsification; (b) premix membrane emulsification. SPG, Shirasu porous glass.

membranes are preferable for producing O/W emulsions. The surface wettability can be changed by reacting with organic silanes, such as octadecyltrichlorosilane (Fujihara, 2008), making the hydrophilic membrane surface hydrophobic. Hydrophobically surface-treated membranes as well as inherently hydrophobic membranes are useful for producing W/O emulsions.

In direct membrane emulsification, droplet detachment from the membrane was normally determined by the balance between the drag force on the droplet from the cross-flowing continuous phase, the buoyancy of the droplet, the interfacial tension forces, and the driving pressure (Schröder *et al.*, 1998). It is worth noting that Vladisavljević *et al.* (2007) observed successful detachment of the oil droplets from the SPG membrane pores in the absence of the cross-flowing continuous phase. The continuous phase velocity influences the mean droplet size when the drag force from the cross-flowing continuous phase is large enough. The degree of coalescence at the membrane surface and in the bulk solution could also influence the mean droplet size and size distribution.

Suzuki *et al.* (1996) reported premix membrane emulsification in which a coarse premixed emulsion is pressed through the membrane to reduce the droplet size of the dispersed phase (Figure 5.5b). A major advantage of premix membrane emulsification is a very high emulsification flux (e.g., about 10000 L/m^2/h at an operating pressure of 1.5 MPa) and a mean droplet size similar to the membrane pore size (Vladisavljević *et al.*, 2012b). In contrast, direct membrane emulsification has considerably lower flux values of 10–100 L/m^2/h (Joscelyne and Trägårdh, 2000). Research on premix membrane emulsification has been reviewed by Nazir *et al.* (2010), who focused on droplet breakup mechanisms and the influence of governing process parameters.

5.5 PREPARATION OF MONODISPERSE EMULSIONS USING MICROFABRICATED EMULSIFICATION DEVICES

Micromachining technology has seen remarkable development over the recent decades, leading to the fabrication of miniature embedded systems involving many micromachined components and structures, the so-called micro-electro-mechanical systems (MEMS) (Bhushan, 2004). Micro-/nano-fabrication techniques have enabled the precise formulation of microchannels (MCs) available for flowing fluid(s) on a chip. Microfluidics, which handles liquid-based systems in MCs fabricated on a chip, is an emerging multidisciplinary technology. The materials used as microfluidic devices, their surface properties, and fabrication techniques are presented in Table 5.2 (Santos, 2005). Emulsification techniques using microfabricated devices have been proposed for producing monodisperse emulsions within the past two decades (Table 5.3). Advantages of these emulsification techniques are monodispersity of the resulting emulsions, superior controllability of the droplet size, and *in situ* microscopic monitoring that enables on-demand fine tuning of the process parameters during emulsification.

5.5.1 Microfluidic emulsification

A T-junction is one of the most frequently used microfluidic geometries to produce monodisperse emulsions in the presence of the cross-flowing continuous phase (Figure 5.6a). Droplet generation using a T-junction is driven by shear force imposed by an immiscible cross-flow fluid. Droplet generation regimes can be predicted by the dimensionless capillary number (Ca) for the continuous phase, defined as

$$Ca_c = \eta_c u_c / \gamma \tag{5.2}$$

where η_c is the viscosity of the continuous phase (Pa s), u_c is the velocity of the continuous phase (m/s), and γ is the interfacial tension between the oil and water phases (N/m) (Teh *et al.*, 2008). The droplet generation regimes consisted of a squeezing regime at a Ca_c of < 0.002, a dripping regime at a Ca_c of 0.01–0.3, and a transient regime at a Ca_c of 0.002 to 0.01 (Xu *et al.*, 2008). Droplets produced with T-junctions are usually several microns to several hundred microns in size, with a CV < 5%. In general, the size of droplets generated using a T-junction is directly proportional to the flow rate of the dispersed phase, and inversely proportional to that of the continuous phase.

Table 5.2 The material properties of microfluidic devices and microfabrication techniques. (Source: data from Santos, 2005.)

Material property	Microfabrication technique
Oxidation of silicon[a]	Anisotropic wet/dry etching
Silicon nitride[a]	Anisotropic dry etching
Quartz glass[a]	Micromechanical cutting
Pyrex glass[a]	Isotropic wet etching
Silicone resin[b]	Soft lithography
Urethane resin[b]	Soft lithography
Poly(methyl methacrylate) (PMMA)[b]	Injection molding,lithography, electroplating and molding (LIGA) process
Stainless steel	Micromechanical cutting
Nickel[b]	Lithography, electroplating and molding (LIGA) process

[a,b] Surface properties of each material: a, hydrophilic; b, hydrophobic.

Table 5.3 Comparison of microfluidics emulsification and microchannel (MC) emulsification. (Source: Data from Steegmans *et al.*, 2009; Anna *et al.*, 2003; Utada *et al.*, 2007; Kobayashi *et al.*, 2007b; Kobayashi *et al.*, 2002; van Dijke *et al.*, 2009a.)

Techniques		**Monodispersity**	**Droplet size (μm)**	**Throughput (mL/h)**	**Device fabrication**
Microfluidics emulsification	T-/Y-junctions	Good	50–500	1–300	Simple
	Flow-focusing	Good	~100	1–50	Elaborate
	Microcapillaries	Good	~10	1–10	Elaborate
MC emulsification	Groove microchannels	Excellent	1–500	1–10	Simple
	Asymmetric straight-through microchannels	Good	5–2000	1–1000	Elaborate
	EDGE	Good	5–10	1–10	Simple

EDGE: edge-based droplet generation system.

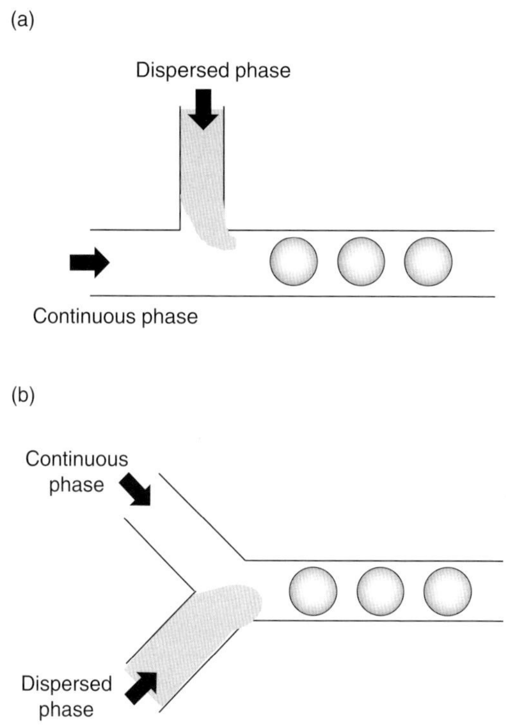

Figure 5.6 Microfluidic devices can produce monodisperse droplets. (a) T-juction; (b) Y-junction. Adapted from Nisisako *et al.* (2004) and Steegmans *et al.* (2009).

A Y-junction has often been used for emulsification (Kawai *et al.*, 2003; Kubo *et al.*, 2006; Steegmans *et al.*, 2009). Variations in the flow rate and viscosity of the dispersed phase do not affect the size of droplets generated using a Y-junction. At Y-junctions, droplets snap off when the nascent droplets are elongated by the surrounding continuous phase, which is known as flow-focusing or extensional flow (Figure 5.6b). Kawai *et al.* (2003) mass-parallelized the Y-junction and operated up to 1500 Y-junctions simultaneously, in which relatively monodisperse droplets were produced with a diameter of 85 μm and CV of about 10%.

Cross-junctions have been used as one-droplet or two-droplet generation units, depending on the location of the two phases. When the dispersed phase flows in the middle MC, each cross-junction works as a one-droplet generation unit, where droplets are generated by extension of Reyleigh's approach (Reyleigh, 1879). When the dispersed phase flows in the side MCs, each cross-junction works as two-droplet generation units. Droplet generation in this case is similar to that at a T-junction.

Emulsification using a flow-focusing geometry was first reported by Anna *et al.* (2003). In the flow-focusing geometry, three coaxial flows consisting of a central dispersed-phase flow and two outside continuous-phase flows are forced to pass through a narrow orifice (Figure 5.7). Droplet generation takes place inside or downstream of the orifice, depending on the flow rates of liquid phases. There are four types of droplet generation processes: squeezing, dripping, jetting, and tip streaming (Anna and Mayer, 2006). Flow-focusing devices typically produce monodisperse emulsions with sizes of a few microns to a few 100 μm. Tip streaming, which occurs in the presence of surfactants and at very high flow-rate ratios of a continuous phase to a dispersed phase can produce droplets as small as 1/20 of the orifice size. In flow-focusing geometries, as well as different types of junctions, the flow of the continuous and dispersed phases must be controlled precisely in order to obtain monodisperse emulsions with a desired droplet size, as the resulting droplet size is sensitive to the flow rate of each phase. Li *et al.* (2009) developed parallelized flow-focusing devices that are capable of producing monodisperse emulsions with a maximum droplet throughput of about 50 mL/h.

In addition to planar microfluidic emulsification devices, research groups at Harvard University have proposed three-dimensional asymmetric microfluidic emulsification devices (Takeuchi *et al.*, 2005; Utada *et al.*, 2005, 2007). Figure 5.8 shows a glass capillary device consisting of two round glass

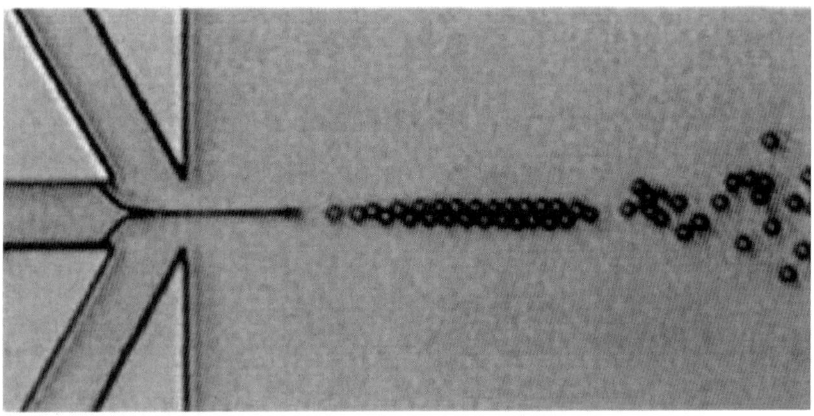

Figure 5.7 Microfluidic devices can produce monodisperse droplets. Flow-focusing (Xu and Nakajima, 2004).

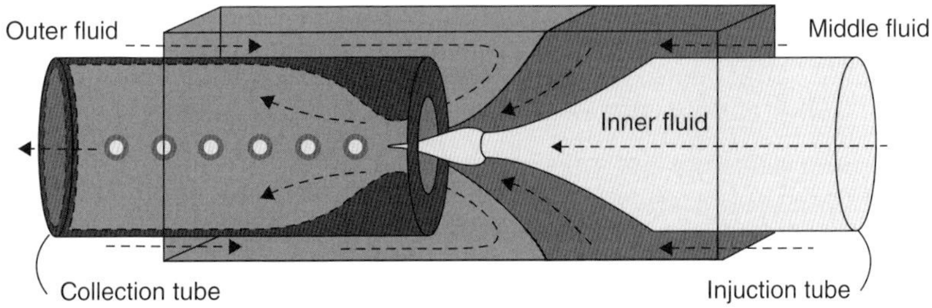

Figure 5.8 Schematic illustration of a microcapillary device. (Utada *et al.*, 2007. Reproduced with permission from Cambridge University Press.)

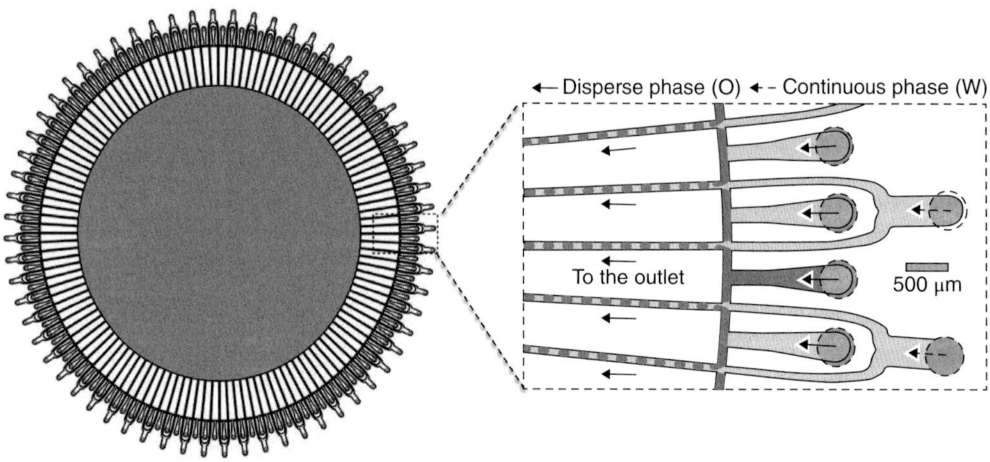

Figure 5.9 Schematic of the 128 cross-junctions on a chip. (Nisisako and Torii, 2008. Reproduced with permission from The Royal Society of Chemistry.)

capillaries lying end to end within a square glass tube. When the two fluids enter the collection tube, emulsion droplets are generated without chemical surface modification, in which the droplets obtained are smaller than the size of the collection tube.

Microfluidic emulsification devices are also advantageous for producing monodisperse multiple emulsions with controlled size and number of internal droplets. For instance, Chu *et al.* 2007 used sequential glass-capillary devices for producing highly controlled monodisperse double and triple emulsions. Additionally, Abate and Weitz (2009) produced high-order multiple emulsions (triple, quadruple, and quintuple emulsions) using a series of cross-junctions. Highly controlled multiple emulsions are useful templates for producing new classes of microcapsules with complex internal morphology.

As described above, microfluidic emulsification devices generally have throughput values up to one or several milliliters per hour. Even higher droplet throughputs have been required for practical applications. Nisisako and Torii (2008) developed large-scale parallelized cross-flow junctions with 256 shear-based droplet generation units and formulated monodisperse droplets with a size of around 100 μm at a throughput of 320 mL/h (Figure 5.9). In these microfluidic devices, however, it is rather difficult to keep a constant and uniform flow rate in all the channels for successful long-term emulsification. Moreover, regular droplet generation may be affected by bubble formation or presence of impurities in the channels. It is also difficult to clean such large microfluidic devices, as their long channels are often chemically bonded to flat plates.

Taking into account the aforementioned unfavorable conditions for uniformly sized and stable droplet formation, the following technology was developed, aiming to minimize such problems.

5.5.2 Microchannel emulsification

Microchannel emulsification was first proposed by Kawakatsu *et al.* (1997), which can produce uniform sized droplets with a CV < 5%. Microchannel array devices have been used for droplet formation, in which hundreds or hundreds of thousands of parallel MCs can be fabricated on a single crystal silicon plate using photolithography and anisotropic wet etching (Kawakatsu *et al.*, 1999, 2000; Kobayashi *et al.*, 2005a). The MC arrays can be fabricated onto the surface of a silicon wafer as micro-grooves (Kobayashi *et al.*, 2002). Modules with microgrooves can be either dead-end (Figure 5.10a) or cross-flow (Figure 5.10b). In a typical dead-end module, MCs are fabricated on terraces arranged on four sides of a silicon plate. Currently, MC plates with 1–100 μm depth are available. In operation,

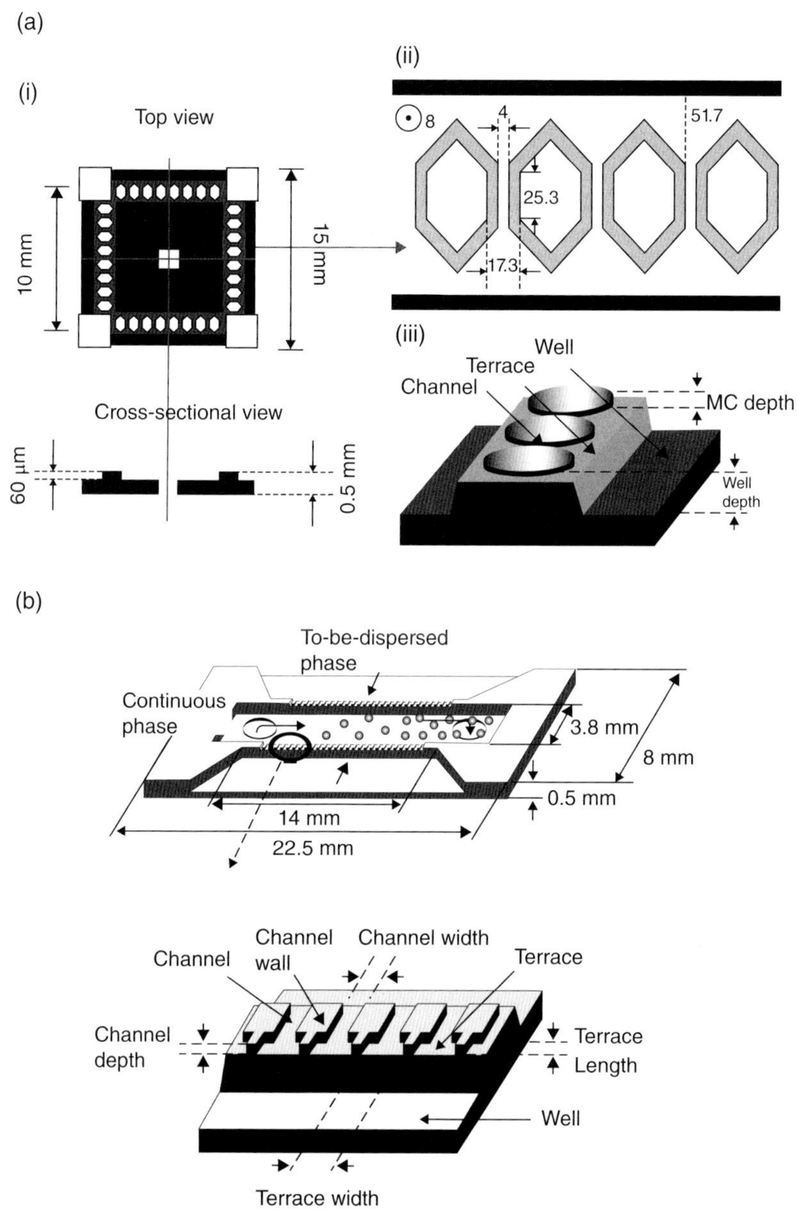

Figure 5.10 Grooved-type MC plates. (a) Dead-end groove MC plate: (i), top and cross-sectional views; (ii), top; (iii), three-dimensional views of part of a MC array. (Fujiu et al., 2011. Reproduced with permission from Springer Science and Business Media.). (b) Cross -low groove MC plate. (Kobayashi *et al.*, 2007b. Reproduced with permission from Elsevier.).

the MC plate is tightly sealed with a transparent glass plate. The dispersed phase is supplied through a central hole and flows out through MCs on all four sides. Compared with other microfluidic devices, the MC array device can be easily cleaned because each device can be disassembled after operation.

Microchannel emulsification exploits the interfacial tension as a driving force for droplet formation. The dispersed phase exiting the channels becomes a disc-like shape on the terrace, which causes high

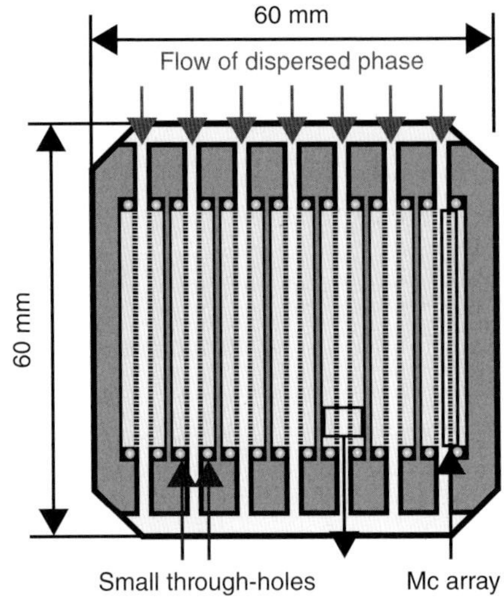

Figure 5.11 Schematic top view of large scale of MC emulsification chip consisting of 14 MC arrays. (Kobayashi *et al.*, 2010. Reproduced with permission from Springer Science and Business Media.)

Laplace pressure due to interfacial tension, resulting in hydrodynamic instability. This instability is a driving force for spontaneous transformation of the dispersed phase into spherical droplets (Sugiura *et al.*, 2002). Droplet formation behavior drastically changes above the critical velocity, due to the transition from a dripping regime to a continuous outflow regime. The critical velocity can be predicted from the physical properties of the dispersed and continuous phases, interfacial tension, and system geometry. We also analyzed the generation of soybean oil-in-water emulsion droplets using CFD with an asymmetric straight-through MC array (Kobayashi *et al.*, 2012b), and showed that successful droplet generation was observed below a critical dispersed-phase velocity.

When the dead-end modules with grooved MC arrays were used, flow rates of the dispersed phase were lower than 0.1 mL/h for vegetable oils, due to the limited number of MCs (100–1500). Large-scale cross-flow modules with grooved MC arrays are more suited for higher production rates, because 11,900 microgrooves can be incorporated onto a 60 × 60 mm single MC plate with 14 parallel arrays. This MC plate, shown in Figure 5.11, was capable of producing monodisperse emulsions with fine droplets of soybean oil of around 10 μm with a throughput of 1.5 mL/h for soybean O/W emulsions (Kobayashi *et al.*, 2010).

Grooved-type modules have a limited droplet throughput, due to insufficient utilization of the MC-plate surface, since MCs are arranged on the plate surface in longitudinal direction, and feed channels for dispersed and continuous phase must be provided on the plate surface. A vertical array of straight-through MCs allows much better utilization of the plate surface, resulting in significantly higher throughputs (Figure 5.12). For example, a 40 × 40 mm straight-through MC plate has 211,248 MCs, that is, 132 MCs per square millimeter and a soybean-oil flow rate can exceed 30 mL/h (Kobayashi *et al.*, 2012c).

Straight-through MCs can have either symmetric or asymmetric structure. Symmetric MCs are of the same size and shape (e.g., circular or rectangular) along the whole cross-section of the plate. Rectangular MCs provide better performance than circular MCs, and the aspect ratio (ratio between slot length to slot width) should be at least 3–3.5, in order to ensure stable production of uniformly sized droplets (Kobayashi *et al.*, 2004). An asymmetric MC plate typically contains circular channels on the upstream (inlet) side and slots on the downstream (outlet) side. Asymmetric structure is particularly useful for generation of uniform droplets when the dispersed phase viscosity is <1 mPa/s,

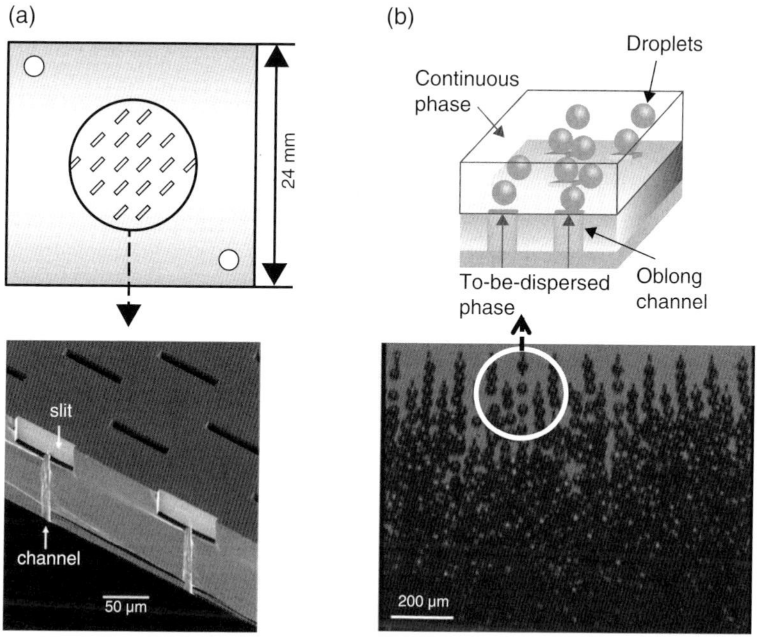

Figure 5.12 Straight-through microchannels (MC). (a) top view and SEM image, (b) droplet formation from straight-through MC. (Kobayashi *et al.*, 2002. Reproduced with permission from Wiley.)

for example when the dispersed phase is a volatile hydrocarbon (C_6-C_{10}), such as decane (Kobayashi *et al.*, 2005b). Asymmetric straight-through MCs have also been used successfully for production of W/O emulsions (Kobayashi *et al.*, 2009), polyunsaturated fatty acids (PUFA)-loaded O/W emulsions (Neves *et al.*, 2008). The size of droplets produced using asymmetric silicon MCs can range from several microns (Kobayashi *et al.*, 2008b) to several hundred microns (Kobayashi *et al.*, 2009). Using asymmetric MCs micromachined in stainless steel plates, the droplet size can reach several millimeters (Kobayashi *et al.*, 2008b). A maximum throughput of *n*-tetradecane droplets in an asymmetric MC device with a standard size (24 × 24 mm) was found to be 270 mL/h (Vladisavljević *et al.*, 2011). Kobayashi *et al.* (2012a) have recently achieved a throughput of *n*-tetradecane droplets of 1400 mL/h using a large 40 × 40 mm asymmetric MC plate with four arrays. Saito *et al.* (2006) prepared O/W emulsions stabilized by bovine serum albumin using straight-through MC and mechanical homogenizers. They reported that emulsions prepared by MC emulsification had more uniform diameters and higher stability to thermal processing, compared to those obtained using a mechanical homogenizer.

Compared with conventional mechanical homogenizers, the MC emulsification may have much higher energy efficiencies at around 30–60% (Sugiura *et al.*, 2001; Kobayashi *et al.*, 2008a), defined as:

$$\varepsilon = \left(\Delta E_{\text{in,min}} / \Delta E_{\text{in,exp}} \right) \times 100 \quad (5.3)$$

where ε is energy efficiency, $\Delta E_{\text{in,min}}$ is theoretical minimum energy input necessary for emulsions (J/m^3), $\Delta E_{\text{in,exp}}$ is actual energy input for emulsions (J/m^3). On the other hand, the energy efficiency of a high-pressure homogenizer was estimated to be below 0.1%, since most of the energy supplied is converted into heat (McClements, 2004). Also, most recently, an MC device has been used to produce monodispersed emulsions continuously for up to 1 week of operation.

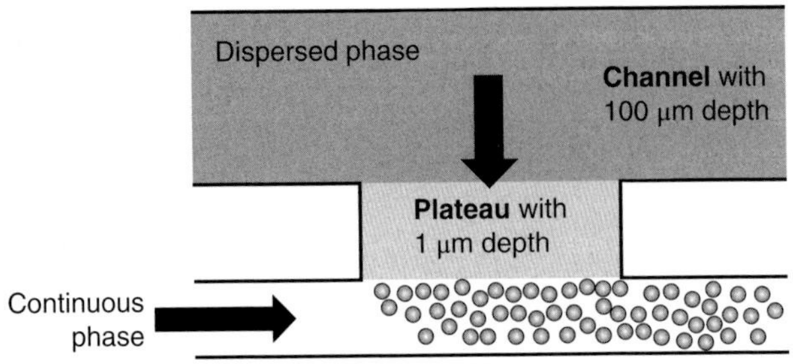

Figure 5.13 Schematic of the microchip with the parallelized edge-based droplet generation system (EDGE).

5.5.3 Edge-based droplet generation emulsification

Van Dijke *et al.* (2009a) proposed the edge-based droplet generation (EDGE) technique, where simultaneous formation of many oil droplets in a single EDGE module occurs. In an EDGE module, multiple droplets are formed at the edge of a shallow but rather wide rectangular plateau (Figure 5.13). The plateau is typically 1–3 μm deep, 200 μm long and 500–10,000 μm wide (van Dijke *et al.*, 2009b). Unlike the grooved-type MC arrays, where the droplet volume is of the same order of magnitude as the volume of the terrace where they are formed, the volume of the droplets formed in an EDGE device is only a small fraction of the volume of the plateau. The ratio of the droplet size to the plateau depth is typically 5.5–6.5 (van Dijke *et al.*, 2010a). Van Dijke *et al.* (2009a) have fabricated 196 plateaus (200 × 500 × 1.2 μm) on a 1.5 × 1.5 cm chip to produce up to 300,000 droplets per second per chip. The plateaus are arranged in 14 parallel arrays, with 14 plateaus per array, and a channel for the continuous phase meandering around the plateau lines. Other EDGE devices have been developed, with 14 plateaus per chip and a single 9500 μm wide plateau per line and with trapezoidal plateaus (van Dijke *et al.*, 2009a). These devices have been tested for production of various food-grade emulsions and foams (van Dijke *et al.*, 2010b).

5.6 EMULSION PROPERTIES AND APPLICATIONS

A variety of physicochemical mechanisms may be responsible for emulsion properties, such as chemical and physical stability. For instance, different physical mechanisms may be responsible for the instability of food emulsions, among which creaming, flocculation, and coalescence are the most common. Moreover, various chemical, biochemical, and microbiological processes that occur in food emulsions can also adversely affect their shelf life and quality: for example, lipid oxidation, enzyme hydrolysis, and bacterial growth (McClements, 2000). Here, we will mainly focus on lipid oxidation and introduce the relationship between droplet formulation and oxidative stability.

Antioxidant foods and ingredients play major roles in the food industry. The oxidation mechanism depends on the reactive species present in the system and their physicochemical environments (Coupland *et al.*, 1996). In the O/W emulsions, the common cause for oxidative instability is the interaction between the lipid hydroperoxides at the droplet surface and transition metals originating from the aqueous phase (Mei *et al.*, 1998). In most previous studies about the influence of droplet sizes on oxidative stabilities of lipids in emulsions, the emulsions were prepared using mechanical techniques. For example, Lethuaut *et al.* (2002) reported the immediate development of oxidation in sunflower oil-in-water emulsions compared to nonemulsified systems. They mentioned the increase in the interface area, but also other factors could be involved. Additionally, oxidation developed faster in high-pressure emulsification, compared to low-pressure emulsification and rotor–stator emulsification.

The conventional techniques usually provide undesirable oxidative stress to unsaturated lipids during preparation of emulsions prior to subjecting the emulsions to the oxidation process. An oxidative stress-free technique is therefore necessary when unsaturated oil emulsions are prepared for the purpose of evaluating the effects of droplet size on oxidative stability. Azuma *et al.* (2009) prepared an O/W emulsion using soybean oil and fish oil by MC emulsification and ultrasonic homogenizer, and investigated their oxidative stabilities. Emulsions produced by MC emulsification were found to have higher oxidative stability compared with an ultrasonic homogenizer.

Many food products can be found in the form of emulsions (Garti, 1997). An example of a naturally occurring food emulsion is milk, which contains globules of milk fat (cream) dispersed in water. The whitish color of milk is due to light scattering as it strikes the microscopic fat particles (Simpson, 2012). Salad dressings, beverages, whipped dessert toppings, peanut butter, and ice cream are also examples of emulsions of various edible fats and oils (Friberg *et al.*, 2007). Food emulsions are subdivided into different types, depending on their end-uses (Dalgleish, 2006). Some emulsions are end products themselves, for example, coffee creamers and cream liqueurs are simple emulsions that remain stable towards creaming and coalescence during their production and shelf life. Emulsions can also be used as ingredients that participate in forming the structures of more complex products. For example, yoghurts and other gelled systems contain emulsion droplets that must interact with other food ingredients, but that must not be destabilized in the process. Finally, the emulsion droplets may create new structures, as in ice cream or whipped products.

As for the potential applications of emulsions, they are commonly used in many major industries, such as chemistry (Oh *et al.*, 2002), pharmaceutics (Okochi *et al.*, 2000), and cosmetics (Lee *et al.*, 2004). These emulsions have been also investigated for controlled-release drug delivery systems (DDS) (Okochi *et al.*, 2000), microcapsules loaded with functional food ingredients (Weiss and Vladisavljević, 2006), and the formulation of reduced-calorie food emulsions (De Cindio *et al.*, 1995). Other applications include the use of multiple emulsions as intermediate products for the preparation of inorganic particles (Liu *et al.*, 2004), lipid nanoparticles (Garcia-Fuentes *et al.*, 2003), polymeric microspheres (Ma *et al.*, 2004), biodegradable microspheres (Herrmann and Bodmeier, 1995), and gel microbeads (Kawakatsu *et al.*, 2001).

In the pharmaceutical industry, emulsions may be used to turn medicines more palatable, to improve effectiveness by controlling dosage of active ingredients, or to provide improved aesthetics for topical drugs such as ointments. Emulsions can be delivered by oral, topical, or parenteral routes. When administered by oral route, O/W emulsions can efficiently improve the absorption and bioavailability of poorly water-soluble drugs. For parenteral drug delivery, both W/O and O/W emulsions have been investigated but O/W emulsions are predominantly used. The W/O/W emulsions have a very attractive potential as prolonged release systems for hydrophilic drugs, with the oil phase and surfactant layer acting as a release barrier (Bjerregaard *et al.*, 1999).

In the agricultural industry, emulsions have been used as delivery vehicles for insecticides, fungicides, and pesticides (Sawyer, 1982). These water insoluble biocides must be applied to crops at very low levels, usually by spraying through mechanical equipment. Emulsion technology allows these chemicals to be effectively diluted and provides improved sprayability. Nonionic emulsions are often used in this regard due to their low foaming properties and lack of interaction with the biocidal agents they are carrying (Mulqueen, 2003).

In cosmetics, emulsions are the delivery vehicle of choice for many hair and skin conditioning agents (Chandy *et al.*, 2011). Many cosmetic formulations are used to impart a pleasant odor, make the skin feel smooth and provide moisturizing agents, provide protection against sunburn, etc. In many cases, cosmetic formulations are designed to provide a protective, occlusive surface layer, which prevents the penetration of unwanted foreign matter and moderates the loss of water from the skin. Novel cosmetic formulations have also been developed, such as sunscreens, liposomes, and other ingredients that may maintain healthy skin and provide protection against drying, irritation, among others (Tadros, 2008).

5.7 CONCLUSIONS

This chapter has reviewed the production of emulsion droplets using various techniques. For instance, conventional mechanical homogenizers can be used to prepare emulsions in considerably small droplets, but they are generally polydisperse and require high-energy input. A quasi-monodisperse emulsion can be prepared by membrane emulsification, in which CV values of the droplet size are around 10%. On the other hand, microfluidic devices such as T- and Y-junctions, cross-junctions and flow focusing of quasi-two/three dimensions can produce monodisperse emulsion droplets, but their productivities have been below 1 mL/h in most cases. Recently, large microfluidic devices have been designed and used to formulate monodisperse emulsions with productivity of a few hundred milliliters per hour. However, successful long-term emulsification using this device was not possible, because of the difficulty in controlling the flow rate uniformly in all the channels. Microchannel emulsification can produce monodisperse emulsions using parallel MCs with a slit-like terrace or symmetric and asymmetric structure on a silicon chip. The droplet formation mechanism for the MC device is based on spontaneous transformation into spherical droplets, by action of the interfacial tension between the continuous and dispersed phases. The resulting droplet size depends mainly on the channel depth and terrace length and is not sensitive to the flow rates of the two phases below the critical values. Another major advantage of MC devices over other microfluidic devices is that the MC module can be easily assembled and disassembled; therefore, all the components can be appropriately cleaned after detachment. Most recently, the development of a large MC emulsification devices has enabled a maximum productivity of about 1 L/h. Further scale-up research on MC emulsification devices is in progress, aiming to achieve industrial-scale production of monodisperse emulsions.

References

Abate, A.R. and Weitz, D.A. (2009) High-order multiple emulsions formed in poly(dimethylsiloxane) microfluidics. *Small*, 5, 2030–2032.

Abismail, A., Canselier, J.R., Wilhelm, A.M., Delmas, H. and Gourdon, C. (1999) Emulsification by ultrasound: drop size distribution and stability. *Ultrasonics Sonochemistry*, 6, 75–83.

Anna, S.L. and Mayer, H.C. (2006) Microscale tipstreaming in a microfluidic flow focusing device. *Physics of Fluids*, 18, 1–13.

Anna, S.L., Bontoux, N. and Stone, H.A. (2003) Formation of dispersions using "flow focusing" in microchannels. *Applied Physics Letters*, 82: 364–366.

Aserin, A. (2008) *Multiple Emulsions: Technology and Applications*, John Wiley and Sons, Inc., Hoboken, NJ; 1–27.

Azuma, G., Kimura, N., Hosokawa, M. and Miyashita, K. (2009) Effect of droplet size on the oxidative stability of soybean oil TAG and fish oil TAG in oil-in-water emulsion. *Journal of Oleo Science*, 58(6), 329–338.

Bhushan, B. (2004) *Springer Handbook of Nanotechnology*, Springer-Verlag, Berlin; 147–317.

Bjerregaard, S., Söderberg, I., Vermehren, C. and Frokjaer, S. (1999) The effect of controlled osmotic stress on release and swelling properties of a water-in-oil emulsion. *International Journal of Pharmaceutics*, 183, 17–20.

Bouaouina, H., Desrumaux, A., Loisel, C. and Legrand, J. (2006) Functional properties of whey proteins as affected by dynamic high-pressure treatment. *International Dairy Journal*, 16, 275–284.

Burgess, D.J. and Yoon, J.K. (1995) Influence of interfacial properties on perfluorocarbon/aqueous emulsion stability. *Colloids and Surfaces B*, 4, 297–308.

Canselier, J.R., Delmas, H., Wilhelm, A.M. and Abismail, A. (2002) Ultrasound emulsification – an overview. *Journal of Dispersion Science and Technology*, 23, 333–349.

Chandy, A., Mandal, S. and Yadav, R. S. (2011) Current concept of cosmetic creams intended for scalp and hair. *International Journal of Current Trends in Science and Technology*, 2(1), 220–231.

Chappat, M. (1994) Some applications of emulsions. *Colloids and Surfaces A: Physicochemical and Engineering Aspects*, 91, 57–77.

Charcosset, C. (2009) Preparation of emulsions and particles by membrane emulsification for the food processing industry. *Journal of Food Engineering*, 92, 241–249.

Chu, L.Y., Utada, A.S., Shah, R.K., Kim, J.W. and Weitz, D.A. (2007) Controllable monodisperse multiple emulsions. *Angewandte Chemie International Edition* 46, 8970–8974.

Coupland, J.N. and McClements, D.J. (1996) Lipid oxidation in food emulsion. *Trends in Food Science and Technology*, 7, 83–91.

Dalgleish, D.G. (2006) Food emulsions – their structures and structure-forming properties. *Food Hydrocolloids*, 20, 415–422.

Davis, H.T. (1994) Factors determining emulsion type: hydrophile-lipophile balance and beyond. *Colloids and Surfaces A: Physicochemical and Engineering Aspects*, 91: 9–24.

De Cindio, B. and Cacace, D. (1995) Formulation and rheological characterization of reduced-calorie food emulsions. *International Journal of Food Science and Technology*, 30, 505–514.

Dickinson, E. (1992) *An Introduction to Food Colloids*. Oxford Science Publications, Oxford, UK; 85–125.

Friberg, S., Larsson, K. and Sjoblom, J. (2007) *Food Emulsions*, 4th edn. Marcel Dekker, Inc., New York; 485–572,

Fujihara, M. (2008) Applications of Shirasu porous glass. *New Glass*, 23(1), 28–33.

Fujiu, K.B., Kobayashi, I., Uemura, K. and Nakajima, M. (2011) Temperature effect on microchannel oil-in-water emulsification. *Microfluidics and Nanofluidics*, 10, 773–783.

Garcia-Fuentes, M., Torres, D. and Alonso, M.J. (2003) Design of lipid nanoparticles for the oral delivery of hydrophilic macromolecules. *Colloids and Surfaces B*, 27, 159–168.

Garti, N. (1997) Progress in stabilization and transport phenomena of double emulsions in food applications. *Lebensmittel-Wissenschaft und Technologie (Food Science and Technology)*, 30, 222–235.

Griffin, W.C. (1949) Classification of surface-active agents by HLB. *Journal of the Society of Cosmetic Chemists*, 1, 311–326.

Herrmann, J. and Bodmeier, R. (1995) Somatostatin containing biodegradable microspheres prepared by a modified solvent evaporation method based on W/O/W-multiple emulsions. *International Journal of Pharmaceutics*, 126, 129–138.

Joscelyne, S. M. and Trägårdh, G. (2000) Membrane emulsification-a literature review. *Journal of Membrane Science*, 169, 107–117.

Karbstein, H. and Schubert, H. (1995) Developments in the continuous mechanical production of oil-in-water macro-emulsions. *Chemical Engineering and Processing*, 34, 205–211.

Kawai, A., Matsumoto, S., Kiriya, H., *et al.* (2003) Development of a microreactor for manufacturing gel particles without class selection of diameter. *TOSOH Research and Technology Review*, 47, 3–9.

Kawakatsu, T., Kikuchi, Y. and Nakajima, M. (1997) Regular sized cell creation in microchannel emulsification by visual microprocessing method. *Journal of the American Oil Chemists' Society*, 74, 317–321.

Kawakatsu, T., Komori, H., Nakajima, M., Kikuchi, Y., Komori, H. and Yonemoto, Y. (1999) Production of monodispersed oil-in-water emulsion using crossflow-type silicon microchannel plate. *Journal of Chemical Engineering of Japan*, 32, 241–244.

Kawakatsu, T., Tragardh, G, Kikuchi, Y., Nakajima, M., Komori, H. and Yonemoto, T. (2000) Effect of microchannel structure on droplet size during crossflow microchannel emulsification. *Journal of Surfactants and Detergents*, 3, 295–302.

Kawakatsu, T., Tragardh, G. and Tragardh, C. (2001) Production of W/O/W emulsions and S/O/W pectin microcapsules by microchannel emulsification. *Colloid Surfaces A*, 189, 257–264.

Kobayashi, I., Nakajima, M., Chun, K., Kikuchi, Y. and Fujita, H. (2002) Silicon array of elongated through-holes for monodisperse emulsion droplets. *American Institute of Chemical Engineers Journal*, 48(8), 1639–1644.

Kobayashi, I., Mukataka, S. and Nakajima, M. (2004) Effect of slot aspect ratio on droplet formation from silicon straight-through micro-channels. *Journal of Colloid and Interface Science*, 279, 277–280.

Kobayashi, I., Mukataka, S. and Nakajima, M. (2005a) Production of monodisperse oil-in-water emulsions using a large silicon straight-through microchannel plate. *Industrial and Engineering Chemistry Research*, 44, 5852–5856.

Kobayashi, I., Mukataka, S. and Nakajima, M. (2005b) Novel asymmetric through-hole array microfabricated on a silicon plate for formulating monodisperse emulsions. *Langmuir*, 21, 7629–7632.

Kobayashi, I., Uemura, K. and Nakajima, M. (2007a) CFD analysis of generation of soybean oil-in-water emulsion droplets using rectangular straight-through microchannels. *Food Science and Technology Research*, 13, 187–192.

Kobayashi, I., Uemura, K. and Nakajima, M. (2007b) Formulation of monodisperse emulsions using submicron-channel arrays. *Colloid Surfaces A*, 296, 285–289.

Kobayashi, I., Takano, T., Maeda, R., Wada, Y., Uemura, K. and Nakajima, M. (2008a) Straight-through microchannel devices for generating monodisperse emulsion droplets several microns in size. *Microfluidics and Nanofluidics*, 4, 167–177.

Kobayashi, I., Wada, Y., Uemura, K. and Nakajima, M. (2008b) Generation of uniform drops via through-hole arrays micromachined in stain-less-steel plates. *Microfluidics and Nanofluidics*, 5, 677–687.

Kobayashi, I., Neves, M.A., Yokota, T., Uemura, K. and Nakajima, M. (2009) Generation of geometrically confined droplets using microchannel arrays: effects of channel and step structure. *Industrial and Engineering Chemistry Research*, 48, 8848–8855.

Kobayashi, I., Wada, Y., Uemura, K. and Nakajima, M. (2010) Microchannel emulsification for mass production of uniform fine droplets: integration of microchannel arrays on a chip. *Microfluidics and Nanofluidics*, 8, 255–262.

Kobayashi, I., Ichikawa, S., Neves, M. A., Kuroiwa, T. and Nakajima, M. (2011) Formulation of lipid micro/nanodispersion systems. In *Lipids in Nanotechnology*, Ahmad, M. (ed.). AOCS Press, Urbana, IL; 95–133.

Kobayashi, I., Neves, M. A., Wada, Y., Uemura, K. and Nakajima, M. (2012a) Large microchannel emulsification system for producing monodisperse droplets on a liter scale. *IMRET 12, The International Conferences on Microreaction Technology. Proceedings*, Lyon; 445–446.

Kobayashi, I., Vladisavljević, G.T., Uemura, K. and Nakajima, M. (2012b) CFD analysis of microchannel emulsification: Droplet generation process and size effect of asymmetric straight flow-through microchannels. *Chemical Engineering Science*, 66, 5556–5565.

Kobayashi, I., Neves, M.A., Wada, Y., Uemura, K. and Nakajima, M. (2012c) Large microchannel emulsification device for mass producing uniformly sized droplets on a liter per hour scale. *Green Processing and Synthesis*, 1, 353–362.

Kubo, A., Shinmori, H. and Takeuchi, T. (2006) Atrazine-imprinted microspheres prepared using a microfluidic device. *Chemistry Letters*, 35, 588–589.

Lee, J.S., Kim, J.W., Han, S.H., *et al.* (2004) The stabilization of l-ascorbic acid in aqueous solution and water-in-oil-in-water double emulsion by controlling pH and electrolyte concentration, *Journal of Cosmetic Science*, 55, 1–12.

Lethuaut, L., Métro, F. and Genot, C. (2002) Effect of droplet size on lipid oxidation rates of oil-in-water emulsions stabilized by protein. *Journal of the American Oil Chemists' Society*, 79(5), 425–430.

Li, W., Greener, J., Voicu, D. and Kumacheva, E. (2009) Multiple modular microfluidic (M3) reactors for the synthesis of polymer particles. *Lab on a Chip*, 9, 2715–2721.

Liu, H.J., Ni, Y.H., Fei, W., *et al.* (2004) Fabrication of submicron Cu_2O hollow spheres in an O/W/O multiple emulsions. *Colloid Surfaces A*, 235, 79–82.

Ma, G.H., Sone, H. and Omi, S. (2004) Preparation of uniform sized polystyrene–polyacrylamide composite microspheres from a W/O/W emulsion by membrane emulsification technique and subsequent suspension polymerization. *Macromolecules*, 37, 2954–2964.

McClements, D.J. (2000) Comments on viscosity enhancement and depletion flocculation by polysaccharides. *Food Hydrocolloids*, 14, 173–177.

McClements, D.J. (2004) *Food Emulsions*, 2nd edn. CRC Press, Boca Raton, FL; 632 pp.

Mei, L.Y., Decker, E.A. and McClements, D.G. (1998) Evidence of iron association with emulsion droplets and its impact on lipid oxidation. *Journal of Agricultural and Food Chemistry*, 46, 5072–5077.

Mulqueen, P. (2003) Recent advances in agrochemical formulation. *Advances in Colloid and Interface Science*, 106, 83–107.

Myers, D. (2006) Emulsions. In: *Surfactant Science and Technology*, 3rd edn, Myers, D. (ed.). John Wiley & Sons, Inc., Hoboken, NJ; 280–322.

Nakashima, T. and Kurki, Y. (1981) Effects of composition and heat treatment on the phase separation of Na2O–B2O3–SiO2–Al2O3–CaO glass prepared from volcanic ashes. *Nippon Kagaku Kaishi*, 8, 1231–1238.

Nakashima, T. and Shimizu, M. (1988) Preparation of O/W emulsion by porous glass membrane. *The 21st Autumn Meeting of the Society of Chemical Engineering*, Japan; 86.

Nakashima, T., Shimizu, M. and Kukizaki, M. (1991) Membrane emulsification by microporous glass. *Key Engineering Materials*, 513, 61–62.

Nazir, A., Schroën, K. and Boom, R. (2010) Premix emulsification: a review. Journal of Membrane Science, 362, 1–11.

Neves, M.A., Ribeiro, H.S., Fujiu, K.B., Kobayashi, I. and Nakajima, M. (2008) Formulation of controlled size PUFA-loaded oil-in-water emulsions by microchannel emulsification using beta-carotene-rich palm oil. *Industrial and Engineering Chemistry Research*, 47, 6405–6411.

Nisisako, T. and Torii, T. (2008) Microfluidic large-scale integration on a chip for mass production of monodisperse droplets and particles. *Lab on a Chip*, 8, 287–293.

Nisisako, T., Torii, T. and Higuchi T. (2004) Novel microreactors for functional polymer beads. *Chemical Engineering Journal*, 101, 23–29.

Oh, C., Chung, S.C., Shin, S.I., Kim, Y.C., Im, S.S. and Oh, S.G. (2002) Distribution of macropores in silica particles prepared by using multiple emulsions. *Journal of Colloid and Interface Science*, 254, 79–86.

Okochi, H. and Nakano, M. (2000) Preparation and evaluation of W/O/W type emulsions containing vancomycin. *Advanced Drug Delivery Reviews*, 45, 5–26.

Perrier-Cornet, J.M. and Gervais, P.M.P. (2005) Comparison of emulsification efficiency of protein-stabilized oil-in-water emulsions using jet, high pressure and colloid mill homogenization. *Journal of Food Engineering*, 66, 211–217.

Reyleigh, L. (1879) On the capillary phenomena of jets. *Proceedings of the Royal Society of London* (1854–1905), 29, 71–97.

Rhodes, M. (1998) *Introduction to Particle Technology*. John Wiley and Sons, Ltd, Chichester, UK; 320 pp.

Ribeiro, H.S., Janssen, J., Kobayashi, I. and Nakajima, M. (2010) Membrane emulsification for food applications. In *Membrane Technology*, Vol. 3, *Membranes for Food Treatment*, Peinemann, K.V., Nunes, S.P. and Giorno, L. (eds). Wiley-VCH, Weinheim, Germany; 149–171.

Roach, A. and Harte, F. (2008) Disruption and sedimentation of casein micelles and casein micelle isolates under high-pressure homogenization. *Innovative Food Science and Emerging Technologies*, 9, 1–8.

Saito, M., Yin, L.J., Kobayashi, I. and Nakajima, M. (2006) Comparison of stability of bovine serum albumin-stabilized emulsions prepared by microchannel emulsification and homogenization. *Food Hydrocolloids*, 20, 1020–1028.

Santos, H. (2005) *Principles and Applications of NanoMEMS Physics*. Springer, Dordrecht, Netherlands; 1–39.

Sawyer, E.W. (1982) Stabilizing agents for agricultural suspensions and emulsions. *Industrial and Engineering Chemistry Product Research and Development*, 21(2), 285–290.

Schröder, V., Behrend, O. and Schubert, H. (1998) Effect of dynamic interfacial tension on the emulsification process using microporous ceramic membranes. *Journal of Colloid and Interface Science*, 202, 334–340.

Shundo, A. (2009) A high-speed mixer enabling particle design. *International Journal of Cosmetic Science*, 31(1), 77.

Simpson, B.K. (2012) *Food Biochemistry and Food Processing*, Part 4, 2nd edn. John Wiley and Sons, Inc., Hoboken, NJ; 427–491.

Steegmans, M.L.J., Schroën, K.G.P.H. and Boom, R.M. (2009) Characterization of emulsification at flat microchannel Y junctions. *Langmuir*, 25, 3396–3401.

Strawbridge, K.B., Ray, E., Hallett, F.R., Tosh, S.M. and Dalgleish, D.G. (1995) Measurement of the particle size distributions in milk homogenized by a microfluidizer-estimation of populations of particles with radii less than 100 nm. *Journal of Colloid and Interface Science*, 171, 392–398.

Suzuki, K., Shuto, I. and Hagura, Y. (1996) Characteristics of the membrane emulsification method combined with preliminary emulsification for preparing corn oil-in-water emulsions. *Food Science and Technology International, Tokyo*, 2(1), 43–47.

Sugiura, S., Nakajima, M., Iwamoto, S. and Seki, M. (2001) Interfacial tension driven monodispersed droplet formation from microfabricated channel array. *Langmuir*, 17, 5562–5566.

Sugiura, S., Nakajima, M., Kumazawa, N., Iwamoto, S. and Seki, M. (2002) Characterization of spontaneous transformation-based droplet formation during microchannel emulsification. *The Journal of Physical Chemistry B*, 106, 9405–9409.

Tadros, T.F. (2008) *Colloids in Cosmetics and Personal Care*, Vol. 4. Wiley-VCH GmbH and Co. KGaA, Weinheim, Germany; 1–34.

Tadros, T.F. (2009) *Emulsion Science and Technology*. Wiley-VCH Verlag GmbH and Co. KGaA, Weinheim, Germany; 1–3.

Takagi, K. (2013) Production of nanoscale foods using high-pressure emulsification technology. In: *Bio-Nanotechnology. A Revolution in Food, Biomedical and Health Sciences*, Bagchi, D. (ed.). Wiley-Blackwell, Ltd, Chichester, UK; 530–541.

Takeuchi, S., Garstecki, P., Weibel, D.B. and Whitesides, G.M. (2005) An axisymmetric flow-focusing microfluidic device. *Advanced Materials*, 17, 1067–1072.

Teh, S.Y., Lin, R., Hung, L.H. and Lee, A.P. (2008) Droplet microfluidics. *Lab on a Chip*, 8, 198–220.

Utada, A.S., Lorenceau, E., Link, D.R., Kaplan, P.D., Stone, H.A. and Weitz, D.A. (2005) Monodisperse double emulsions generated from a microcapillary. *Science*, 308, 537–541.

Utada, A.S., Chu, L.Y., Fernandez-Nieves, A., Link, D.R., Holtze, C. and Weitz, D.A. (2007) Dripping, jetting, drops, and wetting: the magic of microfluidics. *Materials Research Society Bulletin*, 32(9), 702–708.

Van Dijke, K.C., Veldhuis, G., Schroën, K. and Boom, R.M. (2009a) Parallelized edge-based droplet generation (EDGE) devices. *Lab on a Chip*, 9(19), 2824–2830.

Van Dijke, K.C., Veldhuis, G., Schroën, K. and Boom, R. (2009b) Simultaneous formation of many droplets in a single microfluidic droplet formation unit. *American Institute of Chemical Engineers Journal*, 56, 833–836.

Van Dijke, K, de Ruiter, R., Schroën, K. and Boom, R. (2010a) The mechanism of droplet formation in microfluidic EDGE systems. *Soft Matter*, 6, 321–330.

Van Dijke, K, Schroën, K., van der Padt, A. and Boom, R. (2010b) EDGE emulsification for food-grade dispersions. *Journal of Food Engineering*, 97, 348–354.

Vladisavljević, G.T. and Schubert, H. (2002) Preparation and analysis of oil-in-water emulsions with a narrow droplet size distribution using Shirasu-porous-glass (SPG) membranes. *Desalination*, 144, 167–172.

Vladisavljević, G.T. and Williams, R.A. (2005) Recent developments in manufacturing emulsions and particulate products using membranes. *Advances in Colloid and Interface Science*, 113, 1–20.

Vladisavljević, G. T., Kobayashi, I., Nakajima, M., Williams, R.A., Shimizu, M. and Nakashima, T. (2007) Shirasu porous glass membrane emulsification: Characterisation of membrane structure by high-resolution X-ray microtomography and microscopic observation of droplet formation in real time. *Journal of Membrane Science*, 302, 243–253.

Vladisavljević, G.T., Kobayashi, I. and Nakajima, M. (2011) Effect of dispersed phase viscosity on maximum droplet generation frequency in microchannel emulsification using asymmetric straight-through channels. *Microfluidics and Nanofluidics*, 10, 1199–1209.

Vladisavljević, G.T., Kobayashi, I. and Nakajima, M. (2012a) Production of uniform droplets using membrane, microchannel and microfluidic emulsification devices. *Microfluidics and Nanofluidics*, 13, 151–178.

Vladisavljević, G.T., Duncanson, W.J., Shum, H.C. and Weitz, D.A. (2012b) Emulsion templating of poly(lactic acid) particles: droplet formation behavior. *Langmuir*, 28, 12948–12954.

Weiss, J. and Vladisavljević, G. (2006) Microencapsulation systems with emphasis on successful applications in foods. In: Encyclopedia of Agricultural, Food, and Biological Engineering, Heldman, D.R. (ed.). Marcel Dekker, New York.

Wilbey, A. (1992) Homogenization. *Journal of the Society of Dairy Technology*, 45(2), 31–32.

Xu, J.H., Li, S.W., Tan, J. and Luo, G.S. (2008) Correlations of droplet formation in T-junction microfluidic devices: from squeezing to dripping. *Microfluidics and Nanofluidics*, 5, 711–717.

Xu, Q.Y. and Nakajima, M. (2004) The generation of highly monodisperse droplets through the breakup of hydrodynamically focused microthread in a microfluidic device. *Applied Physics Letters*, 85, 3726–3728.

6 Delivery systems for food applications: an overview of preparation methods and encapsulation, release, and dispersion properties

Qixin Zhong, Huaiqiong Chen, Yue Zhang, Kang Pan, and Wan Wang

Department of Food Science and Technology, University of Tennessee, Knoxville, Tennessee, USA

6.1 INTRODUCTION

Encapsulation has a long history of use in the food industry. In the past decade, encapsulation has evolved to goal-oriented approaches by enclosing a target compound/object in structures that have physical, chemical, or biological characteristics fulfilling the goal of encapsulation. This is largely due to the introduction of a delivery systems concept from disciplines such as pharmaceutical sciences, where studies of polymeric and colloidal systems are focused on the inclusion of a compound/structure to deliver targeted functions. Functions significant to food systems can be unique physical properties such as dispersibility, dispersion stability, turbidity, viscosity, color and gelation, protection of compounds from degradation during processing and storage, masking of unpleasant flavors, release of the encapsulated compound at a controlled rate, triggered release by stimuli of acidity, ionic strength, temperature, and enzymes, enhanced physical or biological binding with microorganisms, enhanced transcellular activity, and improved uptake from the adsorption sites to the circulatory system – that is, bioavailability. These functions are typically enabled using physical structures of particulates, fibers, or coatings/films. Physical structures can be formed by processes such as homogenization and spray drying, reversible or irreversible association of molecules (self-assembly) driven by thermodynamics to minimize the system free energy, nonspecific binding due to attractive electrostatic and hydrophobic interactions or hydrogen bonds, or specific binding between groups that recognize each other, for example, base pairs of deoxyribonucleic acid (DNA), antigen and antibody, and bacteria and bacteriophages.

Compounds requiring delivery systems include those important to flavor, appearance (color and turbidity), and texture, and those with biological activities that are important to the quality, safety, satiety, nutrition, and healthfulness of food products (McClements *et al.*, 2009). The goal of delivery systems varies significantly for these compounds. Variations in physicochemical properties of these compounds may require different strategies to achieve the goal. The scope of this chapter is to review physical properties of delivery systems, including dimension and geometry of structures commonly studied, entrapment efficiency, release properties of the encapsulated compounds, and dispersion stability in aqueous

Nanotechnology and Functional Foods: Effective Delivery of Bioactive Ingredients, First Edition.
Edited by Cristina M. Sabliov, Hongda Chen, and Rickey Y. Yada.

foods. The dimension of delivery systems is important to sensory (texture and visual appearance) and stability in aqueous dispersions. Entrapment efficiency is the first parameter to evaluate delivery systems. Release profiles are used to ensure a delivery system functions according to the design goal. Eventually, delivery systems are to be incorporated in foods that mostly have an aqueous continuous phase, and their stability against precipitation and aggregation during storage is important to product quality.

6.2 METHODS OF FABRICATING DELIVERY SYSTEMS AND THEIR TYPICAL DIMENSIONS

The dimension of delivery systems is one of the starting parameters when designing structures to achieve the study goal. Microparticles are commonly referred to as those with a diameter greater than 1 μm, but particles larger than ~10 μm are usually undesirable because they may cause a sandy texture (Nickerson, 1954; Walstra, 2002). The definition of nanomaterials, however, varies among countries and disciplines. According to the definition of the National Nanotechnology Initiative of the United States (www.nano.gov), nanomaterials have at least one dimension smaller than approximately 100 nm and have unique properties different to bulk materials. Therefore, the phrase nano can be used for spheres, rods, and fibers with a diameter/radius smaller than 100 nm and shells, sheets, and coatings/films with a thickness less than 100 nm. Submicroparticles are then used for those with a dimension between 100 nm and 1 μm (Fryd and Mason, 2012).

For food applications, fabrication of micro- and nanostructures can be practiced by two routes: bottom-up or top-down. A top-down approach is achieved by applying mechanical energy in unit operations, such as homogenization and milling to disrupt droplets or particles until most particles have a micrometer or nanometer dimension (Sozer and Kokini, 2009; Thies, 2012). Because the application of intense mechanical forces consumes a significant amount of energy, these methods are called high-energy methods. In contrast, bottom-up approaches refer to the formation of micro- and nanostructures from molecules or smaller particles (Thies, 2012) by physical, chemical, and biological means. The bottom-up methods typically do not require high mechanical energy and therefore are called low-energy methods. The two methods can also be used in combination. Several top-down and bottom-up methods are reviewed below.

6.2.1 Top-down methods

Size reduction of dry matter by ball and jet millings is a common method used in the food industry to produce particles of ≤ 1–10 μm (Sanguansri and Augustin, 2006). The discussion here concerns "soft" particles that are suspended in water and can be disrupted by mechanical forces, especially those with a liquid dispersed phase, like oil-in-water (O/W) emulsions. Generally, a coarse emulsion with large droplets is produced by blending two immiscible phases that are kinetically stabilized by surfactants, followed by reduction of droplet dimension by strong mechanical forces. The forces breaking emulsion droplets can be shear or normal forces, dependent on the motion of particles with respect to the continuous phase (Walstra, 2002), and have to overcome the Laplace pressure (equation 6.1) at the oil–water interface (McClements, 2010). The Laplace pressure becomes larger as emulsion droplets are reduced to a smaller dimension, which defines the limit of reducing droplet dimension by a high-energy method:

$$\Delta P = \frac{2\gamma}{R} \tag{6.1}$$

where ΔP is the Laplace pressure, γ is the oil–water interfacial tension, and R is the radius of curvature. The break up of droplets is accompanied by the creation of new droplet surfaces that are not covered by surfactants (Walstra, 2002). Fine emulsion droplets are produced if surfactants in the continuous

Table 6.1 Principles of common top-down processes and the typical dimension of processed particles.

Process	Principle	Typical particle diameter
Rotor–stator homogenization (colloid mills)	Coarse emulsion droplets/particles are sheared in the rotor–stator space. The dominant droplet disruption mechanism is laminar shear flow for smooth rotor and stator surfaces, while it is turbulent for corrugated surfaces.	1–5 μm (McClements, 2005; Urban *et al.*, 2006)
High-shear disc homogenization	A combination of longitudinal, rotational, and radial shear forces is generated by the rotating disc which disrupts the droplets.	As small as 1 μm (Zuidam and Shimoni, 2010) or below (Urban *et al.*, 2006)
High-pressure homogenization	The premix is passed through a narrow valve/orifice/nozzle under high pressure. Size reduction is enabled by the combination of turbulent and laminar flows and cavitation.	At small as about 0.1 μm (McClements, 2005)
Microfluidization	Two liquid streams of separate dispersed and continuous phases or coarse emulsions are forced into contact at a high velocity, creating fine droplets or reducing droplet dimension. The disruption involves both turbulent fluid flow and cavitation.	As small as < 0.1 μm (McClements, 2005)
Membrane or microchannel emulsification	The dispersed phase is extruded through the pores in a membrane or microchannels into the flowing continuous phase with surfactants.	2–10 times of the membrane pore/microchannel dimension (Joscelyne and Trägårdh, 2000); 0.3 μm (McClements, 2005)
High-power ultrasonication	Intense shear and pressure gradients are formed by ultrasonic waves and the cavitational and turbulent effects disrupt droplets.	As small as 0.2–0.25 μm (Urban *et al.*, 2006)

phase can quickly adsorb onto and subsequently stabilize the new surfaces, otherwise the newly formed droplets can coalesce to large droplets (Walstra, 2002). Common high-energy methods are briefly discussed below, with the principles and typical particle dimensions summarized in Table 6.1.

6.2.1.1 High-shear homogenization

High-shear homogenization is used to process dispersions by shear forces, typically at ambient pressure. These homogenizers apply a rotor–stator or disc set-up (Urban *et al.*, 2006). In a rotor-stator setup like colloid mills and gear-rim dispersing machine, the coarse dispersion is being sheared in the space between the rotor and the stator (Urban *et al.*, 2006). The surfaces of the rotor and stator can be fabricated to be smooth or toothed, generating laminar or turbulent flows, respectively, in colloid mills, but only turbulent flow in gear–rim dispersing systems (Urban *et al.*, 2006). The lower particle diameter limit processed by a laminar flow in colloid mills is about 1 μm (Urban *et al.*, 2006).

In the disc set-up, the dispersed phase is usually directly blended into a continuous phase in a suitable vessel by a disc head rotating at a high speed (Urban *et al.*, 2006). This discontinuous process only generates a turbulent flow with a low to medium stress and is mostly used for dispersing solid particles and emulsification of high-viscous products (50–10,000 mPa-s) or stress-sensitive ingredients (Urban *et al.*, 2006; El Kinawy *et al.*, 2012).

6.2.1.2 High-pressure homogenization

The high-pressure homogenization (HPH) technique is currently the most popular method of creating fine emulsions/dispersions in the food industry. The HPH systems have three categories, each with various configurations, as comprehensively reviewed elsewhere (Schultz *et al.*, 2004). In the most

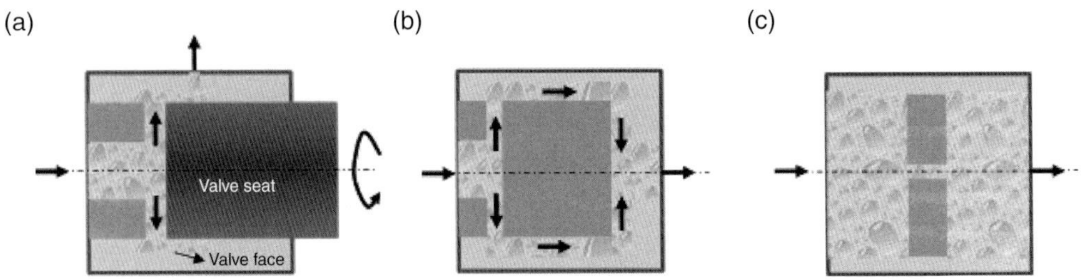

Figure 6.1 Three categories of high-pressure homogenizers: (a) a radial diffuser, (b) a counterjet dispergator in the example of a microfluidizer, and (c) an axial flow nozzle system. (Adapted from Schultz *et al.*, 2004.)

commonly used radial diffuser set-up (Figure 6.1a), also known as homogenizing valve, an incoming fluid is diffused radially and deflected for 90° by the valve face, and is sheared by the mobile valve seat. In counterjet dispergators, with a microfluidizer, as illustrated in Figure 6.1b, an incoming dispersion is diverged into two streams with much reduced dimensions in order to increase the velocity of fluid flow before bringing together to disrupt particles at a high velocity. Both shear force in a turbulent flow and cavitation are involved in breaking up particles during microfluidization. In axial-flow nozzle systems (Figure 6.1c), a dispersion is forced through a controllable orifice in a valve. Configurations in HPH systems can be designed and controlled to generate a pressure commonly in the range from 5 to 400 MPa and the dispersions can be homogenized for multiple passes to reduce the particle dimension (Schultz *et al.*, 2004). Generally, a higher pressure during operation and an increased number of passes enable finer particles to be produced.

6.2.1.3 Membrane/microchannel emulsification

This method has received growing attention due to its simplicity, relatively low energy requirement, and possibility of producing monodispersed droplets (Chuah *et al.*, 2014; El-Abbassi *et al.*, 2013). The dispersed phase is forced through pores in a membrane or straight-through channels with uniform diameters into a flowing continuous phase with surfactants (McClements, 2005). The relatively low stress applied during emulsification makes it appealing to process sensitive compounds. The pore/channel dimension, the O/W interfacial tension, surfactant chemistry, the transmembrane pressure, and the flow rate of the continuous phase affect droplet dimension (McClements, 2005). A hydrophobic membrane is used to produce a W/O emulsion, while an O/W emulsion is enabled with a hydrophilic membrane (McClements, 2005). The droplet size is normally 2–10 times that of the membrane pore size, with details provided elsewhere (Joscelyne and Trägårdh, 2000).

6.2.1.4 Ultrasonication

Ultrasonication offers another method of producing micro-/nanoscale particles by utilizing high-intensity ultrasonic waves with a frequency higher than 20 kHz (McClements, 2010). High-power ultrasound has been adopted by the food industry (Thies, 2012) and can be used to directly homogenize a dispersed phase into a continuous phase, or to reduce the droplet size of a coarse emulsion (Souza *et al.*, 2013). High-power ultrasound is also used for depolymerization or reduction of the size of solid particles (Souza *et al.*, 2013). The exact mechanism of this method is given in the review article by McClements (2010). One drawback of ultrasonication is that cavitation and high-intensity shear are restricted in a small zone, which limits the applicable sample volume to a few hundred milliliters (Leong *et al.*, 2009).

6.2.2 Bottom-up methods

In bottom-up methods, larger structures are formed from molecules dissolved or particles dispersed in a solvent. This category of methods includes processes that are already used in industrial productions, such as spray drying, and many others that are still practiced only in laboratories. Physical chemistry is at the center of many bottom-up methods as physical forces are used to bring molecules together to form structures. Chemical methods are less appealing in food systems because of implications on the safety and regulation of materials. Biological methods are also studied, if there are suitable substrates and enzymes. Numerous molecules and preparation methods have been studied to fabricate structures, with common ones discussed in this section.

6.2.2.1 Precipitated particles

In this group of methods, one principle is to dissolve a structure-forming molecule in a solvent and disperse the solution into another miscible solvent that is an antisolvent for the structure-forming molecule. For food systems, one solvent is usually water, and the other is a polar organic solvent like ethanol (Patel *et al.*, 2010a), methanol (Patel *et al.*, 2010a), isopropanol (Patel *et al.*, 2010a), and dimethylsulfoxide (Thies, 2012). One of the commonly studied food biopolymers is prolamins that are soluble in a binary solvent mixture of water and alcohol with appropriate proportions. For example, dispersing corn prolamins, zein, dissolved in aqueous ethanol into deionized water results in the precipitation of zein to nanoparticles with a diameter ranging from <100 nm to ~400 nm, which can be controlled by parameters such as concentrations of zein and ethanol and dispersion conditions (Zhong and Jin, 2009b). Aqueous alcohol in prolamin solutions can be used as a solvent to co-dissolve many lipophilic bioactive compounds that can co-precipitate with prolamins and become simultaneously encapsulated. This has been demonstrated for polyunsaturated fatty acids (Zhong *et al.*, 2009b), curcumin (Patel *et al.*, 2010a, b), essential oils (Parris *et al.*, 2005), among others.

Another principle of precipitation methods is based on biopolymers that have a solubility dependent on environmental conditions such as acidity and temperature. Hydroxypropyl methylcellulose phthalate (HPMCP) is insoluble below a critical pH but becomes soluble at higher pH, and so can be used to form particles upon changes in pH (Wege *et al.*, 2008). Chitosan is another biopolymer that is soluble at pH below 6.5 but precipitates to particles at neutral and alkaline pH (Jia *et al.*, 2001). Casein micelles are supramolecular structures that are dissociated at alkaline pH (e.g., pH 10) but can reassociate into nanoparticles upon neutralization (Huppertz *et al.*, 2008). Casein micelles also dissociate in warm (≥60°C) aqueous ethanol (≥35%) but become reassociated upon cooling (O'Connell *et al.*, 2001). If a compound has solubility characteristics similar to biopolymers, it can be encapsulated in biopolymer structures formed upon changes in environmental conditions. One example is a recent study of nanoencapsulating a lipophilic pigment, bixin, based on warm aqueous ethanol solution with both caseins and bixin (Zhang and Zhong, 2013).

6.2.2.2 Biopolymer complex particles

Coacervation, formation of nano- and microparticles between proteins and negatively charged polysaccharides, is an interesting topic. The principles were reviewed previously (Jones and McClements, 2010) and are summarized briefly herein. At a pH way above the isoelectric point (pI) of proteins, proteins are highly negatively charged and no complexes are formed. As pH is reduced to below a critical value (pH_s) but above the critical value of coacervation (pH_c), polysaccharides bind with positively charged patches of overall weakly negatively charged proteins. Soluble complexes are formed and systems are transparent or slightly turbid. As pH is reduced to below pH_c but above the precipitation pH (pH_p), in a narrow pH range below pI, proteins have a small amount of net positive charges, and complexes (coacervates) are formed at a dimension of ~100–10000 nm. Systems become turbid and separate into two phases, with the bottom being the viscous coacervate phase with > 70% solvent. Coacervates are open, dynamic, and reversible structures. At pH below pH_p, precipitates of complexes are expected, containing much less solvent than the coacervate phase. At a pH well below the p*K*a of polysaccharides, polysaccharides have few charges and no complexation is expected. Coacervates have been used to encapsulate various compounds (Yeo *et al*., 2005).

6.2.2.3 Colloidal structures self-assembled from small molecular surfactants and biopolymers

Small molecular surfactants like polysorbates and phospholipids have distinct regimes of hydrophilic and lipophilic structures. When dissolved above the critical micelle concentration, the portion with polarity opposite to the continuous phase associates in order to minimize the contact with the continuous phase and thus the system free energy. The association of surfactants is described in the surfactant number (N_s, equation 6.2) (Israelachvili, 1992). With the increase of N_s from 1/3 to 3 and the transition from a polar solvent to a nonpolar solvent, the self-assembled surfactant structures take spherical normal micelles, cylindrical micelles, planar structures, reverse cylindrical micelles, and reverse spherical micelles, with the last two systems having the lipophilic group facing the continuous nonpolar phase (Lawrence and Rees, 2000). The self-assembled structures are commonly studied to encapsulate compounds in micelles, microemulsions, and vesicles. Micelles and characteristic structures in microemulsions are typically smaller than 100 nm, which produces transparent dispersions that are thermodynamically stable (Bagwe *et al*., 2001). Liposomes are vesicular structures with a dimension and structure dependent on the molecular structure of phospholipids and preparation conditions. Multilamellar vesicles typically have a dimension of 0.1–10 μm, while unilamellar vesicles are further divided into three groups based on the dimensions: small (<100 nm), large (100–500 nm) and giant (over 1 μm) (Ulrich, 2012).

$$N_s = \frac{v}{l_c a_o} \tag{6.2}$$

where ν is the volume of the lipophilic segment, l_c is the chain length of lipophilic segment, and a_o is the area of the hydrophilic head group (Joscelyne and Trägårdh, 2000).

Likewise, proteins like β-casein (Esmaili and Ghaffari, 2011), polysaccharides after substituting hydroxyl groups with hydrophobic side chains (Yu and Huang, 2010), and proteins glycated with saccharides (Pan *et al*., 2006, 2007) can self-assemble to form micelles, which can be applied to encapsulate bioactive compounds. Caseins also self-assemble into nanoparticles by hydrophobic interactions and ionic bridging through divalent calcium ions (Walstra, 2002). In addition, globular proteins from dairy proteins – α-lactalbumin (Ipsen and Otte, 2007) and β-lactoglobulin (Mezzenga *et al*., 2005; Nicolai *et al*., 2011), soy proteins (Wang *et al*., 2011), kidney bean phaseolin (Tang *et al*., 2010), and egg white proteins (Cao *et al*., 2009) undergo steps of denaturation, hydrolysis to oligopeptides, and assembly of amyloid-like fibrils from oligopeptides by β-sheet alignment during heating at pH 2.0 and above the denaturation temperature (Jones and Mezzenga, 2012).

6.2.2.4 Emulsions and nanoemulsions

Emulsions are systems with a liquid phase dispersed in another liquid phase. When the droplets of the dispersed phase are smaller than about 200 nm, the term nanoemulsion is used (McClements and Rao, 2011). Most emulsions have opposite polarity between continuous and dispersed phases, with the interface being surfactants that reduce O/W interfacial tension. Water-in-water emulsions can also be produced if the dispersed water droplets can form distinct particles by gelation or adsorption of biopolymers on surfaces (McClements and Rao, 2011). Compared to thermodynamically stable microemulsions that possibly have coexisting spherical and nonspherical particles, emulsions and nanoemulsions are not thermodynamically stable and have spherical droplets with a defined interface (McClements, 2012). Nanoemulsions can be prepared from coarse emulsions after reducing the emulsion droplet dimension by using the above high-energy methods, such as high-pressure homogenization, microfluidization and high-power ultrasound (McClements and Rao, 2011).

Low-energy methods have also been studied to prepare nanoemulsions by controlling interfacial properties. Self-emulsifying nanoemulsions can be prepared by gradually dispersing an oil phase with water-soluble surfactants into an aqueous phase, and the dissolution of surfactants out of the oil phase into the continuous aqueous phase is simultaneously accompanied by the inclusion of oil in surfactant shells (McClements and Rao, 2011). Another group of technologies is based on phase-inversion principles that are due to changes in composition or temperature (McClements and Rao, 2011). In the phase-inversion temperature method, an O/W emulsion is transformed to a W/O emulsion upon exceeding a critical temperature, which can be transformed back to an O/W emulsion upon cooling. The redistribution of surfactants and oil during thermal processes may enable the transformation of a turbid coarse emulsion into a transparent nanoemulsion. Analogously, gradual increases in the volume fraction of the continuous phase can invert an O/W emulsion to a W/O one, or vice versa, and the dimension of droplets can be reduced after phase inversion.

Higher orders of emulsion architectures can be fabricated. This includes multilayered emulsions that are produced with a charged surface, enabling the sequential adsorption of multiple layers of molecules with charges opposite the outer surface of emulsion droplets (McClements *et al.*, 2007). The increased layer thickness and possibly strengthened repulsive forces enhance the interfacial stability. Multiple emulsions can also be designed to disperse another phase in primary emulsion droplets such as water-in-oil-in-water emulsions that enable the dissolution of multiple compounds with opposite polarity in each phase of emulsion droplets (McClements *et al.*, 2007). Various methods of controlling emulsion properties are described in more detail in other chapters.

6.2.2.5 Solid particles coated with biopolymers

Delivery systems can also be fabricated by physically coating solid particles. For example, a composite layer of HPMCP and medium chain fatty acids was coated on solid particles of glutamine by spray-drying (Zhong and Shah, 2013). The majority of water-soluble glutamine was kept in the coating in a simulated gastric fluid at pH 1.2 up to 4 h but was completely released at pH 6.8, resulting from the solubility characteristics of HPMCP. Probiotics have a dimension typically of 1–5 μm (Burgain *et al.*, 2011). Coating probiotics with food biopolymers using processes of spray drying and spray-freeze drying improved the viability (Burgain *et al.*, 2011). Precipitating or gelling alginate (by calcium ions) on probiotics or the combination of both principles to entrap probiotics using a biopolymer blend have also been studied to protect probiotics from inactivation after ingestion (Liu *et al.*, 2006).

6.2.2.6 Structures formed as a result of processing

Many structures can be formed from solutions or dispersions by processing to obtain functions desired in a delivery system. This includes particles produced using spray drying, spray freeze-drying, and supercritical antisolvent precipitation (Gouin, 2004). For example, with appropriate formulations, sustained release of lysozyme was observed from spray-dried (Zhong and Jin, 2009a) and supercritical antisolvent precipitated (Zhong *et al.*, 2009a) zein capsules. Another example is nanofibers produced by electrospinning of biopolymer solutions that can also include microemulsions (Kriegel *et al.*, 2009, 2010).

6.3 ENCAPSULATION EFFICIENCIES OF VARIOUS DELIVERY SYSTEMS

Encapsulation efficiency (equation 6.3) is usually estimated as the percentage of a compound included in a structure with respect to the total amount used in encapsulation:

$$\text{encapsulation efficiency}\left(\%\right)=\frac{\text{mass of a compound enpsulated in a structure}}{\text{total mass of the compound used before entrapment}}. \tag{6.3}$$

While this definition is commonly used, there are several points to be considered based on the nature of delivery systems and properties of the encapsulation process and encapsulated compound. In some microemulsion systems, there is no defined boundary between the two phases, and the encapsulation efficiency concept may not be applicable. Some systems undergo finishing steps, for example, preparation into the powdered form, which can cause the loss, or molecular structural changes, of the encapsulated compound. Examples can be the loss of volatile components and enzyme activity during spray drying. In addition, in laboratory scale spray-drying experiments, the overall mass yield can be low, which may cause an inaccurate evaluation of encapsulation efficiency if the content in spray-dried powder is used in equation 6.3. There also can be a considerable time lag before encapsulation efficiency is determined, and inconsistency can be observed if the initially encapsulated fraction is lost or deactivated during the time lag. As each delivery system is fabricated under different conditions and the encapsulation efficiency may be determined differently, the comparison of different literature studies should be examined with care. Furthermore, many studies apply a solvent to "extract" the "unencapsulated" fraction, which, however, can also extract the fraction that is initially entrapped.

Nevertheless, encapsulation efficiency is discussed in this section based on water solubility of the compound to be encapsulated, as most consumer products have an aqueous continuous phase (McClements *et al*., 2007). Many bioactive compounds are insoluble in water or marginally soluble in water. Encapsulation of lipophilic bioactive compounds is typically enabled in emulsion/microemulsion or biopolymer particles. Lipophilic compounds are used as the oil body of emulsions or are dissolved in the oil body to prepare emulsions. For biopolymer particles, lipophilic compounds are usually dissolved in an organic solvent that can also dissolve biopolymers before going through process(es) enabling encapsulation. Because of insolubility, encapsulation efficiency in emulsion droplets or biopolymer is expected to be high, unless lipophilic compounds separate into a cream or a precipitated phase. However, as listed in Table 6.2, very low encapsulation efficiency has also been reported in the literature.

Some studies are summarized in Table 6.3 for compounds with good water solubility. Physically, when encapsulation is conducted in an aqueous phase only, a bioactive compound can be attracted to the structure-forming molecules by opposite charges or hydrophobic interactions, and the encapsulation efficiency depends on the partition coefficient between the particle matrix and continuous phase. In addition to molecular structures, pH, ionic strength, temperature, and co-solutes can impact the partition coefficient. Multiple emulsions enable the incorporation of an aqueous phase in the dispersed oil phase, and the complete encapsulation is expected before emulsions are destabilized (McClements *et al*., 2007). Alternatively, water-soluble compounds can be prepared to solid particles that can be coated with a structure impermeable to the continuous water phase (Zhong and Shah, 2013).

There are also many compounds that are overall lipophilic but can be dissolved in water to a limited extent, for example, below ~1% w/v. These include essential oils including those used as flavoring agents (Table 6.4). Encapsulation of these compounds can be enabled similarly to those with no or marginal solubility. When prepared as emulsions/nanoemulsions, the slight water solubility of these compounds can cause the destabilization of emulsions by Ostwald ripening. Physically, compounds have an increased solubility when present in smaller particles and the dissolved fraction joins and

Table 6.2 Examples of efficiencies of encapsulating bioactive compounds with water insolubility or marginal solubility.

Category	Compound	Nature of particles	Entrapment efficiency (%)	Particle size	Reference
Fat-soluble vitamins	Vitamin A	Gliadin nanoparticles	75	ND	Duclairoir *et al.*, 1999
		Silica microparticles	7.3–44.0	2.4–6.6 μm	Lee *et al.*, 2001; Hwang *et al.*, 2005
		Methoxy poly(ethylene glycol)-grafted chitosan nanoparticles	≥80	50–200 nm	Jeong *et al.*, 2006
		Glyceryl behenate solid lipid nanoparticles	ND	ND	Jenning and Gohla, 2001
		Poly(lactic acid-co-glycolic acid) particles	27.1–54.9	ND	Jeong *et al.*, 2003
		Nanoemulsion	99.8	60–330 nm	Zhong *et al.*, 2011
	Vitamin E	Gliadin nanoparticles	77–79	900 nm	Renard *et al.*, 2002; Duclairoir *et al.*, 2003
		Poly(lactic acid-co-glycolic acid) nanoparticles	89.6–95.6	100–200 nm	Zigoneanu *et al.*, 2008
		Medium chain triglyceride nanoemulsion	99.7	78 nm	Laouini *et al.*, 2012
	Vitamin K_1	Solid lipid nanoparticles	≥85	125 nm	Liu *et al.*, 2010
	Vitamin D_2	Chitosan-ethylcellulose complexes	>95	2–20 μm	Shi and Tan, 2002
Polyunsaturated fatty acids	Docosahexaenoic acid	Casein nanoparticles	62.3–67.6	50 nm, 289 nm	Zimet *et al.*, 2011
		β-Lactoglobulin-pectin nanocomplexes	64	100 nm	Zimet and Livney, 2009
	Fish oil	Emulsion	97	≤1.3 μm	Kosaraju *et al.*, 2009; Patten *et al.*, 2009
		Zein nanoparticles	ND	350–540 nm	Zhong *et al.*, 2009b
	Linoleic acid (flaxseed oil)	Zein microspheres	59.6–93.3	ND	Quispe-Condori *et al.*, 2011
		Spray-dried emulsions	62.3–95.7	~20 μm	Carneiro *et al.*, 2013
Carotenoids	β-Carotene	Caseinate nanoparticles	ND	17 nm	Chu *et al.*, 2007
		Nanoparticles of casein glycated with dextran	54	110 nm	Pan *et al.*, 2007
		Nanoemulsions	ND	<200 nm	Qian *et al.*, 2012
	Lutein	Solid lipid nanoparticles or nanoemulsions	89.3–101	150–350 nm	Mitri *et al.*, 2011
	Curcumin	Nanoemulsions	98.2	132 nm	Ganta and Amiji, 2009
		Poly(lactic acid-co-glycolic acid) nanoparticles	84.6	242 nm	Shaikh *et al.*, 2009
		Alginate–chitosan–pluronic composite nanoparticles	10–13	100 nm	Das *et al.*, 2010
		Solid lipid nanoparticles	35–70	450 nm	Tiyaboonchai *et al.*, 2007
		Solid lipid nanoparticles	81.9	134.6 nm	Kakkar *et al.*, 2011
Flavonoids	Quercetin	Poly(lactic acid-co-glycolic acid) nanoparticles	32.7	139.5 nm	Song *et al.*, 2008

ND, not determined.

Table 6.3 Excmples of efficiencies of entrapping water soluble compounds.

Category	Compound	Nature of particles	Entrapment efficiency (%)	Particle size	Reference
Water-soluble vitamins	Vitamin C	Tripolyphosphate cross-linked chitosan microspheres	45.0–68.6	6.1–9.0 μm	Desai and Park, 2005, 2006
	Vitamin C	Oleic acid micelles	ND	150 nm	Abraham and Narine, 2009
	Vitamin B_2	Alginate-whey protein microspheres	>80	90–885 μm	Chen and Subirade, 2007
Carbohydrate	Blue dextran	Poly(lactic acid-co-glycolic acid) microspheres	12.5–37.0	6.2–16.2 μm	Ito *et al.*, 2008
Glycosamino-glycan	Heparin	Chitosan /carboxymethyl-β-cyclodextrin / tripolyphosphate nanoparticles	69.3–70.6	366–574 nm	Krauland and Alonso, 2007
Oligonu-cleotide	17-mer	Chitosan nanoparticles	78–93	128–162 nm	Janes *et al.*, 2001
Proteins	Insulin	Biodegradable polymers/soybean phosphatidylcholine nanoparticles	>90	200 nm	Cui *et al.*, 2006
		Vitamin B12-dextran nanoparticles	45–71	160–250 nm	Chalasani *et al.*, 2007
		Lectin-modified solid lipid nanoparticles	>60	57–75 nm	Zhang *et al.*, 2006
		Poly(lactic acid) microspheres	>95	ND	Uchida *et al.*, 1997
		Chitosan/carboxymethyl-β-cyclodextrin/ tripolyphosphate nanoparticles	85.5–93.3	377–613 nm	Krauland and Alonso, 2007
		Solid lipid nanoparticles	72.7	146 nm	Xie *et al.*, 2008
	Lysozyme	Solid lipid nanoparticles	43.2–59.2	549–644 nm	Almeida *et al.*, 1997
		Solid lipid nanoparticles	75.7	125 nm	Xie *et al.*, 2008
	Immunoglobulin G	W/O/W emulsions	61.2–65.8	ND	Chen *et al.*, 1999
	Bovine serum albumin	Nanoemulsion	>90	21.8 nm	Sun *et al.*, 2012
		Solid lipid nanoparticles	83.8	156 nm	Xie *et al.*, 2008
		Microspheres of poly(oxyethylene)–poly(lactic acid) diblock copolymers	93	33–44 μm	Bouillot *et al.*, 1999

ND, not determined.

Table 6.4 Examples of efficiencies of entrapping bioactive compounds with some water solubility.

Category	Compound	Nature of particles	Entrapment efficiency (%)	Particle size	Reference
Pharmaceuticals	Vincristine	Microemulsions	91.3–94.6	127–138 nm	Wang *et al.*, 2003
		Poly(lactic acid-co-glycolic acid) nanoparticles	92.8	139.5 nm	Song *et al.*, 2008
	Aclacinomycin A	Emulsions	96.3	123 nm	Wang *et al.*, 2002
	Ketoprofen	Alginate microparticles	93	350 μm	Del Gaudio *et al.*, 2009
		Eudragit RL polymer microparticles	60–67	<500 μm	El-Kamel *et al.*, 2001
		Solid lipid nanoparticles	97	75	Kheradmandnia *et al.*, 2010
Essential oil	Linalool/linalyl acetate	Gliadin nanoparticles	82.4	930 nm	Duclairoir *et al.*, 2003
	Artemisia arborescens L essential Oil	Solid lipid nanoparticles	87–92	100–500 nm	Lai *et al.*, 2006
	Caraway essential oil	Spray-dried emulsions emulsified with whey protein concentrate	78.8	ND	Bylaitë *et al.*, 2001
		Spray-dried emulsions emulsified with skimmed milk powder	74.1	ND	Bylaitë *et al.*, 2001
		Spray-dried emulsions emulsified with whey protein concentrate and co-dissolved maltodextrins	68–85	ND	Bylaitë *et al.*, 2001
	Cardamom essential oil	Spray-dried emulsions emulsified by mesquite gum	83.6	ND	Beristain *et al.*, 2001
	Orange oil	Spray-dried emulsion emulsified by whey protein isolate	72.7	40 μm	Kim and Morr, 1996
		Spray-dried emulsion emulsified by soy protein isolate	85.7	36 μm	Kim and Morr, 1996
		Spray-dried emulsion emulsified by gum arabic	75.9	33 μm	Kim and Morr, 1996
		Spray-dried emulsion emulsified by sodium caseinate	81.5	41 μm	Kim and Morr, 1996
		Wet emulsion emulsified by whey protein isolate	99.3	ND	Kim and Morr, 1996
		Wet emulsion emulsified by soy protein isolate	91.7	ND	Kim and Morr, 1996
		Wet emulsion emulsified by gum arabic	96.3	ND	Kim and Morr, 1996
		Wet emulsion emulsified by sodium caseinate	95.7	ND	Kim and Morr, 1996

ND, not determined.

becomes a part of the larger particles, eventually resulting in the disappearance of small particles (Walstra, 2002). Ostwald ripening can be prevented by preparing emulsions with an oil body of these compounds dissolved in other oils that are insoluble in water (Ziani *et al.*, 2011). Furthermore, essential oils are composed of volatile components and loss due to vaporization is common during spray drying and shelf-life storage (Zhong and Jin, 2009a; Xiao *et al.*, 2011a, b).

Similarities exist between pharmaceuticals and dietary supplements utilizing tablets and capsules as delivery vehicles. For food applications, the meaningful encapsulation efficiency should be evaluated based on the targeted function of delivery systems in food matrices (McClements *et al.*, 2007). Some products like cereals have low moisture contents and a powdered form of delivery system is needed, either by adding powders directly or being dried together with other materials during food manufacturing. Spray drying is a dominant process used to prepare powdered samples, with and without intentionally applying structure-design principles to achieve the functions (Tables 6.2–6.4). For moist or fluidic foods like beverages, the shelf life of these products can be weeks, months, and possibly years. The encapsulation should then be evaluated based on the amount remaining in a structure after the delivery system reaches the target location (Ulrich, 2012). If a delivery system is designed to enhance bioavailability, the meaningful entrapment efficiency should be the amount that reaches the absorption sites in the digestive system, which would require the entrapment of bioactive compounds in particles and the protection of their degradation during shelf-life storage and before reaching absorption sites after ingestion. For delivery systems to function in food matrices, for example, preservatives and antioxidants, the entrapment efficiency should be evaluated by the delivery modes, examples of which are sustained release at a controlled rate or a burst release upon triggers of environmental stresses, as discussed in the next section. These scenarios are due to the nature of delivery systems that are different from traditional encapsulation systems, which focus on how much is entrapped after the encapsulation process (equation 6.3), without examining how systems function in food matrices and after ingestion.

6.4 RELEASE PROPERTIES OF ENCAPSULATED COMPOUNDS

The release of encapsulated compounds can be enabled by the concentration (including vapor pressure) gradient, gradual loss of particle structures, for example, by dissolution or acid/alkaline and/or enzymatic hydrolysis, or changes in the solubility characteristics of particle components upon changes in environmental conditions. These release properties are briefly discussed below.

6.4.1 Diffusion-controlled release mechanism

Considering delivery systems with a defined geometry such as slabs, spheres and cylinders, release properties of an encapsulated compound can be described by typical time-dependent mass transport phenomena, where two diffusion coefficients exist for the process diffusing from the interior to the particle surface, followed by diffusion to the continuous bulk phase. As concentration gradient is the driving force of diffusion/release rate, the gradual loss of an encapsulated compound to the continuous phase causes a lowered release rate after longer incubation, eventually reaching a release equilibrium (Kim *et al.*, 2001). The typical Fickian diffusion principle can be applied in modeling, by considering particle structures of core–shell or matrix types (Siepmann and Siepmann, 2011). For spheres with a continuous matrix, the time-dependent concentration in particles is given in the second-order differential equation

$$\frac{\partial c}{\partial t} = D\frac{1}{r^2}\frac{\partial}{\partial r}\left(r^2\frac{\partial c}{\partial r}\right). \tag{6.4}$$

The release rate can then be calculated, with an example given in the equation

$$\frac{d\left(M_{\text{rel}}/M_{\text{eqm}}\right)}{dt} = \frac{6D}{R^2}\sum_{n=0}^{\infty}\exp\left(-\frac{Dn^2\pi^2 t}{R^2}\right), \tag{6.5}$$

where c is the compound concentration at a radial distance r from the sphere center, M_{rel} and M_{eqm} are the released and equilibrium concentrations, respectively, t is time, D is the diffusion coefficient of the compound in the matrix, and R is the radius of spheres (Kim *et al.*, 2001).

The purely Fickian diffusion typically is based on an assumption of uniform mass transfer resistance throughout a particle and typically follows the first-order release kinetics. However, the internal particle structure can change during the release process, due to the loss of compounds to the continuous phase, the infusion of water and other co-solutes, and possible impacts of water and other compounds on the morphology of structure-forming molecules (Zhong and Jin, 2009a). This can cause the formation of "channels" inside particles that change release characteristics. Interactions between the encapsulated compound and particle matrix play a critical role in controlling the release rate: strong repulsive forces can cause complete release at the first time point of a realistic release study, strong attractive forces correspond to a low release rate, while comparable strengths of repulsive and attractive forces enable the gradual release of an encapsulated compound (Zhong and Jin, 2009a). These interactions can be controlled by the inclusion of appropriate compounds, such as glycerol and polysorbates, to change hydrophobic or electrostatic interactions within particles (Xiao *et al.*, 2011a, b).

For the core–shell type structure, the shell curvature radius is typically much larger than molecules and Fick's law for a planar geometry may be applied:

$$\left(\frac{\partial c}{\partial t}\right)_x = D\left(\frac{\partial^2 c}{\partial x^2}\right)_t. \tag{6.6}$$

In addition to the partitioning properties of the encapsulated compound, the shell structure and composition are important for controlling the release rate.

6.4.2 Release properties in evolving particle structures

The dimension and structure of delivery systems may not be constant during the release of an encapsulated compound. The infusion of water can cause swelling and even disintegration of particles (Wani *et al.*, 2008). Particles made with water-soluble compounds can be gradually dissolved. Acid and alkaline hydrolysis can cause the dissolution or degradation of particles or changes of particle structure. If particles are made of substrates of enzymes such as amylases and proteases, particles can be gradually hydrolyzed, releasing the encapsulated compound. The modeling may use time-evolving boundary conditions and diffusion coefficients in mass transfer models and include the release due to the hydrolyzed/dissolved fraction. Solutions to the differential equations may then become complicated, and numerical simulations may become an option to simulate release profiles.

6.4.3 Triggered release

Triggered release is the burst release of an encapsulated compound responding to changes in environmental conditions such as acidity, ionic strength, temperature, enzymes, electromagnetic radiation such as UV and visible light, and external magnetic and electric fields (Esser-Kahn *et al.*, 2011). Triggers can be chemical and biological reactions or changes in physical properties, which can be used to create food products with novel quality, develop indicators of inappropriate heating or storage, or enhance the delivery of bioactive compounds to the adsorption/function sites.

Based on acidity differences in sequential stages of the gastrointestinal digestion system, designing particles with multiple layers of polyelectrolytes using layer-by-layer deposition is one of the most studied systems responding to environmental conditions. This includes the deposition of polyelectrolytes on charged emulsion droplets to both protect emulsion stability during storage and control the digestion of oil (Weiss *et al.*, 2006, Hu *et al.*, 2011). Tumor cells have an acidity slightly lower than that of normal cells, and nanoparticles with multiple layers of polyelectrolytes were demonstrated to protect drugs from degradation *in vivo* and release naked particles nearby tumor cells to enter the cells to release drugs (Poon *et al.*, 2011). Structured beads were produced by extruding a pectin solution containing bovine serum albumin (BSA) into a zein solution in aqueous ethanol and with calcium ions (Liu *et al.*, 2006). The gelled pectin beads were coated with a layer of zein, which enabled limited release of BSA after 2 h at pH 3.5 and 4 h at pH 7.4 and gradual release afterwards. The release profile is applicable to deliver compounds to the colon. Solubility characteristics of some biopolymers such as HPMCP can be used to protect the degradation at gastric conditions and release the encapsulated compound at intestinal acidity (Zhong and Shah, 2013), as discussed previously. Finally, multiple triggers can be used in combination based on *in vivo* conditions. In a study, insulin solid particles were coated with alternate layers of soybean oil and HPMCP (Toorisaka *et al.*, 2005). The combined HPMCP and oil layers prevented the release of insulin at pH 1.2. After increasing pH to 6.8, the HPMCP layer was dissolved and the oil layer was hydrolyzed by lipase, which enabled the release of encapsulated insulin.

In addition to delivery systems improving bioactivity *in vivo*, electrostatic complexes of type B gelatin (hydrolyzed from collagen at alkaline conditions) and gum arabic were studied to encapsulate flavor oil (Yeo *et al.*, 2005). The microencapsulation system showed complete release of flavor oil after heating at 100°C or higher and was able to re-encapsulate the released oil upon cooling. At typical storage temperatures of 4 and −20 °C, no significant release was observed in 4 weeks. The electrostatic complexes also released oil at room temperature when NaCl concentration was increased to 100 mM.

6.5 STABILITY OF MICRO- AND NANOPARTICLES IN AQUEOUS DISPERSIONS

When dispersed in aqueous systems, both particle integrity and the stability against aggregation and precipitation are important to develop successful delivery systems. The structure of particles can be altered by solution acidity or relevant chemical and biological reactions, as discussed previously. These structural changes can also cause dispersion instability, but are beyond the scope of this chapter. This section is focused on maintaining particles against gravitational sedimentation and particle aggregation as they are important to the quality of food systems.

6.5.1 Stability of particles

The stability of particles against gravitational sedimentation in aqueous dispersions is clearly illustrated in the Stokes' equation:

$$v_s = \frac{2\,\Delta\rho\, g\, R^2}{9\eta} \quad (6.7)$$

where g is the acceleration due to gravity, and η is the viscosity of the continuous phase. When a density difference ($\Delta\rho$) exists between particles and the aqueous phase, the sedimentation velocity (v_s) of spheres with a radius R is proportional to the second power of the particle radius (R). To prevent particle sedimentation, fluid flow properties around particles need to be changed, as discussed in the next subsection. In addition to the gravitational force, Brownian motion is a well-known random motion of colloidal particles driven by thermal energy whose magnitude is a product of the Boltzman constant and absolute temperature (Israelachvili, 1992). For particles with a sufficiently small dimension, the thermal energy of particles can be more significant than gravitational energy and Brownian motion becomes significant to particle stability (Israelachvili, 1992). As Brownian motion takes a random direction, the possibility of particle aggregation is controlled by interparticle colloidal forces, as discussed below. The aggregated particles behave as a large particle and precipitation or creaming then again becomes an instability factor. Soft particles with a fluidic core, such as emulsions, can also be destabilized by coalescence and/or Ostwald ripening (Walstra, 2002). The principle of Ostwald ripening, involving a dispersed phase with some solubility in the continuous phase, was discussed above. For coalescence, aggregation of particles is followed by the formation and subsequent disruption of a thin interparticle lamella, forming a large droplet (Walstra, 2002). Therefore, controlling particle aggregation and/or the stability of lamella is required for the stability of emulsions.

6.5.2 Controlling particle precipitation and motion

As indicated in equation 6.7, there are several ways to control particle sedimentation. The density of particles can be controlled by incorporating multiple components with a density higher or lower than that of the continuous phase. In the food industry, brominated oil is used to increase the density of oil droplets to match that of water. Reduction of particle dimension is an efficient approach but may require energy in top-down unit operations. Increasing the viscosity of the continuous phase is a convenient approach, which can be done by dissolving hydrocolloids such as gelatin and polysaccharides that are effective in increasing viscosity at a low concentration. Hydrocolloid solutions are typically shear-thinning, that is, a high viscosity at a low shear rate – the initial stage of particle motion (Walstra, 2002). Some hydrocolloids such as high-acyl gellan can form a network at low biopolymer concentrations, and the network formed creates a yield stress that has to be overcome before creating motion, which is effective in preventing the precipitation and aggregation of colloidal particles (Ikeda *et al.*, 2013). The network and the increased viscosity can be enabled by hydrogen bonds formed at a lowered temperature (Dickinson, 2010). Finally, if a food product can be stored in a freezer, an ice-crystal network is an effective mechanism to arrest the motion of particles.

6.5.3 Controlling particle aggregation

The Derjaguin–Landau–Verwey–Overbeek (DLVO) theory is the classic framework that describes the possibility of particle aggregation using long-range van der Waals and electrostatic forces (Israelachvili, 1992). For food systems, short-range interaction forces such as hydrophobic, hydrogen-bond, steric, and biological-affinity interactions are also significant. Particle stability against aggregation is the overall impact of these repulsive and attractive forces:

$$W_{\text{total}} = W_{\text{vdW}} + W_{\text{hydrophobic}} + W_{\text{electr}} + W_{\text{steric}} + W_{\text{steric}} \tag{6.8}$$

where subscripts from left to right indicate total, van der Waals, hydrophobic, electrostatic, and steric interactions, respectively.

In addition, when a system contains two groups of particles with significantly different dimensions, depletion flocculation can occur (Dickinson, 2010). This is because depletion of small particles, including biopolymers, between two large particles causes a lower osmotic pressure in the interparticle regime than in other regimes around large particles, and the osmotic pressure imbalance induces the flocculation of large particles. Depletion flocculation is reversible upon dilution. Strategies of stabilizing particles against aggregation are discussed below for electrostatic and/or steric mechanisms.

6.5.3.1 Surface charge

Electrostatic repulsion is a long-range force significant in the prevention of particle aggregation. When fabricating food delivery systems, charge properties of materials in a given food matrix environment should receive strong considerations. Common food compounds with charges include ionic surfactants, proteins, and polysaccharides. Given a known molecular structure, charge properties can be estimated based on the p*Ka* of charged groups. Summation of different types and quantity of charges gives the calculated net charge at a given pH. It should be mentioned, however, that charges on the colloidal particle surface control the stability against aggregation (Dickinson, 2003). Surface charge properties are also utilized to engineer the particle surface, for example, by adsorbing polyelectrolytes enabling changes in charge type and density. The adsorbed polyelectrolytes on the particle surface can also function as a polymer brush to provide steric repulsion against particle aggregation (Kleshchanok and Lang, 2007). For multilayered emulsions, the increased thickness and engineered charge properties of interfaces have been demonstrated to improve the physical stability of emulsion droplets against environmental stresses during processing and storage and reduce the chemical degradation of compounds in the oil droplets (McClements *et al.*, 2007).

6.5.3.2 Steric repulsion provided by adsorbed biopolymers

Adsorbing appropriate polymers on particles enables the formation of a polymer brush that stabilizes particle against aggregation by steric repulsion (Kleshchanok and Lang, 2007). This is because a higher solute density and thus osmotic pressure exist in the regime of two overlapping polymer brushes, which induces the influx of solvent to mediate the osmotic pressure gradient and thus repel particles (Israelachvili, 1992). In addition to the example of polyelectrolytes adsorbing onto particles with opposite charges, there are several studies that utilized biopolymers to stabilize colloidal particles. Soluble soybean polysaccharides are conjugates with polypeptide and polysaccharide segments, and the polypeptide fraction can adsorb onto milk protein particles, which orientates the polysaccharide fraction facing the aqueous phase to provide steric repulsion to stabilize milk protein particles at acidic pH (Nakamura *et al.*, 2003; El Kinawy *et al.*, 2012). Another example is caseins that adsorb on zein nanoparticles to prevent the aggregation and precipitation of the hydrophobic nanoparticles (Patel *et al.*, 2010a, b). Glycating proteins with reducing saccharides is another established approach to improve the stability of the fabricated colloidal particles (Akhtar and Dickinson, 2003).

6.6 CONCLUSIONS

In conclusion, delivery systems have great promise for improving the quality, safety, and healthfulness of food products. Various structures can be fabricated using top-down or bottom-up approaches to design systems that deliver the desired functions. Particle dimension is significant for the appearance and stability of dispersions. When particulate structures are used in encapsulation, encapsulation efficiency is an important parameter in evaluating the performance of a delivery system. However, encapsulation efficiency of delivery systems should be defined clearly and evaluated accordingly, for example, by measuring the efficiency of delivering the encapsulated compounds to their targeted functioning location, and not following the encapsulation process. Encapsulation efficiency can be attributed to the ability of delivery systems to control the manner and release rate of encapsulated compounds to perform their functions. Finally, when delivery systems are incorporated in a food matrix, it is not only important to prevent their aggregation and precipitation but also to maintain their physical and chemical stability during processing and storage and to function appropriately after ingestion.

References

Abraham, S. and Narine, S.S. (2009) Self-assembled nanostructures of oleic acid and their capacity for encapsulation and controlled delivery of nutrients. *Journal of Nanoscience and Nanotechnology*, 9, 6326–6334.

Akhtar, M. and Dickinson, E. (2003) Emulsifying properties of whey protein-dextran conjugates at low pH and different salt concentrations. *Colloids and Surfaces B: Biointerfaces*, 31, 125–132.

Almeida, A.J., Runge, S. and Müller, R.H. (1997) Peptide-loaded solid lipid nanoparticles (SLN): Influence of production parameters. *International Journal of Pharmaceutics*, 149, 255–265.

Bagwe, R.P., Kanicky, J.R., Palla, B.J., Patanjali, P.K. and Shah, D.O. (2001) Improved drug delivery using microemulsions: rationale, recent progress, and new horizons. *Critical Reviews in Therapeutic Drug Carrier Systems*, 18, 77–140.

Beristain, C.I., García, H.S. and Vernon-Carter, E.J. (2001) Spray-dried encapsulation of cardamom (*Elettaria cardamomum*) essential oil with mesquite (*Prosopis juliflora*) gum. *Lebensmittel-Wissenschaft und Technologie (Food Science and Technology)*, 34, 398–401.

Bouillot, P., Ubrich, N., Sommer, F., Minh Duc, T., Loeffler, J. and Dellacherie, E. (1999) Protein encapsulation in biodegradable amphiphilic microspheres. *International Journal of Pharmaceutics*, 181, 159–172.

Burgain, J., Gaiani, C., Linder, M. and Scher, J. (2011) Encapsulation of probiotic living cells: From laboratory scale to industrial applications. *Journal of Food Engineering*, 104, 467–483.

Bylaitė, E., Rimantas V. P. and Maþdþierienė, R. (2001) Properties of caraway (*Carum carvi* L.) essential oil encapsulated into milk protein-based matrices. *European Food Research and Technology*, 212, 661–670.

Cao, A., Hu, D. and Lai, L. (2009) Formation of amyloid fibrils from fully reduced hen egg white lysozyme. *Protein Science*, 13, 319–324.

Carneiro, H.C.F., Tonon, R.V., Grosso, C.R.F. and Hubinger, M.D. (2013) Encapsulation efficiency and oxidative stability of flaxseed oil microencapsulated by spray drying using different combinations of wall materials. *Journal of Food Engineering*, 115, 443–451.

Chalasani, K.B., Russell-Jones, G.J., Yandrapu, S.K., Diwan, P.V. and Jain, S.K. (2007) A novel vitamin B_{12}-nanosphere conjugate carrier system for peroral delivery of insulin. *Journal of Controlled Release*, 117, 421–429.

Chen, C., Tu, Y. and Chang, H. (1999) Efficiency and protective effect of encapsulation of milk immunoglobulin G in multiple emulsion. *Journal of Agricultural and Food Chemistry*, 47, 407–410.

Chen, L. and Subirade, M. (2007) Effect of preparation conditions on the nutrient release properties of alginate–whey protein granular microspheres. *European Journal of Pharmaceutics and Biopharmaceutics*, 65, 354–362.

Chu, B.S., Ichikawa, S., Kanafusa, S. and Nakajima, M. (2007) Preparation of protein-stabilized beta-carotene nanodispersions by emulsification-evaporation method. *Journal of the American Oil Chemists Society*, 84, 1053–1062.

Chuah, A.M., Kuroiwa, T., Kobayashi, I. and Nakajima, M. (2014) The influence of polysaccharide on the stability of protein stabilized oil-in-water emulsion prepared by microchannel emulsification technique. *Colloids and Surfaces A: Physicochemical and Engineering Aspects*, 440, 136–144.

Cui, F., Shi, K., Zhang, L., Tao, A. and Kawashima, Y. (2006) Biodegradable nanoparticles loaded with insulin–phospholipid complex for oral delivery: Preparation, *in vitro* characterization and *in vivo* evaluation. *Journal of Controlled Release*, 114, 242–250.

Das, R.K., Kasoju, N. and Bora, U. (2010) Encapsulation of curcumin in alginate-chitosan-pluronic composite nanoparticles for delivery to cancer cells. *Nanomedicine: Nanotechnology, Biology and Medicine*, 6, 153–160.

Del Gaudio, P., Russo, P., Rosaria L. M., Colombo, P. and Aquino, R.P. (2009) Encapsulation of ketoprofen and ketoprofen lysinate by prilling for controlled drug release. *American Association of Pharmaceutical Scientists PharmSciTech*, 10, 1178–1185.

Desai, K.G.H. and Park, H.J. (2005) Encapsulation of vitamin C in tripolyphosphate cross-linked chitosan microspheres by spray drying. *Journal of Microencapsulation*, 22, 179–192.

Desai, K.G. and Park, H.J. (2006) Effect of manufacturing parameters on the characteristics of vitamin C encapsulated tripolyphosphate-chitosan microspheres prepared by spray-drying. *Journal of Microencapsulation*, 23, 91–103.

Dickinson, E. (2003) Hydrocolloids at interfaces and the influence on the properties of dispersed systems. *Food Hydrocolloids*, 17, 25–39.

Dickinson, E. (2010) Food emulsions and foams: Stabilization by particles. *Current Opinion in Colloid and Interface Science*, 15, 40–49.

Duclairoir, C., Irache, J., Nakache, E., Orecchioni, A., Chabenat, C. and Popineau, Y. (1999) Gliadin nanoparticles: formation, all- trans-retinoic acid entrapment and release, size optimization. *Polymer International*, 48, 327–333.

Duclairoir, C., Orecchioni, A.M., Depraetere, P., Osterstock, F. and Nakache, E. (2003) Evaluation of gliadins nanoparticles as drug delivery systems: a study of three different drugs. *International Journal of Pharmaceutics,* 253, 133–144.

El-Abbassi, A., Neves, M.A., Kobayashi, I., Hafidi, A. and Nakajima, M. (2013) Preparation and characterization of highly stable monodisperse argan oil-in-water emulsions using microchannel emulsification. *European Journal of Lipid Science and Technology,* 115, 224–231.

El-Kamel, A.H., Sokar, M.S., Al Gamal, S.S. and Naggar, V.F. (2001) Preparation and evaluation of ketoprofen floating oral delivery system. *International Journal of Pharmaceutics,* 220, 13–21.

El Kinawy, O.S., Petersen, S., Helmdach, L. and Ulrich, J. (2012) Parameter selection of emulsification processes: conditions for nano-and macroemulsions. *Chemical Engineering and Technology*, 35, 1604–1608.

Esmaili, M. and Ghaffari, S. M. (2011) Beta casein-micelle as a nano vehicle for solubility enhancement of curcumin; food industry application. *Lebensmittel-Wissenschaft und Technologie (Food Science and Technology),* 44, 2166–2172.

Esser-Kahn, A.P., Odom, S.A., Sottos, N.R., White, S.R. and Moore, J.S. (2011) Triggered release from polymer capsules. *Macromolecules,* 44, 5539–5553.

Fryd, M.M. and Mason, T.G. (2012) Advanced nanoemulsions. *Annual Review of Physical Chemistry,* 63, 493–518.

Ganta, S. and Amiji, M. (2009) Coadministration of paclitaxel and curcumin in nanoemulsion formulations to overcome multidrug resistance in tumor cells. *Molecular Pharmaceutics,* 6, 928–939.

Gouin, S. (2004) Microencapsulation: industrial appraisal of existing technologies and trends. *Trends in Food Science and Technology,* 15, 330–347.

Hu, M., Li, Y., Decker, E.A., Xiao, H. and McClements, D.J. (2011) Impact of layer structure on physical stability and lipase digestibility of lipid droplets coated by biopolymer nanolaminated coatings. *Food Biophysics,* 6, 37–48.

Huppertz, T., Vaia, B. and Smiddy, M.A. (2008) Reformation of casein particles from alkaline-disrupted casein micelles. *Journal of Dairy Research,* 75, 44–47.

Hwang, Y., Oh, C. and Oh, S. (2005) Controlled release of retinol from silica particles prepared in O/W/O emulsion: The effects of surfactants and polymers. *Journal of Controlled Release,* 106, 339–349.

Ikeda, S., Gohtani, S., Nishinari, K. and Zhong, Q. (2013) High acyl gellan networks probed by rheology and atomic force microscopy. *Food Science and Technology Research*, 19(2), 201–210.

Ipsen, R. and Otte, J. (2007) Self-assembly of partially hydrolysed α-lactalbumin. *Biotechnology Advances,* 25, 602–605.

Israelachvili, J. N. (1992) *Intermolecular and Surface Forces*. Academic Press, London, UK.

Ito, F., Fujimori, H. and Makino, K. (2008) Factors affecting the loading efficiency of water-soluble drugs in PLGA microspheres. *Colloids and Surfaces B: Biointerfaces,* 61, 25–29.

Jenning, V. and Gohla, S.H. (2001) Encapsulation of retinoids in solid lipid nanoparticles (SLN). *Journal of Microencapsulation,* 18, 149–158.

Janes, K.A., Calvo, P. and Alonso, M.J. (2001) Polysaccharide colloidal particles as delivery systems for macromolecules. *Advanced Drug Delivery Reviews,* 47, 83–97.

Jeong, Y., Song, J., Kang, S., *et al.* (2003) Preparation of poly(dl-lactide-co-glycolide) microspheres encapsulating all-trans retinoic acid. *International Journal of Pharmaceutics,* 259, 79–91.

Jeong, Y., Kim, S., Jung, T., *et al.* (2006) Polyion complex micelles composed of all-trans retinoic acid and poly (ethylene glycol)-grafted-chitosan. *Journal of Pharmaceutical Sciences,* 95, 2348–2360.

Jia, Z., Shen, D. and Xu, W. (2001) Synthesis and antibacterial activities of quaternary ammonium salt of chitosan. *Carbohydrate Research,* 333, 1–6.

Jones, O.G. and McClements, D.J. (2010) Functional biopolymer particles: Design, fabrication, and applications. *Comprehensive Reviews in Food Science and Food Safety,* 9, 374–397.

Jones, O.G. and Mezzenga, R. (2012) Inhibiting, promoting, and preserving stability of functional protein fibrils. *Soft Matter,* 8, 876–895.

Joscelyne, S.M. and Trägårdh, G. (2000) Membrane emulsification—a literature review. *Journal of membrane science,* 169, 107–117.

Kakkar, V., Singh, S., Singla, D. and Kaur, I.P. (2011) Exploring solid lipid nanoparticles to enhance the oral bioavailability of curcumin. *Molecular Nutrition and Food Research,* 55, 495–503.

Kheradmandnia, S., Vasheghani-Farahani, E., Nosrati, M. and Atyabi, F. (2010) Preparation and characterization of ketoprofen-loaded solid lipid nanoparticles made from beeswax and carnauba wax. *Nanomedicine: Nanotechnology, Biology and Medicine,* 6, 753–759.

Kim, S.Y., Kim, J.H., Kim, D., An, J.H., Lee, D.S. and Kim, S.C. (2001) Drug-releasing kinetics of MPEG/PLLA block copolymer micelles with different PLLA block lengths. *Journal of Applied Polymer Science,* 82, 2599–2605.

Kim, Y.D. and Morr, C.V. (1996) Microencapsulation properties of gum arabic and several food proteins: Spray-dried orange oil emulsion particles. *Journal of Agricultural and Food Chemistry,* 44, 1314–1320.

Kleshchanok, D. and Lang, P.R. (2007) Steric repulsion by adsorbed polymer layers studied with total internal reflection microscopy. *Langmuir,* 23, 4332–4333.

Kosaraju, S.L., Weerakkody, R. and Augustin, M.A. (2009) *In vitro* evaluation of hydrocolloid-based encapsulated fish oil. *Food Hydrocolloids,* 23, 1413–1419.

Krauland, A.H. and Alonso, M.J. (2007) Chitosan/cyclodextrin nanoparticles as macromolecular drug delivery system. *International Journal of Pharmaceutics,* 340, 134–142.

Kriegel, C., Kit, K.M., McClements, D.J. and Weiss, J. (2009) Nanofibers as carrier systems for antimicrobial microemulsions. Part I: fabrication and characterization. *Langmuir,* 25, 1154–1161.

Kriegel, C., Kit, K.M., McClements, D.J. and Weiss, J. (2010) Nanofibers as carrier systems for antimicrobial microemulsions. II. Release characteristics and antimicrobial activity. *Journal of Applied Polymer Science,* 118, 2859–2868.

Lai, F., Wissing, S.A., Müller, R. and Fadda, A. (2006) *Artemisia arborescens* L essential oil-loaded solid lipid nanoparticles for potential agricultural application: Preparation and characterization. *American Association of Pharmaceutical Scientists PharmSciTech,* 7, E10–E18.

Laouini, A., Fessi, H. and Charcosset, C. (2012) Membrane emulsification: A promising alternative for vitamin E encapsulation within nano-emulsion. *Journal of Membrane Science,* 423–424, 85–96.

Lawrence, M.J. and Rees, G.D. (2000) Microemulsion-based media as novel drug delivery systems. *Advanced Drug Delivery Reviews,* 45, 89–121.

Lee, M., Oh, S., Moon, S. and Bae, S. (2001) Preparation of silica particles encapsulating retinol using O/W/O multiple emulsions. *Journal of Colloid and Interface Science,* 240, 83–89.

Leong, T.S.H., Wooster, T.J., Kentish, S.E. and Ashokkumar, M. (2009) Minimising oil droplet size using ultrasonic emulsification. *Ultrasonics Sonochemistry,* 16, 721–727.

Liu, C., Wu, C. and Fang, J. (2010) Characterization and formulation optimization of solid lipid nanoparticles in vitamin K1 delivery. *Drug Development and Industrial Pharmacy,* 36, 751–761.

Liu, L.S., Fishman, M.L., Hicks, K.B., Kende, M. and Ruthel, G. (2006) Pectin/zein beads for potential colon-specific drug delivery: Synthesis and in vitro evaluation. *Drug Delivery,* 13, 417–423.

McClements, D.J. (2005) *Food Emulsions: Principles, Practices, and Techniques.* CRC Press, Boca Raton, FL.

McClements, D.J. (2010) Edible nanoemulsions: fabrication, properties, and functional performance. *Soft Matter,* 7, 2297–2316.

McClements, D.J. (2012) Nanoemulsions versus microemulsions: terminology, differences, and similarities. *Soft Matter,* 8, 1719–1729.

McClements, D.J. and Rao, J. (2011) Food-grade nanoemulsions: Formulation, fabrication, properties, performance, biological fate and potential toxicity. *Critical Reviews in Food Science and Nutrition,* 51, 285–330.

McClements, D.J., Decker, E.A. and Weiss, J. (2007) Emulsion-based delivery systems for lipophilic bioactive components. *Journal of Food Science,* 72, R109–R124.

McClements, D.J., Decker, E.A., Park, Y. and Weiss, J. (2009) Structural design principles for delivery of bioactive components in nutraceuticals and functional foods. *Critical Reviews in Food Science and Nutrition,* 49, 577–606.

Mezzenga, R., Schurtenberger, P., Burbidge, A. and Michel, M. (2005) Understanding foods as soft materials. *Nature Materials,* 4, 729–740.

Mitri, K., Shegokar, R., Gohla, S., Anselmi, C. and Müller, R.H. (2011) Lipid nanocarriers for dermal delivery of lutein: Preparation, characterization, stability and performance. *International Journal of Pharmaceutics,* 414, 267–275.

Nakamura, A., Furuta, H., Kato, M., Maeda, H. and Nagamatsu, Y. (2003) Effect of soybean soluble polysaccharides on the stability of milk protein under acidic conditions. *Food Hydrocolloids,* 17, 333–343.

Nickerson, T.A. (1954) Lactose crystallization in ice cream. *I. Control of crystal size by seeding. Journal of Dairy Science,* 37, 1099–1105.

Nicolai, T., Britten, M. and Schmitt, C. (2011) β-Lactoglobulin and WPI aggregates: Formation, structure and applications. *Food Hydrocolloids,* 25, 1945–1962.

O'Connell, J.E., Kelly, A.L., Fox, P.F. and De Kruif, K.G. (2001) Mechanism for the ethanol-dependent heat-induced dissociation of casein micelles. *Journal of Agricultural and Food Chemistry,* 49, 4424–4428.

Pan, X., Mu, M., Hu, B., Yao, P. and Jiang, M. (2006) Micellization of casein-graft-dextran copolymer prepared through Maillard reaction. *Biopolymers,* 81, 29–38.

Pan, X., Yao, P. and Jiang, M. (2007) Simultaneous nanoparticle formation and encapsulation driven by hydrophobic interaction of casein-graft-dextran and β-carotene. *Journal of Colloid and Interface Science,* 315, 456–463.

Parris, N., Cooke, P.H. and Hicks, K.B. (2005) Encapsulation of essential oils in zein nanospherical particles. *Journal of Agricultural and Food Chemistry,* 53, 4788–4792.

Patel, A., Hu, Y., Tiwari, J.K. and Velikov, K.P. (2010a) Synthesis and characterisation of zein–curcumin colloidal particles. *Soft Matter,* 6, 6192–6199.

Patel, A.R., Bouwens, E.C.M. and Velikov, K.P. (2010b) Sodium caseinate stabilized zein colloidal particles. *Journal of Agricultural and Food Chemistry,* 58, 12497–12503.

Patten, G.S., Augustin, M.A., Sanguansri, L., Head, R.J. and Abeywardena, M.Y. (2009) Site dpecific delivery of microencapsulated fish oil to the gastrointestinal tract of the rat. *Digestive Diseases and Sciences,* 54, 511–521.

Poon, Z., Chang, D., Zhao, X. and Hammond, P. T. (2011) Layer-by-layer nanoparticles with a pH-sheddable layer for *in vivo* targeting of tumor hypoxia. *ACS Nano,* 5, 4284–4292.

Qian, C., Decker, E.A., Xiao, H. and McClements, D.J. (2012) Nanoemulsion delivery systems: Influence of carrier oil on β-carotene bioaccessibility. *Food Chemistry,* 135, 1440–1447.

Quispe-Condori, S., Saldaña, M.D.A. and Temelli, F. (2011) Microencapsulation of flax oil with zein using spray and freeze drying. *Lebensmittel-Wissenschaft und Technologie (Food Science and Technology),* 44, 1880–1887.

Renard, D., Robert, P., Lavenant, L., *et al.* (2002) Biopolymeric colloidal carriers for encapsulation or controlled release applications. *International Journal of Pharmaceutics,* 242, 163–166.

Sanguansri, P. and Augustin, M.A. (2006) Nanoscale materialsdevelopment – a food industry perspective. *Trends in Food Science and Technology,* 17, 547–556.

Schultz, S., Wagner, G., Urban, K. and Ulrich, J. (2004) High-pressure homogenization as a process for emulsion formation. *Chemical Engineering and Technology,* 27, 361–368.

Shaikh, J., Ankola, D.D., Beniwal, V., Singh, D. and Kumar, M.N.V.R. (2009) Nanoparticle encapsulation improves oral bioavailability of curcumin by at least 9-fold when compared to curcumin administered with piperine as absorption enhancer. *European Journal of Pharmaceutical Sciences,* 37, 223–230.

Shi, X. and Tan, T. (2002) Preparation of chitosan/ethylcellulose complex microcapsule and its application in controlled release of Vitamin D_2. *Biomaterials,* 23, 4469–4473.

Siepmann, J. and Siepmann, F. (2011) Modeling of diffusion controlled drug delivery. *Journal of Controlled Release*, 161(2), 351–362.

Song, X., Zhao, Y., Hou, S., *et al.* (2008) Dual agents loaded PLGA nanoparticles: Systematic study of particle size and drug entrapment efficiency. *European Journal of Pharmaceutics and Biopharmaceutics,* 69, 445–453.

Souza, H.K.S., Campiña, J.M., Sousa, A.M.M., Silva, F. and Gonçalves, M.P. (2013) Ultrasound-assisted preparation of size-controlled chitosan nanoparticles: Characterization and fabrication of transparent biofilms. *Food Hydrocolloids,* 31, 227–236.

Sozer, N. and Kokini, J.L. (2009) Nanotechnology and its applications in the food sector. *Trends in Biotechnology,* 27, 82–89.

Sun, H.W., Liu, K.Y., Liu, W., *et al.* (2012) Development and characterization of a novel nanoemulsion drug-delivery system for potential application in oral delivery of protein drugs. *International Journal of Nanomedicine,* 7, 5529–5543.

Tang, C.H., Zhang, Y.H., Wen, Q.B. and Huang, Q. (2010) Formation of amyloid fibrils from kidney bean 7S globulin (phaseolin) at pH 2.0. *Journal of Agricultural and Food Chemistry,* 58, 8061–8068.

Thies, C. (2012) Nanocapsules as delivery systems in the food, beverage and nutraceutical industries. In: *Nanotechnology in the Food, Beverage and Nutraceutical Industries*, Huang, Q. (ed.). Woodhead Publishing Ltd, Cambridge, UK.

Tiyaboonchai, W., Tungpradit, W. and Plianbangchang, P. (2007) Formulation and characterization of curcuminoids loaded solid lipid nanoparticles. *International Journal of Pharmaceutics,* 337, 299–306.

Toorisaka, E., Hashida, M., Kamiya, N., Ono, H., Kokazu, Y. and Gato, M. (2005) An enteric-coated dry emulsion formation for oral insulin delivery. *Journal of Controlled Release,* 107, 91–96.

Uchida, T., Nagareya, N., Sakakibara, S., *et al.* (1997) Preparation and characterization of polylactic acid microspheres containing bovine insulin by a w/o/w emulsion solvent evaporation method. *Chemical and Pharmaceutical Bulletin,* 45, 1539–1543.

Ulrich, A.S. (2012) Biophysical aspects of using liposomes as delivery vehicles. *Bioscience Report,* 22, 129–150.

Urban, K., Wagner, G., Schaffner, D., Röglin, D. and Ulrich, J. (2006) Rotor-Stator and Disc Systems for Emulsification Processes. *Chemical engineering and technology,* 29, 24–31.

Walstra, P. (2002) *Physical Chemistry of Foods*, Marcel Dekker, Inc., New York.

Wang, J., Maitani, Y. and Takayama, K. (2002) Antitumor effects and pharmacokinetics of aclacinomycin A carried by injectable emulsions composed of vitamin E, cholesterol, and PEG-lipid. *Journal of Pharmaceutical Sciences,* 91, 1128–1134.

Wang, J., Takayama, K., Nagai, T. and Maitani, Y. (2003) Pharmacokinetics and antitumor effects of vincristine carried by microemulsions composed of PEG-lipid, oleic acid, vitamin E and cholesterol. *International Journal of Pharmaceutics,* 251, 13–21.

Wang, J.M., Yang, X.Q., Yin, S.W., Yuan, D.B., Xia, N. and Qi, J.R. (2011) Growth Kinetics of amyloid-like fibrils derived from individual subunits of soy β-conglycinin. *Journal of Agricultural and Food Chemistry,* 59, 11270–11277.

Wani, M.S., Dehghan, M., Polshettiwar, S.A., *et al.* (2008) Controlled released system – a review. *Pharmainfo. net,* 6.

Wege, H., Kim, S., Paunov, V. N., Zhong, Q. and Velev, O.D. (2008) Long-term stabilization of foams and emulsions with *in situ* formed microparticles from hydrophobic cellulose. *Langmuir,* 24, 9245–9253.

Weiss, J., Takhistov, P. and McClements, D.J. (2006) Functional materials in food nanotechnology. *Journal of Food Science,* 71, R107–R116.

Xiao, D., Davidson, P.M. and Zhong, Q. (2011a) Spray-dried zein capsules with co-encapsulated nisin and thymol as antimicrobial delivery systems for enhanced antilisterial properties. *Journal of Agricultural and Food Chemistry,* 59, 7393–7404.

Xiao, D., Gömmel, C., Davidson, P. M. and Zhong, Q. (2011b) Intrinsic Tween 20 improves release and antilisterial properties of co-encapsulated nisin and thymol. *Journal of Agricultural and Food Chemistry,* 59, 9572–9580.

Xie, S., Wang, S., Zhao, B., Han, C., Wang, M. and Zhou, W. (2008) Effect of PLGA as a polymeric emulsifier on preparation of hydrophilic protein-loaded solid lipid nanoparticles. *Colloids and Surfaces B: Biointerfaces,* 67, 199–204.

Yeo, Y., Bellas, E., Firestone, W., Langer, R. and Kohane, D. S. (2005) Complex coacervates for thermally sensitive controlled release of flavor compounds. *Journal of Agricultural and Food Chemistry,* 53, 7518–7525.

Yu, H. and Huang, Q. (2010) Enhanced in vitro anti-cancer activity of curcumin encapsulated in hydrophobically modified starch. *Food Chemistry,* 119, 669–674.

Zhang, N., Ping, Q., Huang, G., Xu, W., Cheng, Y. and Han, X. (2006) Lectin-modified solid lipid nanoparticles as carriers for oral administration of insulin. *International Journal of Pharmaceutics,* 327, 153–159.

Zhang, Y. and Zhong, Q. (2013) Encapsulation of bixin in sodium caseinate to deliver the colorant in transparent dispersions. *Food Hydrocolloids*, 33, 1–9.

Zhong, C., Yinghe, H., Lixian, S. and Xueqiang, C. (2011) Preparation and characterization of vitamin a palmitate-loaded nanostructured lipid carriers as delivery systems for food products. *Advanced Materials Research,* 236–238, 1818–1823.

Zhong, Q. and Jin, M. (2009a) Nanoscalar structure of spray-dried zein microcapsules and in vitro release kinetics of the encapsulated lysozyme as affected by formulations. *Journal of Agricultural and Food Chemistry,* 57, 3886–3894.

Zhong, Q. and Jin, M. (2009b) Zein nanoparticles produced by liquid–liquid dispersion. *Food Hydrocolloids,* 23, 2380–2387.

Zhong, Q. and Shah, B. (2013) Surface coating of L-glutamine solid microparticles for enteric delivery. *Journal of Food Engineering,* 114, 538–544.

Zhong, Q., Jin, M., Davidson, P.M. and Zivanovic, S. (2009a) Sustained release of lysozyme from zein microcapsules produced by a supercritical anti-solvent process. *Food Chemistry,* 115, 697–700.

Zhong, Q., Tian, H. and Zivanovic, S. (2009b) Encapsulation of fish oil in solid zein particles by liquid-liquid dispersion. *Journal of Food Processing and Preservation,* 33, 255–270.

Ziani, K., Chang, Y.H., Mclandsborough, L. and McClements, D.J. (2011) Influence of surfactant charge on antimicrobial efficacy of surfactant-stabilized thyme oil nanoemulsions. *Journal of Agricultural and Food Chemistry,* 59, 6247–6255.

Zigoneanu, I.G., Astete, C.E. and Sabliov, C.M. (2008) Nanoparticles with entrapped alpha-tocopherol: synthesis, characterization, and controlled release. *Nanotechnology,* 19(10), 105606.

Zimet, P. and Livney, Y.D. (2009) Beta-lactoglobulin and its nanocomplexes with pectin as vehicles for ω-3 polyunsaturated fatty acids. *Food Hydrocolloids,* 23, 1120–1126.

Zimet, P., Rosenberg, D. and Livney, Y.D. (2011) Re-assembled casein micelles and casein nanoparticles as nanovehicles for ω-3 polyunsaturated fatty acids. *Food Hydrocolloids,* 25, 1270–1276.

Zuidam, N.J. and Shimoni, E. (2010) Overview of microencapsulates for use in food products or processes and methods to make them. In: *Encapsulation Technologies for Active Food Ingredients and Food Processing,* Zuidam, N.J. and Nedovic, V.A. (eds). Springer, New York.

7 Characterization of nanoscale delivery systems

Rohan V. Tikekar

Department of Nutrition and Food Science, University of Maryland, College Park, Maryland, USA

7.1 INTRODUCTION

Characterization of nanoparticles in terms of their size, surface charge, appearance, distribution of encapsulants in the matrix, and their micro-environment is crucial for development of optimized delivery systems that can enhance the stability and bioavailability of the material and enable their controlled release. Researchers constantly employ innovative design strategies to control the properties of nanoparticles. To evaluate the success and effectiveness of these approaches, there is a need to develop novel characterization tools. Additionally, these characterization tools are also essential for quality control purposes. In the present chapter, we will discuss some of the most commonly employed nanoparticle characterization techniques and their applications.

7.2 PARTICLE-SIZE MEASUREMENT

Size of the nanoscale delivery system has a significant impact on its stability, uptake, deposition and eventual clearance from the body (Jani *et al.*, 1990; Chanamai and McClements, 2000; Win and Feng, 2005; Chithrani *et al.*, 2006). Despite its tremendous significance in nanotechnology research, only a few methods can accurately measure particle size directly in solution. One of the most commonly used method for *in situ* measurement of particle size is based on dynamic light scattering (DLS) or photon correlation spectroscopy (PCS). In this technique, nanoparticle solution is illuminated using laser. The incident laser light is scattered by the nanoparticles, which is detected by a detector set either at a fixed or a variable scattering angle. Due to Brownian motion of the nanoparticles in solution, the intensity of scattered light fluctuates as a function of time. This fluctuation in scattering intensity is a manifestation of diffusion of nanoparticles in the solution, which in effect is a function of its size. These intensity fluctuations are captured several times over a period of short time (microseconds) and analysed using an autocorrelation function. The correlogram thus generated measures decay in autocorrelation function as a function of time and reveals information regarding the translational diffusion coefficient (D) of the particle in solution through

Nanotechnology and Functional Foods: Effective Delivery of Bioactive Ingredients, First Edition.
Edited by Cristina M. Sabliov, Hongda Chen, and Rickey Y. Yada.

equation (7.1). The translational diffusion coefficient (D) is related to the hydrodynamic diameter of the particle through the Stokes–Einstein equation:

$$\tau = q^2 D \tag{7.1}$$

where τ is the exponential decay rate, q is the modulus of the scattering vector and D the translational diffusion

$$d = \frac{kT}{3\pi\eta D} \tag{7.2}$$

where d is the hydrodynamic diameter, k the Boltzmann constant, T the absolute temperature and η is the absolute zero-shear viscosity of the medium.

The DLS technique is more accurate in determination of size in samples with a relatively narrow size distribution. In multimodal particle distribution, the results tend to be less accurate due to difficulties associated with interpretation of more complex autocorrelation decay curves (McClements, 1999a). The average particle size distribution can be depicted as number-averaged, surface-area averaged or volume averaged:

$$\text{number average particle size} = \sum n_i d_i / \sum n_i \tag{7.3}$$

$$\text{surface average particle size} = \sqrt{\sum n_i d_i^2 / \sum n_i} \tag{7.4}$$

$$\text{volume average particle size} = \sqrt[3]{\sum n_i d_i^3 / \sum n_i} \tag{7.5}$$

Each representation has its own advantages and provides unique information regarding particle size (McClements, 1999b).

Apart from particle-size distribution, the DLS technique can provide valuable information regarding the stability of nanoparticles and their response to environmental factors such as changes in pH, temperature, salt concentration, and presence of enzymes (Tokle and McClements, 2011; Teng *et al.*, 2012; Tikekar *et al.*, 2013).

7.3 ζ-POTENTIAL MEASUREMENT

7.3.1 Principle of the technique

ζ-potential measures the magnitude and nature of electric charge present on the surface of a nanoparticle. Similar to particle size, ζ-potential measurement also has a significant role in determining the stability and uptake of a nanoparticle delivery system. In ionic solutions, nanoparticles with surface charge have a strong layer of oppositely charged counter-ions attached to their surface, called a Stern layer. A subsequent layer of ions attached to these counter-ions is relatively loosely associated with the Stern layer. When the nanoparticle moves in solution due to either Brownian motion or applied electric potential, ions in the diffuse layer partition into two types. One type of ions moves with the nanoparticles and the other stays within the bulk solution. This creates an electrostatic potential at the diffuse layer or a "slipping plane," which is termed ζ-potential. Nanoparticles with ζ-potential of more than 30 mV (positive or negative) are considered strongly charged, while those with ζ-potential of less than 10 mV (positive or negative) are considered weakly charged (Patri and Clogtson, 2011).

7.3.2 Applications of ζ-potential measurement

Since ζ-potential measures the magnitude and nature of surface charge on the nanoparticles, this technique can be used in variety of ways, such as (i) confirmation of deposition of surfactant at the interface of nanoparticles, (ii) validation of the fabrication method, and (iii) measurement of the behavior and stability of nanoparticles in response to environmental conditions.

The magnitude of surface charge is a reasonably good indicator of stability of nanoparticles against destabilizing mechanisms such as coalescence and flocculation. Typically, a high magnitude of ζ potential (>30 mV) imparts good physical stability to nanoparticles against destabilization mechanisms (Freitas and Muller, 1998). The strong repulsive forces exerted by like charges localized at the interface of nanoparticles prevent nanoparticle aggregation. Since ζ-potential can be altered by the presence of (i) salts that can cause an electrostatic screening effect and (ii) changes in pH that can alter the electrostatic charge density, the stability of nanoparticles is significantly affected by these two factors (Binks *et al.*, 2006). A dramatic change in ζ-potential is also responsible for destabilization of emulsion and hydrolysis of encapsulated lipids during the digestion process. In a previous study, it was shown that nano-emulsion droplets stabilized by silica nanoparticles partially destabilized during simulated gastric digestion and completely destabilized during simulated intestinal digestion. Measurements showed that the ζ-potential decreased from 50 mV to –60 mV during simulated gastric digestion, while it decreased to 13 mV during intestinal digestion (Tikekar *et al.*, 2013). It is possible that due to such low surface charge, emulsion droplets did not show electrostatic repulsion and coalesced. Similar results have also been reported in other studies (Hu *et al.*, 2010; Sarkar *et al.*, 2010).

The surface charge on nanoparticles can significantly affect the stability of encapsulated materials. Previous studies have shown that oil undergoes oxidation at a faster rate in emulsions stabilized by negatively charged surfactants compared to positively charged surfactants (Yoshida and Niki, 1992; Mei *et al.*, 1998). This is possibly due to the ability of positively charged surfactants to repel positively charged transition metal ions such as ferrous ions, which can play a catalytic role in the generation of oxidative species such as free radicals that trigger oxidation processes within emulsions.

A layer-by-layer arrangement of oppositely charged surfactant molecules at the interface of nanoparticles has been hypothesized to improve the stability of encapsulated material (Guzey and McClements, 2006). This process is accomplished by stabilization of nanoparticles by a primary surfactant, followed by addition of oppositely charged surfactant to the nanoparticle dispersion. The excess surfactant is separated from the solution by centrifugation. The process is repeated until a suitable numbers of layers are deposited on the interface. ζ-potential measurement is possibly the fastest and easiest method to demonstrate the success of layer-by-layer arrangement of surfactants on nanoparticles. After addition of each layer of surfactant, the ζ-potential of nanoparticles is measured. If the process is successful, ζ-potential on the nanoparticles changes from negative to positive or vice versa upon addition of surfactant (Table 7.1) (Thanasukarn *et al.*, 2006).

Table 7.1 Influence of chitosan concentration on the properties of secondary emulsions (5 wt% hydrogenated palm oil, 10 mmol/kg SDS, 100 mM acetic acid buffer, pH 3.0) after 24 h storage. (Source: adapted from Thanasukarn *et al.*, 2006.)

Chitosan (wt%)	ζ potential (mV)
0	–68.4 ± 1.3
0.1	–8.95 ± 8.6
0.2	+20.6 ± 14
0.3	+50.2 ± 1.7

7.4 SCANNING AND TRANSMISSION ELECTRON MICROSCOPY

7.4.1 Principle of techniques

In electron microscopy, electron beam is bombarded on the sample. Interaction of these electrons with the sample surface results in generation of secondary electrons knocked out from the sample material, back-scattered electrons from the source and electrons that are transmitted through the sample. Based on the type of electrons detected, it is possible to evaluate structure, topography and composition of material. Backscattered electrons reveal details regarding topography and composition, while secondary electrons reveal topography alone. Scanning electron microscopy (SEM) detects either back-scattered or secondary electrons and is therefore frequently used for surface evaluation. In contrast to this, transmission electron microscopy (TEM) is based on the signal from electrons transmitted through the sample and therefore can be used to evaluate a three-dimensional structure and composition of the sample (Myhra and Riviere, 2012). Extensive discussion regarding the principle and equipment design considerations of SEM and TEM is beyond the scope of this chapter.

7.4.2 Applications of SEM and TEM

Sample preparation is crucial in SEM and is also a source of potential artifacts. Since SEM is carried out under vacuum, it is desirable to remove water from the sample. In addition to this, the sample must be conductive in nature. This poses a significant challenge as a large proportion of nanoparticles are composed of soft-matter or organic polymers. Therefore, these samples are typically coated with gold, platinum or other metals before imaging to improve their conductivity. However, as a result, some of the native structure of the nanoparticles is lost. This is probably the most significant limitation of SEM imaging. However, one of the most significant advantages of electron microscopy imaging is the much higher resolution (of the order of a few nanometer) that can be achieved as compared to optical or fluorescence microscopy.

Scanning electron microscopy imaging is frequently used to evaluate the surface characteristics of the nanoparticles. Dinsmore *et al.* (2002) used SEM to demonstrate the ability of polystyrene beads to occupy an oil–water interface, resulting in the formation of colloidosome. Many studies have subsequently used SEM to demonstrate change in the permeability of interface of colloidosomes (Miguel and Behrens, 2011; Thompson *et al.*, 2010; Wang *et al.*, 2011). Scanning electron microscopy has been used to demonstrate localization of nanoparticles at the oil–water interface to form Pickering emulsion, a significant class of encapsulation systems (Binks, 2002; He and Yu, 2007) and visualize surfaces of regular surfactant-stabilized emulsions (Dubes *et al.*, 2003; Palanuwech and Coupland, 2003). Scanning electron microscopy can also be used to visualize aggregation of nanoparticles, however, care must be taken to ensure that aggregation did not occur due to sample preparation that involves drying of the sample.

7.5 ELECTRON SPIN RESONANCE SPECTROSCOPY

7.5.1 Principle of ESR spectroscopy

Electron spin resonance (ESR) or electron paramagnetic resonance (EPR) is a widely used spectroscopic technique that relies on measurement of properties of paramagnetic molecules containing unpaired electrons (including free radicals) under the externally applied magnetic field. In some ways, ESR or EPR are similar to nuclear magnetic resonance (^{1}H-NMR) spectroscopy except that it detects the behavior of an unpaired electron rather than a proton. Material is exposed to an external magnetic field that distributes the population of unpaired electrons between lower and higher energy states. A second, variable magnetic field is applied through electromagnetic radiation such as microwaves to achieve resonance conditions for the unpaired, electrons resulting in transition of electrons from lower to higher energy states, thus producing an ESR signal (Lund *et al.*, 2011). Detailed discussion of the principle of ESR is beyond the scope of this chapter but Lund *et al.* provide a comprehensive source for fundamental understanding of this technique.

7.5.2 Applications of the ESR technique using nanoscale particles

One of the significant applications of ESR involves developing an understanding of the localization of the encapsulant in nanoparticles. Jores *et al.* (2003) used the ESR technique to demonstrate the differences in the distribution of encapsulated materials within nanoemulsions, solid lipid nanoparticles (SLNs), and nanostructured lipid carriers (NLCs). It is widely hypothesized that SLNs expel the encapsulated molecules during the processing crystallization, resulting in exposure of encapsulants to the aqueous environments. The NLCs are considered to be better encapsulation systems than SLNs due to presence of liquid oil fractions that retain the encapsulant within the liquid lipid core surrounded by solid lipid matrix. Based on this understanding, it is expected that there may be fractions of encapsulants experiencing different molecular environment within SLNs and NLCs. To demonstrate these differences Jores *et al.* (2003) used 4-hydroxy-2,2,6,6-tetramethylpiperidin-1-oxyl (TB) as an ESR-responsive encapsulant. Based on the spectral shape and hyperfine coupling constant of the probe, it is possible to quantify the relative polarity and the mobility of the molecular environment surrounding this probe. Based on this analysis, it was observed that all the TB molecules encapsulated within nanoemulsions experienced identical, highly mobile and hydrophobic environment. This was expected since TB is highly hydrophobic and readily dissolves in the oil phase. Since the oil phase was in the liquid state, relatively high mobility was observed. However, in both SLNs and NLCs three distinct populations of TB were observed. Based on their ESR characteristics, these populations were characterized as: Type 1, localized within the lipid compartment; Type 2, with a very hydrophilic environment, indicating TB dissolved in the water phase; Type 3, with a moderately hydrophilic environment, indicating TB localized at the nanoparticle surface. It was also evident that the fraction of TB in the Type 1 environment was significantly higher in NLCs compared to SLNs, indicating that presence of liquid oil enables enhanced solubilization of the encapsulant. Concurrently, the fraction of the population in the Type 2 and 3 environments was also significantly lower in NLCs than SLNs, indicating that the extent of expulsion of encapsulant was lower in NLCs compared to SLNs. These results are consistent with another study performed using fluorescence imaging to characterize the distribution of encapsulants within SLNs and NLCs (Tikekar and Nitin, 2012) indicating a direct correlation between the two techniques. The paramagnetic properties of TB can be quenched by reduction by antioxidants such as ascorbic acid (AA). By adding AA to the aqueous phase of these nanoparticles, they measured relative susceptibility of TB to reduction. They observed that the rate of loss of signal was slowest in nanoemulsions, followed by NLCs. The rate was fastest in SLNs. They also observed that the rate of loss of signal was faster in Type 2 and Type 3 species compared to Type 1 species, indicating susceptibility of encapsulant to reactions based on its localization within nanoparticles.

Yucel *et al.* (2013) investigated the effect of carbon chain length of lipid and the type of emulsifier on the distribution of encapsulant within nanoemulsions and SLNs. To accomplish this, they used 4-phenyl-2,2,5,5-tetramethyl-3-imidazoline-1-oxyl nitroxide (PTMIO) as an ESR probe that was encapsulated within emulsions or SLNs containing either C14 or C20 alkane as the oil phase stabilized by either sodium caseinate or lecithin + bile salt. By measuring the broadening of the hyperfine peak and magnitude of the hyperfine coupling constant (aN, high in an aqueous environment and low in a lipid environment) they measured the relative polarity of the environment experienced by PTMIO in these nanoparticles. It was observed that in caseinate stabilized SLNs containing C14 alkane, close to 70% of the probe was encapsulated in the lipid phase while 30% remained in the aqueous phase. When the lipid was replaced with C20 alkane, close to 100% of the probe was in the aqueous phase with a minute fraction remaining localized in the caseinate phase. The SLNs containing C14 alkane and stabilized by lecithin + bile salt showed a significantly different trend, with three distinct populations of PMTIO, as in the study by Jores *et al.* (2003). Approximately 60% lipid was encapsulated in the lipid phase, approximately 20% of the probe was expelled and 20% showed intermediate polarity. However, in SLNs containing C20 alkane and stabilized by lecithin + bile salt, the population resembling mobility in the lipid phase disappeared, and only two distinct populations resembling intermediate mobility and mobility in the aqueous phase remained. These results demonstrate that the ability of SLNs to retain the encapsulants depended on the nature of lipid phase and the type of emulsifier. The rate of quenching of the paramagnetic signal from PTMIO by iron/ascorbate was also compared between SLNs stabilized by caseinate and lecithin + bile salt. As expected, regardless of the type of emulsifier, the rate of quenching was higher in C20 SLNs as compared to C14 SLNs. However, interestingly, the rate of quenching was higher in lecithin + bile salt stabilized SLNs, which was counterintuitive considering the distribution of the probe in these systems.

Berton-Carabin *et al.* (2012) recently explored the effect of surfactant type and electrostatic charge on them on the ability of their micelles to encapsulate small molecular weight compounds. To accomplish this, they used PTMIO as an ESR probe that was encapsulated within the micelles of sodium dodecyl sulfate (SDS), an anionic surfactant, and dodecyl trimethylammonium bromide (DTAB), a cationic surfactant. It was observed that below the critical micellar concentration (CMC) of the surfactant, PTMIO was mainly localized within the polar aqueous phase, however, above the CMC the fraction of PTMIO encapsulated within micelles gradually increased with the concentration of the surfactant. The rate of quenching of the paramagnetic signal from PTMIO by ascorbate was also compared between SDS and DTAB micelles. The rate of quenching was significantly higher in micelles stabilized by DTAB than those stabilized by SDS. These results highlight the role of surface charge in controlling the permeation of molecular species across the nanoparticle interfaces.

Haag *et al.* (2011) used (2,2,6,6-tetramethyl-1-piperidinyloxy) (TEMPO) as an ESR probe and a model encapsulant to compare *ex vivo* and *in vivo* the differences in the dynamics of skin permeation of the encapsulant between NLCs and invasomes, a type of phospholipid vesicle. By comparing the changes in the hyperfine coupling constant and the upward/downward shift of the spectra they found that both NLC and invasomes were better than aqueous solution in reducing the rate of loss of intensity from TEMPO, indicating a more gradual release of the probe into skin compared to delivery through solution.

7.6 FLUORESCENCE SPECTROSCOPY AND IMAGING

7.6.1 Principle of fluorescence spectroscopy and imaging

Fluorescence is a phenomenon where a molecule absorbs light, resulting in excitation of electrons to a higher energy state. As these electrons return to their lower energy state, they emit photons of slightly lower energy, resulting in fluorescence. Since the emitted photon is of lower energy than the absorbed photon, the emission wavelength is higher than the excitation wavelength (Lacowicz, 2006). Both fluorescence spectroscopy and imaging are powerful tools that rely on either measurement of fluorescence from specific indicator probes (spectroscopy) or observance of fluorescence from these indicator probes (imaging). These techniques have been used extensively to characterize the distribution of the encapsulant in nanoparticles, understand the microenvironment of the encapsulant, measure transport of materials across the nanoparticles, and visualize the distribution of nanoparticles and their cargo in cells, tissues and other *in vivo* systems. We will briefly discuss some studies that used fluorescence spectroscopy and imaging to enable characterization of properties of nanoparticles.

7.6.2 Applications of fluorescence spectroscopy and imaging

Maintaining the stability of encapsulants is a critical challenge that limits the shelf-life of nanoparticle-based encapsulation systems. The problem is particularly exacerbated in emulsion-based nanoparticles that encapsulate hydrophobic, degradation-sensitive materials such as unsaturated fish oil (Frankel *et al.*, 2002). These encapsulants typically undergo oxidation or degradation induced by free radicals that are generated in the vicinity of the encapsulation system due to the interaction of metal ions, high temperature and other minor impurities (Coupland and McClements, 1996; Chaiyasit *et al.*, 2000). Upon generation, these free radicals diffuse into the hydrophobic core of nanoparticles and trigger degradative reactions. Therefore, the ability of encapsulation system to prevent such transport of free radicals is a highly desirable attribute. Various approaches based on the selection of optimized emulsifier or core material have been employed to improve the stability of encapsulants (McClements *et al.*, 2007). The effectiveness of these approaches is typically evaluated based on the measurement of the stability of a unique encapsulant (Klinkesorn *et al.*, 2005; Souto and Muller, 2005). Although such an approach is direct and provides insight into the systems under investigations, it has significant limitations such as: (i) difficulty in extrapolating results to other combinations of encapsulant–encapsulation systems; (ii) resource and time intensive analytical measurements; (iii) difficulty in comparing various nanoparticle-based encapsulation systems for their barrier properties. Therefore, there is a need to develop a rapid method that enables measurement of barrier properties of encapsulation systems against transport of species such as oxygen and free radicals. To enable rapid measurement of transport of oxygen and free radical species within nanoparticles, techniques were developed based on fluorescence spectroscopy. These methods are based on encapsulation of fluorescent molecules within nanoparticles that show changes in their fluorescence intensity upon interaction with oxygen and free radicals transported from the aqueous phase.

7.6.3 Transport of hydroxyl radicals

To measure transport of hydroxyl radicals across the oil-in-water emulsion interface, the hydroxyl radical sensitive dye carboxy-H_2DFFDA (5-(and-6) carboxy-2′,7′-difluorodihydro-fluorescein diacetate) was dispersed in the oil phase of the emulsion. Hydroxyl radicals were generated in the aqueous phase by the addition of Fenton's reagent (ferrous chloride and hydrogen peroxide) (Tikekar *et al.*, 2011a). The fluorescent dye carboxy-H_2DFFDA is activated only upon reaction with hydroxyl radicals, that is, when it is converted from a nonfluorescent to a fluorescent state. Thus, the rate of increase of fluorescence of the dye can be a good measure of the barrier properties of the emulsion interface. Figure 7.1a shows the fluorescence values of the dye as a function of time after addition of Fenton's reagent in the aqueous phase of whey protein emulsion. To complement spectroscopic measurements, we also performed fluorescence imaging of the emulsion droplets as a function of time after addition of Fenton's reagent (Figure 7.1b). These results further validate that the increase in fluorescence was indeed due to transport of hydroxyl radicals across the emulsion interface. To demonstrate the sensitivity of the measurement technique, the rate of increase of fluorescence in the presence of metal ion quencher and antioxidant added to the aqueous phase was also measured. The results show that addition of these molecules reduced the rate of increase of fluorescence, indicating quenching of hydroxyl radicals before they are transported into the oil phase.

7.6.4 Transport of peroxyl radicals

Similar to hydroxyl radicals, peroxyl radicals are frequently generated in food systems and lead to degradation of compounds. To measure the barrier properties of an encapsulation system against peroxyl radicals, a fluorescence-based method was developed (Tikekar and Nitin, 2011, 2012). Similar to the hydroxyl radical transport-measurement technique, a hydrophobic fluorescent dye,

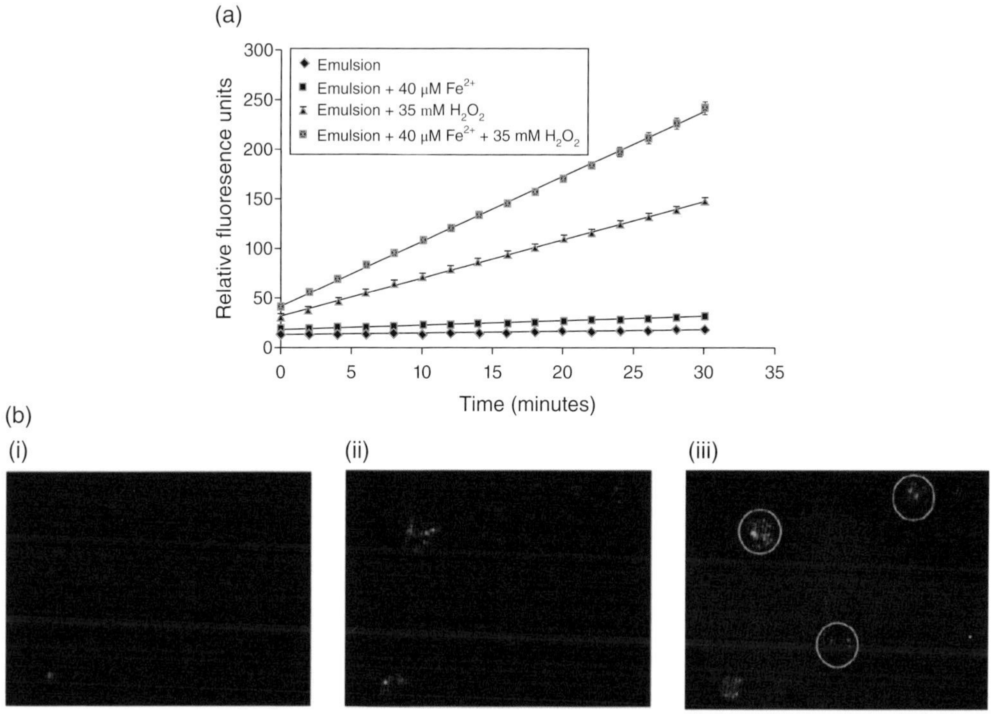

Figure 7.1 (a) Increase in fluorescence of free-radical-sensitive dye encapsulated in the oil phase of whey protein isolate (WPI) emulsion (3% oil, 1% WPI, pH 7.0, and 50 μg/g of fluorescent dye) as a function of time after addition of Fe^{2+} (40 μM), H_2O_2 (35 mM) ,and 40 μM Fe^{2+} + 35 mM H_2O_2 respectively to the aqueous phase of an emulsion. Each data point is an average of triplicate measurements ± standard deviation. (b) Time-series white imaging data based on activation of free-radical-sensitive dye encapsulated in the oil phase of a WPI emulsion upon exposure to Fenton's reagent (40 μM Fe^{2+} + 35 mM H_2O_2). Images were collected at discrete time points: (i) 0, (ii) 30, and (iii) 60 min after exposure of WPI emulsion to Fenton's reagent. (Adapted from Tikekar *et al.*, 2011a. Reproduced with permission from Elsevier.)

BODIPY 665/676, was encapsulated within the oil phase of the emulsion. This dye loses its fluorescence intensity upon interaction with peroxyl radicals. Peroxyl radicals were generated in the aqueous phase of the emulsion by addition of AAPH (2,2'-azobis-2-methyl-propanimidamide, dihydrochloride), a compound known to generate peroxyl radicals over a long period of time (half-life 175 h) through spontaneous decomposition. The loss in fluorescence intensity of the encapsulated peroxyl-radical-sensitive dye was measured as a function of time after addition of AAPH in Tween 20 and lecithin emulsion (Figure 7.2a) (Tikekar *et al.*, 2013). The results clearly show a significant difference in the rate of loss of fluorescence within the Tween 20 and lecithin stabilized emulsions. Lecithin emulsion showed a shoulder where for up to 400 min, there was no significant change in the fluorescence intensity. This was in contrast to the Tween 20 stabilized emulsion where the fluorescence decay was continuous. These results suggest excellent barrier properties of lecithin against permeation of peroxyl radicals compared to Tween 20. This can be attributed to the antioxidant activity of lecithin. Direct evidence of this was shown by comparing the peroxyl radical transport rates in emulsions stabilized by lecithin and oxidized lecithin (Figure 7.2b) (Tikekar *et al.*, 2013). The results clearly showed that the peroxyl radical permeation was significantly slower in emulsions stabilized by lecithin compared to that stabilized by oxidized lecithin. Additionally, the results of peroxyl radical measurements in lecithin and Tween 20 emulsions were in agreement with

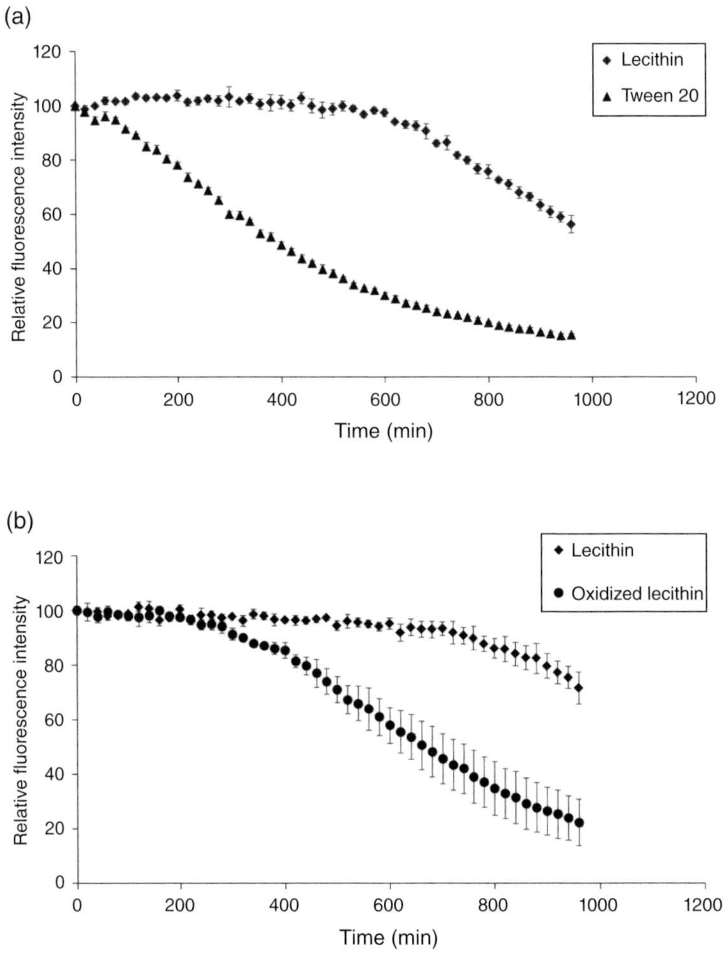

Figure 7.2 (a) Permeation of peroxyl radicals from the aqeuous phase to the oil phase of Tween 20 and lecithin emulsions. Radicals were generated using 20 mM of AAPH in the aqueous phase of emulsions. The radical permeation rate was measured based on loss in fluorescence of a peroxyl radical sensitive dye (C11 BODIPY665/676) encapsulated in the oil phase of the emulsion. (b) Permeation of peroxyl radicals from the aqeuous phase to the oil phase of lecithin and oxidized lecithin emulsions. Radicals were generated using 20 mM of AAPH in the aqueous phase of emulsions. The radical permeation rate was measured based on loss in fluorescence of a peroxyl-radical-sensitive dye (C11-BODIPY665/676) encapsulated in the oil phase of the emulsion. Each data point represents an average of three independent measurements ± standard deviation. (Adapted from Pan *et al.*, 2013. Reproduced with permission from Elsevier.)

measurement of the stability of a model bioactive compound, curcumin, encapsulated within these emulsions (Figure 7.3) (Tikekar *et al.*, 2013). Peroxyl radical transport rates have been used to demonstrate the impact of interfacial thickness and interfacial mobility in limiting permeation of free radicals across the films (Bricarello and Nitin, 2012). In another study, peroxyl radical measurements were used to demonstrate that encapsulations within solid lipid nanoparticles (SLNs) and liquid emulsion may be equally susceptible to oxidative degradation. This was possibly due to exclusion of the dye molecules encapsulated within SLNs towards the interface during crystallization of the lipid phase (Tikekar and Nitin, 2011).

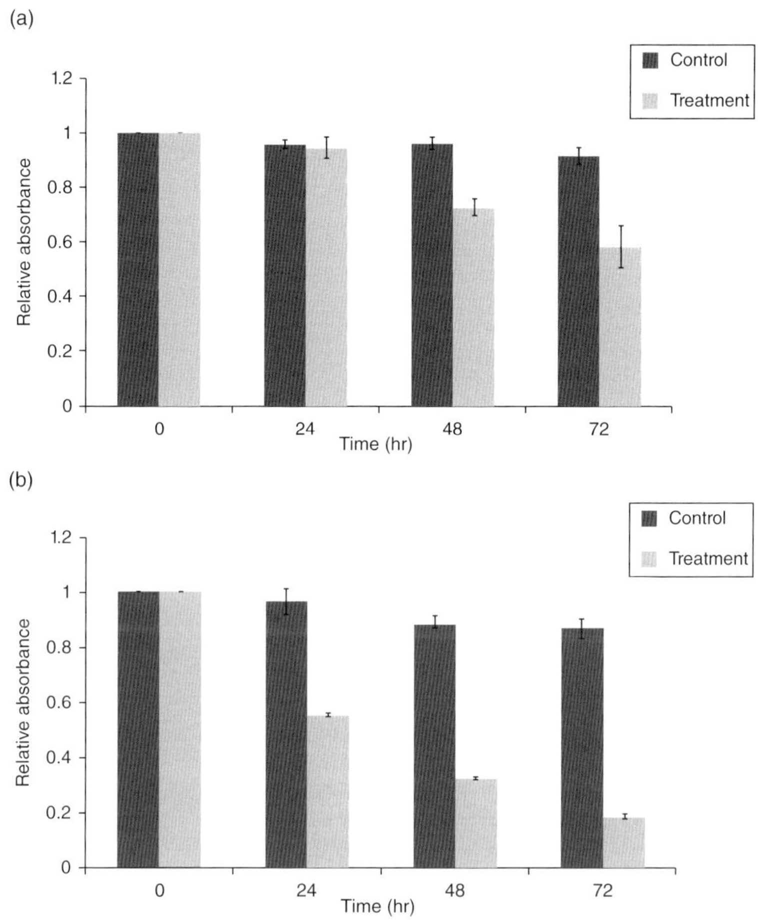

Figure 7.3 (a) Oxidative stability of encapsulated curcumin in Tween 20 stabilized emulsions with and without AAPH treatment. Both the control and the treatment emulsion samples were stored in the dark at room temperature for 72 h. (b) Oxidative stability of encapsulated curcumin in lecithin stabilized emulsions with and without AAPH treatment. Both the control and the treatment emulsion samples were stored in the dark at room temperature for 72 h. (Adapted from Pan *et al.*, 2013. Reproduced with permission from Elsevier.)

7.6.5 Permeation of oxygen within nanoparticles

To measure permeation of oxygen within the encapsulation system, an oxygen sensitive dye tris (4,7-diphenyl-1,10-phenanthroline) ruthenium(II) bis (hexafluorophosphate) complex was encapsulated within the oil phase of the emulsion (Tikekar *et al.*, 2011a). This dye shows reversible changes in fluorescence intensity as a function of concentration of oxygen in its surroundings. The fluorescence intensity of the dye is inversely proportional to the surrounding oxygen concentration. The emulsion encapsulating this dye was purged with nitrogen to remove traces of oxygen and subsequently exposed to atmosphere. A decrease in the fluorescence intensity of the dye as a function of time was measured until equilibrium values were reached, thereby indicating oxygen diffusion from atmospheric oxygen to the oil phase was complete (Figure 7.4). By converting the fluorescence decay rate into the fraction of equilibrium concentration of oxygen diffused into the emulsions, it was possible to calculate the diffusion coefficient of oxygen within emulsions stabilized by various emulsifiers. It was evident that, despite significantly low values of diffusion coefficients for various emulsions, oxygen transport

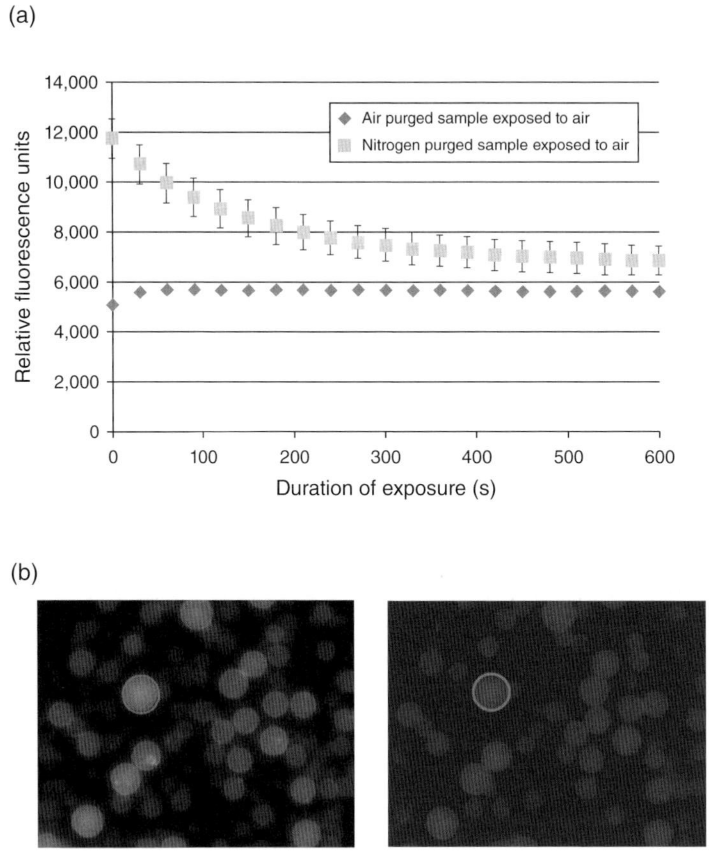

Figure 7.4 (a) Measurement of changes in fluorescence intensity of nitrogen purged whey protein isolate (WPI) emulsion (2.5% oil, 1% WPI, 50 lg/g oil of the dye, pH 7.0) upon exposure to air. (b) Fluorescent images of WPI emulsion when purged with nitrogen (t = 0 min, left) and when exposed to oxygen for 10 min (right). (Magnification 200×). (Adapted from Tikekar *et al.*, 2011b. Reproduced with permission from Elsevier.)

occurred efficiently. Various strategies such as cross-linking of interfacial protein matrix and layer-by-layer assembly of emulsifiers have been proposed to reduce the interfacial transport. However, based on the results obtained through quantitative measurement of oxygen transport, it was evident that these strategies were only marginally successful in limiting oxygen transport. This study demonstrates that limiting oxygen transport across the emulsion interface is a challenging task and new approaches maybe necessary to mitigate this challenge. However, using the novel approach based on fluorescence spectroscopy, it was possible to highlight this challenge.

Fluorescence spectroscopy can also be used to determine the localization of encapsulant within a nanoparticle. Jores *et al.* (2005) demonstrated that by comparing the changes in the fluorescence emission spectrum of Nile-red as a function of the polarity of its surrounding environment (solvatochromism), they were able to identify the location of Nile-red within solid lipid nanoparticles and nanostructured lipid carriers. It was evident that Nile-red was mainly localized on the surface of solid nanoparticles, possibly due to exclusion during crystallization of the lipid phase. Addition of liquid lipid created two populations of Nile-red, one dispersed in liquid lipid (hydrophobic environment) and the other one localized on the surface of the nanoparticles (more polar environment). When the liquid oil fraction increased to > 20% of the total lipid composition, the fraction of Nile-red localized on the surface of the nanoparticles disappeared. Such differences may have a significant impact on the stability of encapsulants within nanoparticles. The results of this study were further validated by another that used fluorescence imaging to show similar variations in the distribution of encapsulated Nile-red (Tikekar and Nitin, 2012). It was further shown that these variations led to differences in susceptibility of encapsulated material to oxidizing free radicals generated in the aqueous phase. Material encapsulated in SLNs showed higher susceptibility to oxidation by these free radicals compared to that encapsulated within NLC. In another study, Nile-red was used as a model encapsulant to demonstrate the differences in the release rate of cargo from SLNs and NLCs in a skin model system (Borgia *et al.*, 2005). It was found that SLNs showed the higher rate of release compared to NLCs, possibly due to surface localization of Nile-red in SLNs that may enhance its release and uptake compared to NLCs, where the dye may be localized within the core of the nanoparticle. Additionally, it was observed that the release rate can also be affected by interaction of the encapsulant with the nanoparticle matrix.

7.6.6 Fluorescence imaging

Fluorescence imaging is an important tool to enable visualization of target molecule in a nanoscale environment. As discussed earlier, fluorescence imaging has played a significant role in the determination of localization of the encapsulant within the nanoparticles. Imaging has also been used to measure uptake and develop fundamental understanding of the distribution of nanoparticles within *in vivo* systems. Wang *et al.* (2010) developed magnetic-fluorescent magnetite/poly(dl-lactic acid-co-α,β-malic acid) composite nanoparticles that were conjugated with fluorescein isothiocyanate (FITC) on the surface to enable their visualization within human mesenchymal stem cells using fluorescence imaging. As observed by fluorescence imaging, these nanoparticles appeared to have aggregated within the cell cytoplasm. In a recent study, to enable simultaneous visualization of encapsulation system and its cargo within cells and tissue, a novel approach was developed to conjugate the encapsulant as well as carrier material with fluorescent probes. Figure 7.5 shows distinctly different distributions of the micelle-based encapsulation system (dark gray fluorescence) and its cargo, curcumin (light gray fluorescence), indicating the micelles were disrupted and cargo was released (Luo *et al.*, 2014). Such an approach aids in investigating the fate of the nanoparticle-based encapsulation systems, evaluate their stability and integrity in a cellular environment, and visualize their distribution. Such fundamental understanding of the system can enable engineering of nanoparticles with the desirable stability and distribution within a tissue environment.

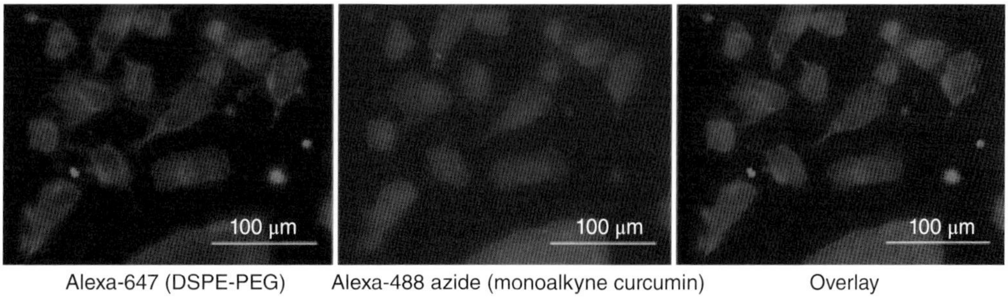

Figure 7.5 Simultaneous imaging of the intracellular distribution of DSPE-PEG and monoalkyne curcumin in a two-dimensional cell culture model. The DSPE-PEG-amine micelles were labeled with Alexa-647 ester (dark gray), while monoalkyne curcumin was labeled with Alexa-488 azide (light gray). (Adapted from Luo *et al.*, 2014. Reproduced with permission from American Chemical Society.)

7.7 ATOMIC FORCE MICROSCOPY

7.7.1 Principle of atomic force microscopy

The main sensing element of atomic force microscopy (AFM) consists of a microscale (100–400 μm in length) cantilever beam with a pyramidal tip. The beam and the tip typically consist of silicon or silicon nitride due to their low cost and ease of production. Analysis of a sample is typically performed by moving the cantilever beam along the surface, and as the AFM tip interacts with the surface of the material, it leads to bending of the cantilever beam. The extent of bending depends on the force applied to the beam and the nature of the surface. The extent of bending of the beam is detected by the extent of deflection of laser light reflected from the surface of the beam (Figure 7.6). This iterative process can occur in *x*, *y* and *z* dimensions, resulting in a three-dimensional topography of the material. Scanning of material can be performed in contact mode, intermittent contact mode (tapping mode), or phase-imaging mode. In contact mode, the tip is in constant contact with the material and the beam deflection is used to plot the topography of the material. However, due to the large amount of stress that this process generates, some features of the samples may be damaged. To address this issue, a variation of this method has been developed, where the height of the AFM sample stage is altered to accomplish a constant deflection of the beam. In intermittent contact mode, the cantilever beam is oscillated using a piezoelectric crystal. When the beam interacts with sample surface, the oscillations are dampened. This dampening is used as a feedback to alter the stage height in order to restore the set oscillation frequency, which results in generation of the three-dimensional topography of the material. Phase-imaging mode is based on measurement of changes in the phase of oscillations of the beams due to interaction of the tip with material. Since the phase change of oscillation is a function of nature of the material (elasticity vs. softness), spatial variations in surface materials can be measured using AFM (Sitterberg *et al.*, 2010).

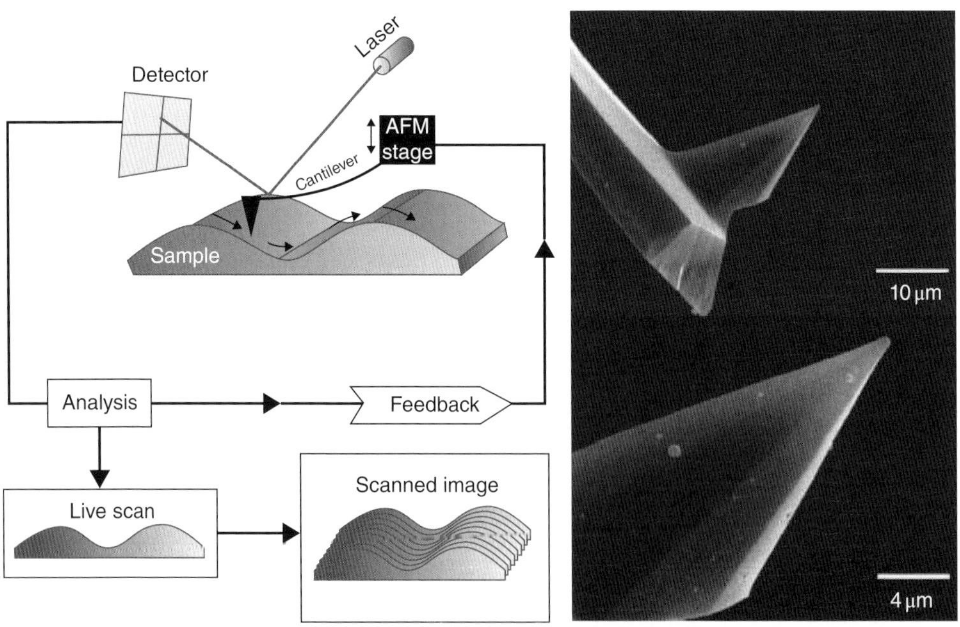

Figure 7.6 Schematic representation of AFM microscopy. The figure on the right shows the typical cantilever beam with tip. (Adapted from Sitterberg *et al.*, 2010. Reproduced with permission from Elsevier.)

7.7.2 Applications of AFM

Atomic force microscopy is a technique commonly used for a three-dimensional visualization of nanoparticles. This is a distinct advantage over techniques such as SEM, which only allow a two-dimensional visualization of the particle. In addition to this SEM requires significant sample preparation such as surface coating with platinum or palladium and may need to be performed in vacuum to enhance visualization. An extensive sample preparation has several limitations such as: (i) destabilization of nanoparticles, (ii) introduction of artifacts, and (iii) structures visualized through SEM may be significantly different from the native structure. The AFM technique is able to address these limitations and provides a more accurate visualization of the topography and three-dimensional structure of the material. In this section, we will consider specific examples that highlight the advantages of AFM in improving the characterization of nanoparticles. Although DLS is an effective tool for measuring average particles size, it has limitations in measuring multimodal distribution of nanoparticles and does not provide accurate particle-size distribution if the samples contain mixtures of relatively large and small particles (Mahl *et al.*, 2011). In such a situation, AFM may be able to measure particle size more effectively. Zur Mühlen *et al.* (1996) used AFM to image drug-loaded solid lipid nanoparticles. Using AFM, they were able to demonstrate that the nanoparticles were disc shaped and crystalline and extended storage of these nanoparticles in aqueous medium resulted in these particles appearing flatter compared to freshly prepared nanoparticles. They also found that the particles were surrounded by a layer of material relatively softer than lipid, indicating possibly a noncrystalline surface. However, interaction of the AFM tip with nanoparticles can result in flattening of particles, which may explain the disc-like appearance of particles (Montasser *et al.*, 2002; Sitterberg *et al.*, 2010). In some cases at least, the flattening of nanoparticles during AFM can be reduced by addition of 25% glutaraldehyde to the sample before imaging and a good correlation between average size measured by the AFM and DLS techniques was observed, indicating that AFM could serve to measure particle size as well (de Assis *et al.*, 2008).

Atomic force microscopy can also be used to visualize changes in the structure of nanoparticles due to surface modifications, changes in the surrounding environment, and release of encapsulants. In one study, insulin-loaded dialkylaminoalkyl-amine-poly(vinyl alcohol)-g-poly(lactide-co-glycolide) nanoparticles were prepared that could aggregate upon interaction with isotonic buffer to form a semi-solid reserve of insulin at the injection site. The nanoparticles were stable under synthesis conditions, but immediately aggregated upon insertion into isotonic buffer solution. The macroscopic change in the nanoparticles was validated by performing AFM imaging of nanoparticles and aggregated nanoparticles (Packhaeuser and Kissel, 2007). In another study, stability of poly(L-lactide-co-glycolide) (PLGA) nanoparticles in phosphate buffer saline was evaluated over a period of 15 weeks using the phase-imaging mode of AFM. Changes in the elastic properties of PLGA nanoparticles surfaces were evident. Freshly prepared nanoparticles had a smooth surface. With increase in duration of storage, however, surface pitting was observed, leading to increased porosity of the material. The depth of pores increased over time, but the nanoparticle surface appeared smooth again after 15 weeks (Weiss *et al.*, 2006) (Figure 7.7). Wang *et al.* (2006) reported use of AFM imaging to demonstrate localization of bovine serum albumin protein on silica nanoparticles.

Based on the research examples presented earlier, it is apparent that AFM complements other characterization techniques such as DLS, SEM, and TEM, and in some cases address their limitations to enhance the fundamental understanding of nanoparticles. A more comprehensive review of applications of AFM for nanoparticle characterization has been published recently by Sitterberg *et al.* (2010).

7.8 CONCLUSIONS

Various characterization techniques, their applications and limitations have been discussed. It is apparent that each technique has unique advantages and limitations. Therefore, to enhance the fundamental understanding of nanoparticle-based delivery systems, it is recommended that multiple

8 Impact of delivery systems on the chemical stability of bioactive lipids

Ketinun Kittipongpittaya, Lorena Salcedo, David Julian McClements, and Eric Andrew Decker

Department of Food Science, University of Massachusetts, Amherst, Massachusetts, USA

8.1 INTRODUCTION

There is growing interest in incorporating various kinds of bioactive lipids into commercial food and beverage products. This can be a challenge since bioactive lipids such as ω-3 fatty acids, carotenoids, and antioxidants are highly susceptible to chemical degradation by heat, reactive oxygen species, metals, and light. In addition, incorporation of bioactive lipids into aqueous-based foods is often problematic because of their low water-solubility, for example beverages, infant formula, dressings, sauces, desserts, and soups. Numerous researchers have suggested that incorporation of bioactive lipids into colloidal delivery systems such as conventional emulsions, multiple emulsions, nano-emulsions, solid lipid nanoparticles, and filled hydrogel particles could be an effective method for their physical and chemical stabilization. However, bioactive lipids are susceptible to different chemical destabilization pathways so it would not be expected that all delivery systems would be effective for all bioactive lipids. This chapter will examine the causes of chemical destabilization of bioactive lipids and how the physical properties of different delivery systems could protect these important nutrients.

8.2 PATHWAYS OF DEGRADATION OF BIOACTIVE LIPIDS

8.2.1 Free radicals

Free radicals are molecules with unpaired electrons. Depending on the free radical species, the energy and thus ability to attack other molecules varies greatly. Reactivity of free radicals can be determined by measuring their *reduction potential*. Free radicals such as the hydroxyl radical have such high energy that they will oxidize almost any molecules they encounter, and therefore the reaction is diffusion-limited. However, other radicals have very low energy and thus do not normally promote oxidation on their own, such as the bi-radicals of atmospheric, triplet oxygen.

Free radicals can be formed by several processes. For example, superoxide anion is produced by enzymes, the release of oxygen from oxymyoglobin to produce metmyoglobin, and by cells such as

Nanotechnology and Functional Foods: Effective Delivery of Bioactive Ingredients, First Edition.
Edited by Cristina M. Sabliov, Hongda Chen, and Rickey Y. Yada.

phagocytes. Free radicals can also be produced from hydroperoxides. Hydrogen peroxide can be found in foods when it is used as a processing aid and via its production by superoxide dismutase. In addition, lipid hydroperoxides are found in essentially all commercial food oils, meats, dairy products, food ingredients (e.g. Tween emulsifiers and phospholipids) and essential oils (Mancusco *et al.*, 2000; Nuchi *et al.*, 2001, 2002). Hydroperoxides can decompose into high-energy free radicals such as hydroxyl, peroxyl, and alkoxyl radicals. Hydroperoxide decomposition into free radicals is accelerated by light, heat, and transition metals.

The molecules most susceptible to free radical attack contain labile hydrogen atoms that can be easily abstracted from the parent molecule. In foods, this includes the hydroxyl groups of phenolics and molecules with double bonds such as unsaturated fatty acids and carotenoids. In polyunsaturated fatty acids, there is a methylene-interrupted carbon, which is a carbon between two double bonds. The electrons of the carbon-hydrogen bond of this methylene-interrupted carbon are drawn towards the double bonds, making it easily abstracted by free radicals. This is why polyunsaturated fatty acids will be oxidized much more quickly than monounsaturated fatty acids.

Once free radicals abstract a hydrogen from a lipid, a free radical is formed on the bioactive lipid. This free radical can react with the bi-radical of triplet oxygen through a series of pathways to form additional hydroperoxides on bioactive lipids such as unsaturated fatty acids, phytosterols, and carotenoids. The bioactive lipid hydroperoxides can also decompose into free radicals, resulting in the further destruction of the molecule and loss of bioactivity. In addition, these decomposition reactions can form off-flavors (fatty acid decomposition products), loss of color (carotenoid bleaching), and formation of potentially toxic compounds (unsaturated aldehydes).

8.2.2 Transition metals

Transition metals are found in all foods since they are common constituents of the biological tissues we use for food as well as water, processing equipment, and packaging materials. Transition metals can promote oxidation by decomposing hydroperoxides into free radicals (Halliwell and Gutteridge, 1990; Girotti, 1998) or directly interacting with the bioactive lipid. Interactions between metals and hydrogen or lipid hydroperoxides occur through the following redox cycling pathway.

$$\mathrm{Mn}^{n} + \mathrm{LOOH\ or\ HOOH} \rightarrow \mathrm{Mn}^{n+1} + \mathrm{LO\bullet\ or\ HO\bullet} + \mathrm{OH}^{-}$$
$$\mathrm{Mn}^{n+1} + \mathrm{LOOH} \rightarrow \mathrm{Mn}^{n} + \mathrm{LOO\bullet} + \mathrm{H}^{+}$$

Where Mn^{n} and Mn^{n+1} are transition metals in their reduced and oxidized states, respectively; LOOH and HOOH are lipid and hydrogen peroxide; and LO•, HO•, and LOO• are alkoxyl, hydroxyl, and peroxyl radicals, respectively. Hydroxyl radicals are produced from hydrogen peroxide while alkoxyl radicals are produced from lipid hydroperoxides.

The rate of hydroperoxide decomposition is dependent on the type of metal and its oxidation state. Copper and iron are the most abundant hydroperoxide-decomposing transition metals in foods, with iron generally being found at greater concentrations than copper. Copper is more reactive, with cuprous ions (Cu^{1+}) decomposing hydrogen peroxide over 50-fold faster than ferrous ions (Fe^{2+}). Ferrous decomposes hydrogen peroxide over 100 times faster than Fe^{3+} (ferric) indicating that the reduced state of the metal is more reactive. Ferrous is also more water-soluble, resulting in even greater hydroperoxide decomposing activity than ferric in aqueous food systems (Halliwell and Gutteridge, 1990; Girotti, 1998).

The higher hydroperoxide reducing activity of the reduced state of transition metals means that compounds capable of promoting the redox cycling of transition metals can promote lipid oxidation. In foods, these prooxidative reducing compounds include superoxide anion (O_2^{-}), ascorbic acid, carotenoids, and certain flavonoids. The superoxide anion is produced by the addition of an electron to triplet oxygen, with this electron being able to reduce metals. The redox cycling of iron by the superoxide anion to promote free-radical generation is known as the Haber–Weiss reaction and is shown below.

$$\begin{aligned} &Fe^{3+} + O_2^- \rightarrow Fe^{2+} + O_2 \\ &Fe^{2+} + H_2O_2 \rightarrow Fe^{3+} + \bullet OH + OH^- \\ &\text{Net}: O_2^- + H_2O_2 \rightarrow O_2 + \bullet OH + OH^- \end{aligned}$$

Ascorbic acid and certain flavonoids can also participate in Haber–Weiss-like reactions, however, unlike the superoxide anion, these reducing agents can also act as antioxidants. This is because at high concentrations, their ability to scavenge free radicals may outweigh their metal-reducing prooxidant activity, resulting in a net antioxidant effect.

Transition metals associated with proteins such as the heme group in myoglobin, hemoglobin, peroxidases, and catalase can also promote both hydrogen peroxide and lipid hydroperoxide decomposition. Thermal denaturation of these proteins can expose the heme iron to increased prooxidant activity. Thermal denaturation of myoglobin may be one of the factors that accelerates lipid oxidation in cooked meats, a quality defect known as warmed-over flavor.

8.2.3 Light promoted oxidation

Decomposition of hydroperoxides to produce free radicals can be accelerated by UV and visible light. This means that packaging systems to decrease light exposure can reduce lipid oxidation rates. Light can also accelerate lipid oxidation by aiding in the production of singlet oxygen. Triplet oxygen ($^3O_2$) is a bi-radical because two of its electrons have the same spin direction and thus cannot exist in the same electron orbital. Triplet oxygen does not readily react with electrons in another molecule because its electrons do not have matching spin directions. Triplet oxygen can be converted to singlet oxygen if the direction of one of its unpaired electrons is reversed (Min and Boff, 2002). Singlet oxygen can react directly with double bonds since its electrons match the spin direction of the electrons in double bonds. Therefore it can produce lipid hydroperoxides in unsaturated fatty acids more than 1500 times faster than triplet oxygen.

Singlet oxygen is most commonly produced in the presence of light and photosensitizers such as chlorophyll, riboflavin, and myoglobin. This occurs when the photosensitizer becomes excited by light and then this energy is transferred to triplet oxygen to produce singlet oxygen. During this process, some superoxide anions can also be formed, so in some cases the acceleration of metal-promoted lipid oxidation is also observed (Lee and Decker, 2011).

8.2.4 Lipoxygenase

Lipoxygenases (LOXs) are naturally occurring enzymes found in numerous plant tissues and selective animal tissues. This enzyme directly produces lipid hydroperoxides at very specific carbon positions, usually at carbon 9 or 13 of linoleic acid (for review see Zhuang *et al.*, 2002). In plants, LOXs contain nonheme iron and the enzymes primarily react with free fatty acids. Lipoxygenases can also been found in animal tissues such as those highly associated with the circulatory system (e.g. fish gills) (German and Creveling, 1990). However, lipoxygenases are not usually an important prooxidant in muscle foods.

8.3 BIOACTIVE LIPID DELIVERY SYSTEMS AS A MEANS TO CONTROL LIPID OXIDATION

Some of the bioactive lipids that the food industry would like to incorporate into commercial products are highly susceptible to oxidation, for example ω-3 fatty acids and carotenoids. Research articles commonly state that numerous nanotechnology-based delivery systems can be used to improve the oxidative stability of bioactive lipids. However, it is important to recognize that there are often additional challenges associated with stabilizing bioactive lipids when using delivery systems. The major problem with nanoscale delivery systems is that they create more surface area than traditional food dispersions. As many oxidation reactions occur at oil–water interfaces, the huge surface area in nanoscale delivery systems could decrease the oxidative stability of bioactive lipids. This section will look at different delivery systems and nanostructures and correlate their properties to the factors that impact lipid oxidation.

8.3.1 Conventional emulsions

In most food dispersions, transition metals are the major prooxidants (Yoshida and Niki, 1992; Mei *et al.*, 1998a, b; Mancuso *et al.*, 2000; Dimakou *et al.*, 2007). Since lipid hydroperoxides are surface active (Nuchi *et al.*, 2001), transition metals primarily form free radicals at the emulsion droplet interface.

Most studies report that emulsion droplet size is not an important factor in oxidation rates in conventional emulsions (Hegenauer *et al.*, 1979; Coupland *et al.*, 1996; Shimada *et al.*, 1996; Gohtani *et al.*, 1999; Lethuaut *et al.*, 2002; Osborn and Akoh, 2004; Nakaya *et al.*, 2005; Dimakou *et al.*, 2007; Kiokias *et al.*, 2007; Paraskevopoulou *et al.*, 2007; Imai *et al.*, 2008; Sun and Gunasekaran, 2009). The fact that lipid oxidation rates do not change dramatically in oil-in-water emulsions with different surface areas suggests that the extremely large surface area of all emulsions is such that it never limits reaction rates. Alternately, conventional emulsions are polydisperse systems with large variations in droplet size, meaning that they can have significant amounts of small droplets (Figure 8.1). Since lipid oxidation can be detected in food dispersion even at very low lipid levels, it is possible that oxidation could start in the smallest droplets. If this was the case, differences in oxidation rates in two conventional emulsions that have different mean droplet sizes might not be different if both emulsions have similar populations of small droplets. More studies are needed to evaluate differences in oxidation rates in monodispersed emulsions to determine if size really matters in oxidation kinetics. Comparison of emulsions with very large differences in particle size (e.g. nanoemulsions vs. conventional emulsions) could provide more insight into how surface area impacts oxidation rates.

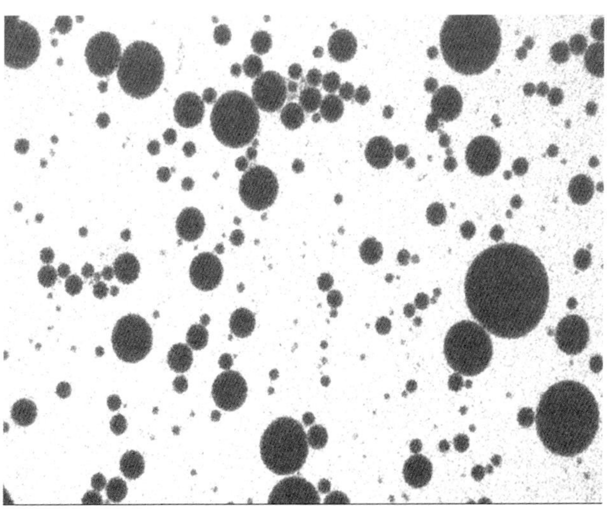

Figure 8.1 A micrograph of a conventional emulsion. Note the large variation in emulsion droplet size.

The physical properties of the oil–water interface of emulsion droplets are an important factor in oxidation reactions. The interfacial layer is chemically and structurally complex since it contains emulsifiers and other surface-active materials such as antioxidants, minor lipid components (e.g. sterols and triacylglycerol hydrolysis products), biopolymers, and mineral ions (McClements and Decker, 2000; Chaiyasit *et al.*, 2007). The interfacial layer is usually 1 to 40 nm thick and can be modified by altering emulsifier type. The physical properties of the interfacial layer can strongly influence oxidation kinetics since it can influence the ability of aqueous phase prooxidants (i.e. metals) to interact with lipid-phase oxidizable substrates (i.e. hydroperoxides).

Since transition metals are such important prooxidants in conventional emulsions, the oxidative stability of oil-in-water emulsions is often related to the ability of the interfacial layer to either attract or repel cationic metals (Mei *et al.*, 1998a, b; Mancuso *et al.*, 1999, 2000; Silvestre *et al.*, 2000; Boon *et al.*, 2008; Haahr and Jacobsen, 2008). For example, emulsions stabilized with anionic surfactants can oxidize faster than emulsions stabilized with cationic surfactants since the anionic surfactant attracts cationic metal ions to the emulsion droplet surface (Mei *et al.*, 1998a, b; Mancuso *et al.*, 1999, 2000; Silvestre *et al.*, 2000; Boon *et al.*, 2008). In addition, factors such as the presence of free fatty acids in the oil phase of emulsions can increase oxidation rates by increasing the negative charge of the oil droplets (Waraho *et al.*, 2011).

The dependence of oil droplet charge on oxidation rates is also seen in protein-stabilized emulsions. In this case, the rate of lipid oxidation is faster when the pH is greater than the pI of the protein and the protein-coated oil droplets are negatively charged (Donnelly *et al.*, 1998; Mei *et al.*, 1998a, b; Mancuso *et al.*, 1999, 2000; Hu *et al.*, 2002, 2003, 2004; Trunova *et al.*, 2007; Djordjevic *et al.*, 2008). However, in the case of emulsions stabilized with different types of proteins, the extent of the droplet charge does not always correlate with oxidative stability. For instance, casein-stabilized emulsions with a lower positive charge were reported to be more oxidatively stable than whey-protein-stabilized emulsions (pH 3.0). This is because proteins have additional properties that impact oxidation, such as the ability to scavenge free radicals, chelate transition metal ions, and alter interfacial properties (Hu *et al.*, 2002; Villiere *et al.*, 2005).

Emulsion droplet interfacial thickness is dependent on the nature of the emulsifier used, and depends on head-group size for surfactants, and molecular weight and conformation for biopolymers (Klinkesorn *et al.*, 2005a, b; Djordjevic *et al.*, 2007; Shaw *et al.*, 2007). Emulsifiers with large molecular dimensions can form a barrier that decreases interactions between labile lipids and aqueous phase prooxidants (Chaiyasit *et al.*, 2000; Silvestre *et al.*, 2000). This was observed in salmon oil-in-water emulsions stabilized with Brij 76 or Brij 700, where the larger hydrophilic head group of Brij 700 (10 times larger than Brij 76) resulted in slower lipid oxidation rates. Large proteins such as casein are more antioxidative than small proteins such as whey proteins, despite the fact that casein-stabilized emulsion droplets have a lower positive charge than whey-protein-stabilized emulsion droplets. This could be due to the thicker interfacial layer produced in casein-stabilized emulsions. Increasing surfactant tail-group size only slightly decreases oxidation rates (Chaiyasit *et al.*, 2000).

8.3.2 Multilayer emulsions

The physical and chemical properties of conventional emulsions can be altered by coating their droplet interface with one or multiple layers of biopolymers. Multilayer emulsions could inhibit lipid oxidation by producing a thicker interfacial layer that could limit interactions between lipid- and aqueous-phase components or by making the emulsion droplet positively charged so that it can repel transition metals. Shaw *et al.* (2007), Klinkesorn *et al.* (2005a, b) and Djordjevic *et al.* (2007) have all shown that lipid oxidation can be inhibited when the exterior layer of the nanolaminated oil droplets are cationic. The thickness and permeability of the interfacial layer may also be important in decreasing oxidation since studies have shown that oil droplets coated with anionic β-lactoglobulin-pectin layers had a similar oxidative stability as those coated by a cationic β-lactoglobulin layer (Katsuda *et al.*, 2008).

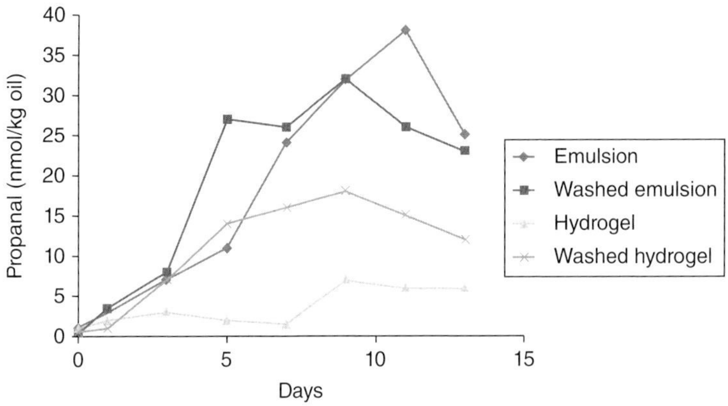

Figure 8.2 Formation of propanal in casein- and pectin-stabilized algae oil-in-water emulsions and hydrogels prepared with 1% oil at pH 7.0. Emulsions were stored at 55°C. Washing the emulsion and hydrogels by centrifugation and removal of the aqueous phase a total of three times was performed in order to remove continuous-phase casein and pectin.

8.3.3 Filled hydrogel particles

Filled hydrogels are produced by trapping oil droplets within biopolymer microspheres to produce oil-in-water-in-water (O/W/W) emulsions (McClements *et al.*, 2007). There are several different methods by which filled hydrogel particles can be produced, including phase separation of biopolymer solutions, physical fragmentation of preformed biopolymer gels, and formation of double emulsions followed by biopolymer gelation (Pich and Adler, 2007; Matalanis *et al.*, 2011). Filled hydrogel particles have been proposed to inhibit oxidation because they present an opportunity to concentrate antioxidants in the same physical location at the emulsion droplets. The ability of hydrogels to protect ω-3 fatty acids was reported by Lamprecht *et al.* (2001) using gelatin and acacia, and by Wu *et al.* (2005) using hydroxylpropyl methylcellulose. Weinbreck *et al.* (2004) reported that hydrogel particles filled with whey protein and gum arabic could stabilize flavor oils. Recent work in our laboratory found that filled hydrogels prepared with casein and pectin could stabilize ω-3 fatty acids more effectively than Tween 20 stabilized emulsions (Matalanis *et al.*, 2012). However, further work showed that the majority of the antioxidant activity was due to the presence of casein and pectin in the continuous phase since the oxidation of algal oil was much faster in washed hydrogels (Figure 8.2). The ability of the continuous phase casein and pectin to inhibit lipid oxidation was likely due to their ability to act as a metal chelator.

8.3.4 Microemulsions and nanoemulsions

Microemulsions and nanoemulsions are both colloidal dispersions that consist of lipid nanoparticles dispersed in water, however, microemulsions are thermodynamically stable whereas nanoemulsions are only kinetically stable (McClements, 2012). In most cases, microemulsions are formed using small molecule surfactants, whereas nanoemulsions can be stabilized by small molecule surfactants, phospholipids, or amphiphilic biopolymers. The main difference between nanoemulsions ($r < 100$ nm) and conventional emulsions ($r > 100$ nm) is their particle size (Mason *et al.*, 2006; McClements, 2012). Microemulsions and nanoemulsions have gained much attention in the beverage industry because of the potential to deliver bioactive lipids without scattering light, an important property for preparing clear beverages.

Microemulsions containing free fatty acids have very unusual oxidation tendencies, with the most unsaturated fatty acids (e.g. 22:6) having the highest oxidative stability (Miyashita *et al.*, 1993, 1994). This phenomenon has been suggested to be due to differences in the physical orientation of the unsaturated double bonds in the free fatty acids, such that they are shielded from interactions with aqueous phase prooxidants. This trend was not observed for cholesterol esters, where oxidative stability decreased with increasing unsaturation (Almeida *et al.*, 2010).

Qian *et al.* (2012) found that oxidation of β-carotene in nanoemulsions stabilized by Tween 20 or β-lactoglobulin were inhibited by both EDTA and ascorbic acid, indicating that transition metals and free radicals were active prooxidants in nanoemulsions. Lee and Decker (2011) reported that menhaden oil in nanoemulsions stabilized by whey protein isolate (WPI) was less stable than menhaden oil in conventional WPI emulsions, presumably due to the larger surface area. Research in this area is important since it is possible that the extremely high surface area of nanoemulsions could cause rapid oxidation, which would severely limit the application of this technology for delivery of oxidizable bioactive lipids into foods. Another possible limitation of nanoemulsions is that the rate of light-induced oxidation may be higher than in conventional emulsions due to their higher optical clarity.

8.3.5 Solid lipid particles

Solid lipid particles have the potential to inhibit lipid oxidation if unsaturated fatty acids or other bioactive lipids can be entrapped in a matrix of solid lipid, thus decreasing their exposure to oxygen and aqueous-phase prooxidants (McClements *et al.*, 2007). To prepare suspensions of solid lipid particles, the lipids are first melted and then homogenized with water to form an emulsion, which is then cooled to solidify the droplets (Wissing *et al.*, 2004). In a study where methyl linolenate was incorporated into octadecane oil-in-water emulsions, oxidation was found to occur more rapidly in droplets containing solid than liquid octadecane (Okuda *et al.*, 2005). This increased oxidation rate was suggested to be due to the solid octadecane expelling the methyl linolenate to the emulsion droplet surface, where it interacts more readily with continuous-phase prooxidants. Yucel *et al.* (2012) recently confirmed this by showing that upon crystallization of lipids in an emulsion droplet a lipid soluble electron paramagnetic resonance probe was either expelled to the droplet surface or into the continuous phase. Both the above studies used hydrocarbons as a model of solid lipid. The crystals formed by these lipids have highly organized molecular packing that does not enable bioactive components to be internalized within the solid phase. If solid lipid particles are going to be an effective strategy to stabilize lipids, then food oils that do not form highly organized crystals might be more effective at trapping bioactive compounds in the center of the fat droplet. Alternately, surfactants that can crystallize on the surface of the emulsion droplet might also provide some protection from the continuous-phase prooxidants (Helgason *et al.*, 2009).

8.4 ANTIOXIDANTS IN BIOACTIVE LIPID DELIVERY SYSTEMS

In most delivery systems it is unlikely that the physical properties of the droplet will be able to completely protect bioactive lipids, especially those that are particularly prone to oxidation such as ω-3 fatty acids and carotenoids. Therefore most delivery systems will also need to utilize antioxidants, either free-radical scavengers or metal chelators, to optimize stability. The activity of antioxidant delivery systems will depend on factors such as concentration, reactivity, location, interactions with other food components, and environmental conditions such as pH and temperature (Frankel, 1998).

The antioxidant "polar paradox" has been used to try to predict how the physical properties of free-radical scavengers impact their ability to inhibit lipid oxidation (Porter *et al.*, 1989; Frankel, 1998). According to the antioxidant polar paradox, nonpolar antioxidants are more effective in oil-in-water emulsions. This was originally proposed to be due to the ability of nonpolar antioxidants to be retained in the emulsion droplet where oxidation is most prevalent. The hypothesis was supported by evidence showing that in oil-in-water emulsion systems, nonpolar antioxidants (α-tocophorol, ascorbyl palmitate, carnosol) were more effective than their polar counterparts (Trolox, ascorbic acid, carnosic acid and rosmarinic acid) (Frankel *et al.*, 1994, 1996a, b; Hopia *et al.*, 1996; Huang *et al.*, 1996a, b).

Unfortunately, the antioxidant polar paradox does not predict antioxidant activity in oil-in-water emulsions with 100% accuracy. For example, esterification of a hydrocarbon chain onto a free radical scavenger to alter the polarity of the antioxidant can cause unpredicted changes in antioxidant activity (Yuji *et al.*, 2007; Hunneche *et al.*, 2008; Laguerre *et al.*, 2009; Medina *et al.*, 2009; Sasaki *et al.*, 2010). This is especially true when the antioxidant is esterified with large hydrocarbons of more than 20 carbons. If the antioxidant polar paradox was accurate, this would be the most nonpolar antioxidant and thus would be expected to be the most effective. However, the opposite is true, as the activity of the most nonpolar antioxidant esters drops dramatically compared to esters with 8–12 hydrocarbon tails (Laguerre *et al.*, 2009). This is thought to be due to the most nonpolar antioxidants partitioning into the center of the emulsion droplet core, where they are unable to scavenge free radicals generated on the surface of the emulsion droplet (Panya *et al.*, 2012). Surfactant micelles that are able to solubilize the most nonpolar esters out of the emulsion core will dramatically increase antioxidant activity (Panya *et al.*, 2012). This suggests that any factors in delivery systems that increase the partitioning of free-radical scavenging antioxidants at the oil–water interface have the potential to increase the oxidative stability of dispersed lipids.

Since iron is an important prooxidant in O/W emulsions, chelating agents can be very effective antioxidants. This can occur by the chelator decreasing metal reactivity or by physically partitioning the metal away from the lipid (Mei *et al.*, 1998b; Mancuso *et al.*, 1999; Jacobsen *et al.*, 2001; Cho *et al.*, 2003; Nielsen *et al.*, 2004). Potential chelators include EDTA, organic acid, polyphosphates, and proteins. Since many delivery systems are stabilized with proteins, these ingredients can have a dual role in providing both physical stability and metal chelation capacity. This is especially true when excess protein is present in the emulsion, such that the oil droplet surface becomes saturated with protein and the remaining protein partitions into the aqueous phase. This aqueous phase protein can partition metals away from the emulsion droplet and in some cases decrease iron reactivity (Faraji *et al.*, 2004; Elias *et al.*, 2006). This is often an overlooked factor in the oxidative stability of protein-stabilized delivery systems, with researchers sometimes attributing increased oxidative stability to the physical properties of the emulsion when in fact it is due to proteins in the continuous phase. Of course proteins will only inhibit lipid oxidation by metal chelation when the pH is above the pI of the protein, and thus the protein can bind metals. Proteins can also scavenge free radicals, so it is possible that they can contribute to the antioxidant capacity of a delivery system even when they are positively charged (Faraji *et al.*, 2004; Elias *et al.*, 2006).

8.5 CONCLUSIONS

Lipid oxidation is a major problem leading to deterioration of polyunsaturated lipids. These oxidative reactions need to be controlled because they cause undesirable changes in flavor, texture, appearance, and nutritional quality of food products. Since many lipid-containing foods are in emulsified form, a thorough understanding of lipid oxidation mechanisms in emulsions should be developed in order to develop innovative technologies to solve this problem. Numerous techniques can be applied to inhibit lipid oxidation in oil-in-water emulsions, including interfacial engineering to control the composition, thickness, or charge of the interfacial layer that separates the encapsulated lipids from the surrounding aqueous phase. The selection of antioxidants is also important to ensure that they are located at the major site of the lipid oxidation reaction. Due to the high susceptibility of emulsified lipids to oxidation, it might be necessary to use combinations of the techniques mentioned in this chapter to effectively retard lipid oxidation and improve the shelf-life, utilization, and quality of food emulsion systems. For example, use of a combination of cationic interfacial membrane, emulsion droplet antioxidant (tocopherols), and metal chelators (EDTA) can be an extremely effective way to protect ω-3 fatty acids in emulsions (Djordjevic *et al.*, 2004). Structured emulsions, such as filled hydrogel particles, multilayer emulsions, or solid lipid particles, may also be useful in protecting lipids against chemical degradation but further work is needed in this area to evaluate their effectiveness. On the other hand, microemulsions and nanoemulsions might not be good for protecting oxidatively labile bioactive compounds, due to their large surface area and the fact that they form clear solutions where light could readily interact with the emulsion droplets. An example of nanoencapsulation to improve folate stability is presented in Chapter 9.

References

Almeida, C., Vital, C., Contente, T., Maria, D. and Maranhao, R. (2010) Modification of composition of a nanoemulsion with different cholesteryl ester molecular species: Effects on stability, peroxidation, and cell uptake. *International Journal of Nanomedicine*, 5, 679–686.

Boon, C., Xu, Z., Yue, X., McClements, D., Weiss, J. and Decker, E. (2008) Factors affecting lycopene oxidation in oil-in-water emulsions. *Journal of Agricultural and Food Chemistry*, 56, 1408–1414.

Chaiyasit, W., Silvestre, M., McClements, D. and Decker, E. (2000) Ability of surfactant tail group size to alter lipid oxidation in oil-in-water emulsions. *Journal of Agricultural and Food Chemistry,* 48, 3077–3080.

Chaiyasit, W., Elias, R., McClements, D. and Decker, E. (2007) Role of physical structures in bulk oils on lipid oxidation. *Critical Reviews in Food Science and Nutrition*, 47, 299–317.

Cho, Y.-J., Alamed, J., McClements, D. and Decker, E. (2003) Ability of chelators to alter the physical location and prooxidant activity of iron in oil-in-water emulsions. *Journal of Food Science,* 68, 1952–1957.

Coupland, J., Zhu, Z., Wan, H., McClements, D., Nawar, W. and Chinachoti, P. (1996) Droplet composition affects the rate of oxidation of emulsified ethyl linoleate. *Journal of the American Oil Chemists Society*, 73, 795–801.

Dimakou, C., Kiokias, S., Tsaprouni, I. and Oreopoulou, V. (2007) Effect of processing and storage parameters on the oxidative deterioration of oil-in-water emulsions. *Food Biophysics,* 2, 38–45.

Djordjevic, D., Cercaci, L., Alamed, J., McClements, D. and Decker, E. (2007). Chemical and physical stability of citral and limonene in sodium dodecyl sulfate chitosan and gum arabic-stabilized oil-in-water emulsions. *Journal of Agricultural and Food Chemistry,* 55, 3585–3591.

Djordjevic, D., Kim, H.-J., McClements, D. and Decker, E. (2004) Oxidative stability of whey protein-stabilized oil-in-water emulsions at pH 3: potential ω-3 fatty acid delivery systems (part B). *Journal of Food Science,* 69, C356–362.

Djordjevic, D., Cercaci, L., Alamed, J., McClements, D. and Decker, E. (2008). Stability of citral in protein- and gum arabic-stabilized oil-in-water emulsions. *Food Chemistry,* 106, 698–705.

Donnelly, J., Decker, E. and McClements, D. (1998) Ability of emulsifiers to influence iron-catalyzed oxidation of emulsified menhaden oil. *Journal of Food Science,* 63, 997–1000.

Elias, R., Bridgewater, J., Vachet, R., Waraho, T., McClements, D. and Decker, E. (2006) Antioxidant mechanisms of enzymatic hydrolysates of β-lactoglobulin in food lipid dispersions. *Journal of Agricultural and Food Chemistry,* 54, 9565–9572.

Faraji, H., McClements, D. and Decker, E. (2004) Role of continuous phase protein on the oxidative stability of fish oil-in-water emulsions. *Journal of Agricultural and Food Chemistry,* 52, 4588–4564.

Frankel, E. (1998) Free radical oxidation. In: *Lipid Oxidation.* (ed. E.N. Frankel), The Oil Press, Dundee, Scotland,13–22.

Frankel, E., Huang, S., Kanner, J. and German, J. (1994) Interfacial phenomena in the evaluation of antioxidants: bulk oils vs. emulsions. *Journal of Agricultural and Food Chemistry,* 42, 1054–1059.

Frankel, E., Huang, S., Aeschbach, R. and Prior, E. (1996a) Antioxidant activity of a rosemary extract and its constituents, carnosic acid, carnosol and rosmarinic acid, in bulk oil and oil-in-water emulsion. *Journal of Agricultural and Food Chemistry,* 44, 131–135.

Frankel, E., Huang, S., Prior, E. and Aeschbach, R. (1996b) Evaluation of antioxidant activity of rosemary extracts, carnosol and carnosic acid in bulk vegetable oils and fish oil and their emulsions. *Journal of the Science of Food and Agriculture,* 72, 201–208.

German, G. and Creveling, R. (1990) Identification and characterization of a 15-lipoxygenase from fish gills. *Journal of Agricultural and Food Chemistry*, 18, 2144–2147.

Girotti, W. (1998) Lipid hydroperoxide generation, turnover, and effector action in biological systems. *Journal of Lipid Rearch*, 39, 1529–1542.

Gohtani, S., Sirendi, M., Yamamoto, N., Kajikawa, K. and Yamamo, Y. (1999) Effect of droplet size on oxidation of Docosahexaenoic acid in emulsion system. *Journal of Dispersion Science and Technology*, 20, 1319–1325.

Halliwell, B. and Gutteridge, J. (1990) Role of free radicals and catalytic metal ions in human disease: an over view. *Methods in Enzymology*, 186, 1–85.

Haahr, A. and Jacobsen, C. (2008) Emulsifier type, metal chelation and pH affect oxidative stability of n-3-enriched emulsions. *European Journal of Lipid Science and Technology*, 110, 949–961.

Hegenauer, J., Saltman, P., Ludwig, D., Ripley, L. and Bajo, P. (1979) Effect of supplemental iron and lipid oxidation in milk. 1. Comparison of metal complexes in emulsified and homogenized milk. *Journal of Agricultural and Food Chemistry*, 27, 860–867.

Helgason, T., Awad, T., Kristbergsson, K., Decker, E., McClements, D. and Weiss, J. (2009) Impact of surfactant properties on oxidative stability of beta-carotene encapsulated within solid lipid nanoparticles. *Journal of Agricultural and Food Chemistry*, 57, 8033–8040.

Hopia, A., Huang, S., Schwarz, K., German, J. and Frankel, E. (1996) Effect of different lipid systems on antioxidant activity of rosemary constituents carnosol and carnosic acid with and without alpha-tocopherol. *Journal of Agricultural and Food Chemistry*, 44, 2030–2036.

Hu, M., McClements, D. and Decker, E. (2002) Lipid oxidation in corn oil-in-water emulsions stabilized by casein, whey protein isolate, and soy protein isolate. *Journal of Agricultural and Food Chemistry*, 51, 1696–1700.

Hu, M., McClements, D. and Decker, E. (2003) Impact of whey protein emulsifiers on the oxidative stability of salmon oil-in-water emulsions. *Journal of Agricultural and Food Chemistry*, 51, 1435–1439.

Hu, M., McClements, D. and Decker, E. (2004) Impact of chelators on the oxidative stability of whey protein isolate-stabilized oil-in-water emulsions containing ω-3 fatty acid. *Food Chemistry*, 88, 57–62.

Huang, S. W., Frankel, E., Schwarz, K. and German, J. (1996a) Effect of pH on antioxidant activity of α-tocopherol and Trolox in oil-in-water emulsions. *Journal of Agricultural and Food Chemistry*, 44, 2496–2502.

Huang, S., Hopia, A., Schwarz, K., Frankel, E. and German, J. (1996b) Antioxidant activity of α-tocopherol and Trolox in different lipid substrates: bulk oils vs. oil-in-water emulsions. *Journal of Agricultural and Food Chemistry*, 44, 444–452.

Hunneche, C., Lund, M., Skibsted, L. and Nielsen, J. (2008) Antioxidant activity of a combinatorial library of emulsifier-antioxidant bioconjugates. *Journal of Agricultural and Food Chemistry*, 56, 9258–9268.

Imai, H., Maeda, T., Shima, M. and Adachi, S. (2008) Oxidation of methyl linoleate in oil-in-water micro- and nanoemulsion systems. *Journal of the American Oil Chemists Society*, 85, 809–815.

Jacobsen, C., Timm, M. and Meyer, A. (2001) Oxidation in fish oil enriched mayonnaise: ascorbic acid and low pH increase oxidative deterioration. *Journal of Agricultural and Food Chemistry*, 49, 3947–3956.

Katsuda, M., McClements, D., Miglioranza, L. and Decker, E. (2008) Physical and oxidative stability of fish oil-in-water emulsions stabilized with β-lactoglobulin and pectin. *Journal of Agricultural and Food Chemistry*, 56, 5926–5931.

Kiokias, S., Dimakou, C. and Oreopoulou, V. (2007) Effect of heat treatment and droplet size on the oxidative stability of whey protein emulsions. *Food Chemistry*, 105, 94–100.

Klinkesorn, U., Sophanodora, P., Chinachoti, P., Decker, E. and McClements, D. (2005a) Encapsulation of emulsified tuna oil in two-layered interfacial membranes prepared using electrostatic layer-by-layer deposition. *Food Hydrocolloids*, 19, 1044–1053.

Klinkesorn, U., Sophanodora, P., Chinachoti, P., McClements, D. and Decker, E. (2005b) Increasing the oxidative stability of liquid and dried tuna oil-in-water emulsions with electrostatic layer-by-layer deposition technology. *Journal of Agricultural and Food Chemistry*, 53, 4561–4566.

Laguerre, M., Lopez Giraldo, L., Lecomte, J., Figueroa-Espinoza, M., Barea, B., Weiss, J., Decker, E. and Villeneuve, P. (2009) Chain length affects antioxidant properties of chlorogenate esters in emulsion: the cutoff theory behind the polar paradox. *Journal of Agricultural and Food Chemistry*, 57, 11335–11342.

Lamprecht, A., Schafer, U. and Lehr, C. (2001) Influences of process parameters on preparation of microparticle used as a carrier system for Omega-3 unsaturated fatty acid ethyl esters used in supplementary nutrition. *Journal of Microencapsulation*, 18, 347–357.

Lee, J. and Decker, E. (2011) Effects of metal chelator, sodium azide, and superoxide dismutase on the oxidative stability in riboflavin-photosensitized oil-in-water emulsion systems. *Journal of Agricultural and Food Chemistry*, 59, 6271–6276.

Lethuaut, L., Metro, F. and Genot, C. (2002) Effect of droplet size on lipid oxidation rates of oil-in-water emulsions stabilized by protein. Journal of the American Oil Chemists Society, 79, 425–430.

Mancuso, J., McClements, D. and Decker, E. (1999) The effects of surfactant type, pH, and chelators on the oxidant of salmon oil-in-water emulsions. *Journal of Agricultural and Food Chemistry,* 47, 4112–4116.

Mancuso, J., McClements, D. and Decker, E. (2000) Iron accelerated cumene hydroperoxide decomposition in hexadecane and trilaurin emulsions. *Journal of Agricultural and Food Chemistry,* 48, 213–219.

Mason, T., Wilking, J., Meleson, K., Chang, C. and Graves, S. (2006) Nanoemulsions: formation, structure, and physical properties. *Journal of Physics: Condensed Matter*, 18, R635-R666.

Matalanis, A., Jones, O. and McClements, D. (2011) Structured biopolymer-based delivery systems for encapsulation, protection, and release of lipophilic compounds. *Food Hydrocolloids,* 25, 1865–1880.

Matalanis, A., Decker, E.and McClements, D. (2012) Inhibition of lipid oxidation by encapsulation of emulsion droplets within hydrogel microspheres. *Food Chemistry*, 132, 766–772.

McClements, D. (2012) Nanoemulsions versus microemulsions: terminology, differences, and similarities. *Soft Matter,* 8, 1719–1729.

McClements, D. and Decker, E. (2000) Lipid oxidation in oil-in-water emulsions: impact of molecular environment on chemical reactions in heterogeneous food systems. *Journal of Food Science,* 65, 1270–1282.

McClements, D., Decker, E. and Weiss J. (2007) Emulsion-based delivery systems for lipophilic bioactive components. *Journal of Food Science,* 72, R109-R124.

Medina, I., Lois, S., Alcantara, D., Luca, R. and Morales, J. (2009) Effect of lipophilization of hydroxytyrosol on its antioxidant activity in fish oils and fish oil-in-water emulsions. *Journal of Agricultural and Food Chemistry,* 57, 9773–9779.

Mei, L., McClements, D., Wu, J. and Decker, E. (1998a) Iron-catalyzed lipid oxidation in emulsions as affected by surfactant, pH and NaCl. *Food Chemistry,* 61, 307–312.

Mei, L., Decker, E. and McClements, D. (1998b) Evidence of iron association with emulsion droplets and its impact on lipid oxidation. *Journal of Agricultural and Food Chemistry,* 46, 5072–5077.

Min, B. and Boff, M. (2002) Chemistry and reaction of singlet oxygen in foods. *Comprehensive Reviews in Food Science and Food Safety*, 1, 58–72.

Miyashita, K., Nara, E. and Ota, T. (1993) Oxidative stability of free fatty acids mixtures from soybean, linseed and sardine oils in aqueous solution. *Bioscience Biotechnology and Biochemistry,* 60, 315–318.

Miyashita, K., Tateda, T. and Ota, T. (1994) Oxidative stability of polyunsaturated fatty acids in aqueous solution. *Bioscience Biotechnology and Biochemistry,* 57, 1638–1640.

Nakaya, K., Ushio, H., Matsukawa, S., Shimizu, M. and Ohshima, T. (2005) Effects of droplet size on the oxidative stability of oil-in-water emulsions. *Lipids*, 40, 501–507.

Nielsen, N., Petersen, A., Meyer, A., Timm-Heinrich, M. and Jacobsen, C. (2004) Effects of lactoferrin, phytic acid, and EDTA on oxidation in two food emulsions enriched with long-chain polyunsaturated fatty acids. *Journal of Agricultural and Food Chemistry,* 52, 7690–7699.

Nuchi, C., McClements, D. and Decker, E. (2001) Impact of Tween 20 hydroperoxides and iron on the oxidation of methyl linoleate and salmon oil dispersions. *Journal of Agricultural and Food Chemistry,* 49, 4912–4916.

Nuchi, C., Hernandez, D., McClements, D. and Decker, E. (2002) Ability of lipid hydroperoxides to partition into surfactant micelles and alter lipid oxidation rates in emulsions. *Journal of Agricultural and Food Chemistry,* 51, 5522–5527.

Okuda, S., McClements, D. and Decker, E. (2005) Impact of lipid physical state on the oxidation of methyl linoleate in oil-in-water emulsions. *Journal of Agricultural and Food Chemistry,* 53, 9624–9628.

Osborn, H. and Akoh, C. (2004) Effect of emulsifier type, droplet size, and oil concentration on lipid oxidation in structured lipid-based oil-in-water emulsions. *Food Chemistry,* 84, 451–456.

Panya, A., Laguerre, M., Bayrasy, C., *et al.* (2012) An investigation of the versatile antioxidant mechanisms of action of rosmarinate alkyl esters in oil-in-water emulsions. *Journal of Agricultural and Food Chemistry,* 60, 2692–2700.

Paraskevopoulou, D., Boskou, D. and Paraskevopoulou, A. (2007) Oxidative stability of olive-lemon juice salad dressings stabilized with polysaccharides. *Food Chemistry,* 101, 1197–1204.

Pich, A. and Adler, H. (2007) Composite aqueous microgels: an overview of recent advances in synthesis, characterization and application. *Polymer International,* 56, 291–307.

Porter, W., Black, E. and Drolet, A. (1989) Use of polyamide oxidative fluorescence test on lipid emulsion: contrast in relative effectiveness of antioxidants in bulk versus dispersed systems. *Journal of Agricultural and Food Chemistry,* 37, 615–624.

Qian, C., Decker, E., Xiao, H. and McClements, D. (2012) Physical and chemical stability of beta-carotene-enriched nanoemulsions: influence of pH, ionic strength, temperature, and emulsifier type. *Food Chemistry*, 132, 1221–1229.

Sasaki, K., Alamed, J., Weiss, J., *et al.* (2010) Relationship between the physical properties of chlorogenic acid esters and their ability to inhibit lipid oxidation in oil-in-water emulsions. *Food Chemistry*, 118, 830–835.

Shaw, L., McClements, D. and Decker, E. (2007) Spray-dried multilayered emulsions as a delivery method for ω-3 fatty acids into food systems. *Journal of Agricultural and Food Chemistry*, 55, 3112–3119.

Shimada, H., Okada, K., Matsuo, K. and Yoshioka, S. (1996) Involvement of chelating action and viscosity in the antioxidative effect of xanthan in an oil/water emulsion. *Bioscience Biotechnology and Biochemistry*, 60, 125–127.

Silvestre, M., Chaiyasit, W., Brannan, R., McClements, D. and Decker, E. (2000) Ability of surfactant head group size to alter lipid and antioxidant oxidation in oil-in-water emulsions. *Journal of Agricultural and Food Chemistry*, 48, 2057–2061.

Sun, C. and Gunasekaran, S. (2009) Effect of protein concentration and oil-phase volume fraction on the stability and rheology of menhaden oil-in-water emulsions stabilized by whey protein isolate with xanthan gum. *Food Hydrocolloids*, 23, 165–174.

Trunova, N., Kartasheva, Z., Maksimova, T., Bogdanova, Y. and Kasaikina, O. (2007) Decomposition of cumene hydroperoxide in the systems of normal and reverse micelles formed by cationic surfactants. *Colloid Journal*, 69, 655–659.

Villiere, A., Viau, M., Bronnec, I., Moreau, N. and Genot, C. (2005) Oxidative stability of bovine serum albumin- and sodium caseinate-stabilized emulsions depends on metal availability. *Journal of Agricultural and Food Chemistry*, 53, 1514–1520.

Waraho, T., McClements, D.and Decker, E. (2011) Mechanisms of lipid oxidation in food dispersions. *Trends in Food Science and Technolonogy*, 22, 3–13.

Weinbreck, F., Minor, M. and De Kruif, C. (2004) Microencapsulation of oils using whey protein/gum arabic coacervates. *Journal of Microencapsulation*, 21, 667–679.

Wissing, S., Kayser, O. and Muller, R. (2004) Solid lipid nanoparticles for parenteral drug delivery. *Advanced Drug Delivery Reviews*, 56, 1257–1272.

Wu, K., Chai, X. and Chen, Y. (2005) Microencapsulation of fish oil by simple coacervation of hydroxypropyl methylcellulose. *Chinese Journal of Chemistry*, 23, 1569–1572.

Yoshida, Y. and Niki, E. (1992) Oxidation of methyl linoleate in aqueous dispersions induced by copper and iron. *Archives of Biochemistry and Biophysic*, 295, 107–114.

Yucel, U., Elias, R. and Coupland, J. (2012) Solute distribution and stability in emulsion-based delivery systems: an EPR study. *Journal of Colloid and Interface Science*, 377, 105–113.

Yuji, H., Weiss, J., Villeneuve, P., Lopez Giraldo, L., Figueroa-Espinoza, M. and Decker, E. (2007) Ability of surface-active antioxidants to inhibit lipid oxidation in oil-in-water emulsion. *Journal of Agricultural and Food Chemistry*, 55, 11052–11056.

Zhuang, H., Barth, M. and Hildebrand, D. (2002) Fatty acid oxidation in planttissues. In: *Food lipids: Chemistry, Nutrition, and Biotechnology*, 2nd edn (ed. C.C. Akoh and D.B. Min). Marcel Dekker, New York, 413–464.

9 Encapsulation strategies to stabilize a natural folate, L-5-methyltetrahydrofolic acid, for food fortification practices

David D. Kitts and Yazheng Liu

Food Science, Food, Nutrition and Health, The University of British Columbia, Vancouver, British Columbia, Canada

9.1 INTRODUCTION

Folates, or natural folates, are pteroylpolyglutamates, which exist in mixtures of reduced, one-carbon substituted forms of up to eight glutamyl moieties bound to *p*-aminobenzoic acid (*p*-ABA) (Figure 9.1). Dietary folate is absorbed at the jejunum segment of the small intestine, where it is enzymatically reduced to dihydrofolic acid and can be further reduced to the active form, tetrahydrofolic acid. There are many different forms of folate made available through metabolism, however, the monoglutamate, the alkyl (5-methyl-tetrahydrofolic acid) and the formyl forms are the most common species absorbed. The synthetic pathways for *de novo* folate synthesis are absent in mammals, thus the human diet, which includes beef liver, leafy green vegetables, citrus fruits and lentils, are the major sources of folates. Natural folates are unstable, prone to oxidation and usually undergo significant losses in foods that undergo thermal processing or exposure to moisture. Folic acid (Figure 9.2), or pteroylmonoglutamin acid, is a fully oxidized, synthetic form of the vitamin B9 and often used for vitamin supplements, and as a food fortificant. Folic acid contains a monoglutamate molecule, while naturally occurring folates contain additional glutamate residues, rendering them less bioavailable compared to folic acid.

9.2 FOLATE FORTIFICATION MANDATES

Due to the limited bioavailability and large losses that occur with natural folates during food processing, storage, cooking, or formulation of ingredient-based foods, it is challenging to reach the recommended intake levels for consumers. The requirement for folate during the peri-conception and pregnancy periods is to prevent folate deficiency anemia and neural tube defects (MRC Vitamin Study Research Group, 1991). In 1998, both Health Canada and the US Food and Drug Administration issued regulations requiring mandatory fortification of foodstuffs, especially grains used in breakfast cereals, rice, and noodles (Neuhouser and Beresford, 2001). Today, Health Canada has in place mandatory food fortification programs with folic acid as the fortifcant in cereals and grains, which are the primary

Nanotechnology and Functional Foods: Effective Delivery of Bioactive Ingredients, First Edition.
Edited by Cristina M. Sabliov, Hongda Chen, and Rickey Y. Yada.

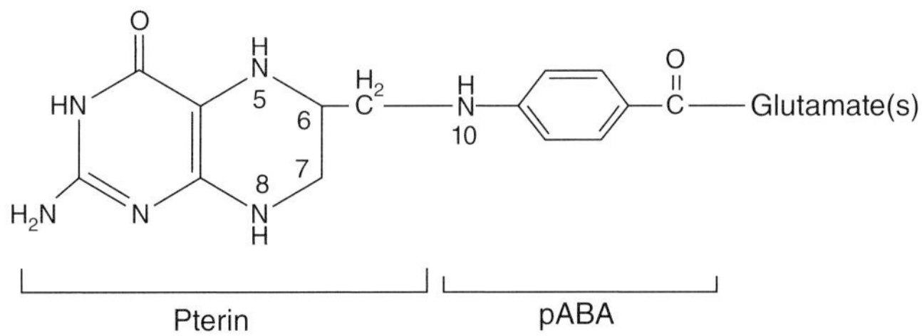

Figure 9.1 Chemical structure of natural folates. Folates are tripartite molecules comprising pterin, *p*-aminobenzoate (pABA) and glutamate moieties. In natural folates, the pterin ring exists in tetrahydro form (as shown) or in 7,8-dihydro form, and a linked polyglutamyl tail of up to eight moieties bound to pABA. One-carbon units (formyl, methyl, etc.) can be coupled to the N5 and/or N10 positions. The pterin ring is fully oxidized in folic acid, which is not a natural folate.

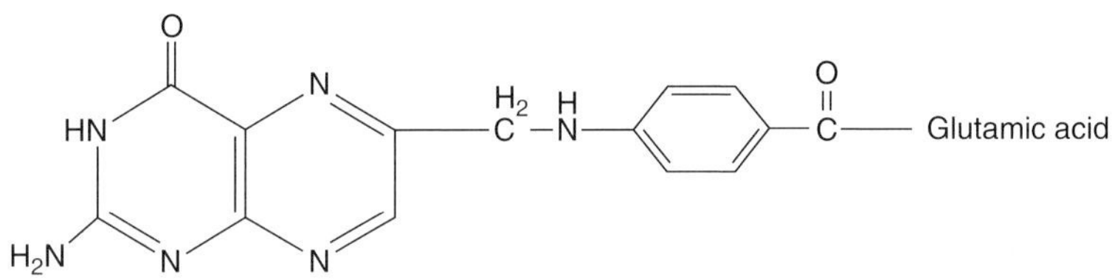

Figure 9.2 Chemical structure of folic acid.

sources for flour (e.g. 150 μg folic acid/100 g flour), pasta (200 μg folic acid/100 g pasta) and cornmeal-derived products (Health Canada, 2004). The choice of folic acid as the fortification of choice is based on the fact that it is inexpensive to synthesize and it is also relatively stable and resistant to destruction, thus it exhibits better bioavailability compared to naturally occurring folates.

Folates are an essential cofactor for *de novo* biosynthesis of purines and thymidylate that is required for DNA synthesis and replication. It is therefore especially important during the peri-conception period, as mentioned above, where supplementation of women with folic acid has a particular importance to reduce the risk of neural tube diseases, such as spina bifida and anencephaly, initiated at around 24 days after conception. Many health organizations recommend that women take 400 μm synthetic folic acid daily in addition to consuming naturally occurring folates from a varied diet. For pregnant women, this increases to 600 μg folic acid and 500 μg folic acid for lactating women (Health Canada, 2010). Mandatory folic acid fortification has most importantly been shown to successfully increase folate status, as indexed by increased whole folate concentrations in blood serum, plasma and red blood cells, and a decrease in plasma homocysteine levels (Jacques *et al.*, 1999: Venn *et al.*, 2003). Folate deficiency has been implicated in cardiovascular disease, which is associated with elevated homocysteine concentrations (Brattström and Wilcken, 2000, Vasan, 2003). Supplementation with folic acid or fully reduced folate has shown improved endothelial function and reduced risk for atherosclerosis due to hypercholesterolemia (Houghton *et al.*, 2006).

However, concerns also exist that the increased dietary intakes of folic acid may also result in masked symptoms of pernicious anemia caused by vitamin B_{12} deficiency (Ross *et al.*, 1948, Hoffbrand and Jackson, 1993, Savaria Morris *et al.*, 2007). This is a particular relevant for seniors where 10–15% of the population over the age of 60 has been diagnosed with vitamin B_{12} deficiency (Food and Nutrition Board, 1998). High intakes of folic acid are reduced to tetrahydrofolic acid that produces an influx of tetrahydrofolic acid, which in turn avoids the requirement for 5-methytetrahydrofolic acid to be recycled back to tetrahydrofolic acid; a reaction step that involves vitamin B_{12}. Hence by avoiding the apparent need for folate recycling that requires vitamin B_{12}, a potential masking of vitamin B_{12} can occur. Symptoms of this deficiency appear much later on as neurological lesions. There is also some concern that excess folic acid intakes may have a role in enhancing the development and progression of already existing, but undiagnosed, premalignant lesions (Kim, 2004). To overcome these potential health concerns, it has been recommended that the current folate fortificant, folic acid, be replaced with L-5-methyltetrahydrofolic acid (L-5-MTHF), which is equally effective at preventing neural tube defects and does not pose a risk for masking vitamin B_{12} deficiency (Scott, 2001; Venn *et al.*, 2003; Pietrzik *et al.*, 2010).

9.3 LIMITATIONS OF FORTIFYING FOODS WITH NATURAL FOLATES

L-5-MTHF is a natural folate and abundantly present in plant and animal food systems. However, L-5-MTHF is sensitive to oxygen, heat, and light, and is rapidly degraded during cooking and food processing (Hawkes and Villota, 1989; Dang *et al.*, 2000; Nguyen and Indrawati, 2003; Liu *et al.*, 2012). L-5-MTHF exhibits a half-life of 20 min at 50°C at pH 6.8 and 80 min at pH 4.0 in buffer solutions, respectively (Figure 9.3). L-5-MTHF is relatively more stable in an acidic environment, compared to more neutral pH environments, where it exists in its anionic form, thus favoring the process of electron movement and ultimate oxidation to L-5-methyl-5,6-dihydrofolate. The latter compound can be reduced back to L-5-MTHF by reducing agents, such as ascorbate and β-mercaptoethanol (Liu *et al.*, 2012).

An example of the relative stabilities of free L-5-MTHF compared to folic acid is shown with bread and noodles; two food products that involve heat process treatments and exposure of nutrients to moisture. L-5-MTHF retention in both baked breads and noodles is considerably lower than folic acid during these food preparations, cooking process and storage (Tables 9.1 and 9.2). The average retention of L-5-MTHF in freshly made breads and after 3 and 7 days storage at room temperature was about 53%, 21% and 10%, respectively, while retention of folic acid was about 90% upon 7 days storage. A previous

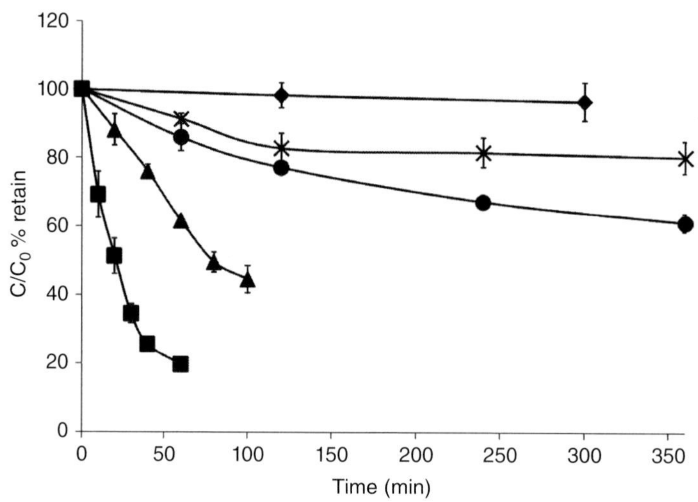

Figure 9.3 Degradation of folic acid and L-5-MTHF in buffer, milk and soy milk at 50°C. ◆, folic acid in pH 6.8 buffer; ■, L-5-MTHF in pH 6.8 buffer; ▲, L-5-MTHF in pH 4.0 buffer; ✳, L-5-MTHF in milk; ●, L-5-MTHF in soy milk.

Table 9.1 Recovery of L-5-MTHF or folic acid in breads after baking and storage at room temperature.[1]

Folate	Day 0 (fresh)	Day 3	Day 7
L-5-MTHF	53.1 ± 2.5^{ax}	21.2 ± 2.7^{ay}	9.8 ± 3.2^{az}
Folic acid	88.8 ± 1.2^{bx}	75.9 ± 3.2^{by}	69.0 ± 2.3^{bz}

[1] Values are means ± SD of three baking processes. Breads were baked at 365°C for 25 min. Data are expressed as per cent of L-5-MTHF or folic acid spiked to the bread.
[ab] Mean values within each column not sharing a common superscript were significantly different and [xy] mean values within each row not sharing a common superscript were significantly different (two-way ANOVA, followed by Bonferroni post-test, $p < 0.05$).

Table 9.2 Recovery of L-5-MTHF or folic acid in fresh and cooked noodles.[1]

Folate	Fortification level mg/Kg	Fresh noodle mg/Kg	Cooked noodle mg/Kg
L-5-MTHF	5.2 ± 0.4^{x}	3.7 ± 0.3^{y}	1.6 ± 0.2^{z}
Folic acid	4.6 ± 0.4^{x}	4.3 ± 0.9^{x}	4.0 ± 0.7^{x}

[1] Measured values are means ± SD for four noodle batches. Noodles were made by a home style noodle maker and 100g of the noodles was boiled at 92°C for 13 min. Data are expressed as mg L-5-MTHF or folic acid/kg flour.
[xyz] In the same row indicate significant difference (one-way ANOVA, followed by Bonferroni post-test, $p < 0.05$).

study showed that reduced loaf weight, thus, shortened baking time, actually resulted in an increase in the retention of L-5-MTHF (Ohrvik *et al.*, 2010). Some success has been made by adding ascorbic acid to protect folates from oxidation throughout the processing of a variety of foods (Sotiriadis and Hoskins, 1982; Tomiuk *et al.*, 2012), but this practice has limited success.

9.4 ENCAPSULATION STRATEGIES

Encapsulation technologies, whether they involve micro- or nanocapsules, have in common the combination of a bioactive core material that is encased by a protective shell, or aggregate structure, and which acts to enable the protection and controlled delivery of the bioactive core agent (Favaro-Trindale and Grosso, 2002, Champagne and Fustier, 2007, Oliver and Augustin, 2009). This process has been very useful to protect bioactive and functional food ingredients, such as probiotics, flavors, and vitamins (Jackson and Lee, 1991). In addition to protecting the bioactive core material from interacting with other reactive components in the food matrix, encapsulation using surface-active biopolymers can enhance efficacy by improving solubility and increasing the active lifetime of the active agent. The use of spray drying, liposome entrapment, coacervation, and extrusion coating are some examples of encapsulation methods used by the food industry (Oliver and Augustin, 2009). In more recent years, the application of nanotechnology has been used to develop nanodosage forms of herbal drugs, which have been shown to have advantages over more conventional formats (Ajazuddin and Saraf, 2010). Nanoparticles that vary in size, ranging from 10 to 1000 nm, are referred to as both nanospheres and nanocapsules (Ajazuddin and Saraf, 2010). Nanoparticles can be recovered using spray drying of solutions or emulsions that have undergone a rigorous process to reduce particle size, which can be achieved using ultrasonication or microfluidization. Nanospheres disperse the bioactive ingredient throughout the particle, whereas nanocapsules contain it within a core. Some of the many advantages of nanoparticles include improved bioavailability, solubility, stability, and pharmacological activity, as well as sustained delivery protection from chemical and physical gradation. Nanocarriers can also be utilized in delivery of bioactive ingredients or nutrients by employing liposomal or encapsulated particles. Similarly, microemulsions with droplets that range from 5 to 50 μm in diameter have been shown to be more thermodynamically stable than both conventional emulsions and nanoemulsions (McClements and Rao, 2011). Studies have shown that encapsulated materials diffuse quickly from microemulsion droplets, due to the greater surface-to-mass ratio (130–1300 m^2/g), compared to nanoemulsions (70–330 m^2/g).

9.5 COATING MATERIALS FOR ENCAPSULATION OF L-5-MTHF

Encapsulation of natural folates using micro-sized particles have recently been used to stabilize and protect L-5-MTHF for food fortification purposes that involve thermal treatment or exposure to moisture (Tomiuk *et al.*, 2012; Liu *et al.*, 2013). A number of surface-active biopolymers were examined in our laboratory to determine the relative yield for optimal recovery of the natural folate, L-5-MTHF, from the coating materials for usage as potential fortificants (Table 9.3). An inherent problem that requires a solution before being able to claim that the encapsulation functions as a microparticle or a nanoparticle is also the aggregation tendency characteristics, which are specific to the coating materials. In our studies, we were relatively unsuccessful to achieve encapsulation at the nanoscale. Instead, microcapsules prepared by spray drying with particle sizes ranging from 10 to 50 μm diameter were used to encapsulate the natural folate.

Pectin and alginate are materials often used for encapsulation of bioactives due to relative biocompatibility and low cost (Nazzaro *et al.*, 2009). We chose pectin, due to the high affinity to form a viscous gel, which would act to strongly encase the natural folate in a durable matrix. Pectin gels were formed by adding divalent calcium ions which enabled a cross-linking or an "egg-box" structure that facilitated the required coating strength. This coating material would be expected to enhance particle solubility and thus bioavailability by affecting optimal release of the bioactive compound contained within the carrier. Using 3 and 4% pectin as coating materials, respectively, proved to produce small (e.g., < 1%) losses during the encapsulation process, but greater (29% and 13%) losses occurred after the slurry material was vacuum dried at room temperature. Other workers, who used an 80:20 pectin:alginate ratio reported approximately 60% recovery after spray drying (Madziva *et al.*, 2005). It is likely that the poor recovery of the natural folate in this example is due to the instability of the L-5-MTHF in an aqueous dispersion system and the exposure to heat that was incurred during the spray-drying process. We also observed that this encapsulation process did not provide uniform mixing of the encapsulated natural folate material into the rest of the formulation, due a tendency to aggregate into even larger, nonuniform-sized particles. This result was viewed as an additional disadvantage, due to the complication that only a very small absolute amount of the folate is required for fortification at the levels required to meet daily requirements.

Table 9.3 Recovery of L-5-MTHF using different coating material.

Coating material[1]	Coating ratio[2]	Per cent recovery[3]
Pectin	0.4%	70
Pectin	0.2%	87
Stearate	15%	65
Skimmed milk powder	1%	90
Modified starch	10%	89
Modified starch	1%	91

[1] Pectin coated L-5-MTHF was vacuum dried and all the others were spray dried.

[2] Coating ratio = weight of L-5-MTHF/(weight of L-5-MTHF + weight of coating material).

[3] Per cent recovery = concentration of L-5-MTHF in the final encapsulation product/concentration of L-5-MTHF initially added to the dry mixture.

A second attempt to encapsulate the natural folate involved using the long chain saturated fatty acid, stearate, which was derived from soybean oil. An emulsion contained two immiscible liquids, which consisted of the natural folate dissolved in an aqueous fluid dispersed into stearate. The mixture was subsequently spray dried to produce a powder that embedded both the water-soluble and lipid-soluble materials (Figure 9.4). Encapsulated material was free flowing with a melting point of 66.8°C and a mean particle size distribution, measured by laser diffraction, of 10.9 μm. Encapsulating folates with 15% stearate produced lower L-5-MTHF recoveries than the pectin gels, which was attributed mostly to the spray-drying process. The final material also had a course texture and like the pectin powders was found to aggregate, which disabled it from being mixed into the flour as a potential vehicle for the folate fortificant. Nevertheless, good signs of release of both the natural folate and folic acid were obtained from the stearate coating material *in vitro*, using a simulated intestinal fluid at 37°C (Figure 9.5). The release patterns of L-5-MTHF and folic acid were similar, albeit that the relative percentage release of L-5-MTHF in the simulated intestinal

Figure 9.4 Image of microencapsulated L-5-MTHF with stearate coating using a scanning electron microscope. 2500× magnification; scale bar 10 μm.

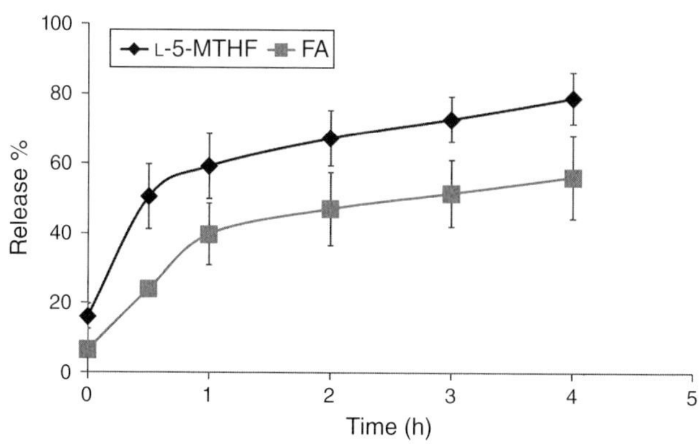

Figure 9.5 Release time-course of microencapsulated L-5-MTHF and folic acid in simulated intestinal fluid at 37°C. Microencapsulated L-5-MTHF contains 15% L-5-MTHF and 85% stearate; microencapsulated folic acid (FA) contains 25% folic acid and 75% stearate.

Table 9.4 Recovery of L-5-MTHF or folic acid with stearate coating in baked breads.[1]

Treatment[2]	Day 0 (fresh)	Day 2	Day 4
Free L-5-MTHF	76.9 ± 0.5[bx]	40.5 ± 1.4[by]	27.7 ± 2.9[bz]
Encap L-5-MTHF	70.4 ± 2.3[ax]	31.3 ± 1.2[ay]	22.9 ± 0.4[az]
Free folic acid	89.8 ± 1.0[cx]	88.5 ± 1.2[dxy]	87.3 ± 1.2[dy]
Encap Folic acid	84.4 ± 3.3[cx]	70.6 ± 1.1[cy]	64.6 ± 3.3[cz]

[1]Results are expressed as μg L-5-MTHF or folic acid per bread (mean ± SD, $n = 3$). Breads were baked at 365°C for 25 min and stored at room temperature.
[abcd]Mean values within each column not sharing a common superscript were significantly different and [xyz]mean values within each row not sharing a common superscript were significantly different (two-way ANOVA, followed by Bonferroni post-test, $p < 0.05$).

Figure 9.6 Scanning electron microscope image of microencapsulated L-5-MTHF with skimmed-milk powder. 2500x magnification; scale bar 10 μm.

fluid was greater than that of folic acid. The relatively faster initial release of L-5-MTHF was likely due to the greater amount of the L-5-MTHF that diffused close to the surface of the microcapsules. After 1 h in the simulated intestinal fluid, a similar release rate of both L-5-MTHF and folic acid was observed. The ability of the stearate to protect the natural folate present in food products that were to undergo a thermal processing procedure, however, could not be achieved. Table 9.4 shows this with the high losses of L-5-MTHF and folic acid that occurred in baked breads over 4 days of storage. The baking temperatures exceeded the melting point of the stearate, thus destroying the capsule and likely resulting in leakage of the L-5-MTHF and folic acid, which lead to subsequent thermal oxidation and degradation.

A skimmed-milk powder was also used to encapsulate L-5-MTHF (Tomiuk *et al.*, 2012). Dairy components, casein, and whey proteins have the characteristics to effectively solubilize bioactives, form gels and glassy matrices, and to stabilize interfaces. They have been used successfully to microencapsulate fish oils (Kagami *et al.*, 2003) and vitamin D (Semo *et al.*, 2007) in attempts to prevent oxidation. Baranauskienė *et al.* (2006) microencapsulated flavoring compounds such as oregano, citronella and marjoram using spray drying in the wall materials of skimmed-milk powder and whey-protein concentrate. Our previous work showed that the use of skimmed-milk powder produced excellent recovery of the natural folate after the spray-drying process was applied (Table 9.3). The scanning electron microscope (SEM) image of microencapsulated L-5-MTHF with skimmed-milk powder (SMP) is shown in Figure 9.6. Encapsulation was obtained and likely involved the globular proteins, such as β-lactoglobulin, which form covalent cross-links and a two-dimensional network to microencapsulate the natural folate. Encapsulation, however, was again at the microencapsulation scale, as nanoparticles were not recovered sufficiently to be used to provide the structure required to protect L-5-MTHF against external conditions.

The success of encapsulating L-5-MTHF using different coating materials can be effectively measured by using time-of-flight second ion mass spectrometry (ToF-SIMS). This analytical tool makes use of an Au^+ ion beam that reaches the sample surface and ionizes its components before returning back to the detector. The result is generation of mass-spectra data on the elemental and molecular composition of individual compounds and their distribution on the surface material. In addition, this analysis is also obtained at shallow depths of 2 nm (e.g., the few topmost atomic layers are sampled) from the surface. The combination of both measures provides convincing data on the success of encapsulated core materials into micro- or nanoparticles. Using ToF-SIMS in static mode, we confirmed the encapsulation of L-5-MTHF within the SMP (Figure 9.7): Figure 9.7a represents the surface spectra of the natural folate (m/z = 159), which is distinctly different from that of the skimmed-milk coating material used to microencapsulate the L-5-MTHF (Figure 9.7b). It is important to note that the surface spectra of microencapsulated L-5-MTHF shows a similar pattern to that of the skimmed-milk coating and no denoting peaks of L-5-MTHF (e.g., peaks m/z 206, 209, 211, 225, 237, 239, and 269) are present on the surface spectra of the microencapsulated L-5-MTHF (Figure 9.7c). This result indicates a successful microencapsulation of the natural folate and confirms that SMP can be a useful coating material for micro- or nanoencapuslation of L-5-MTHF.

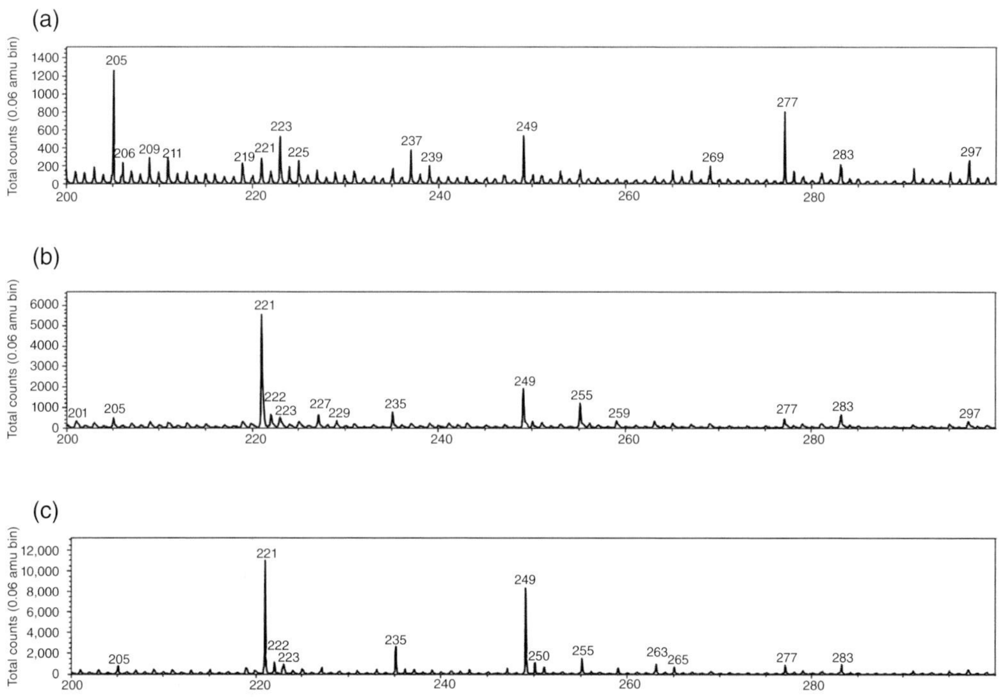

Figure 9.7 Negative ToF-SIMS spectra of the (a) L-5-MTHF, (b) skimmed-milk powder (SMP) and (c) microencapsulated L-5-MTHF with SMP. *x*-axis is m/z (mass to charge ratio); *y*-axis is total counts of pixels (0.04 amu bin). Selected spectra m/z range from 200 to 300. The ToF-SIMS surface spectrum of microencapsulated L-5-MTHF (L-5-MTHF:SMP = 1:99) shows similar pattern as that of the SMP.

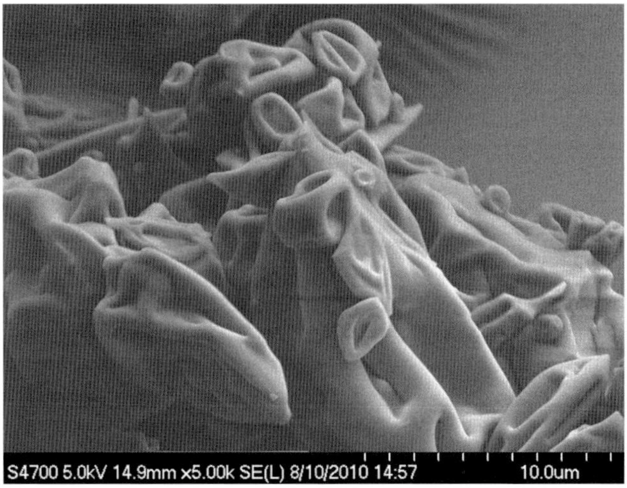

Figure 9.8 Scanning electron microscope image of microencapsulated L-5-MTHF with modified starch coating. 2500× magnification; scale bar 10 μm.

Our last coating material examined to test for feasibility of encapsulating L-5-MTHF was a modified starch, derived from waxy maize and characterized by its excellent resistance to oxidation (Partanen *et al.*, 2002a, b). The customized starch (HiCap 100) used by Liu *et al.* (2013) was a chemically modified food starch refined from waxy maize sodium octenyl succinate derivative, with a particle size of generally 30–50 μm. Other workers have used this starch for various micro- and nanoencapsulation applications that include different core materials, such as caraway (Partanen *et al.*, 2002a), sea buckthorn kernel oil (Partanen *et al.*, 2002b), black pepper oleoresin (Shaikh, 2006), and fish oil (Jafari *et al.*, 2007). It has a characteristic high thermal stability to protect core material and is also used in nanotechnologies to coat fish oil and protect the labile, highly unsaturated fatty acids. Coating L-5-MTHF with the modified starch produced microcapsules that had a high recovery of L-5-MTHF (Table 9.3) and a smooth surface (Figure 9.8). This facilitated a relatively easy and uniform blending of the encapsulated folate into the flour. The surface mass spectra of the L-5-MTHF (Figure 9.9a) and the modified starch as the coating material (Figure 9.9b) is shown using ToF-SIMS. The primary peak denoting the L-5-MTHF (m/z 162) is present along with the starch (*m/z* 159) when a relatively small ratio of modified starch was used to encapsulate the L-5-MTHF (e.g., Encap 1, MTHF: modified starch = 1:9) (Figure 9.9c). Increasing the amount of modified starch relative to the natural folate (e.g. Encap 2, L-5-MTHF: modified starch = 1:99) resulted in no sign of the L-5-MTHF on the starch surface (Figure 9.9d). Mass resolved ToF-SIMS images of Encap 1 and Encap 2 also showed an even distribution of the core material within the starch coating (Encap 1; Figure 9.10a), but an absence of L-5-MTHF in the outer layer (Encap 2; Figure 9.10b). These findings indicate that there was successful encapsulation of the L-5-MTHF in the modified starch, with a high percentage of the coating material. When baked in bread, Encap 2 had a significantly ($p < 0.05$) higher recovery of L-5-MTHF compared to breads that were fortified with Encap 1, and the free L-5-MTHF, during both bread baking and following storage (Table 9.5). The modified starch provided an excellent retention of L-5-MTHF during the encapsulation process and improved the stability of L-5-MTHF in baked breads. It had low viscosity and high solubility in water and produced the desired microcapsules, which possessed a smooth surface and evenly distributed core material. It is also low in cost and bland in taste. Thus, modified starch proved to be a good coating material for the microencapsulation of L-5-MTHF.

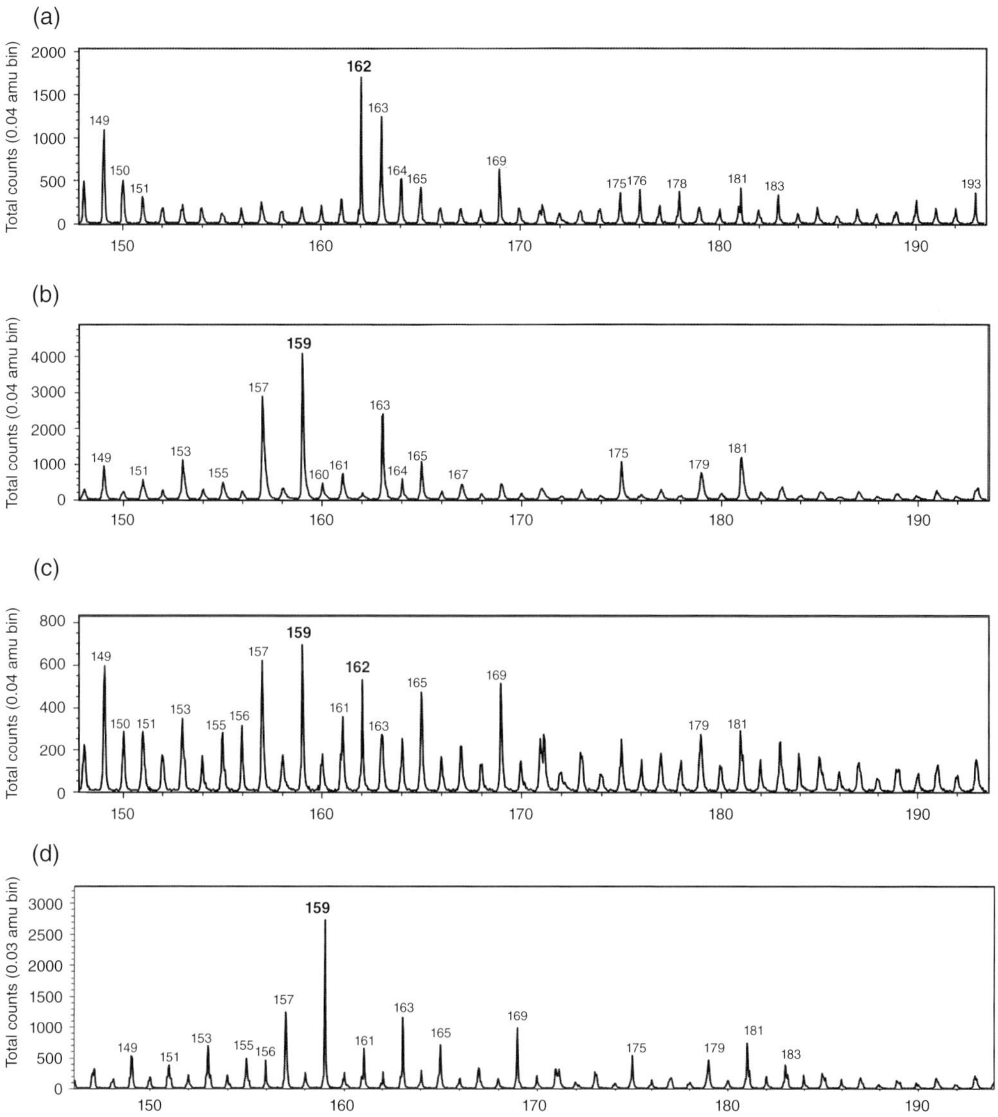

Figure 9.9 Negative ToF-SIMS spectra of (a) the L-5-MTHF, (b) modified starch, (c) Encap 1 and (d) Encap 2. *x*-axis is *m/z* (mass to charge ratio); *y*-axis is total counts of pixels (0.04 amu bin). Peak *m/z* 159 and peak *m/z* 162 are characteristic fragments denoting modified starch and L-5-MTHF, respectively. Peak *m/z* 159 and peak *m/z* 162 were detected in the outer layer of the Encap 1; the Encap 2 spectrum only has an intensive peak *m/z* 159. Encap1 (L-5-MTHF:starch = 1:9); Encap 2 (L-5-MTHF:starch = 1:99). (Liu *et al.*, 2013. Reproduced with permission from American Chemical Society.)

9.6 CO-MICROENCAPSULATION OF L-5-MTHF WITH ANTIOXIDANT

Our studies demonstrated that encapsulating L-5-MTHF produced defined benefits for protecting the labile vitamin from degradation; however, the lack of a significant carry-over effect of this technology to ensure stability of the reduced folates over time and post-processing, remains to be solved. As mentioned above,

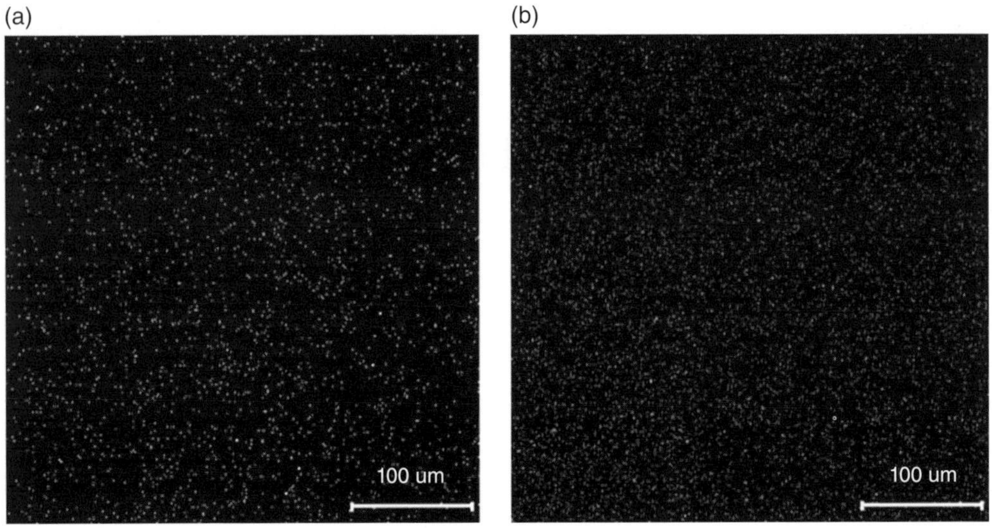

Figure 9.10 Mass resolved images (chemical maps) of (a) Encap 1 and (b) Encap 2. Peak m/z 159 and peak m/z 162 are characteristic fragments denoting modified starch and L-5-MTHF, respectively. Red spots represent m/z 159 (modified starch) and green spots represents m/z 162 (L-5-MTHF). Brighter color corresponds to higher intensity of fragments, black to the absence of fragments. Encap 1 (L-5-MTHF:starch = 1:9); Encap 2 (L-5-MTHF:modified starch = 1:99). (Liu *et al.*, 2013. Reproduced with permission from American Chemical Society.)

Table 9.5 Recovery of L-5-MTHF with modified starch coating in baked breads.[1]

Treatment [2]	Day 0 (fresh)	Day 3	Day 7
Free L-5-MTHF	51.5 ± 7.1^{ax}	21.6 ± 1.7^{ay}	12.1 ± 1.1^{az}
L-5-MTHF + NaAsc	70.4 ± 3.5^{bx}	58.6 ± 5.5^{cy}	39.6 ± 6.2^{cz}
Encap 1	54.5 ± 9.4^{ax}	23.7 ± 3.9^{ay}	13.1 ± 2.0^{az}
Encap 2	68.5 ± 5.2^{bx}	41.1 ± 3.7^{by}	22.3 ± 1.9^{bz}
Encap 3	79.3 ± 1.5^{cx}	70.8 ± 0.5^{dy}	46.7 ± 1.1^{cz}
Encap 4	91.9 ± 7.7^{dx}	89.2 ± 3.9^{ex}	73.7 ± 1.7^{dy}

[1] Values are means ± SD of three baking processes. Data are expressed as % of L-5-MTHF added to the bread. Breads were baked at 365 °C for 25min and stored at room temperature.

[2] Free L-5-MTHF, flour fortified with free L-5-MTHF; MTHF + NaAsc, flour fortified with free L-5-MTHF and NaAsc; Encap 1, flour fortified with microencapsulated L-5-MTHF (L-5-MTHF:modified starch = 1:9); Encap 2, flour fortified with microencapsulated L-5-MTHF (L-5-MTHF:modified starch = 1:99); Encap 3, flour fortified with NaAsc co-encapsulated L-5-MTHF (L-5-MTHF:NaAsc:modified starch = 0.1:1:99); Encap 4: flour fortified with NaAsc co-microencapsulated L-5-MTHF (L-5-MTHF:NaAsc:modified starch = 0.1:10:99).

[abcde] Mean values within each column not sharing a common superscript were significantly different and [xyz] mean values within each row not sharing a common superscript were significantly different (two-way ANOVA, followed by Bonferroni post-test, $p < 0.05$).

oxidation degradation of the L-5-MTHF is the major source of loss of this natural folate. Reducing agents, such as sulfites and ascorbate are known to stabilize L-5-MTHF against oxidative degradation in different foods and buffered systems (Pizzoferrato, 1992; Donnelly-Vanderloo *et al.*, 1994; Oley *et al.*, 2006; Liu *et al.*, 2012). Different nutrients present in the breads, such as iron, may also influence the oxidation rate of the folate. It therefore stands to reason that if natural folates such as L-5-MTHF can be microencapsulated or co-microencapsulated with a reducing agent, such as ascorbate, then an improved stability of the L-5-MTHF would be realized during cooking, and subsequently afterwards during storage of the food product.

Recently, Liu *et al.* (2013) successfully microencapsulated L-5-MTHF with sodium ascorbate (NaAsc), using modified starch as the coating material. The relative success at co-microencapsulating L-5-MTHF with ascorbate in baked breads is shown in Table 9.5. Fortifying flour with L-5-MTHF and NaAsc in the free form improved the stability of the L-5-MTHF during the bread baking process; where NaAsc also worked to stabilize the L-5-MTHF during storage. This can be supported by the significantly ($p < 0.05$) higher recovery of L-5-MTHF in breads fortified with free L-5-MTHF and NaAsc, compared to those spiked with Encap 2 after 3 and 7 days of storage (Table 9.5). Co-microencapsulation of L-5-MTHF with NaAsc in Encap 3 (L-5-MTHF: NaAsc: modified starch = 0.1:1:99) gave greater ($p < 0.05$) stability for L-5-MTHF and enhanced the stability of L-5-MTHF in the bread (Encap 4, L-5-MTHF: NaAsc: modified starch = 0.1:10:99) following baking and during storage. Repeating the ToF-SIMS with NaAsc co-microencapsulated L-5-MTHF enabled us to see that NaAsc was present on the surface of microcapsules, especially when present at a higher proportion (Figure 9.11). The fact that the ToF-SIMS spectra for Encap 4 (Figure 9.11e) produced a similar profile to that of the NaAsc (Figure 9.11c), indicates that NaAsc residues were on the outer layer of the surface and not taken up along with the L-5-MTHF to the same degree inside the microcapsule. No L-5-MTHF was present on the surface of these two microencapsulated products (Encap 3 and 4), which confirms that it was successfully microencapsulated. In a subsequent human clinical trial, Green *et al.* (2013) demonstrated high bioavailability of the encapsulated L-5-MTHF when formulated as part of a baked bread product. The improved stability of L-5-MTHF in Encap 4 was also seen with cooked noodles (Table 9.6). Although wet heat (boiling) was applied when noodles were cooked, thus producing unfavorable conditions for the labile vitamin, we observed an improved retention and stability of L-5-MTHF in the cooked noodles when co-encapsulated with ascorbic acid. Losses of free L-5-MTHF were almost 70%, compared to only 25–30% loss when co-microencapsulated with NaAsc (Encap 4). The retention of the L-5-MTHF in noodles fortified with Encap 4 reached values that were close to that found for folic acid in boiled fortified noodles.

9.7 SAFETY AND EFFICACY OF MICRO- AND NANOENCAPUSLATION OF FOLATES

In our work with reduced folates, we were unable to utilize nanoparticles to successfully encapsulate the labile vitamin. There are a few potential drawbacks that involve the use of nanoparticles and liposomes for the general purpose of providing a nutrient carrier. These can include poor stability, low embedded nutrient ratios in particles, and vulnerable leakage during storage periods (Li *et al.*, 2009; McNeil, 2011). Moreover, long-term testing to determine cumulative effects on humans from particles that have degraded into component parts has created an additional concern (Popov *et al.*, 2010). Ironically, the same properties of nanomaterials that make them attractive as a component for use in nutrient delivery systems, such as the increased surface area to volume ratio, also raise some concerns on safety attributes. In employing a number of different coating materials for microencapsulating L-5-MTHF, we also observed poor stability with the stearate emulsion, while considerable loss of L-5-MTHF occurred due to leakage of pectin microcapsules. In both cases, very low recoveries of the vitamin were obtained with these coating materials used to construct the microencapsulation. The use of skimmed-milk powder and modified starch to encapsulate L-5-MTHF was more successful and did not have the same limitations observed with other materials. In both cases, the choices of materials for constructing microparticles for L-5-MTHF were linked to functionality and not so much food safety, since there is little concern as to potential toxicity consequences. In contrast, *in vivo* and *in vitro* studies are needed to determine safety regulations for nanotechnology that include nanoencapsulation of vitamins, since no standardized regulations for the production and use of nanotechnology in food products currently exist (Rashidi and Khosravi-Darani, 2011). This situation has been hampered by

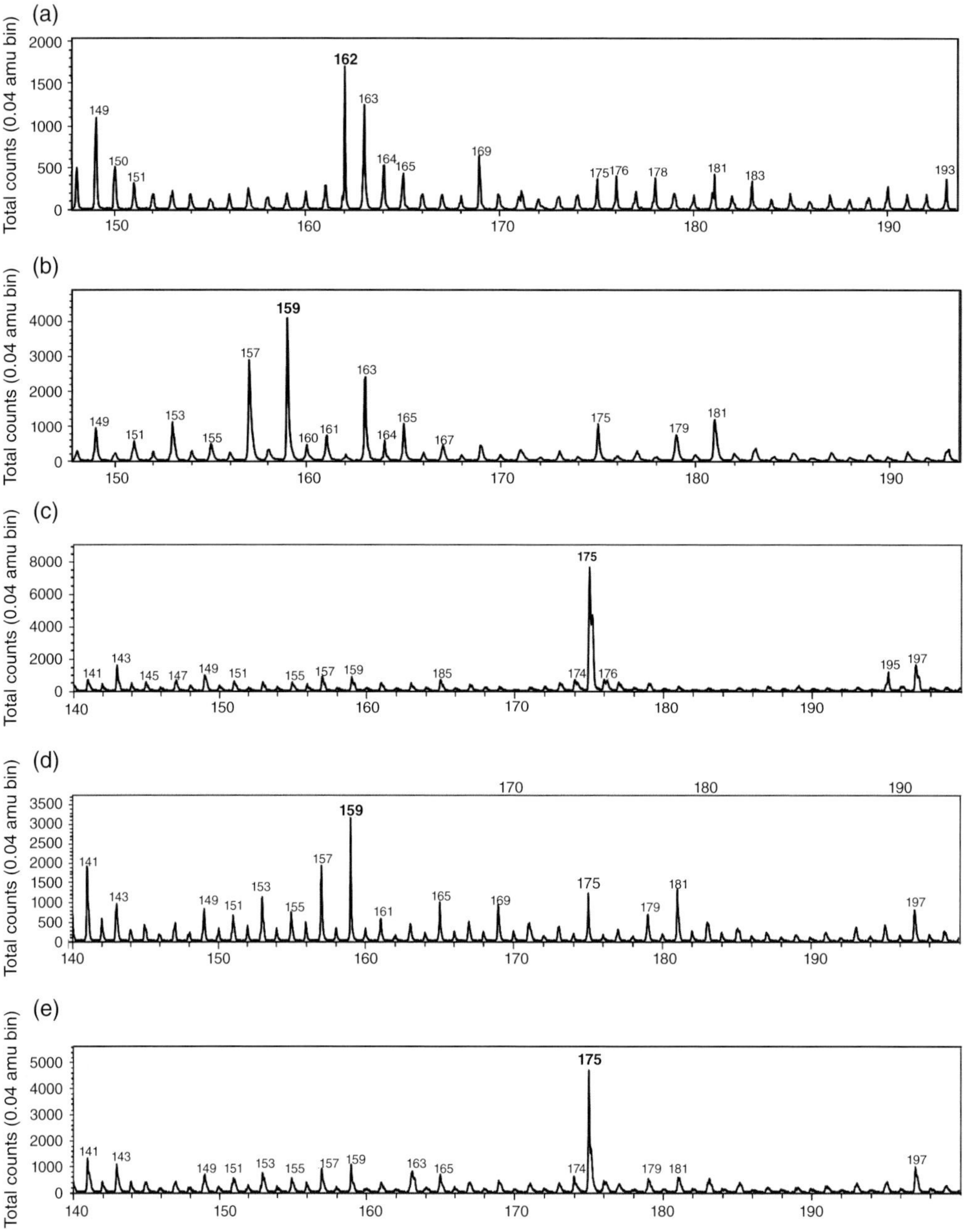

Figure 9.11 Negative ToF-SIMS spectra of (a) the L-5-MTHF, (b) modified starch, (c) NaAsc, (d) Encap 3 and (e) Encap 4. *x*-axis is *m/z* (mass to charge ratio); *y*-axis is total counts of pixels (0.04 amu bin). Peak *m/z* 162 and peak *m/z* 159 are characteristic fragments denoting L-5-MTHF and modified starch, respectively. Peak *m/z* 175 denotes NaAsc. Encap 3 (L-5-MTHF:NaAsc:starch = 0.1:1:99); Encap 4 (L-5-MTHF:NaAsc:starch = 0.1:10:99). (Liu *et al.*, 2013. Reproduced with permission from American Chemical Society.)

Table 9.6 Recovery of L-5-MTHF with modified starch coating in fresh and cooked noodles.[1]

Treatment	Fortification level mg/Kg	Fresh noodle mg/Kg	Cooked noodle mg/Kg	Fresh recovery (%)	Cooked recovery (%)
Free L-5-MTHF	5.2 ± 0.4^{x}	3.7 ± 0.3^{y}	1.6 ± 0.2^{z}	71	31
Encap 4	4.8 ± 0.4^{x}	3.8 ± 0.2^{y}	3.4 ± 0.3^{z}	79	71
Free folic acid	4.6 ± 0.4^{x}	4.3 ± 0.9^{x}	4.0 ± 0.7^{x}	93	87

[1] Measured values are means ± SD for four noodle batches. Noodles were made by a homestyle noodle maker and 100 g of the noodles was boiled at 92°C for 13 min.
[xyz] In the same row indicate significant difference (one-way ANOVA, followed by Bonferroni post-test, $p < 0.05$).

the fact that few studies have reported results of stability and toxicity of different polymer materials (Dobrovolskaia and McNeil, 2007). Though the use of nanotechnology can potentially improve human health by enhancing bioavailability of bioactive phytochemicals or nutrients (Weiss *et al.*, 2006; Rashidi and Khosravi-Darani, 2011), the consumer is more likely to prefer foods that they perceive as natural (Siegrist *et al.*, 2008). Thus, there is a need for well-defined methods of analysis of the components of engineered nanoparticles to ascertain safety prior to clinical use (Tiede *et al.*, 2008). Currently there are no regulations specified by Health Canada that are specific to nanotechnology-based health and food products, nor are there existing legislative and regulatory frameworks currently available to mitigate potential health risks (Health Canada, 2011). As such, more research is required to address the public concerns over the risk–benefits of using nanotechnology to encapsulate vitamins for human consumption. For more information on consumer perception, safety assessment, and regulation of nanomaterials for use in foods, please consult Chapters 20–23.

References

Ajazuddin and Saraf, S. (2010) Applications of novel drug delivery system for herbal formulations. *Fitoterapia*, 81, 680–689.

Baranauskienė, R., Venskutonis, P.R., Dewettinck, K. and Verhé, R. (2006) Properties of oregano (*Origanum vulgare* L.), citronella (*Cymbopogon nardus* G.) and marjoram (*Majorana hortensis* L.) flavors encapsulated into milk-protein-based matrices. *Food Research International*, 39, 413–425.

Brattström, L. and Wilcken, D.E. (2000) Homocysteine and cardiovascular disease: cause or effect? *The American Journal of Clinical Nutrition*, 72, 315–323.

Champagne, C.P. and Fustier, P. (2007) Microencapsulation for the improved delivery of bioactive compounds into foods. *Current Opinion in Biotechnology*, 18, 184–190.

Dang, J. Arcot, J. and Shrestha, A. (2000) Folate retention in selected processed legumes. *Food Chemistry*, 68, 295–298.

Dobrovolskaia, M.A. and Mcneil, S.E. (2007) Immunological properties of engineered nanomaterials. *Nature Nanotechnology*, 2, 469–478.

Donnelly-Vanderloo, M., O'Connor, D.L. and Shoukri, M. (1994) Impact of pasteurization and procedures commonly used to rethermalize stored human milk on folate content. *Nutrition Research*, 14, 1305–1316.

Favaro-Trindale, C.S. and Grosso, C.R.F. (2002) Microencapsulation of L-acidophilus (La-05) and B-lactis (Bb-12) and evaluation of their survival at the pH values of the stomach and in bile. *Journal of Microencapsulation*, 19, 485–494.

Food and Nutrition Board. (1998) *Dietary Reference Intakes for Thiamin, Riboflavin, Niacin, Vitamin B_6, Folate, Vitamin B_{12}, Pantothenic Acid, Biotin, and Choline: A Report of the Standing Committee on the Scientific Evaluation of Dietary Reference Intakes and its Panel on Folate, Other B Vitamins, and Choline and Subsommittee on Upper Reference Levels of Nutrients*. National Academy Press, Washington, DC.

Green, T.J., Liu, Y., Dadgar, S., Li, W., Böhni, R. and Kitts, D.D. (2013). Wheat rolls fortified with microencapsulated l-5-Methylteetrahydrofolic acid or equimolar folic acid increase blood folate concentrations to a similar extent in healthy men and women. *Journal of Nutriton*, 143, 867–871.

Hawkes, J.G. and Villota, R. (1989) Folates in foods-reactivity, stability during processing and nutritional implications. *Critical Reviews in Food Science and Nutrition*, 28, 439–538.

Health Canada. (2010) *Dietary Reference Intakes: Reference Values for Vitamins*. [online] Available: http://www.hc-sc.gc.ca/fn-an/nutrition/reference/table/ref_vitam_tbl-eng.php (accessed 29 November 2010).

Hoffbrand, A.V. and Jackson, B.F.A. (1993) Correction of the DNA synthesis defect in vitamin B_{12} deficiency by tetrahydrofolate: evidence in favour of the methyl-folate trap hypothesis as the cause of megaloblastic anaemia in vitamin B_{12} deficiency. *British Journal of Haematology*, 83, 643–647.

Houghton, L.A., Sherwood, K.L., Pawlosky, R., Ito, S. and O'connor, D.L. (2006) 6S -5-Methyltetrahydrofolate is at least as effective as folic acid in preventing a decline in blood folate concentrations during lactation. *American Journal of Clinical Nutrition*, 83, 842–850.

Jackson, L.S. and Lee, K. (1991) Microencapsulation and the food industry. *Food Science and Technology*, 24, 289–297.

Jacques, P.F., Selhub, J., Bostom, A.G., Wilson, P.W.F. and Rosenberg, I.H. (1999) The effect of folic acid fortification on plasma folate and total homocysteine concentrations. *New England Journal of Medicine*, 340, 1449–1454.

Jafari, S.M., Assadpoor, E., Bhandari, B. and He, Y. (2008). Nano-particle encapsulation of fish oil by spray drying. *Food Research International*, 41, 172–183.

Kagami, Y., Sugimura, S., Fujishima, N., Matsuda, K., Kometani, T. and Matsumura, Y. (2003) Oxidative stability, structure, and physical characteristics of microcapsules formed by spray drying of fish oil with protein and dextrin wall materials. *Journal of Food Science*, 68, 2248–2255.

Kim, Y.I. (2004) Will mandatory folic acid fortification prevent or promote cancer? *American Journal of Clinical Nutrition*, 80, 1123–1128.

Li, D., Zhong, X., Zeng, Z., *et al.* (2009) Application of targeted drug delivery system in Chinese medicine. *Journal of Controlled Release*, 138, 103–112.

Liu, Y., Tomiuk, S., Rozoy E., *et al.* (2012) Thermal oxidation studies on reduced folate, L-5-methyltetrahydrofolic acid (L-5-MTHF) and strategies for stabilization using food matrices. *Journal of Food Science*, 77, 236–243.

Liu, Y., Green, T.J., Wong, P. and Kitts, D.D. (2013) Microencapsulation of l-5-methyltetrahydrofolic acid with ascorbate improves stability in baked bread products. *Journal of Agricultural and Food Chemistry*, 61, 247–254.

Loksuwan, J. (2007) Characteristics of microencapsulated β-carotene formed by spray drying with modified tapioca starch, native tapioca starch and maltodextrin. *Food Hydrocolloids* 21, 928–935.

Madziva, H., Kailasapathy, K. and Phillips, M. (2005) Alginate–pectin microcapsules as a potential for folic acid delivery in foods. *Journal of Microencapsulation* 22, 343–351.

McClements, D.J. and Rao, J. (2011) Food-grade nanoemulsions: formulation, fabrication, properties, performance, biological fate, and potential toxicity. *Critical Reviews in Food Science and Nutrition*, 51, 285–330.

McNeil, S.E. (2011) Challenges for nanoparticle characterization. In*: Characterization of Nanoparticles Intended for Drug Delivery* (ed. S.E. McNeil). Springer, New York, 9–16.

MRC Vitamin Study Research Group. (1991) Prevention of neural tube defects: Results of the Mdical Resarch Council Vitamin Study. *Lancet*, 338, 131–137.

Nazzaro, F., Fratianni, F., Coppola, R., Sada, A. and Orlando, P. (2009) Fermentative ability of alginate-prebiotic encapsulated Lactobacillus acidophilus and survival under simulated gastrointestinal conditions. *Journal of Functional Foods*, 1, 319–323.

Neuhouser, M.L. and Beresford, S.A. (2001) Folic acid: Are current fortification levels adequate? *Nutrition*, 17, 868–872.

Nguyen, M.T. and Indrawati, H.M. (2003) Model studies on the stability of folic acid and 5-methyltetrahydrofolic acid degradation during thermal treatment in combination with high hydrostatic pressure. *Journal of Agricultural and Food Chemistry*, 51, 3352–3357.

Ohrvik, V., Ohrvik, H., Tallkvist, J. and Witthoft, C. (2010) Folates in bread: retention during bread-making and in vitro bioaccessibility. *European Journal of Nutrition*, 49, 365–372.

Oley, I. Verlinde, P. Hendrickx, M. and Van Loey, A. (2006) Temperature and pressure stability of L-ascorbic acid and/or 6s 5-methyltetrahydrofolic acid: A kinetic study. *European Food Research and Technology*, 223, 71–77.

Oliver, C.M. and Augustin, M.A. (2009) Using dairy ingredients for encapsulation. In*: Woodhead Publishing in Food Science, Technology and Nutrition*. Woodhead Publishing Ltd, Cambridge, 565–588.

Partanen, R., Yoshii, H., Kallio, H., Yang, B. and Forssell, P. (2002a) Encapsulation of sea buckthorn kernel oil in modified starches. *Journal of the American Oil Chemists Society*, 79, 219–223.

Partanen, R., Ahro, M., Hakala, M., Kallio, H. and Forssell, P. (2002b) Microencapsulation of caraway extract in ß-cyclodextrin and modified starches. *European Food Research and Technology*, 214, 242–247.

Pietrzik, K. Bailey, L. and Shane, B. (2010) Folic acid and L-5-methyltetrahydrofolate comparison of clinical pharmacokinetics and pharmacodynamics. *Clinical Pharmacokinetics*, 49, 535–548.

Pizzoferrato, L. (1992) Examples of direct and indirect effects of technological treatments on ascorbic acid, folate and thiamine. *Food Chemistry*, 44, 49–52.

Popov, K.I., Filippov, A.N. and Khurshudyan, S.A. (2010) Food nanotechnologies. *Russian Journal of General Chemistry (Translation of Zhurnal Obshchei Khimii)*, 80, 630–642.

Rashidi, L. and Khosravi-Darani, K. (2011) The applications of nanotechnology in food industry. *Critical Reviews in Food Science and Nutrition*, 51, 723–730.

Ross, J.F., Belding, H. and Paegel, B.L. (1948) The development and progression of subacute combined degeneration of the spinal cord in patients with pernicious anemia treated with synthetic pteroylglutamic (folic) acid. *Blood*, 3, 68–90.

Savaria Morris, M., Jacques, P.F., Rosenberg, I.H. and Selhub, J. (2007) Folate and vitamin B-12 status in relation to anemia, macrocytosis, and cognitive impairment in older Americans in the age of folic acid fortification. *American Journal of Clinical Nutrition*, 85, 193–200.

Scott, J. (2001) Methyltetrahydrofolate: the superior alternative to folic acid. In*: Oxidative Stress and Disease*, Vol. 6. Marcel Dekker, Inc., New York.

Semo, E., Kesselman, E., Danino, D. and Livney, Y.D. (2007) Casein micelle as a natural nano-capsular vehicle for nutraceuticals. *Food Hydrocolloids*, 21, 936–942.

Shaikh, J., Bhosale, R. and Singhal, R. (2006). Microencapsulation of black pepper oleoresin. *Food Chemistry*, 94, 105–110.

Siegrist, M., Stampfli, N., Kastenholz, H. and Keller, C. (2008) Perceived risks and perceived benefits of different nanotechnology foods and nanotechnology food packaging. *Appetite*, 51, 283–290.

Sotiriadis, P.K. and Hoskins, F.H. (1982) Vitamin retention during storage of processed foods. I. Effect of ascorbic acid on folates in cowpeas, okra and tomatoes. *Scientia Horticulturae*, 16, 125–130.

Tang, H.G., Xia, W.S. and Yuan, S.L. (2005) Study on microencapsulation of vitamin E. *Food and Machinery*, 21, 4–6.

Tiede, K., Boxall, A.B.A., Tear, S.P., Lewis, J., David, H. and Hassellov, M. (2008) Detection and characterization of engineered nanoparticles in food and the environment. *Food Additives and Contaminants Part A – Chemistry Analysis Control Exposure and Risk Assessment*, 25, 795–821.

Tomiuk, S., Liu, Y.Z., Green, T.J., King, M.J., Finglas, P.M. and Kitts, D.D. (2012) Studies on the retention of microencapsulated L-5-methyltetrahydrofolic acid in baked bread using skim milk powder. *Food Chemistry*, 133, 249–255.

Vasan, R.S.B., D'agostino, R.B., Levy, D., *et al.* (2003) Plasma homocysteine and risk for congestive heart failure in adults without prior myocardial infarction. *Journal of the American Medical Association*, 289, 1251–1257.

Venn, B.J., Green, T.J., Moser, R. and Mann, J.I. (2003) Comparison of the effect of low-dose supplementation with L-5-methyltetrahydrofolate or folic acid on plasma homocysteine: a randomized placebo-controlled study. *American Journal of Clinical Nutrition*, 77, 658–662.

Weiss, J., Takhistov, P. and Mcclements, D.J. (2006) Functional materials in food nanotechnology. *Journal of Food Science*, 71, 107–116.

10 The application of nanoencapsulation to enhance the bioavailability and distribution of polyphenols

Alison Kamil, C-Y. Oliver Chen, and Jeffrey B. Blumberg*

Antioxidants Research Laboratory, Jean Mayer USDA Human Nutrition Research Center on Aging, Tufts University, Boston, Massachusetts, USA

10.1 INTRODUCTION

Polyphenols are secondary metabolites produced by plants, in which they play an important role in growth and reproduction as well as resistance to pathogens, predators, and diseases (Beckman, 2000). These phytochemicals also contribute importantly to the color and organoleptic properties of plant foods. While they were once considered as antinutrients, research conducted during the past 30 years has revealed that polyphenols contribute to the promotion of health and reduction in the risk of common chronic diseases (Liu, 2003). This evidence is derived from a body of *in vitro* experiments, animal models, observational studies, and clinical trials. For example, prospective cohort studies show an inverse correlation between the consumption of polyphenol-rich fruits and vegetables and the risk of major noncommunicable diseases such as cancer, cardiovascular disease, type 2 diabetes mellitus, neurodegenerative diseases, and osteoporsis (Knekt *et al.*, 2002; Tapiero *et al.*, 2002; Bazzano *et al.*, 2003; Smith-Warner *et al.*, 2003; He *et al.*, 2004; Hung *et al.*, 2004; Dauchet *et al.*, 2006; Esmailzadeh *et al.*, 2004; McCullough *et al.*, 2012; Wedick *et al.*, 2012). Thousands of polyphenols have been characterized, with data on the composition of the major classes of these phytochemicals now available in nutrient databases, including the US Department of Agriculture Database for the Flavonoid Content of Selected Foods, Phenol Explorer (Scalbert *et al.*, 2011), and the electronic BioActive Substances Information System (eBASIS). Dietary polyphenols are typically categorized into four classes: flavonoids (including the subclasses anthocyanidins, catechins, flavones, flavonols, flavanones, and isoflavones), coumarins, stilbenes, and tannins, though other dietary constituents, for example, chalcones and lignans, also have polyphenolic structures (Scalbert and Williamson, 2000). The main sources of polyphenols in the human diet are fruits, plant-based beverages (especially tea, coffee, wine, and fruit juices), chocolate, and, to a lesser extent, vegetables, cereals, legumes, and nuts (Bravo, 1998; Scalbert *et al.*, 2005).

Polyphenols can act as potent effectors of biologic processes associated with the pathogenesis of human disease. Polyphenols are recognized for their direct free-radical scavenging activity and reducing power *in vitro*. However, the more relevant of their mechanisms for bioactivity appear to be

*The authors declare no conflicts of interest.

Nanotechnology and Functional Foods: Effective Delivery of Bioactive Ingredients, First Edition.
Edited by Cristina M. Sabliov, Hongda Chen, and Rickey Y. Yada.

associated with their ability to beneficially affect inflammation, upregulate detoxification pathways, and modulate cell-signal transduction (Nijveldt *et al.*, 2001; Fraga *et al.*, 2010). While polyphenols appear to have a dynamic interaction with gut microbiota, their efficacy in promoting health and reducing the risk of chronic diseases is dependent on their systemic bioavailability and metabolism. Generally, polyphenols have a low bioavailability due to a number of intrinsic and extrinsic factors, including their chemical structure and molecular weight, low hydrosolubility, binding within food matrices, low stability in the gastrointestinal (GI) environment, extensive phase II metabolism, and rapid elimination (Scalbert and Williamson, 2000; D'Archivio *et al.*, 2010).

Using data from the US National Health and Examination Survey, Chun *et al.* (2007) estimated mean daily total flavonoid intake of 189.7 mg/day, a level of consumption below the intermediate range of intake shown to be associated with lower risk of death related to cardiovascular disease (McCullough *et al.*, 2012). Polyphenol intake is low in the United States because fewer than 10% of people report eating the number of servings of fruits or vegetables recommended by the *Dietary Guidelines for Americans* (Kimmons *et al.*, 2009). While increased consumption of polyphenol-rich fruits, vegetables, and whole grains (as well as beverages derived from these foods) would serve to enhance their concentrations in blood and tissues, additional strategies are being considered to obtain their health benefits via increased bioavailability. For example, co-consumption of polyphenols with selected fats, proteins, or inhibitors of phase II metabolism (such as piperine) (D'Archivio *et al.*, 2010; Scheepens *et al.*, 2010) may act to increase their uptake across the gut wall. Encapsulation of polyphenols with nanomaterials presents an alternate approach with several advantages over that of using food synergies to increase bioavailability and bioactivity (Acosta, 2009; Nair *et al.*, 2010).

10.2 BIOAVAILABILITY OF POLYPHENOLS

In pharmacology, bioavailability is a subcategory of absorption and is defined as the fraction of an administered dose of unchanged drug that reaches the systemic circulation (Chen *et al.*, 2001). In nutrition science, which covers the intake of essential nutrients and other dietary constituents, the concept of bioavailability lacks the well-defined standards commonly used for drugs. In contrast to most drugs, the absorption, distribution, and utilization of nutrients (including phytochemicals) are subject to the influence of the steady-state status of the nutrient and physiological state of the subject (Heaney, 2001). Therefore, bioavailability for nutrients can be defined as the proportion of the consumed substance being absorbed and available for use or storage (Srinivasan, 2001). Nevertheless, nutrikinetic parameters overlap substantially with pharmacokinetic parameters such as area under the plasma time curve (AUC), peak plasma concentration (C_{max}), time to reach C_{max} (T_{max}), and time required for the concentration to reach half its original value or elimination half-life ($T_{1/2}$) (Raghuram and Krishnaswamy, 1992), so the latter term will be used throughout this review. The relative oral bioavailability between two formulations of the same compound can be measured by comparing the bioavailability (estimated as AUC) per dose of each formulation. The pharmacokinetics of a compound can be divided into four processes: absorption, distribution, metabolism, and excretion (ADME) (Thomas *et al.*, 2006). Intrinsic and extrinsic factors affecting the ADME and, thereby, the consequent bioavailability of a nutrient include the nutrient and its dietary source, its physicochemical properties alone and within the food matrix, gastric emptying rate, permeability across intestinal membranes, and extent of metabolism and distribution throughout the body. In addition to pharmacokinetic parameters, measures of bioactivity, including intermediary biomarkers of physiological function or disease pathogenesis, can be employed for making inferences about bioavailability (Verhagen *et al.*, 2004).

The bioavailability of dietary polyphenols is relatively low, ranging from ≈1% to ≈20% for anthocyanins (Milbury *et al.*, 2010) and isoflavones (Hendrich, 2002), respectively. The low bioavailability of most polyphenols has been attributed mainly to low absorption in the GI tract following consumption, extensive biotransformation within the gut as well as following systemic distribution, and rapid

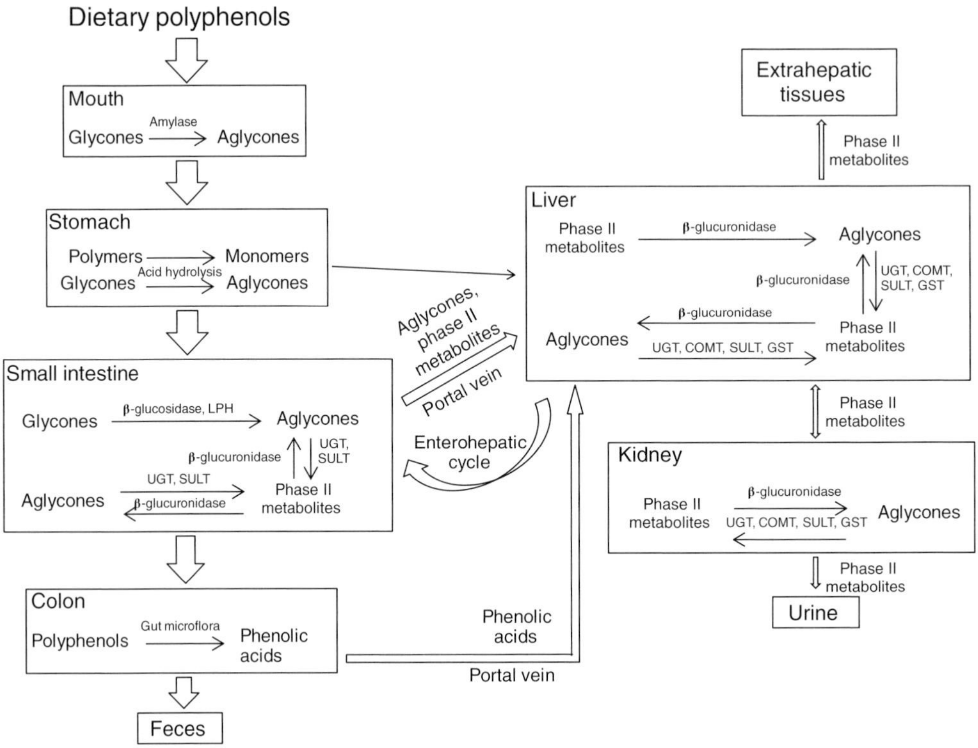

Figure 10.1 Absorption, distribution, metabolism, and excretion of dietary polyphenols. Size of arrows is proportional to per cent distribution from consumption to excretion. Abbreviations: UGT, UDP-glucuronosyltransferase; COMT, catechol-O-methyltransferase; SULT, sulfotransferase; GST, glutathione S-transferase; LPH, lactase-phlorizin hydrolase.

clearance from the body (see Figure 10.1 for a schematic of the ADME of polyphenols). Many dietary polyphenols are found as glycosylated compounds, and this form may diminish their diffusion across barriers in the GI tract (Scalbert and Williamson, 2000). Many polyphenols are labile in the acidic condition of the stomach and alkaline condition of the small intestine, so potentially much less than consumed may actually reach enterocyte absorption sites (Bermúdez-Soto *et al.*, 2007). For example, the flavanol epigallocatechin-3-gallate (EGCG) is stable in acidic pH but easily oxidized in neutral or basic pH conditions; in a pH 7.4 solution, half an EGCG concentration is lost within 2 h (Krook and Hagerman, 2012). During absorption, polyphenols are extensively transformed via phase II pathways, predominately methylation, glucuronidation, and sulfation, in enterocytes of the small intestine and then further metabolized in the liver (Scalbert and Williamson, 2000). For example, absorption of quercetin glucosides is quite high (≈50%), whereas its excretion in urine is low (<1%), indicating extensive transformation of quercetin via phase II metabolism (Manach *et al.*, 2005). Such biotransformation facilitates quick polyphenol excretion from the enterocytes via the phase III transporters, the liver via the bile, and the body through urine. Generally, polyphenols, including most flavonoids, have a $T_{1/2}$ of ≈1–3 h (Manach *et al.*, 2005).

As nutrient bioavailability must precede its bioactivity at tissue sites in the body, a number of strategies have been used to increase the chemical stability or permeability of nutrients. These approaches usually rely upon the addition of chemical additives such as reducing agents to maintain chemical structure, dissolving agents to increase solubility (Ader *et al.*, 2000; Dube *et al.*, 2010a), inhibitors of phase I and/or II enzymes to escape biotransformation (Shoba *et al.*, 1998; Brand *et al.*, 2010),

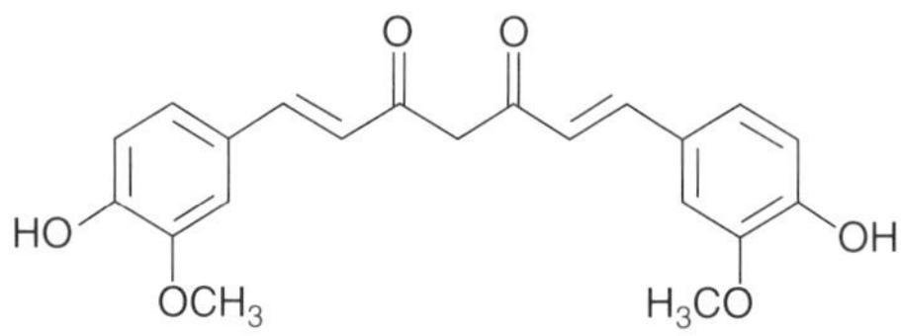

Curcumin

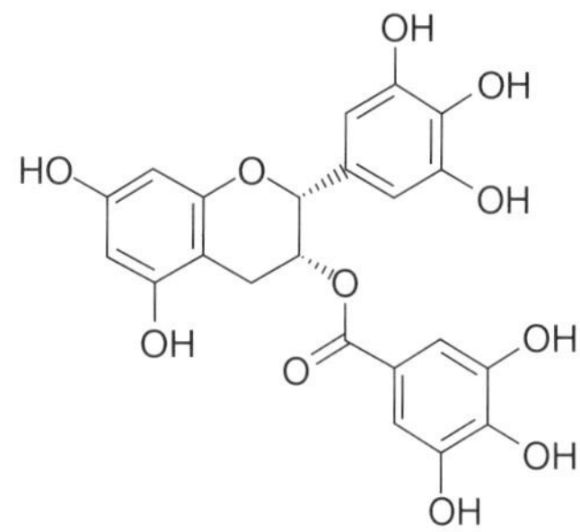

Epigallocatechin 3-gallate

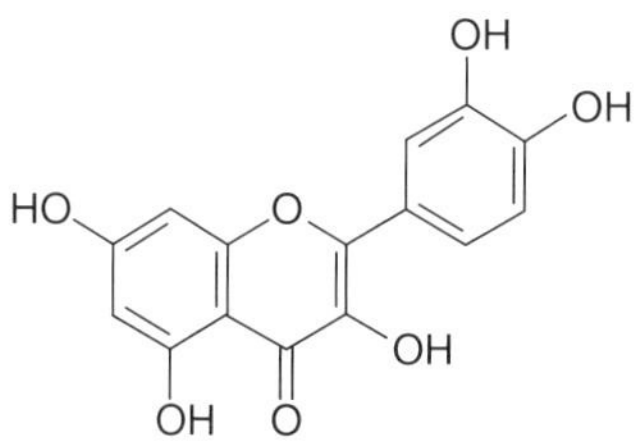

Quercetin

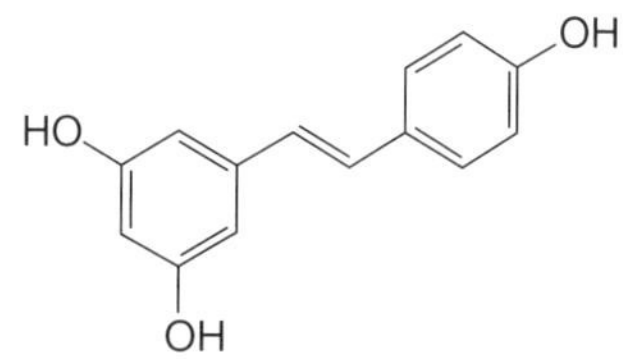

Resveratrol

Figure 10.2 Chemical structure of selected polyphenols.

modulation of the food matrix or addition of complementary ingredients such as lipids or protein (Lesser *et al.*, 2004), and/or manipulation through conjugation with promoiety groups (Lam *et al.*, 2004). More recently, encapsulation with food-grade or related Generally Recognized As Safe (GRAS) materials that exhibit controlled-release behavior have been used to increase the bioavailability and bioactivity of polyphenols. Encapsulation and controlled-release of active food ingredients can be attained through novel formulations such as microemulsions, matrix systems, solid dispersions, and liposomes (Pralhad and Rajendrakumar, 2004; Yuan *et al.*, 2006; Cui *et al.*, 2009; Musthaba *et al.*, 2009). However, as there are concerns and/or limitations associated with the safety and stability of these approaches (Frijlink *et al.*, 1991; Windrum *et al.*, 2005; Mu and Zhong, 2006), nanoencapsulation of nutrients, especially polyphenols such as curcumin, EGCG, quercetin, and reseveratrol (see Figure 10.2 for images of their chemical structure), has emerged as a novel strategy to improve nutrient delivery, distribution, and/or bioactivity.

10.3 NANOENCAPSULATION: CHARACTERISTICS AND FORMULATIONS

Nanoencapsulation is defined as a technology involving the formation of active loaded particles with diameters ranging from 1 to 1000 nm and for which the end product exhibits properties or phenomena attributable to its dimensions (FDA, 2011). The term nanoparticle (NP) is a collective name for colloidal systems including nanospheres and nanocapsules (Gurley, 2011). A nanosphere is a matrix structure in which the active ingredient is dispersed throughout the particles, whereas nanocapsules have a membrane and an active ingredient in the core. Some examples of natural food-derived or GRAS materials that can fulfill these requirements include polysaccharides of plant (e.g., carrageenan) or microbial (e.g., dextran) origin, food proteins (e.g., whey proteins), and emulsifiers (e.g., lecithin). Polymeric NPs have been adopted as a preferred method for improving phytochemical bioavailability and distribution because of their stability and diversity. They have proven more stable in the GI tract than other colloidal carriers, such as liposomes, so that they protect encapsulated compounds from the conditions of the GI environment (Brannon-Peppas, 1995). The use of various polymeric materials enables the modulation of physiochemical characteristics (e.g., hydrophobicity, zeta potential), compound release properties (e.g., delayed, prolonged, triggered), and biological behavior (targeting, bioadhesion, cellular uptake) of NPs (Soppimath *et al.*, 2001). Moreover, the submicron size of NPs and their large surface area favor their absorption compared to larger carriers (Jani *et al.*, 1990; Desai *et al.*, 1996).

Polymeric materials used for the formulation of NPs include synthetic poly(lactic acids) (PLA), poly(lactic-co-glycolytic acid) (PLGA), poly(ε-caprolactone) (PCL), poly(methyl methacrylates), and poly(alkyl cyanoacrylates)) or natural polymers (albumin, gelatin, alginate, collagen, or chitosan). Synthetic polymers with precise chemical composition display highly predictable physical properties such as solubility, permeability, degradation, and erosion, as well as targeting behavior. Natural polymers, besides being less predictable, are also often mildly immunogenic (des Rieux *et al.*, 2006). Among the different synthetic polymers developed to formulate polymeric NPs, PLGA has attracted considerable attention due to its high biodegradability (Kumari *et al.*, 2010). Decomposition of PLGA yields lactic acid ("milk acid") and glycolic acid ("fruit acid") monomers. Because these two compounds are present endogenously and easily metabolized via the Krebs cycle, systemic toxicity is presumed to be unlikely with the use of PLGA for nutrient encapsulation. In addition, the polymers are commercially available with different molecular weights and copolymer compositions allowing the degradation to vary from several months to several years. Poly(lactic acid) has also been employed for nutrient-based NPs, but to a lesser extent than PLGA due to its lower degradation rate.

10.4 TRANSPORT OF NPs ACROSS INTESTINAL MUCOSA

Besides protection from the environment in the GI lumen, strategies have been developed to improve the bioavailability of compounds encapsulated in polymeric NPs, focusing principally on approaches to facilitating transport across the intestinal mucosa and enhancing uptake into intestinal cells. For food-related applications, two types of intestinal cells are most involved in bioavailability: enterocytes, the absorptive cells representing the majority of the GI tract lining, and microfold ("M") cells, mainly located in the follicle-associated epithelium of the Peyer's patch. In contrast to normal enterocytes, M cells lack microvilli on their apical surface but possess broader microfolds with the ability to transport antigens from the lumen via endocytosis or phagocytosis to the basolateral side for processing by immune cells. The M cells do not secrete mucus or digestive enzymes and have a small or absent glycocalyx, an extracellular polysaccharide layer found on enterocytes. The M cells represent ≈1% of the intestinal surface area.

The mechanisms by which polymeric NPs are taken up into intestinal epithelial cells are not entirely clear. There is evidence from *ex vivo* studies in rats and rabbits and from *in vitro* studies in human epithelial colorectal adenocarcinoma cells (Caco-2) that translocation of NPs occurs in the villi (McClean *et al.*, 1998; Behrens *et al.*, 2002; Zhang *et al.*, 2006; Reix *et al.*, 2012). However, because of the low endocytic activity of enterocytes, the amount of NPs translocated via this route in Caco-2 cells appears minimal (Pietzonka *et al.*, 2002; Cartiera *et al.*, 2009). Thus, the bulk of NP translocation may occur across M cells in the follicle associated epithelium (des Rieux *et al.*, 2005, 2007). Hillery *et al.* (1994) studied the uptake of 60 nm polystyrene particles by gut epithelial cells in female Sprague Dawley rats after 5 days of oral dosing via gavage and found only 10% of the administered dose to be recovered from the entire GI tract. They estimated that 60% of this uptake occurred through the lymphoid aggregates of the Peyer's patches.

Two major obstacles affecting the transport of compounds across the intestinal mucosa are GI transit time and mucus-layer barriers. Thus, the absorption of a polyphenol could be improved if its residence time in the gut is increased and if its permeability through mucus layers is enhanced. Research efforts have been undertaken to develop mucoadhesive and mucus-penetrating NPs. Electrostatic interaction is one of the more exploited forms of mucoadhesion, as exemplified by use of chitosan, a cationic natural polymer obtained from deacetylation of chitin, for a variety of oral delivery applications (Thanou *et al.*, 2001; Chae *et al.*, 2005). However, chitosan is only soluble in acidic environments, so its effectiveness is diminished in the slight alkaline environment in the rat small intestine and in Caco-2 cells (Sonaje *et al.*, 2009, 2012). A potential solution to this limitation with chitosan is *N*-trimethyl chitosan chloride (TMC), a derivative that is soluble in neutral and basic environments (van der Merwe *et al.*, 2004). Alternatively, mucoadhesion can also be achieved via conjugation with ligands that recognize particular mucin glycoproteins. For example, wheat germ agglutinin (WGA) from *Tricticum vulgare* specifically binds to *N*-acetyl-D-glucosamine and sialic acid on the cell surface. The WGA conjugated PLGA NPs have been found to bind *in vitro* to the surface of Caco-2 cells as well as *ex vivo* and *in vivo* to the intestinal villous epithelium and Peyer's patches of rats; these NPs may also be taken up into cells via receptor-mediated endocytosis involving the epidermal growth factor receptor that is expressed at a considerable density on enterocytes (Weissenboeck *et al.*, 2004; Yin *et al.*, 2007).

Although mucoadhesion originally appeared as a promising approach for increasing absorption, these NPs adsorb to mucin fibers and so are not efficient in reaching the surface of the enterocyte. As GI mucus has a turnover time of 4–5 h, those NPs bound to mucus are excreted rapidly (Galindo-Rodriguez *et al.*, 2005; Bernkop-Schnurch *et al.*, 2005). Hence, this approach will be unsuccessful unless the encapsulated compound is readily released from the NPs at the site of absorption and can directly reach the enterocyte. To achieve a longer residence time of NPs at mucosal surfaces, especially for particle systems that require intact intracellular delivery, attempts have been made to develop mucus-penetrating particles. One approach has been to synthesize particles with a strong hydrophilicity and neutral charge that mimics the essential surface properties of viruses that allow them to avoid mucoadhesion. Densely coating biodegradable particles with poly(ethylene glyocol) (PEG), an uncharged hydrophilic polymer, has been demonstrated to minimize adhesive interactions between NP and mucins (Vila *et al.*, 2002; Lai *et al.*, 2009). This approach appears to allow NP to rapidly penetrate human mucus by moving through openings between mucin mesh fibers (Huang *et al.*, 2000; Tobio *et al.*, 2000; Yoncheva *et al.*, 2007).

A macromolecule or NP can theoretically cross the intestinal epithelium by paracellular (between adjacent cells) and/or transcellular routes. However, under normal physiological conditions, the overall efficiency of the paracellular route is limited because of the small surface area of the intercellular spaces and the tightness of the junctions between the cells (Anderson, 2001). So, the transcellular route has been the most explored for NP translocation. Transport of NPs by the transcellular pathway is especially dependent on factors associated with the physiochemical properties of particles, such as size, zeta potential, and surface hydrophobicity. For example, NP transcytosis increases as the particle diameter decreases (Jani *et al.*, 1990; Desai *et al.*, 1996). However, while the available data from Caco-2 cells and rat models on the effect of charge and hydrophobicity using a variety of NPs are not consistent, there is agreement that NP surface properties are a critical aspect of developing NPs that will enhance oral delivery (Jung *et al.*, 2000; Behrens *et al.*, 2002; Vila *et al.*, 2002; Shakweh *et al.*, 2004). In addition to their mucoadhesive properties, water-soluble polymers such as chitosan and TMC do appear in Caco-2 cells and in rats to be useful strategies to enhance paracellular transport of compounds alone (van der Merwe *et al.*, 2004; Sonaje *et al.*, 2009). The mechanism for paracellular transport appears to be the formation of entanglements and chemical bonds between the negatively charged cell membrane and the positive charges of the polymer or by complexing Ca^{2+} involved in the structure of tight junctions (Smith *et al.*, 2004; Sonaje *et al.*, 2012). The results of previous studies, also conducted in Caco-2 cells and in rats, indicate that the transport is inversely related to molecular weight and reversible (Chae *et al.*, 2005).

With the challenge of overcoming obstacles to the intestinal uptake and transport of NP delivery systems, strategies are also being pursued that will prolong the $T_{1/2}$ to allow for a greater uptake into target tissues. The body generally perceives hydrophobic particles as being foreign. The reticuloendothelial system (RES) eliminates these particles from the circulation and moves them to the liver or spleen. This RES process presents one of the most important biological barriers to NP systemic delivery (Kumari *et al.*, 2010). To address this matter, NPs can be coated with molecules that hide the hydrophobicity by providing a hydrophilic layer on the particle surface. The most common moiety for surface modification is the hydrophilic and non-ionic polymer PEG. In animal models, the attachment of PEG, mostly to injectable polymeric NPs, can increase $T_{1/2}$ by several orders of magnitude (Owens and Peppas, 2006). Surface modification can also be used to target NPs toward specific organs and increase selective cellular binding and internalization through receptor-mediated endocytosis. As noted, surface charges of NP have an important influence on their interaction with and uptake into cells, such as the human dendritic cell line (Foged *et al.*, 2005; Vasir and Labhasetwar, 2008). Positively charged NPs seem to be able to escape the lysosomes after being internalized and exhibit perinuclear localization in a variety epithelial, fibroblastic, endothelial, and blood cell lines (Yue *et al.*, 2011). In addition, targeting ligands can be grafted at the NP surface (Alexis *et al.*, 2008).

10.5 NANOPARTICLES AS POTENTIAL DELIVERY SYSTEMS OF POLYPHENOLS

A growing body of literature reveals a careful characterization of the *in vitro* stability and absorption profile of NP encapsulated polyphenols. A much more limited number of animal model studies have examined the *in vivo* pharmacokinetic profiles of free versus NP polyphenols (Table 10.1). The ADME data from human studies of polyphenols in NP formulations is not available. A discussion of experimental studies using NPs as a potential delivery system for selected polyphenols follows.

10.5.1 Curcumin

Curcumin, found particularly in turmeric (*Curcuma longa*), has been associated with a variety of health benefits, including antioxidant actions to reduce oxidative stress, chemopreventive or cancer therapeutic effects mediated by a capacity for anti-angiogenesis, anti-inflammatory effects associated with the prevention or treatment of chronic conditions including cardiometabolic diseases as well as acute anti-allergic responses, and enhancement of immune responses to modulate autoimmune disorders. However, as a lipophilic polyphenol, curcumin is poorly absorbed due to its low stability and solubility in the GI tract (Tønnesen, 2002). The maximum solubility of curcumin in an aqueous buffer (pH 5.0) is 11 ng/mL. Thus, several efforts have been undertaken to promote the stability and bioavailability of curcumin with NP formulations.

In a pharmacokinetic study with male Sprague Dawley rats, Shaikh *et al.* (2009) found that curcumin-PLGA-NP (size, 242 nm; entrapment efficiency (EE), 76.9%; zeta potential, −4.27 mV) at a dose of 100 mg/kg body weight (BW) increased relative oral bioavailability (AUC/dose) by 25.8-fold as compared to free curcumin at a dose of 250 mg/kg BW. In addition, they found that the T_{max} and C_{max} of the curcumin NP were increased four- and threefold, respectively. Consistent with the results from Shaikh *et al.* (2009), in the same animal model Tsai *et al.* (2011) found the relative oral bioavailability of curcumin-PLGA-NP at a dose of 50 mg/kg BW (size, 158 nm; EE, 46.6%; zeta potential, −12 mV) was 22.8-fold larger than pure curcumin at a, 20-fold larger dose of 1.0 g/kg BW. The greater urinary curcumin excretion and a 50% reduction in curcumin elimination in feces in the rats given curcumin-PLGA-NP as compared to the free form provided further evidence that NPs enhanced curcumin absorption. Interestingly, despite the fact that C_{max} increased by twofold, T_{max} remained essentially the same. In some contrast, using male Sprague Dawley rats, Xie *et al.* (2011) found that 100 mg/kg BW curcumin-PLGA-NP (size, 200 nm; EE, 92.0%), when compared to free curcumin administered at the

Table 10.1 Examples of *in vivo* oral pharmacokinetic studies (nanoparticle vs. free polyphenol) in animal models.[a]

Compound	Studies	Animal	Formulation/dose (per BW)	Time points (h)	C_{max} (μg/mL)	T_{max} (min)	AUC (min*μg/mL)
Curcumin	Shaikh *et al.*, 2009	SD rat	PLGA (50:50) –PVA (1%) NP: 100 mg/kg Free: 250 mg/kg Free + enhancer: 250 mg/kg + 10 mg/kg piperine	0–48	0.26 ± 0.03 vs. 0.09 ± 0.02 vs. 0.12 ± 0.02 $P < 0.001$ for NP-cur $P < 0.01$ for NP-cur+ piperine	120 vs. 30 vs. 45	193.44 ± 19.7 vs. 18.72 ± 0.5 vs. 52.32 ± 2.6 $P < 0.001$ for NP-cur $P < 0.01$ for NP-cur+ piperine
	Tsai *et al.*, 2011	SD rat	PLGA (50:50) –PVA (2%): 50 mg/kg Free: 1 g/kg	0–6	0.044 ± 0.004 vs. 0.022 ± 0.002 $P < 0.05$	~30 vs. ~45	7.32 ± 0.8 vs. 6.44 ± 2.0
	Xie *et al.*, 2011	SD rat	PLGA (50:50) –PVA (1%): 100 mg/kg Free: 100 mg/kg	0–12	6.75 ± 1.54 vs. 1.55 ± 0.21 $P < 0.01$	120 + 0 vs. 102 ± 16 $P < 0.01$	2066 ± 332 vs. 367 ± 21 $P < 0.01$
	Khalil *et al.*, 2013	Wistar rat	PLGA (50:50) –PVA (0.5%): 50 mg/kg PLGA–PEG: 50 mg/kg Free: 50 mg/kg	0–24	0.012 ± 0.0005 vs. 0.030 ± 0.005 vs. 0.004 ± 0.0005 $P < 0.01$	120 vs. 180 vs. 30 $P < 0.01$	8.04 ± 0.21 vs. 26.87 ± 3.84 vs. 0.52 ± 0.11 $P < 0.01$
EGCG	Dube *et al.*, 2011	Swiss outbred mice	CS: 0.76 mg/kg Free: 0.76 mg/kg	0–5	Free EGCG: 0.016 ± 0.004 vs. 0.015 ± 0.002 Total EGCG (free + phase II conjugates): 0.017 ± 0.003 vs. 0.016 ± 0.001	90 for all	Free EGCG: 4.57 ± 0.33 vs. 3.15 ± 0.11 Total EGCG (free + phase II conjugates): 4.93 ± 0.3 vs. 3.19 ± 0.11 $P < 0.05$
Quercetin	Li *et al.*, 2009	Wistar rats	SLN: 50 mg/kg Free: 50 mg/kg	0–48	12.22 ± 2.15 vs. 5.90 ± 1.24	480 vs. 300	19,440 ± 2481 vs. 3403.8 ± 553.8

AUC, area under the curve of time and concentration; C_{max}, maximum concentration; T_{max}, time to C_{max}; EGCG, epigallocatechin-3-gallate; SD, Sprague Dawley; Cur, curcumin; NP, nanoparticle; PLGA, poly(lactic-co-glycolytic acid); PVA, polyvinyl alcohol; PEG, polyethylene glycol; CS, chitosan; SLN, solid lipid nanoparticle.

[a] All data are expressed as mean ± standard deviation.

same dose, provided greater relative oral bioavailability but to a markedly lower amount (5.6-fold) compared to Tsai *et al.* (2011): T_{max} and C_{max} were increased by 1.2- and 4.4-fold, respectively. While the factors accountable for the discrepancy between studies are unclear, it is unlikely attributed to the size of the NP. Recently, Khalil *et al.* (2013) investigated the effect of surface-modified PLGA-PEG NPs on curcumin circulation time in Wistar rats. They observed that both PLGA (size, 161 nm; EE, 77%) and PLGA-PEG (size, 152 nm; EE, 73%) formulations were able to prolong curcumin release in physiological conditions over time. With respect to the efficiency of improving the oral bioavailability of curcumin, PLGA-PEG provided better results in all of the pharmacokinetic parameters compared to PLGA alone. For example, compared to free curcumin in an aqueous solution at a dose of 50 mg/kg BW, the PLGA and PLGA-PEG NP at the same dose increased the relative oral bioavailability of curcumin by 15.6- and 51.6-fold, respectively; the T_{max} was extended by six- and fourfold and the C_{max} was increased by 7.5- and threefold, respectively. One of the potential reasons for the difference in pharmacokinetics between the PLGA and PLGA-PEG NPs may be the longer term of systemic circulation. Further research is required to determine whether the comparative efficacy of PLGA and PLGA-PEG is the same for other polyphenols.

In addition to improving the pharmacokinetic profile, polymeric NPs may increase bioactivity of curcumin. Yallapu *et al.* (2010) found two- and sixfold increases in the cellular uptake of 2–8 μM doses of curcumin-PLGA-NP (size, 76.2 nm; EE, 89%; zeta potential, 0.06 mV) in cisplatin-resistant A2780CP ovarian and metastatic MDA-MB-231 breast cancer cells, respectively, compared to the same dose of free curcumin. The NP-curcumin also showed superior efficacy when compared to the free form in inhibiting proliferation and promoting apoptosis of these two cancer cells *in vitro*. For example, treatment with NP-curcumin resulted in a higher number of cells (five- and eightfold increases in A2780CP and MDA-MB-231 cells, respectively) that were positive for Annexin V staining, an indicator of cell death, compared to free curcumin. Similarly, Anand *et al.* (2010) found that curcumin encapsulated in PLGA-PEG NPs (size, 80.9 nm; EE, 97.5%) was taken up by KBM-5 human chronic myeloid leukemia cells *in vitro* in 5 min as compared to 30 min for the free form. They also noted that NP-curcumin was similarly or more potent than free curcumin in inducing apoptosis in leukemic cells and in suppressing proliferation of breast, colorectal, esophageal, and prostate cancer cell lines. The NP-curcumin was also more potent than free curcumin in suppresing tumor necrosis factor-induced nuclear factor (NF) κB activation and subsequent NFκB-regulated responses, including cell proliferation (cyclin D1), invasion (matrix metallopeptidase-9), and angiogenesis (vascular endothelial growth factor). Similarly, Mukerjee and Vishwanatha (2009) reported that PLGA-NP encapsulation (size, 45 nm; EE, 90%) enhanced curcumin uptake into various prostate cancer cells and arrested cell growth and inhibited NFκB activation as compared to free curcumin. These *in vitro* results demonstrate the potential of PLGA NPs as carriers for the oral delivery of curcumin in cancer therapy. However, *in vivo* research will be necessary to substantiate the therapeutic efficacy of curcumin encapsulated in PLGA-PEG NPs.

10.5.2 Epigallocatechin-3-gallate

The water-soluble flavanol EGCG is found predominately in green tea leaves (*Camellia sinensis*) and has been associated with an array of health benefits including reductions in the risk for some forms of cancer, cardiovascular disease, type 2 diabetes mellitus, and osteoporosis. Epigallocatechin-3-gallate is highly susceptible to degradation in the intestinal milieu and via oxidative processes (Mochizuki *et al.*, 2002); for example, 80% of EGCG was degraded after incubation for 1 h in a simulated intestinal fluid at pH 7.4 (Dube *et al.*, 2010a). Efforts to promote the intake, enhance the stability, and increase bioavailability of EGCG are being explored with the intent to incorporate it into NP for formulation into novel foods. Dube *et al.* (2010a, b, 2011) conducted a series of experiments to evaluate the efficacy of chitosan tripolyphosphate NPs (size, 165 nm; zeta potential, 33 mV) in EGCG stability and bioavailability. Dube *et al.* (2010a) found that EGCG incubated in an alkaline solution (50 mM

KH_2PO_4 buffer, pH 7.4), took 10 and 40 min for pure and chitosan tripolyphosphate EGCG to degrade to 50% of their initial levels, respectively. Using an *ex vivo* mouse jejunum segment mounted in Ussing chambers, Dube *et al.* (2010b) found that chitosan tripolyphosphate EGCG at 6.2 mg/mL (size, 440 nm; zeta potential, 25 mV) exhibited a 1.8-fold greater absorption rate than the same dose of free EGCG. Using mice, these investigators also examined *in vivo* the absorption of 0.76 mg/kg BW chitosan tripolyphosphate EGC (size, 440 nm) and found that the chitosan NP increased the relative oral bioavailability by 1.5-fold compared to the same dose of free EGCG (Dube *et al.*, 2011); the C_{max} increased by onefold but the T_{max} was not different between the NP and pure compound. Further, a 2.3-fold increase in the apparent exposure of encapsulated EGCG was evident in the jejunum tissue at 2.5 h compared to free EGCG, whereas at 1.5 and 5 h the concentrations were similar. Nonetheless, the EE of EGCG in this series of experiments was only 0.05%, a concentration that would not be practical for food applications.

Surface modification with chitosan has been shown to enhance absorption via permeation through the paracellular route (George and Abraham, 2006). Tang *et al.* (2013) developed a green tea flavanol mixture, including EGCG, epicatechin gallate, epigallocatechin, and epicatechin, loaded into a poly-(γ-glutamic acid) γ-(PGA)-chitosan NP (size, 140 nm; zeta potential, 33.5 mV) with an EE of 13.8–23.5% and monitored *in vitro* paracellular permeability through Caco-2 cells. A decrease in transendothelial electrical resistance (TEER), an indicator of monolayer integrity, was utilized as a measure of paracellular permeability. The TEER values of Caco-2 cell monolayers incubated for 2 h with free flavanols showed no significant differences when compared to the control group; however, incubation of the encapsulated tea flavanols on the apical side, at pH 6.6 (intestinal pH) for the same duration led to an immediate and pronounced reduction in TEER. After removal of the incubated NP, a gradual increase in TEER values was observed up to 20 h. Percentages of free tea flavanols transported across the Caco-2 cells were ≤6% but incubation with NPs lead to the enhancement of tea catechins transported to ≈24%. This reduction in TEER followed by an increase in transport with NP incubation compared to free catechins indicate, in an indirect manner, that the NPs lead to an increase in paracellular permeability resulting in an enhancement of transport of the tea flavanols. The gradual increase in TEER after removal of the NPs suggests that the paracellular permeability is reversible.

In addition to enhancing systemic bioavailability, nanoencapsulated EGCG has been shown in a few studies to provide improved bioactivity. Siddiqui *et al.* (2009) evaluated the potential of EGCG encapsulated in PLA-PEG (size, 260 nm; zeta potential, −7.92 mV) for chemoprevention of prostate cancer in athymic nude mice with androgen-responsive 22rv1 cell xenografts in their right and left flanks. The mice received either intraperitoneal administration three times weekly of 100 μg PLA-PEG EGCG, pure EGCG (1 mg) or a control (no treatment). The growth of tumors was monitored over time. At 45 days post-inoculation, the tumor volume in the control mice was larger by 1.8- and 1.5-fold than those administered sNP and free EGCG respectively ($p < 0.01$). In order to achieve a similar extent of tumor growth inhibition, a 10-fold lower dose of the NP EGCG was required compared to free EGCG. The mean serum specific antigen (PSA) level in the control mice was significantly higher, at three- and 11-fold, than that of the NP and free EGCG groups, respectively.

10.5.3 Quercetin

Quercetin, a semi-lipophilic flavonol, is ubiquitous in plant foods. Like most flavonoids, quercetin is extensively transformed via phase II metabolism for elimination from the body (Manach *et al.*, 2008). To date, only one pharmacokinetic study is available which examines the efficacy of polymeric NPs to improve the oral bioavailability of quercetin. Following oral administration of 50 mg/kg BW quercetin to male Wistar rats, Li *et al.* (2009) reported that solid lipid (soya lecithin, Tween-80, and PEG) NPs (size, 155 nm; EE, 91%; zeta potential, −32 mV) increased the relative oral bioavailability of quercetin by 5.7-fold as compared to the free form. The C_{max} and T_{max} of the quercetin NPs increased by two- and 1.6-fold, respectively. The bioactivity of quercetin is largely dependent on its metabolites, so more research is warranted on the effect of quercetin encapsulation on its biotransformation in the GI tract and in other tissues.

While limited information is available on the oral bioavailability of quercetin in polymeric NPs, the effect of nanoencapsulation on its antioxidant and anti-inflammatory action has been examined in several studies. Wu *et al.* (2008) prepared a quercetin NP system with aminoalkyl methacrylate copolymers as a carrier (size, 82 nm; EE, 99%). They found *in vitro* that the radical scavenging activity of quercetin-NP toward DPPH radicals and superoxide anion was enhanced by 883- and 1377-fold as compared to the free form. The quercetin NP was also 59-fold more effective in inhibiting lipid peroxidation reactions. Similarly, Pool *et al.* (2012) reported that quercetin encapsulated within PLGA (1 and 5 μM) displayed a more potent antioxidant action against peroxyl radical-induced lipid peroxidation and a greater capacity for chelating activity toward transient metals than free quercetin. Ghosh *et al.* (2009) evaluated the antioxidant activity of a single pretreatment of 2.71 mg/kg BW quercetin-PLGA administered orally to female rats in protecting against injury from reactive oxygen species (ROS) produced in liver and brain tissue 90 min later by a subcutaneous injection of arsenite. Increased oxidative damage, assessed by glutathione, glutathione peroxidase, glutathione transferase, catalase, glucose-6-dehydrogenase, mitochondrial microviscosity, and lipid peroxidation, was examined 24 h after the arsenite injection. Quercetin-PLGA provided full protection against the arsenite-induced ROS injury while free quercetin was ineffective. The protection of quercetin-PLGA was ascribed mainly to sequestration of ≥60% of the arsenite deposited in mitochondria of the tissues. The same laboratory reported a synergy between quercetin and meso-2,3-dimercaptosuccinic acid (a hydrophilic arsenic chelator) in protection against arsenic-induced damage in the same animal model when they were encapsulated together in PLGA (Gosh *et al.*, 2011).

In a rat model, Chakraborty *et al.* (2012) examined the potency of a single dose of 2.5 mg/kg BW quercetin-PLGA (size, 15 nm; EE, 66%) administered orally prior to alcohol-induced gastric ulcer in protecting against oxidative damage, mitochondrial integrity, and the upregulation of matrix metallopeptidase-9 and inducible nitric oxide synthase. They found that quercetin-PLGA prevented 90% of alcohol-induced ulceration as compared to 20% by free quercetin at the same dose. The quercetin-PLGA maintained the biomarkers of oxidative damage and inflammatory responses in gastric mucosa induced by alcohol at the same level as the negative control. These preclinical results suggest that the potential therapeutic efficacy of quercetin can be enhanced by nanocapsulation. More research is necessary to establish the mechanism(s) by which PLGA nanoencapsulation increases quercetin bioactivity, for example, by protecting quercetin against degradation in acidic environments, increasing overall bioavailability, and/or modifying its metabolite distribution or profile.

10.5.4 Resveratrol

Resveratrol, a lipophilic stilbene found predominantly in grapes and red wine, displays a diverse array of bioactivity, for example, chemoprevention or cancer treatment via roles in cell-cycle arrest, differentiation, and apoptosis, anti-inflammation, putative benefits on risk factors for cardiovascular disease and type 2 diabetes mellitus, and "anti-aging" effects. While resveratrol is poorly absorbed, due in part to its low water solubility and unstable chemical nature, the effect of nanoencapsulation on it pharmacokinetic profile *in vivo* has not yet been tested. Using an *in vitro* simulation of GI conditions with an initial 3-h incubation at 37°C in a pH 1.2 buffer, followed by a 33-h incubation at pH 7.4, Neves *et al.* (2013) tested the concept that resveratrol loaded in lipid NPs (solid lipid NP (cetyl palmitate and polysorbate 60) or nanostructured lipid carriers (cetyl palmitate, polysorbate 60, and the liquid lipid miglyol-812); size, ~170 nm; EE, ~72%; zeta potential, ~ −30 mV) would maintain its chemical integrity in the GI tract environment. They found that ≤24% resveratrol in the NP leached into the GI mimic buffers.

While no pharmacokinetic experiments have been conducted with a NP resveratrol, several studies have examined whether polymeric nanoencapsulation enhances its therapeutic potential. Using lipophilic PCL-PEG as a carrier, Lu *et al.* (2009) examined the effect of a resveratrol-NP (size, 100 nm; EE, 89%) on cell viability, ROS production, and activation of caspase-3 activity (apoptosis) in rat adrenal pheochromocytoma (PC12) cells treated with β-amyloid peptide, an inducer of an Alzheimer's disease model. They found that pretreatment with a resveratrol-NP at 10 μM for 12 h provided a significant protection against β-amyloid peptide-induced toxicity, while free resveratrol exaggerated the toxicity. The authors postulated that nanocapsulation facilitated internalization of resveratrol via endocytosis so that resveratrol would not accumulate in the cell membrane where it could induce lipid peroxidation and its sequelae. Similarly, Shao *et al.* (2009) observed resveratrol encapsulated in a monomethoxypoly-PEG-PCL copolymer NP (size, 87 nm; EE, 90%; zeta potential, −5.6 mV) was more effective to induce cytotoxicity of rat C6 glioma cancer cells than free resveratrol at the same doses (2–32 μM). The induction of cytotoxicity was in parallel with intracellular ROS generation (determined by a H_2DCFDA dye) and was diminished by the co-treatment of vitamin E as its addition decreased ROS generation. It is worth noting the discrepancy in the effect of resveratrol on ROS production with inhibition in the former study and promotion in the latter study, probably due to the use of the ROS inducer in the former study and difference in cell lines. These *in vitro* data appear to support application of nanoencapsulation to magnify therapeutic efficacy of resveratrol and decrease its potential toxicity to normal cells.

10.6 CONCLUSIONS

Research results from efforts to enhance the bioavailability of dietary polyphenols provide encouraging evidence for the feasibility of utilizing nanoencapsulation technology. While there is value in carefully characterizing the transport of nanoparticles *in vitro*, potential applications require substantiation of efficacy and safety *in vivo*. To date, using animal models, only a limited number of *in vivo* studies of nanoencapsulated polyphenols have been completed that characterize their absorption, distribution, metabolism, and elimination. Some studies have demonstrated that increasing the absorption and systemic distribution of polyphenols within nanoparticles is associated with higher concentrations within target cells and/or improved bioactivity. However, the results obtained vary by specific polyphenol and the type and characteristics of the nanoparticle, including its ability to achieve effective entrapment efficiency. Translating concepts for nanoencapsulation developed to improve the transport of drugs and vaccines to food applications has led to modifications of the surface physicochemical properties of polymeric nanoparticles, including the addition of targeting molecules to stabilize the nutrient within the gastrointestinal milieu and to achieve better penetration through the multiple barriers of the gut wall. Further *in vivo* research in animal models and clinical studies is required to demonstrate the putative benefit of nanoparticles of polyphenols and other bioactive constituents of food before we can determine their full range of benefits. Successful "nanonutrient" applications should allow for the development of innovative foods and beverages with a variety of health-promoting and therapeutic functional properties.

References

Acosta, E. (2009) Bioavailability of nanoparticles in nutrient and nutraceutical delivery. *Current Opinion in Colloid and Interface Science,* 14, 3–15.

Ader, P., Wessmann, A. and Wolffram, S. (2000) Bioavailability and metabolism of the flavonol quercetin in the pig. *Free Radical Biology and Medicine,* 28, 1056–1067.

Alexis, F., Pridgen, E., Molnar, L. and Farokhzad, O. (2008) Factors affecting the clearance and biodistribution of polymeric nanoparticles. *Molecular Pharmaceutics,* 5, 505–515.

Anand, P., Nair, H., Sung, B., *et al.* (2010) Design of curcumin-loaded PLGA nanoparticles formulation with enhanced cellular uptake, and increased bioactivity *in vitro* and superior bioavailability *in vivo*. *Biochemical Pharmacology,* 79, 330–338.

Anderson, J. (2001) Molecular structure of tight junctions and their role in epithelial transport. *News in Physiological Sciences,* 16, 126–130.

Bazzano, L., Serdula, M. and Liu, S. (2003) Dietary intake of fruits and vegetables and risk of cardiovascular disease. *Current Atherosclerosis Report*, 5, 492–499.

Beckman, C. (2000) Phenolic-storing cells: keys to programmed cell death and periderm formation in wilt disease resistance and in general defense responses in plants? *Physiological and Molecular Plant Pathology,* 57, 101–110.

Behrens, I., Pena, A., Alonso, M. and Kissel, T. (2002) Comparative uptake studies of bioadhesive and non-bioadhesive nanoparticles in human intestinal cell lines and rats: the effect of mucus on particle adsorption and transport. *Pharmaceutical Research,* 19, 1185–1193.

Bermúdez-Soto, M., Tomás-Barberán, F. and García-Conesa, M. (2007) Stability of polyphenols in chokeberry (*Aronia melanocarpa*) subjected to *in vitro* gastric and pancreatic digestion. *Food Chemistry,* 102, 865–874.

Bernkop-Schnürch, A. (2005) Mucoadhesive polymers: strategies, achievements and future challenges. *Advanced Drug Delivery Reviews,* 57, 1553–1555.

Brand, W., Padilla, B., van Bladeren, P., Williamson, G. and Rietjens, I. (2010) The effect of co-administered flavonoids on the metabolism of hesperetin and the disposition of its metabolites in Caco-2 cell monolayers. *Molecular Nutrition and Food Research,* 54, 851–860.

Brannon-Peppas, L. (1995) Recent advances on the use of biodegradable microparticles and nanoparticles in controlled drug delivery. *International Journal of Pharmaceutics,* 116, 1–9.

Bravo, L. (1998) Polyphenols: chemistry, dietary sources, metabolism, and nutritional significance. *Nutrition Reviews,* 56, 317–333.

Cartiera, M., Johnson, K., Rajendran, V., Caplan, M. and Saltzman, W. (2009)The uptake and intracellular fate of PLGA nanoparticles in epithelial cells. *Biomaterials*, 30(14), 2790–2798.

Chae, S., Jang, M. and Nah, J. (2005) Influence of molecular weight on oral absorption of water soluble chitosans. *Journal of Controlled Release,* 102, 383–394.

Chakraborty, S., Stalin, S., Das, N., Choudhury, S.T., Ghosh, S. and Swarnakar, S. (2012) The use of nano-quercetin to arrest mitochondrial damage and MMP-9 upregulation during prevention of gastric inflammation induced by ethanol in rat. *Biomaterials*, 33, 2991–3001.

Chen, M., Shah, V., Patnaik, R., *et al.* (2001) Bioavailability and bioequivalence: an FDA regulatory overview. *Pharmaceutical Research,* 18, 1645–1650.

Chun, O., Chung, S. and Song, W. (2007) Estimated dietary flavonoid intake and major food sources of U.S. adults. *Journal of Nutrition,* 137, 1244–1252.

Cui, J., Yu, B., Zhao, Y., *et al.* (2009) Enhancement of oral absorption of curcumin by self-microemulsifying drug delivery systems. *International Journal of Pharmaceutics,* 371, 148–155.

D'Archivio, M., Filesi, C., Varì, R., Scazzocchio, B. and Masella, R. (2010) Bioavailability of the polyphenols: status and controversies. *International Journal of Molecular Sciences,* 11, 1321–1342.

Dauchet, L., Amouyel, P., Hercberg, S. and Dallongeville, J. (2006) Fruit and vegetable consumption and risk of coronary heart disease: a meta-analysis of cohort studies. *Journal of Nutrition,* 136, 2588–2593.

Des Rieux, A., Fievez, V., Garinot, M., Schneider, Y.J. and Préat, V. (2006) Nanoparticles as potential oral delivery systems of proteins and vaccines: a mechanistic approach. *Journal of Controlled Release,* 116, 1–27.

Des Rieux, A., Fievez, V., Théate, I., Mast, J., Préat, V. and Schneider, Y. (2007) An improved *in vitro* model of human intestinal follicle-associated epithelium to study nanoparticle transport by M cells. *European Journal of Pharmaceutical Sciences,* 30, 380–391.

Des Rieux, A., Ragnarsson, E., Gullberg, E., Préat, V., Schneider, Y. and Artursson, P. (2005) Transport of nanoparticles across an *in vitro* model of the human intestinal follicle associated epithelium. *European Journal of Pharmaceutical Sciences,* 25, 455–465.

Desai, M., Labhasetwar, V., Amidon, G. and Levy, R. (1996) Gastrointestinal uptake of biodegradable microparticles: effect of particle size. *Pharmaceutical Research,* 13, 1838–1845.

Dube, A., Ng, K., Nicolazzo, J. and Larson, I. (2010a) Effective use of reducing agents and nanoparticle encapsulation in stabilizing catechins in alkaline solution *Food Chemistry,* 122, 662–667.

Dube, A., Nicolazzo, J. and Larson, I. (2010b) Chitosan nanoparticles enhance the intestinal absorption of the green tea catechins (+)-catechin and (−)-epigallocatechin gallate. *European Journal of Pharmaceutical Sciences,* 41, 219–225.

Dube, A., Nicolazzo, J. and Larson, I. (2011) Chitosan nanoparticles enhance the plasma exposure of (−)-epigallocatechin gallate in mice through an enhancement in intestinal stability. *European Journal of Pharmaceutical Sciences*, 44(3), 422–426.

Esmaillzadeh, A., Tahbaz, F., Gaieni, I., Alavi-Majd, H. and Azadbakht, L. (2004) Concentrated pomegranate juice improves lipid profiles in diabetic patients with hyperlipidemia. *Journal of Medicinal Food*, 7, 305–308.

FDA. (2011) *Considering Whether an FDA-Regulated Product Involves the Application of Nanotechnology*. [Online] Available: http://www.fda.gov/RegulatoryInformation/Guidances/ucm257698.htm (accessed on 15 February 2013).

Foged, C., Brodin, B., Frokjaer, S. and Sundblad, A. (2005) Particle size and surface charge affect particle uptake by human dendritic cells in an *in vitro* model. *International Journal of Pharmaceutics*, 298, 315–322.

Fraga, C., Galleano, M., Verstraeten, S. and Oteiza, P. (2010) Basic biochemical mechanisms behind the health benefits of polyphenols. *Molecular Aspects of Medicine*, 31, 435–445.

Frijlink, H., Eissens, A., Hefting, N., Poelstra, K., Lerk, C. and Meijer, D. (1991) The effect of parenterally administered cyclodextrins on cholesterol levels in the rat. *Pharmaceutical Research*, 8, 9–16.

Galindo-Rodriguez, S., Allemann, E., Fessi, H. and Doelker, E. (2005) Polymeric nanoparticles for oral delivery of drugs and vaccines: a critical evaluation of *in vivo* studies. *Critical Reviews in Therapeutic Drug Carrier Systems*, 22, 419–464.

George, M. and Abraham, T. (2006) Polyionic hydrocolloids for the intestinal delivery of protein drugs: alginate and chitosan--a review. *Journal of Controlled Release*, 114, 1–14.

Ghosh, A., Mandal, A., Sarkar, S., Panda, S. and Das, N. (2009) Nanoencapsulation of quercetin enhances its dietary efficacy in combating arsenic-induced oxidative damage in liver and brain of rats. *Life Sciences*, 84, 75–80.

Ghosh, S., Dungdung, S.R., Chowdhury, S.T., *et al.* (2011) Encapsulation of the flavonoid quercetin with an arsenic chelator into nanocapsules enables the simultaneous delivery of hydrophobic and hydrophilic drugs with a synergistic effect against chronic arsenic accumulation and oxidative stress. *Free Radical Biology and Medicine*, 51, 1893–1902.

Gurley, B. (2011) Emerging technologies for improving phytochemical bioavailability: benefits and risks. *Clinical Pharmacology and Therapeutics*, 89, 915–919.

He, K., Hu, F., Colditz, G., Manson, J., Willett, W. and Liu, S. (2004) Changes in intake of fruits and vegetables in relation to risk of obesity and weight gain among middle-aged women. *International journal of Obesity and Related Metabolic Disorders*, 28, 1569–1574.

Heaney, R. (2001) Factors influencing the measurement of bioavailability, taking calcium as a model. *Journal of Nutrition*, 131, 1344S-1348S.

Hendrich, S. (2002) Bioavailability of isoflavones. *Journal of Chromatography. B, Analytical Technologies in the Biomedical and Life Sciences*, 777(1–2), 203–210.

Hillery, A., Jani, P. and Florence, A. (1994) Comparative, quantitative study of lymphoid and non-lymphoid uptake of 60 nm polystyrene particles. *Journal of Drug Targeting*, 2, 151–156.

Huang, Y., Leobandung, W., Foss, A. and Peppas, N. (2000) Molecular aspects of muco- and bioadhesion: tethered structures and site-specific surfaces. *Journal of Controlled Release*, 65, 63–71.

Hung, H., Joshipura, K., Jiang, R., *et al.* (2004) Fruit and vegetable intake and risk of major chronic disease. *Journal of the National Cancer Institute*, 96, 1577–1584.

Jani, P., Halbert, G., Langridge, J. and Florence, A. (1990) Nanoparticle uptake by the rat gastrointestinal mucosa: quantitation and particle size dependency. *Journal of Pharmacy and Pharmacology*, 42, 821–826.

Jung, T., Kamm, W., Breitenbach, A., Kaiserling, E., Xiao, J. and Kissel, T. (2000) Biodegradable nanoparticles for oral delivery of peptides: is there a role for polymers to affect mucosal uptake? *European Journal of Pharmaceutics and Biopharmaceutics*, 50, 147–160.

Khalil, N, do Nascimento, T., Casa, D., *et al.* (2013) Pharmacokinetics of curcumin-loaded PLGA and PLGA-PEG blend nanoparticles after oral administration in rats. *Colloids and Surfaces B: Biointerfaces*, 101, 353–360.

Kimmons, J., Gillespie, C., Seymour, J., Serdula, M. and Blanck, H. (2009) Fruit and vegetable intake among adolescents and adults in the United States: percentage meeting individualized recommendations. *Medscape Journal of Medicine*, 11, 26.

Knekt, P., Kumpulainen, J., Järvinen, R., *et al.* (2002) Flavonoid intake and risk of chronic diseases. *American Journal of Clinical Nutrition*, 76, 560–568.

Krook, M. and Hagerman, A. (2012) Stability of polyphenols epigallocatechin gallate and pentagalloyl glucose in a simulated digestive system. *Food Research International*, 49, 112–116.

Kumari, A., Yadav, S. and Yadav, S. (2010) Biodegradable polymeric nanoparticles based drug delivery systems. *Colloids and Surfaces B: Biointerfaces*, 75, 1–18.

Lai, S., Wang, Y. and Hanes, J. (2009) Mucus-penetrating nanoparticles for drug and gene delivery to mucosal tissues. *Advanced Drug Delivery Reviews,* 61, 158–171.

Lam, W., Kazi, A., Kuhn, D., *et al.* (2004) A potential prodrug for a green tea polyphenol proteasome inhibitor: evaluation of the peracetate ester of (−)-epigallocatechin gallate [(−)-EGCG]. *Bioorganic and Medicinal Chemistry,* 12, 5587–5593.

Lesser, S., Cermak, R. and Wolffram, S. (2004) Bioavailability of quercetin in pigs is influenced by the dietary fat content. *Journal of Nutrition,* 134, 1508–1511.

Li, H., Zhao, X., Ma, Y., Zhai, G., Li, L. and Lou, H. (2009) Enhancement of gastrointestinal absorption of quercetin by solid lipid nanoparticles. *Journal of Controlled Release,* 133, 238–244.

Liu, R. (2003) Health benefits of fruit and vegetables are from additive and synergistic combinations of phytochemicals. *American Journal of Clinical Nutrition,* 78, 517S-520S.

Lu, Z., Ji, C., Xu, H., *et al.* (2009) Resveratrol-loaded polymeric micelles protect cells from Aβ- induced oxidative stress. *International Journal of Pharmaceutics,* 375, 89–96.

Manach, C., Williamson, G., Morand, C., Scalbert, A. and Rémésy, C. (2005) Bioavailability and bioefficacy of polyphenols in humans. I. Review of 97 bioavailability studies. *American Journal of Clinical Nutrition,* 81, 230S-242S.

McClean, S., Prosser, E., Meehan, E., *et al.* (1998) Binding and uptake of biodegradable poly-DL-lactide micro- and nanoparticles in intestinal epithelia. *European Journal of Pharmaceutical Sciences,* 6, 153–163.

McCullough, M., Peterson, J., Patel, R., Jacques, P., Shah, R. and Dwyer, J. (2012) Flavonoid intake and cardiovascular disease mortality in a prospective cohort of US adults. *American Journal of Clinical Nutrition,* 95, 454–464.

Milbury, P., Vita, J. and Blumberg, J. (2010) Anthocyanins are bioavailable in humans following an acute dose of cranberry juice. *Journal of Nutrition,* 140, 1099–1104.

Mochizuki, M., Yamazaki, S.-I., Kano, K. and Ikeda, T. (2002) Kinetic analysis and mechanistic aspects of autoxidation of catechins. *Biochimica et Biophysica Acta (BBA) – General Subjects,* 1569, 35–44.

Mu, X. and Zhong, Z. (2006) Preparation and properties of poly(vinyl alcohol)-stabilized liposomes. *International Journal of Pharmaceutics,* 318, 55–61.

Mukerjee, A. and Vishwanatha, J. (2009) Formulation, characterization and evaluation of curcumin-loaded PLGA nanospheres for cancer therapy. *Anticancer Research,* 29, 3867–3875.

Musthaba, S., Baboota, S., Ahmed, S., Ahuja, A. and Ali, J. (2009) Status of novel drug delivery technology for phytotherapeutics. *Expert Opinion on Drug Delivery,* 6, 625–637.

Nair, H., Sung, B., Yadav, V., Kannappan, R., Chaturvedi, M. and Aggarwal, B. (2010) Delivery of antiinflammatory nutraceuticals by nanoparticles for the prevention and treatment of cancer. *Biochemical Pharmacology,* 80, 1833–1843.

Neves, A., Lúcio, M., Martins, S., Lima, J. and Reis, S. (2013) Novel resveratrol nanodelivery systems based on lipid nanoparticles to enhance its oral bioavailability. *International Journal of Nanomedicine,*. 8, 177–187.

Nijveldt, R., van Nood, E., van Hoorn, D., Boelens, P., van Norren, K. and van Leeuwen, P. (2001) Flavonoids: a review of probable mechanisms of action and potential applications. *American Journal of Clinical Nutrition,* 74, 418–425.

Owens, D. and Peppas N. (2006) Opsonization, biodistribution, and pharmacokinetics of polymeric nanoparticles. *International Journal of Pharmaceutics,* 307, 93–102.

Pietzonka, P., Rothen-Rutishauser, B., Langguth, P., Wunderli-Allenspach, H., Walter, E. and Merkle, H. (2002) Transfer of lipophilic markers from PLGA and polystyrene nanoparticles to caco-2 monolayers mimics particle uptake. *Pharmaceutical Research,* 19, 595–601.

Pool, H., Quintanar, D. and de Dios Figueroa, J. (2012) Antioxidant effects of quercetin and catechin encapsulated into PLGA nanoparticles. *Journal of Nanomaterials,* 2012, 1–12.

Pralhad, T. and Rajendrakumar, K. (2004) Study of freeze-dried quercetin-cyclodextrin binary systems by DSC, FT-IR, X-ray diffraction and SEM analysis. *Journal of Pharmaceutical and Biomedical Analysis*, 34, 333–339.

Raghuram,T. and Krishnaswamy, K. (1992) Basic concepts of clinical pharmacokinetics. *Journal of the Association of Physicians India,* 40, 29–35.

Reix, N., Parat, A., Seyfritz, E., *et al.* (2012) *In vitro* uptake evaluation in Caco-2 cells and *in vivo* results in diabetic rats of insulin-loaded PLGA nanoparticles. *International Journal of Pharmaceutics,* 437, 213–220.

Scalbert, A. and Williamson, G. (2000) Dietary intake and bioavailability of polyphenols. *Journal of Nutrition,* 130, 2073S-2085S.

Scalbert, A., Johnson, I. and Saltmarsh, M. (2005) Polyphenols: antioxidants and beyond. *American Journal of Clinical Nutrition,* 81, 215S-217S.

Scalbert A., Andres-Lacueva C., Arita M., *et al.* (2011) Databases on food phytochemicals and their health-promoting effects. *Journal of Agriculture and Food Chemistry,* 59, 4331–4348.
Scheepens, A., Tan, K. and Paxton, J. (2010) Improving the oral bioavailability of beneficial polyphenols through designed synergies. *Genes and Nutrition,* 5, 75–87.
Shaikh, J., Ankola, D., Beniwal,V., Singh, D. and Kumar, M. (2009) Nanoparticle encapsulation improves oral bioavailability of curcumin by at least 9-fold when compared to curcumin administered with piperine as absorption enhancer. *European Journal of Pharmaceutical Sciences,* 37, 223–230.
Shakweh, M., Ponchel, G. and Fattal, E. (2004) Particle uptake by Peyer's patches: a pathway for drug and vaccine delivery. *Expert Opinion on Drug Delivery,* 1, 141–163.
Shao, J., Li, X., Lu, X., *et al.* (2009) Enhanced growth inhibition effect of resveratrol incorporated into biodegradable nanoparticles against glioma cells is mediated by the induction of intracellular reactive oxygen species levels. *Colloids and Surfaces B: Biointerfaces,* 72, 40–47.
Shoba, G., Joy, D., Joseph, T., Majeed, M., Rajendran, R. and Srinivas, P. (1998) Influence of piperine on the pharmacokinetics of curcumin in animals and human volunteers. *Planta Medica,* 64, 353–356.
Siddiqui, I., Adhami, V., Bharali, D., *et al.* (2009) Introducing nanochemoprevention as a novel approach for cancer control: proof of principle with green tea polyphenol epigallocatechin-3-gallate. *Cancer Research,* 69, 1712–1716.
Smith, J., Wood, E. and Dornish, M. (2004) Effect of chitosan on epithelial cell tight junctions. *Pharmaceutical Research,* 21, 43–49.
Smith-Warner, S., Spiegelman, D., Yaun, S., *et al.* (2003) Fruits, vegetables and lung cancer: a pooled analysis of cohort studies. *International Journal of Cancer,* 107, 1001–1011.
Sonaje, K., Lin, Y., Juang, J., Wey, S., Chen, C. and Sung, H. (2009) *In vivo* evaluation of safety and efficacy of self-assembled nanoparticles for oral insulin delivery. *Biomaterials*, 30, 2329–2339.
Sonaje, K., Chuang, E., Lin, K., *et al.* (2012) Opening of epithelial tight junctions and enhancement of paracellular permeation by chitosan: microscopic, ultrastructural, and computed-tomographic observations. *Molecular Pharmaceutics,* 9, 1271–1279.
Soppimath, K., Aminabhavi, T., Kulkarni, A. and Rudzinski, W. (2001) Biodegradable polymeric nanoparticles as drug delivery devices. *Journal of Controlled Release,* 70, 1–20.
Srinivasan, V. (2001) Bioavailability of nutrients: a practical approach to *in vitro* demonstration of the availability of nutrients in multivitamin-mineral combination products. *Journal of Nutrition,* 131, 1349S–1350S.
Tang, D., Yu, S., Ho, Y., *et al.* (2012) Characterization of tea catechins-loaded nanoparticles prepared from chitosan and an edible polypeptide. *Food Hydrocolloids*, 30, 33–41.
Tapiero, H., Tew, K., Ba, G. and Mathé, G. (2002) Polyphenols: do they play a role in the prevention of human pathologies? *Biomed Pharmacother.* 2002 56, 200–207.
Thanou, M., Verhoef, J. and Junginger, H. (2001) Chitosan and its derivatives as intestinal absorption enhancers. *Advanced Drug Delivery Reviews,* 50, S91–101.
Thomas, V., Bhattachar, S., Hitchingham, L., *et al.* (2006) The road map to oral bioavailability: an industrial perspective. *Expert Opinion on Drug Metabolism and Toxicology,* 2, 591–608.
Tobío, M., Sánchez, A., Vila, A., *et al.* (2000) The role of PEG on the stability in digestive fluids and *in vivo* fate of PEG-PLA nanoparticles following oral administration. *Colloids and Surfaces B: Biointerfaces,* 18, 315–323.
Tønnesen, H. (2002) Solubility, chemical and photochemical stability of curcumin in surfactant solutions. Studies of curcumin and curcuminoids, XXVIII. *Pharmazie*, 57, 820–824.
Tsai, Y., Jan, W., Chien, C. Lee, W., Lin, L. and Tsai, T. (2011) Optimised nano-formulation on the bioavailability of hydrophobic polyphenol, curcumin, in freely-moving rats. *Food Chemistry,* 127, 918–925.
Van der Merwe, S., Verhoef, J., Verheijden, J., Kotzé, A. and Junginger, H. (2004) Trimethylated chitosan as polymeric absorption enhancer for improved peroral delivery of peptide drugs. *European Journal of Pharmaceutics and Biopharmaceutics,* 58, 225–235.
Vasir, J. and Labhasetwar, V. (2008) Quantification of the force of nanoparticle-cell membrane interactions and its influence on intracellular trafficking of nanoparticles. *Biomaterials*, 29, 4244–4252.
Verhagen, H., Coolen, S., Duchateau, G., Hamer, M., Kyle, J. and Rechner, A. (2004) Assessment of the efficacy of functional food ingredients-introducing the concept "kinetics of biomarkers". *Mutation Research,* 551, 65–78.
Vila, A., Sánchez, A., Tobío, M., Calvo, P. and Alonso, M. (2002) Design of biodegradable particles for protein delivery. *Journal of Controlled Release,* 78, 15–24.
Wedick, N., Pan, A., Cassidy, A., *et al.* (2012) Dietary flavonoid intakes and risk of type 2 diabetes in US men and women. *American Journal of Clinical Nutrition,* 95, 925–933.

Weissenboeck, A., Bogner, E., Wirth, M. and Gabor, F. (2004) Binding and uptake of wheat germ agglutinin-grafted PLGA-nanospheres by caco-2 monolayers. *Pharmaceutical Research,* 21, 1917–1923.

Windrum, P., Morris, T., Drake, M., Niederwieser, D. and Ruutu, T. (2005) Variation in dimethyl sulfoxide use in stem cell transplantation: a survey of EBMT centres. EBMT Chronic Leukaemia Working Party Complications Subcommittee. *Bone Marrow Transplantation*, 36, 601–603.

Wu, T., Yen, F., Lin, L., Tsai, T., Lin, C. and Cham, T. (2008) Preparation, physicochemical characterization, and antioxidant effects of quercetin nanoparticles. *International Journal of Pharmaceutics,* 346, 160–168.

Xie, X., Tao, Q., Zou, Y., *et al.* (2011) PLGA nanoparticles improve the oral bioavailability of curcumin in rats: characterizations and mechanisms. *Journal of Agriculture and Food Chemistry,* 59, 9280–9289.

Yallapu, M., Gupta, B., Jaggi, M. and Chauhan, S. (2010) Fabrication of curcumin encapsulated PLGA nanoparticles for improved therapeutic effects in metastatic cancer cells. *Journal of Colloid and Interface Science,* 351, 19–29.

Yin, Y., Chen, D., Qiao, M., Wei, X. and Hu, H. (2007) Lectin-conjugated PLGA nanoparticles loaded with thymopentin: ex vivo bioadhesion and *in vivo* biodistribution. *Journal of Controlled Release,* 123, 27–38.

Yoncheva, K., Guembe, L., Campanero, M. and Irache, J. (2007) Evaluation of bioadhesive potential and intestinal transport of pegylated poly(anhydride) nanoparticles. *International Journal of Pharmaceutics,* 334, 156–165.

Yuan, Z., Chen, L., Fan, L., *et al.* (2006) Liposomal quercetin efficiently suppresses growth of solid tumors in murine models. *Clinical Cancer Research* 12, 3193–3199.

Yue, Z., Wei, W., Lv, P., *et al.* (2011) Surface charge affects cellular uptake and intracellular trafficking of chitosan-based nanoparticles. *Biomacromolecules*, 12, 2440–2446.

Zhang, L., Sun, M., Guo, R., *et al.* (2006) Chitosan surface-modified hydroxycamptothecin loaded nanoparticles with enhanced transport across Caco-2 cell monolayer. *Journal of Nanoscience and Nanotechnology*, 6, 2912–2920.

11 Properties and applications of multilayer and nanoscale emulsions

Moumita Ray, Renuka Gupta, and Dérick Rousseau

Department of Chemistry and Biology, Ryerson University, Toronto, Ontario, Canada

11.1 INTRODUCTION

To respond to increasing consumer demand for healthier foods options, food processors must find new ways of incorporating fat-soluble functional ingredients (e.g., omega-3 fatty acids, phytosterols, and carotenoids) into liquid processed foods such as dairy products, soups, juices, and supplements. Conventional emulsions see extensive use for such applications as they potentially offer: (i) tailored controlled release properties, (ii) the protection of emulsified material (e.g., against oxidation), and (iii) lengthy emulsion kinetic stability (+1 year). Other advantages include effective entrapment of functional ingredients and a single-stage preparation that can easily be scaled up for industrial production.

Nanoemulsions and multilayer emulsions are two recent categories of dispersed systems drawing attention as novel delivery systems for lipophilic compounds. Nanoemulsions possess a number of properties that make them suited for delivery applications, including their optical clarity, small average droplet diameter (<200 nm) and kinetic stability against visual phase separation (months to years) (Wooster *et al.*, 2008; Mao *et al.*, 2009; Wang *et al.*, 2009). Though much of the knowledge base on nanoemulsions has been generated in other fields such as cosmetics and personal-care products, they are rapidly becoming a 'go-to' category for encapsulation and controlled delivery systems in novel food applications (Wang *et al.*, 2008; Liu *et al.*, 2006; Shafiq *et al.*, 2007). In a similar vein, multilayer emulsions, which consist of micron-scale emulsion droplets covered by at least two oppositely charged polyelectrolyte or biopolymer nanoscale layers, have been the subject of extensive food-related research given their ability to yield kinetically stable emulsions that minimize oxidation of polyunsaturated oils of vegetable or animal origin. The objective of this chapter is to provide the reader with insights regarding the formation and stability of nanoemulsions and multilayer emulsions as well as examples of recent applications on the dispersion and incorporation of lipophilic compounds within emulsions. More information on emulsions, specifically on emulsion property control, can be found in Chapter 5.

11.2 CLASSIFICATION OF EMULSIONS

At all length scales, emulsions consist of at least two immiscible fluids where one is dispersed as droplets in the other. In a conventional emulsion, oil is dispersed as micron-scale droplets in an aqueous phase to form an oil-in-water (O/W) emulsion (e.g., milk, cream, and mayonnaise) whereas

Nanotechnology and Functional Foods: Effective Delivery of Bioactive Ingredients, First Edition.
Edited by Cristina M. Sabliov, Hongda Chen, and Rickey Y. Yada.

Table 11.1 Characteristics of different type of emulsions. (Source: McClements, 2011. Reproduced with permission from The Royal Society of Chemistry.)

Emulsion type	Diameter	Thermodynamic stability	Appearance	Surface-to-mass ratio (m^2/g)
Conventional	0.1–100 μm	Unstable	Turbid	0.07–70
Nanoemulsion	10–200 nm	Unstable	Transparent	70–330
Microemulsion	5–100 nm	Stable	Transparent	330–1300

a water-in-oil (W/O) emulsion (e.g., margarine, butter, and spreads) consists of an aqueous phase dispersed as droplets in an oil phase. Conventional emulsions (Table 11.1) are thermodynamically unstable, visually opaque and usually consist of a low surfactant:oil weight ratio (<1:10). Nanoemulsions are typically visually transparent given their small diameter (<200 nm) and consist of a higher surfactant:oil weight ratio (~1:1), given their much greater surface-to-volume ratio compared to conventional emulsions. By contrast, microemulsions are thermodynamically stable dispersions of water and an apolar (oil) phase, stabilized by a surfactant, usually in conjunction with a co-surfactant and/or short chain alcohol. Diverse self-assembled, well-defined and non-birefringent structures, such as droplets (O/W, W/O, and bicontinuous phases) can exist within microemulsions. Their dispersed domains are well-below 100 nm in diameter. They have been investigated as delivery systems for pharmaceuticals, nutraceuticals, and other ingredients as they can improve the solubility of poorly soluble compounds (e.g., sterols) (Garti *et al.*, 2004), increase bioavailability (e.g., of oil-soluble vitamins) (Rozner *et al.*, 2010), and protect labile compounds against light and/or enzyme-induced breakdown (e.g., oxidation of omega-3 fatty acids) (Swenson and Curatolo, 1992; Flanagan and Singh, 2006; Sanguansri and Augustin, 2006). Though they are easily prepared, they normally consist of high surfactant concentrations, which limit their applicability in processed foods.

11.3 EMULSION STABILITY

As both multilayer emulsions and nanoemulsions are thermodynamically unstable, they will break down over time. Hence, they must be stabilized with the use of surfactants or emulsifiers adsorbed to the oil–water interface. The presence of at least one surfactant (e.g., lecithin or synthetic small-molecule surfactants such as polysorbates and sucrose esters) and/or emulsifier (i.e., proteins and surface-active polysaccharides) is essential to form a kinetically stable emulsion. Both surfactants and emulsifiers possess hydrophilic and hydrophobic moieties that can adsorb to the water–oil interface, with their efficacy depending on their molecular composition and spatial arrangement as well as the type of emulsion formed (e.g., O/W vs. W/O). The presence of a surfactant monolayer at the interface normally leads to a reduction in interfacial tension (typically down to < 5 mN/m). Stabilization can be accentuated by the use of charged surfactants that promote charge–charge repulsion between surfactant-coated droplets.

Though emulsifiers can result in a greater reduction in interfacial contact on a molar basis than small-molecule surfactants, their key advantage lies in their ability to form viscoelastic interfacial films that can confer steric (and possibly electrostatic) stabilization to an emulsion. However, conventional emulsions stabilized with emulsifiers often possess limited resistance against stresses associated with local changes in pH, ionic strength, or temperature.

The kinetic stability of an emulsion refers to its ability to resist breakdown as a result of physical processes such as creaming, sedimentation, flocculation, coalescence, Ostwald ripening, and phase inversion. Emulsion stability will depend on a number of factors, including the following.

11.3.1 Strength of the interfacial film

The greatest mechanical stability of an emulsion arises from the presence of a cohesive interfacial film consisting of adsorbed emulsifiers (such as proteins) that provide high film elasticity (Rosen, 2004). Their corollary in W/O emulsions is best represented by interfacial colloidal particles offering full monolayer coverage (Schulman and Leja, 1940; Ford and Furmidge, 1966).

11.3.2 Presence of electrical and steric barriers

A charged interfacial layer is a significant stabilizing factor in O/W emulsions as it improves stability via electrostatic repulsion between adjacent charged droplets thereby retarding the probability of droplet encounters, flocculation, and ultimately coalescence.

Steric stabilization arises from the presence of emulsifier hairy layers at the droplet surface, typically as a result of protein or polysaccharide adsorption. With such layers, work is required to push the droplets closer than their biopolymer layers will allow given that the biopolymer layers overlap and repel. The typical brush layer thickness is ~20 nm for a 1 μm diameter droplet. This mode of stabilization is also important in multilayer emulsions given the presence of multiple biopolymer layers enrobing dispersed droplets.

11.3.3 Droplet size distribution

An emulsion with a fairly uniform narrow size distribution will be more stable than one with the same average particle size but wider distribution. This will be important in regards to emulsion phase separation (creaming and sedimentation), which can be determined using Stokes' law for an isolated spherical particle in a highly diluted emulsion (equation 11.1):

$$v_s = \frac{d^2 \Delta\rho g}{18\eta_c} \tag{11.1}$$

where v_s is the velocity of the droplet, $\Delta\rho$ is the density difference between the dispersed and continuous phases, d is the diameter of the droplet, η_c is the viscosity of the continuous phase and g is the acceleration due to gravity. Creaming/sedimentation in an emulsion can be effectively reduced by increasing aqueous phase viscosity, decreasing droplet size, and/or closely matching the densities of the two phases.

This equation is based on the assumption of infinitely dilute emulsions for a single droplet. In reality, the presence of neighboring droplets will increase emulsion viscosity (Mason, *et al.*, 2006), which in turns will influence their movement thereby resulting in velocities lower than predicted by Stokes' law (Dickinson and Stainsby, 1988). At sufficiently high droplet volume fractions (*i.e.*, close packing of droplets), creaming may eventually be stopped. As well, aggregation increases the effective diameter of the flocs, thereby heightening the rate of creaming or sedimentation. Walstra (2003) combined these factors to estimate the velocity of droplets in an emulsion with a given dispersed phase volume fraction (equation 11.2):

$$v_\phi = v_s(1-\phi)^n \tag{11.2}$$

where $n = 6.55$ for a monodispersed emulsion. The inherent polydispersity in the droplets size distribution of food emulsions results in higher values of n (up to 9) (Dickinson *et al.*, 1988; Walstra, 2003).

11.3.4 Phase volume ratio

Surfactants are most effective at emulsion stabilization when present at their critical micelle concentration (CMC), which results in the monolayer coverage of all droplets in an emulsion. For a given dispersed phase volume and surfactant concentration at the CMC, an increase in surface-to-volume ratio will result in the redistribution of surfactants onto more droplets, thereby yielding a surfactant surface concentration below CMC. In this scenario, the resulting emulsion is less stable as the droplets are not fully covered, thereby exposing bald patches more prone to flocculation and coalescence. A further consideration is the possibility of phase inversion if the dispersed phase volume exceeds the continuous phase volume (Rosen, 2004).

In the case of nanoemulsions, given their small size, the surrounding interfacial layer may appreciably contribute to the overall dispersed-phase volume fraction (Garti *et al.*, 2004; Wilking and Mason, 2007; McClements, 2011). For a nanoemulsion with average droplet radius r and interfacial layer thickness h, the ratio of effective versus core droplet volume fraction can be calculated as $(1 + h/r)$ (Weiss and McClements, 2000; Wilking and Mason, 2007).

11.3.5 Continuous phase viscosity

An increase in continuous phase viscosity reduces the droplet collision frequency and the rate of coalescence, which increase emulsion stability. This is highlighted in the Stoke equation above.

11.3.6 Temperature

Besides its effect on kinetic stability, changes in temperature also affect the interfacial tension between the two liquids as well as the nature and viscosity of the interfacial film surrounding dispersed droplets (Rosen, 2004; McClements, 2005).

11.4 MULTILAYER EMULSIONS

This section describes the low molecular weight surfactants, proteins and polysaccharides that may be used in the formulation of multilayer emulsions. This is followed by a discussion on the formation of multilayer emulsions and the roles of pH, salt, ionic strength, and temperature.

11.4.1 Low molecular weight surfactants

The hydrophile–lipophile balance (HLB) value of a surfactant describes its affinity towards the oil and aqueous phases in an emulsion. A high HLB number indicates that the surfactant contains a higher number of hydrophilic groups than lipophilic groups and vice versa (equation 11.3). In order to calculate the HLB number of a surfactant, the number and type of hydrophilic groups and lipophilic groups must be known:

$$\text{HLB} = 7 + \Sigma(\text{hydrophilic group numbers}) - \Sigma(\text{lipophilic group numbers}). \quad (11.3)$$

In general, low HLB numbers in the range of 3 to 6 indicate that the surfactant has a larger number of lipophilic groups, is oil-soluble and can stabilize W/O emulsions. Longer or more saturated aliphatic chains produce smaller HLB numbers. With a higher HLB number (10–18), a surfactant contains more hydrophilic groups, is soluble in water, and can stabilize O/W emulsions. Surfactants with intermediate HLB numbers between 7 and 9 do not exhibit selectivity between the oil and aqueous

Table 11.2 HLB values and applications of emulsifying agents. (Source: Rosen 2004. Reproduced with permission from Wiley.)

HLB range	Applications
3–6	Water-in-oil emulsions
7–9	Wetting agents
10–18	Oil-in-water emulsion
13–15	Detergent
15–18	Solubilizer

phases. Surfactants with HLB values lower than 3 and higher than 18 are not surface-active and therefore tend to gather in either the oil or aqueous phase. The HLB ranges and their related applications are outlined in Table 11.2.

Surfactants adsorb at the interface and orient their hydrophilic head towards the aqueous phase and the hydrophobic tail towards the nonaqueous phase (Rosen, 2004; McClements, 2005). There are three key surfactant characteristics that enable them to enhance the formation and stability of emulsions: (i) rapid adsorption of the surfactant molecules onto the droplets during homogenization; (ii) monolayer coverage and reduction of the interfacial tension; and (iii) retardation of destabilization phenomena.

11.4.2 Biopolymers as emulsifiers

Biopolymers such as proteins and surface-active polysaccharide-based species can form a thick viscoelastic film at the oil–water interface that provides stability via a combination of electrostatic and steric repulsive forces (McClements, 2005).

11.4.2.1 Proteins

As a result of their amphipathic nature, proteins are often the emulsifier of choice for food processors. From a fundamental perspective, the relationship between protein structure and emulsifying activity depends on a combination of factors such as molecular flexibility, molecular size, surface hydrophobicity, net charge, and amino acid composition. Typically, protein adsorption depends on the surface charge and hydrophobicity as well as the pH and ionic strength of the medium. Adsorbed monolayer thickness has been shown to differ between compact globular protein (e.g., β-lactoglobulin) and disordered flexible protein (e.g., β-casein) (Dickinson, 1995a, b).

11.4.2.2 Gums and thickeners

A great number of polysaccharides are used to gel and thicken food products. In the context of dispersions, emulsion gels consist of an O/W emulsion stabilized by a highly viscous or gelled biopolymer matrix. Such gels have been studied as texture modifiers in processed foods (Yost and Kinsella, 1992; Dickinson and Yamamoto, 1996; Chen and Dickinson, 1998; Cortesi *et al.*, 2002) and more recently as a new class of biomaterials for controlled release applications (Gulsen and Chauhan, 2005; Holtze *et al.*, 2005; Chen *et al.*, 2007; Shingel *et al.*, 2008, 2009; Tokuyama and Kato, 2008).

Thickening/gelling agents are long-chained, straight or branched polysaccharide-based biopolymers that contain hydroxyl groups capable of hydrogen bonding. It is impossible to attribute them a distinct molecular weight as they are polydisperse molecules. The number of monosaccharide units in

Table 11.3 Functional properties of some commonly used hydrocolloids. (Source: Rosen 2004. Reproduced with permission from Wiley.)

Hydrocolloid	Carbohydrate backbone	Function
Starch	D-α-1,4-linked glucose	Thickening agent, stabilizer
Cellulose	D-β-1,4-linked glucose	Gelling agent, thickener, stabilizer, emulsifier
Gum Arabic	Arabino-galactan	Emulsifier, stabilizer, thickener
Guar gum	Galactomannan	Emulsifier, stabilizer, thickener
Pectin	Homogalacturonan	Gelling agent, thickener, stabilizer, emulsifier
Xanthan	D-β-1,4-linked glucose	Stabilizer, thickener, emulsifier, foaming agent
Carrageenan	Sulphated galactans	Thickener, gelling agent, stabilizer, emulsifier
Alginate	Alginic acid	Thickening agent, stabilizer
Carob, guar	Galactomannan	Stabilizer, thickener, emulsifier

a polysaccharide is termed degree of polymerization. The molecular weight of polysaccharides typically varies between approximately 10 and 10,000 kDa (Stauffer, 1999). Their gelling/thickening ability is highly dependent on composition, concentration, interactions with other components present in the formulation, and local environmental conditions such as pH, ionic strength, and temperature. Table 11.3 lists some hydrocolloids that have been used in emulsion systems.

11.4.3 Polyelectrolyte properties

We define a polyelectrolyte as a protein or surface-active polysaccharide that can form physical or chemical crosslinks. A commonly-studied approach to stabilizing O/W emulsions is via polyelectrolyte multilayer deposition at the interface, which is modulated by their concentration, charge status, and the nature of the different polyelectrolytes used for the system. It is advantageous to combine two biopolymers (protein and polysaccharide), as their synergistic effects significantly improve the emulsifying capacity and shelf life over either biopolymer alone (Dickinson, 1993; 1995a, b, 2003; Benichou *et al.*, 2002). Proteins and polysaccharides can associate either via physical interactions (noncovalent interactions such as electrostatic interactions or hydrogen bonding) or covalent interactions. Certain polysaccharides such as gum arabic (Phillips and Williams, 1995, 2003), and soluble soybean polysaccharides can form protein–polysaccharide complexes as they contain a protein moiety representing 2–4% of their total molecular weight (Ray *et al.*, 1995; Nakamura *et al.*, 2003, 2004).

Emulsion stability depends on the concentration of polyelectrolytes or emulsifying agent present at the interface and in the bulk (Xu *et al.*, 2001). When the polyelectrolyte concentration is insufficient to completely saturate the droplet surface, bridging flocculation occurs. Without complete coverage by a secondary polysaccharide layer, dispersed oil droplets may possess both positive and negative patches on their surfaces thereby promoting droplet aggregation upon colliding. Conversely, if the free biopolymer concentration in the continuous phase exceeds a critical value, depletion flocculation occurs as the depletion forces overcome repulsive forces (i.e., electrostatic and steric). Therefore, to produce a stable multilayer emulsion against bridging and depletion flocculation, polyelectrolyte concentrations should permit effective droplet formation while minimizing the presence of free polyelectrolytes in the continuous phase.

11.4.4 pH

pH is an important factor in emulsion stability as it regulates the magnitude and sign of the droplet surface charge status, charge density, and degree of polyelectrolyte adsorption. Depending on surfactant/emulsifier composition, there is often a critical pH value where adsorption first starts, which depends on relative magnitude and the charge status of the different species present (Sukhorukov *et al.*, 2001). Electrostatic interactions between biopolymers will depend on the sign, type, number, and charge distribution along the biopolymer chains (Kato *et al.*, 2002). For example, at pH 7 anionic β-lactoglobulin-coated droplets do not complex with anionic pectin due to their electrostatic repulsion. At pH 3, however, the anionic pectin readily adsorbs onto positively charged β-lactoglobulin-coated droplets (Guzey *et al.*, 2004).

11.4.5 Salt and ionic strength

Electrostatic interactions between charged biopolymers are strongly influenced by changes in ionic strength. The addition of salts such as NaCl results in electrostatic charge screening that decreases the attraction between oppositely charged species and the repulsion between similarly charged species (McClements, 2005). Electrostatic charge screening is strengthened by an increase in the concentration and valency of the counterions present.

11.4.6 Temperature and freeze-thaw stability

During their industrial production, emulsions are usually thermally-treated via pasteurization or sterilization (Srinivasan *et al.*, 2003). In the case of conventional emulsions stabilized with proteins, protein unfolding at higher temperatures increases the likelihood of emulsion destabilization due to droplet flocculation or coalescence. The presence of multilayers around lipid droplets has been shown to slow emulsion breakdown during temperature fluctuations and during freeze–thaw cycling (McClements, 2005; Guzey and McClements, 2006a). For example, a three-component β-Lg-ι-carrageenan–gelatin emulsion showed higher freeze–thaw stability than a β-Lg or mixed β-Lg-ι-carrageenan emulsion during storage at −20°C for 22 h and 40°C for 2 h due to the formation of a thicker interfacial layer around the former emulsion (Gu *et al.*, 2007).

11.4.7 Preparation of multilayer emulsions

The composition of the dispersed phase, continuous phase, and interface as well as unit operations (e.g., homogenization, freezing, sterilization, etc.) play a vital role in ensuring an emulsion's long-term stability (McClements, 2005). It is primarily the role of the surfactant/emulsifier layer that will dictate the stability (i.e., kinetic, oxidative, etc.) of compounds (e.g., bioactive reagents, flavor, antioxidants, antimicrobial agents, etc.) within the dispersed phase (Sukhorukov *et al.*, 2001; Guzey and McClements, 2006b). In this regard, there are a number of design parameters to control and optimize the formation of multilayer emulsions, including the:

1. composition of the emulsifier/surfactant used to form the primary emulsion;
2. composition, order of addition and number of the "layer-forming" polyelectrolytes;
3. type of cross-linking of the polyelectrolyte layers (i.e., chemically or physical);
4. local environment around the droplets (pH, ionic strength, temperature, etc.).

11.4.8 Protein–polysaccharide complexation

There are two approaches to generate protein–polysaccharide complexes for use in emulsion formation and stabilization: the layer-by-layer (LBL) and mixed-layer techniques.

11.4.8.1 Layer-by-layer deposition

With the LBL approach, a primary emulsion stabilized with a surface-active species (usually a protein) is prepared followed by the addition of a second, oppositely charged, polyelectrolyte which is usually a polysaccharide (Guzey *et al.*, 2004). Due to their electrostatic attraction, the second polyelectrolyte adsorbs onto the primary emulsion oil droplets to produce a secondary emulsion now coated with two layers at the water-oil interface. By repeating this procedure, oil droplets may be coated with three or more interfacial layers (Guzey and McCclements, 2006a).

The total number of charges present on the initially adsorbed polyelectrolyte should be less than or equal to that on the second electrolyte so as to neutralize it once adsorbed. If the first polyelectrolyte has a higher surface charge, then an electrostatic repulsive force between nonadsorbed and adsorbed electrolytes may prevent further adsorption. In addition, only an oppositely charged electrolyte can form a dense second layer. Thus, the role of pH in LBL deposition is critical. For instance, in the case of a protein-stabilized emulsion, the electric charge of the primary emulsion would change from negative above the pI to positive below it, thereby rendering LBL deposition ineffective. Such emulsions show improved stability over simple emulsions.

11.4.8.2 Mixed-layer deposition

A concern with LBL-based deposition is the tendency of droplet–droplet flocculation. When oil droplet collisions occur faster than polyelectrolyte adsorption, bridging flocculation is enhanced. In addition, if polyelectrolyte concentration is higher than a critical concentration, there is a greater likelihood of depletion flocculation. It has been experimentally demonstrated that the mixed-layer approach provides better stability compared to the LBL approach (Ganzevles *et al.*, 2008; Jourdain *et al.*, 2008, 2009).

The mixed-layer approach consists of the adsorption of polyelectrolyte complexes formed by two oppositely charged emulsifiers during homogenization. This technique first involves the preparation of a bulk aqueous solution of protein–polysaccharide complexes by controlling the pH and ionic strength of the aqueous phase. The resulting protein–polysaccharide complex is then used as the emulsifying agent during homogenization (Jourdain *et al.*, 2008, 2009).

With both approaches, the thickness and effective density of the coated oil droplets will increase compared to simple emulsions, thus reducing the propensity towards emulsion phase separation, flocculation, coalescence, and Ostwald ripening (McClements, 2010a, b). However, during their preparation, the primary emulsions will be prone to flocculation as mentioned above, while additional layers will require additional ingredients and processing time, making them more expensive to prepare and manipulate for a specific end-use (McClements *et al.*, 2007; McClements, 2010b).

There are a number of approaches to optimize protein–polysaccharide complexation and minimize/remove unwanted (i.e., bulk) polyelectrolyte (Voigt *et al.*, 1999). In the saturation method, the presence of free nonadsorbed electrolyte remaining in the emulsion's aqueous phase following complexation is monitored using zeta potential values. Once sufficient polyelectrolyte is added to permit complete complexation, presence of free biopolymers will result in a change in zeta potential, implying that no further biopolymer should be added (Guzey and McCclements, 2006a). Excess polyelectrolyte beyond what is required for complete saturation of the dispersed droplets is removed via centrifugation. This method is generally quite effective but may promote unwanted droplet/particle aggregation. Finally, nonadsorbed polyelectrolyte may also be removed by membrane filtration of the emulsion.

A large body of research on the incorporation of lipophilic compounds within multilayer emulsions has been generated over the past two decades. Table 11.4 provides a number of examples of biopolymers used in multilayer emulsions as well as why they were formed.

Table 11.4 Recent applications of multilayer emulsions.

Biopolymers	Desired end goal	Reference
Lactoferrin–alginate/low methoxyl pectin/high methoxyl pectin	Heat-stable emulsion	Tokle *et al.*, 2010
SDS–chitosan–pectin	Freeze–thaw stable emulsion	Aoki *et al.*, 2005
Whey protein isolate–Xanthan gum/guar gum	Heat-stable emulsion	Ye and Singh, 2006
Lecithin–Chitosan	Oxidative stability of dispersed tuna oil	Klinkesorn *et al.*, 2005
Gelatin–gum arabic	Oxidative stability of dispersed flaxseed oil	Liu *et al.*, 2010
Maltodextrin–whey protein concentrate or maltodextrin–modified starch	Oxidative stability of dispersed flaxseed oil	Carneiro *et al.*, 2013
Carrageenan–β-lactoglobulin	Shelf-stable emulsion	Gu *et al.*, 2005
Pectin–β-lactoglobulin	Shelf-stable emulsion	Guzy *et al.*, 2004
Lecithin–chitosan–pectin	Heat-stable emulsion	Ogawa *et al.*, 2004
β-lactoglobulin–pectin–chitosan	Shelf-stable emulsion	Guzy and McClements, 2006b

SDS, sodium dodecyl sulfate.

11.5 NANOEMULSIONS

Nanoemulsions have a number of functional properties that can be advantageous over conventional emulsions, including their optical transparency, stability against phase separation, and rheological properties. Though far from exhaustive, these properties exemplify the differences between nanoemulsions and their larger cousins.

In contrast to conventional emulsions, the diameter of nanoemulsion droplets is much smaller than that of visible wavelengths. As a result, they appear optically transparent, despite the significant difference in refractive index between the dispersed and continuous phases. Turbidity shows a maximum value at a particle size where light scattering is strongest (i.e., $r \approx \lambda$). Thus, as the droplet radius increases to values approaching that of the visible portion of the spectrum, nanoemulsions appear cloudy, eventually appearing white due to significant multiple scattering.

In regards to their stability, nanoemulsions have a much larger surface area-to-volume ratio vis-à-vis conventional emulsions. Hence, properties related to their stability (namely Laplace pressure) become more important than in conventional emulsions. According to the Laplace Equation (11.4), the Laplace pressure (p_L) is inversely proportional to the radius (r) of the droplet and twice the interfacial tension (γ) of the droplet:

$$p_{\mathrm{L}} = \frac{2\gamma}{r}. \tag{11.4}$$

As equation (11.4) clearly shows, the inverse dependence of p_L on r implies that molecules in smaller droplets experience a higher pressure than those in larger droplets. Given their small size, nanoemulsions are susceptible to Ostwald ripening, which is the mass transfer of the dispersed phase from smaller droplets to larger droplets via passage through the continuous phase (Kabalnov *et al.*, 1987a,b; 1990; Kabalnov and Shchukin, 1992). Large droplets therefore grow in size whereas small droplets become smaller and may disappear. The basis of Ostwald ripening is that the solubility of the material within a droplet increases as the droplet decreases in size (Kabalnov and Shchukin, 1992). The effect of this process is usually negligible when the mutual solubilities of the oil and water phases are low enough to render the rate of mass transfer insignificant, as is the case in most food applications. However, it becomes more of a factor in nanoemulsion formulations given that the content of nanoscale droplets show greater solubility compared to large droplets with a low radius of curvature (Buffo and Reineccius, 2001). Kabalnov *et al.* (1987a) and Kabalnov and Shchukin (1992) showed that adding low-solubility additives to the dispersed phase of O/W emulsions effectively stopped Ostwald ripening. Similarly, Wooster *et al.* (2008) observed that instability due to Ostwald ripening in hexadecane-in-water nanoemulsions was prevented via addition of long-chained triglycerides.

A key property of nanoemulsions is that they can be diluted with water with no change in average droplet size or breadth of distribution. This last point is a marked advantage over microemulsions, which exist within only a narrow concentration range. Outside of this compositional range, they either break down or evolve towards other structures such as biocontinuous or liquid crystalline arrangements.

Nanoemulsions can show considerably different rheological properties compared to conventional emulsions at a given oil content (Tadros, 2004; Mason *et al.*, 2006). Though they are not particularly viscous at a very low internal phase ratio, their viscosity is appreciably higher than that of conventional emulsions at the same dispersed phase volume, particularly if they contain a thick or electrically charged interfacial layer which increases droplet–droplet repulsion. Weiss *et al.* (2007) demonstrated that hydrocarbon oil-in-water emulsions were liquid-like with droplets larger than 200 nm in diameter but gel-like with droplets smaller than 80 nm in diameter. Hence, it is possible to produce viscous or gel-like nanoemulsions with similar rheological properties to conventional emulsions at higher oil contents. Such properties will greatly depend on whether the droplets interact repulsively or attractively.

Compositionally, though many principles relevant to conventional emulsions apply to nanoemulsions, there are certain nuances to consider. As per conventional emulsions, the oil phase of a nanoemulsion can consist of a variety of nonpolar oils, including common vegetable oils, essential oils, mineral oil, oil-soluble vitamins, lipophilic nutraceuticals, with most studies to date based on vegetable oils containing dissolved compounds of interest within. Critically, the chosen dispersed phase compound(s) should be highly insoluble in the continuous phase so that Ostwald ripening does not readily occur despite the very high Laplace pressures (Sonneville-Aubrun *et al.*, 2004).

The aqueous phase of a nanoemulsion will consist of water, with possibly dissolved components such as carbohydrates, flavours, acids, etc. Their presence should not hinder nanoemulsion formation and kinetic stability (Palamakula and Khan, 2004; Wooster *et al.*, 2008; Do *et al.*, 2009). An important requirement is that the continuous phase has an excess of surfactant that will permit the surface of the newly formed nanodroplets to be rapidly coated during emulsification, thereby inhibiting shear-induced coalescence. Though rapidly growing in scope, little food-related research has been performed on the addition of conventional food ingredients to nanoemulsion formulations, for example, proteins from egg, milk or soy as well as common polysaccharides (e.g., starch, pectin, alginate, guar gum, carrageenan, xanthan, etc.).

11.5.1 Nanoemulsion preparation

Nanoemulsion preparation methods can be classified into high-energy or low-energy approaches (Tadros *et al.*, 2004; Pouton and Porter, 2006; Acosta, 2009; Anton and Vandamme, 2009; Leong *et al.*, 2009). Nanoemulsion preparation can be performed mechanically, which involves an input of high energy generally achieved by methods such as high-pressure homogenization, high-shear mixing, or ultrasound. Given their smaller size and need for significant energy input and high surfactant concentration, such routes can be somewhat unfavorable for industrial applications. Nevertheless, approaches such as high-pressure valve homogenization are well-established in large-scale food processing operations and are capable of generating the disruptive forces necessary to produce nanoscale dispersed droplets. As per conventional emulsion processing, the aqueous (+ surfactant) and oil phases first undergo coarse emulsification, which is followed by fine emulsification. Table 11.5 provides a summary of recent research efforts related to O/W nanoemulsions generated with high-energy methods.

Table 11.5 Recent research related to O/W nanoemulsions made using high-energy methods.

Oil phase	Bioactive component	Diameter (nm)	Reference
MCT	β-Carotene	115–178	Mao *et al.*, 2009
MCT	Curcumin	160	Wang *et al.*, 2008
MCT	Coenzyme Q_{10}	60	Hatanaka *et al.*, 2008
MCT	α-Tocopherol	80–400	Hatanaka *et al.*, 2010
Oleoresin/ethyl acetate	Polycaprolactone	320–460	Choi *et al.*, 2009
Flaxseed oil	–	150	Kentish *et al.*, 2008
Vegetable oils	–	<500	Wulf-Perez *et al.*, 2009
Peanut oil	–	120	Wooster *et al.*, 2008
Corn oil	–	146	Qian and McClements, 2011
MCT and orange oil	β-carotene	140–170	Qian *et al.*, 2012
Corn oil	β-carotene	210–380	Salvia-Trujillo *et al.*, 2013

MCT, medium chain triglycerides.

Following their preparation, the droplet size of nanoemulsions may be altered using solvent displacement and evaporation. After homogenization, the entrapped organic solvent within the droplets is removed by solvent displacement and evaporation, which causes shrinkage (Horn and Rieger, 2001; Tan and Nakajima, 2005a, b; Chu *et al.*, 2007b; Lee and McClements, 2010). For example, Lee and McClements (2010) used this approach to shrink droplets in whey protein isolate-stabilized O/W nanoemulsions from 174 nm to < 70 nm by removing ethyl acetate present in the dispersed phase. In work by Chu *et al.* (Chu *et al.*, 2007a, b), droplet shrinkage from 1000 nm to 18 nm was achieved in emulsions containing β-carotene dispersed in sodium caseinate-stabilized hexane-in-water nanoemulsions.

By contrast, low-energy approaches focus on the spontaneous formation of nanoscale droplets in oil–water–emulsifier mixtures via alteration of environmental and/or solution conditions, for example, phase-inversion and spontaneous-emulsification methods (Bouchemal *et al.*, 2004; Tadros *et al.*, 2004; Freitas *et al.*, 2005; Chu *et al.*, 2007a, b; Anton and Vandamme, 2008; Yin *et al.*, 2009). Initially, the oil and surfactant are mixed followed by slow addition of the aqueous phase, which leads to formation of dispersed nanodroplets. Such an approach is based on the change in the spontaneous curvature of surfactants whose activity is influenced by temperature or other parameters. For example, nonionic surfactants may transition from promoting O/W emulsions at lower temperatures to W/O emulsions at higher temperatures. Upon cooling, the emulsion crosses a point of zero spontaneous curvature and minimal surface tension, promoting formation of nanoscale oil droplets – a process referred to as the phase inversion temperature (PIT) method (Morales *et al.*, 2003). Alternatively, changes in parameters such as salt concentration or pH may be used to promote such inversion (Engels *et al.*, 1995). However, research efforts are still required to optimize the stability of nanoemulsions generated in this way. In a review by Mason *et al.* (2006), it was stated nanoscale droplets created using PIT procedures do not remain stable over long periods of time. Nevertheless, there is interest in this approach in food-based research as the list in Table 11.6 shows.

11.5.2 Applications of nanoemulsions

Though in its infancy, there is currently tremendous interest in the development of nanoemulsions for food use (Garti *et al.*, 2003; Shefer and Shefer, 2003; Wang *et al.*, 2008; Shah *et al.*, 2010; McClements and Hang, 2012; Salvia-Trujillo *et al.*, 2013). By contrast, they are extensively used in cosmetics and the personal-care industry due to the improved penetration efficiency through skin and mucous membranes they provide given their smaller droplet size compared to conventional emulsions. Various

(Mozafari, 2005; Torchilin, 2005). Liposomes are vesicular concentric bilayered structures, usually 0.05–5.0 μm in diameter, composed of relatively biocompatible and biodegradable materials. They consist of one or several lipid bilayers separated by aqueous compartments, and they form spontaneously when certain lipids are hydrated in aqueous media. Drugs with widely varying lipophilicity can be encapsulated in liposomes, either in the phospholipid bilayer, or the entrapped aqueous volume, or at the bilayer interface. A huge number of books and reviews are available and provide information about liposomes as controlled drug delivery systems, but only some of them are mentioned herein (Chonn and Cullis, 1998; Gabizon *et al.*, 1998; Lasic and Papahadjopoulos, 1998; Mozafari, 2005; Gregoriadis, 2007; Suntres, 2011; Mansoori *et al.*, 2012; Xu *et al.*, 2012).

12.2 LIPOSOME CLASSIFICATION

The liposomal system has a major advantage compared to other colloidal carrier systems: it has great flexibility for changing structure and physicochemical characteristics, that is, in terms of size, surface charge, membrane fluidity, and number of lamellae (Maurer *et al.*, 2001; Crommelin and Storm, 2003; Katre, 2004). The versatility of liposomal formulations enables modification of their behavior in both *in vitro* and *in vivo* applications (Mozafari, 2005). As a function of the phospholipid bilayer charge, liposomal vesicles could be neutral, negative or positive (cationic). Exhaustive liposome classifications based on structural parameters, like size and lamellarity, are presented by Torchilin (2005), and also in a recent review by Mansoori *et al.* (2012). The formulations usually used in the DDS area are: multilamellar large vesicles (MLVs), in the size range 500–5000 nm and consisting of several concentric bilayers; oligolamellar vesicles (OLVs), in the range100–500 nm; small unilamellar vesicles (SUVs), around 100 nm in size and formed by a single bilayer; large unilamellar vesicles (LUVs), in the size range 200–800 nm; giant unilamellar vesicles (GUVs) and multivesicular vesicles (MVVs), larger than 1000 nm.

Based upon composition and application, liposomes are conventional, fusogenic, cationic, long-circulating and pH sensitive, and there are also immunoliposomes (Storm and Crommelin, 1998; Costin *et al.*, 2002; Torchilin, 2005; Mansoori *et al.*, 2012; Xu *et al.*, 2012). More attractive for clinical practice are the immunoliposomes (targeted liposomes), which are able to increase drug accumulation in specific tissues or organs where the surface-attached ligands (antibodies) recognize and bind the antigen (Lehtinen *et al.*, 2012). Equally important are the long-circulating liposomes, modified by coating with inert, biocompatible polymers (e.g., poly(ethylene glycol) (PEG) or polysaccharide), which form a protective layer over the liposome surface, slow down liposome recognition by opsonins and, subsequently, prevent liposome clearance (Sihorkar and Vyas, 2001; Moghimi and Szebeny, 2003; Torchilin, 2005). Long-circulating liposomes are now investigated in detail and are widely used for *in vitro* and *in vivo* biomedical studies (Yang *et al.*, 2011; Wong *et al.*, 2013).

12.3 LIPOSOME PROPERTIES

The clinical applications of liposomes are well-known and are based on their attractive biological properties: biocompatibility and ability to entrap water-soluble (hydrophilic) pharmaceutical agents in their internal water compartment and water-insoluble (hydrophobic) pharmaceuticals into their membrane (Lasic and Papahadjopoulos, 1998; Torchilin, 2005). Compared to other carriers, liposomes provide several advantages as delivery systems of bioactive molecules, such as:

- *protection and duration* – ability to protect the encapsulated drug from the external environment and to act as a depot from which the entrapped compound is slowly released over time (Trif *et al.*, 2001, 2007; Williams *et al.*, 2001);

- *direction* – liposomes can target a drug to the intended site of action in the body, thus enhancing its therapeutic efficacy (drug targeting, site-specific delivery) (Martin and Lasic, 1995);
- *internalization* – liposomes can interact with target cells in various modes and, therefore, they are able to promote the intracellular delivery of drug molecules that in their "free" form (i.e., nonencapsulated) would not be able to enter a cell interior due to unfavorable physicochemical characteristics (e.g., DNA molecules) (Chonn and Cullis, 1998; Lasic and Papahadjopoulos, 1998; Miller *et al.*, 1998; Allen *et al.*, 2006; Schwendener and Schott, 2010; Allen and Cullis, 2013).
- *amplification* – if the drug is an antigen, liposomes can act as an immunological adjuvant in vaccine formulations (Copland *et al.*, 2003).

A successful application of liposomes in therapy is based on the combination of two or more of the above characteristics (Storm and Crommelin, 1998; Torchilin, 2005; Crommelin and Zuidam, 2007). Some problems related to leakage or fusion of encapsulated drug, membrane permeability, lipid peroxidation, and liposome stability during storage could be minimized or eliminated by optimization of lipid composition and concentration, lipid to drug ratio, or the preparation method. Solute leakage depends on both membrane permeability and interaction with biological fluids components. Membrane fluidity can be controlled, in turn, by supplementing the lipid bilayer with cholesterol (Chol) (Tsamaloukas *et al.*, 2006) or by manipulating the hydrophobic/lipophobic character of the bilayers. To minimize unsaturated phospholipid oxidation, special precautions must be taken. These include the use of antioxidants, such as α-tocopherol (Hincha, 2008) and deaeration with argon or nitrogen. The degradation of liposomal systems can be prevented or minimized by freeze-drying the liposome suspension to powder (Uhumwangho and Okor, 2005). Lipid composition controls the stability, permeability, and fluidity of the vesicles and these properties control, in turn, the transport and release of the therapeutic agents to the cell or target tissue (Jain *et al.*, 2010). Successful applications of these carrier systems in drug delivery depend on comprehensive understanding of their physical properties, including size, polydispersity, and morphology. Variation in size and shape of the carrier systems indicates their stability and can guide scientists in improving the therapeutic formulations (Chonn and Cullis, 1998; Mozafari, 2005; Gregoriadis, 2007; Samad *et al.*, 2007; Barenholz *et al.*, 2011).

12.4 POTENTIAL USE OF LIPOSOMES AS A DDS IN DIFFERENT ROUTES OF ADMINISTRATION

Owing to their characteristics and many therapeutical advantages, liposomes are considered very promising delivery systems for different routes of administration. Allen and Cullis (2013) recently produced a scientific report focused on almost all the important contributors in the clinical use of liposomes. The *intravenous route* was investigated intensively, especially for the administration of low-solubility and highly toxic drugs, in order to improve their efficiency (Park, 2002; Tang *et al.*, 2010; Barenholz, 2012). Both intravenous and *inhaled routes* for antituberculosis drugs targeting the alveolar macrophages have recently been proposed (Pinheiro *et al.*, 2011). *Intraperitoneal administration* of liposomes may be a beneficial approach to treat tumors in the abdominal cavity. The intraperitoneal pre-targeting method warrants further studies as a potential approach in cancer therapy (Lehtinen *et al.*, 2012).

Topical administration of liposomal formulations was reviewed by Egbaria and Weiner (1990) and Pierre and Costa (2011), and has proven to be an efficient mode of drug administration. The sustained release of superoxide dismutase in burn wounds was promoted by liposomal encapsulation, resulting in enhanced antioxidant effect of superoxide dismutase and reduced edema formation (Vorauer-Uhl *et al.*, 2001). Liposomal CoQ10 is a potential ophthalmic DDS that could ameliorate the oxidative stress in the lenses, thus preventing cataracts (Wang *et al.*, 2011), or cosmetically applied to reduce photoaging and wrinkle depth (Lee and Tsai, 2010). In the latter case, topical application of liposomal CoQ10 enhanced its accumulation in rat skin, compared with a nonencapsulated suspension, prolonging the treatment time (Suntres, 2011). *Transdermal drug delivery* (TDD) also can be a useful route for drug administration, and liposomes have been used to increase the drug penetration rate across the stratum corneum (SC), the main barrier to percutaneous absorption (Touitou *et al.*, 1994; Canto *et al.*, 1999; Maestrelli *et al.*, 2005). Topical delivery of anti-inflammatory drugs could be more desirable than oral/parenteral route of administration in the case of chronic conditions. For this reason, research is focused on topical delivery of anti-inflammatory drugs by encapsulation into liposomes, in order to avoid the side-effects following oral administration, to maintain maximum therapeutic effect for a longer time, and to provide sustained release of drugs through enhancing circulation lifetime (Jithan and Swathi, 2010). It has been shown recently that many drugs have better bioavailability by the *nasal route* than other routes and liposomes can be used as the DDS to enhance intranasal drug absorption (Akhatar *et al.*, 2012).

For many years our expertise has been in the design of a suitable liposome system for efficient *local/intra-articular delivery* of bioactive molecules and, for this reason, we want to highlight this important aspect of liposome technology. Because of the greatly prolonged drug residence time at the administration site, liposomes have been proposed as suitable carriers of bioactive molecules in intra-articular therapy of inflammatory diseases, like rheumatoid arthritis (RA) or osteoarthritis (OA) (Trif *et al.*, 2001, 2007; Zhang and Huang, 2012). The intra-articular route of administration could offer several beneficial advantages for drug uptake due to the presence of phospholipids in both cell membranes and the double-layer of lipid vesicles (liposomes). The only liposomal product available in Germany for intra-articular administration in human patients with OA is Lipotalon (a corticosteroid liposomal formulation containing dexamethasone-21-palmitate) (Gerwin *et al.*, 2006). The potential of liposomes as a DDS in the treatment of arthritis has been thoroughly investigated (Elron-Gross *et al.*, 2009; Vanniasinghe *et al.*, 2009; van den Hoven *et al.*, 2011). Recent reviews have highlighted that liposomes play an important role in joint boundary lubrication, as demonstrated by *in vitro* studies (Barenholz *et al.*, 2011; Goldberg and Klein, 2012; McNary *et al.*, 2012). Verberne *et al.* (2010) screened several liposome types in order to choose the optimal lipid composition and structure that could act as a lubricant for joint dysfunctions. Verberne *et al.* (2010) showed that MLVs composed of 1,2-dimyristoyl-glycero-3-phosphocholine (DMPC) were the best reducers of cartilage wear, and they have been proposed for intra-articular injections to improve joint dysfunctions (Sivan *et al.*, 2010).

Oral delivery of bioactive is the most convenient route especially for poorly water-soluble drugs, or when repeated or routine administration is necessary (Florence and Jani, 1993). In this context, liposomes are the most intensively studied carrier system in order to enhance the oral bioavailability of certain drugs. The oral administration of liposomes requires a high stability and the ability to move from the gut to the blood and to release the drug only when they reach the blood compartment (Rogers and Anderson, 1998). Much data comparing oral administration of liposomal systems to free-form drugs are available but no liposomal formulation that is approved for the oral route is on the market yet due to their instability in the gastrointestinal (GI) tract (Chen *et al.*, 1996; Li *et al.*, 2003; Wu *et al.*, 2004). Effective oral delivery of insulin is increasing in popularity and several researchers have designed and characterized different liposome formulations containing insulin. It was reported that conventional liposomes mainly composed of phospholipids and cholesterol are not efficient carriers for oral drug delivery due to their susceptibility to digestion through the GI tract and poor permeability across the epithelial membrane (Dezim *et al.*, 2004).

There has been great interest to enhance the stability of insulin liposomal formulations in the GI tract and encapsulation efficiency by the modification of the liposomal physicochemical parameters (composition, lipid concentration, lamellarity, size, and surface charge) (Taira *et al.*, 2004; Mozafari *et al.*, 2008; Woitiski *et al.*, 2008; Niu *et al*, 2011). It has been reported that liposomes containing gangliosides, such as GM1 and GM type III, inserted in the lipid bilayer are stable in different biological media and can survive better in the GI tract compared to other liposome formulations (Taira *et al.*, 2004). Niu *et al.* (2011) have shown by *in vitro* experiments that liposomes prepared with sodium glycocholate in the lipidic bilayer have an enhanced stability against enzymatic degradation by pepsin, trypsin, and α-chymotrypsin and enhanced permeation across the GI epithelium. Glycocholate liposomes have potential for the oral delivery of human recombinant insulin or other protein and peptide drugs. *In vivo* studies proved that the incorporation of insulin into liposomes coated with PEG, chitosan, or mucin enhanced its oral absorption and prolonged the therapeutic action in the body. This was possible by the better interaction of polymer-coated liposomes with the mucus layer and better retention of insulin under the aggressive conditions of the GI tract (Iwanaga *et al.*, 1999; Wu *et al.*, 2004; Torchilin, 2005).

In order to evaluate oral bioavailability of the poorly water-soluble drugs, liposomes containing celastrol, an antitumoral drug, were tested using a rat intestinal model. The intestinal absorbtion and antitumoral capacity of celastrol were increased by using liposomes as carriers compared to the free-form drug (Song *et al.*, 2011). Oral administration of a pH-sensitive, encapsulated midazolam liposome (a benzodiazepine, insoluble in water) to rabbits showed both an increased stability in a physiological medium and an efficient release in an acidic environment. The results suggested that it is possible to use the liposome midazolam system for oral administration, especially for sedation in pediatric patients, with better results than the intravenous route (Tomoyasu *et al.*, 2011).

All these findings related to the ability of liposomes to improve the phamacokinetics and the biodistribution of the entrapped bioactive molecule and at the same time to minimize its toxicity have encouraged extensive studies for new liposomal pharmaceutical products. Part of our research is focused on the design of liposomal formulations of chondroitin sulphate (CS) to improve therapeutic efficacy.

12.5 PREPARATION AND PHYSICOCHEMICAL CHARACTERIZATION OF LIPOSOMAL SYSTEMS CONTAINING CS AS THE BIOACTIVE MOLECULE

The most common nutritional and pharmaceutical molecule (nutraceutical) used as a chondroprotective agent is CS, the molecular building block found in articular cartilage (Morelli *et al.*, 2003). CS is an important sulphated glycosaminoglycan from cartilage, consisting of a long, unbranched, polysaccharide chain, composed of repeating disaccharide units of *N*-acetylgalactosamine and glucuronic acid (Iovu *et al.*, 2008). Due to its numerous sulphate groups with negative electrical charges, CS ensures the property of swelling and water retention in cartilage proteoglycans, contributing to the cartilage characteristic property of compressive strength (DuRaine *et al.*, 2011). Moreover, CS is able to quickly and specifically interact with proteins on the cell surface and with enzymes. It has the ability to interact with proteins from the extracellular matrix and to regulate various cellular activities.

In vitro and *in vivo* studies have demonstrated the clinical benefits of CS administration in patients with OA (Gerwin *et al.*, 2006; Henrotin *et al.*, 2010). The chondroprotective action of CS has been attributed to its role in stimulation of extracellular matrix production by chondrocytes, suppression of inflammatory mediators, and inhibition of cartilage degradation (Zhang and Huang, 2012). CS has a beneficial effect on each of the three levels of arthritic joint (i.e., cartilage, synovial membrane, and subchondral bone) and is a well tolerated therapeutic agent *in vivo* (Kobo *et al.*, 2009). CS absorbed through oral applications was found in a high content in the synovial fluid and cartilage and was not rapidly absorbed in the GI tract (Henrotin *et al.*, 2010). Eddington *et al.* (2001) demonstrated the oral bioavailability of CS fragments in horse plasma. CS also showed a tropism for articular cartilage, with synovial fluid levels exceeding plasma concentrations after oral administration (Rodgers, 2006).

12.5.1 General method of liposome preparation and drug loading

For delivery applications, liposomal formulations should have high entrapment efficiency, narrow size distribution, long-term stability, and ideal release properties. These demands require a suitable preparation method, able to produce stable liposomes using a wide range of lipids/phospholipids (Gregoriadis, 2007; Dua *et al.*, 2012). In addition to the above properties, liposomes should be able to protect the incorporated agents from degradation, in order to deliver sensitive compounds, such as proteins or nucleic acids (Chonn and Cullis, 1998; Mozafari, 2005). There are many different strategies for liposome preparation that generate vesicles of different sizes, ranging from 20 nm to several microns in diameter, with a variable number of lipid bilayers, and they are fully compatible with the biological environment, biodegradable, and capable of achieving high encapsulation efficiency of the incorporated compound. Conventional liposome preparation techniques have been discussed extensively in many books and reviews (Lasic and Papahadjopoulos, 1998; Lasic *et al.*, 2001; Mozafari, 2005; Gregoriadis, 2007; Mozafari *et al.*, 2008; Mansoori *et al.*, 2012). Selecting the method of liposome preparation depends on the following parameters: the physicochemical characteristics of the liposomal lipids and those of the entrapped material; the kind of dispersion medium for lipid vesicles; the concentration and potential toxicity of the entrapped material; the processes involved during *in vivo* application of the vesicles; vesicles size, polydispersity and shelf-life intended for each application; batch-to-batch reproducibility; safe and efficient liposome production on a large scale (Mozafari *et al.*, 2008; Mansoori *et al.*, 2012). Depending on their application, a suitable preparation technology should be used and involving an appropriate mechanism of liposome formation, but that is not the scope of this chapter.

12.5.2 Preparation of liposomal systems containing CS

For our purpose, we have selected and adapted those methods of obtaining liposomal systems that reduced the working time and preserved the lipid bilayer structure to eliminate the possibility of membrane phospholipid peroxidation. The methods described in the following section outline the preparation of MLVs and SUVs by sonication and filter extrusion. We favor the use of these techniques and they are recommendable because of their ease, versatility, and high quality of liposomes produced (Schwendener and Schott, 2010). These methods involve the entrapment of active hydrophilic molecules through a passive loading process during the manufacturing procedure (Mozafari, 2005; Mansoori *et al.*, 2012).

Positive liposomes containing CS were prepared by the thin-film hydration method (TFHM) (Mozafari, 2005; Gregoriadis, 2007), the most widely used technique, using a suitable concentration of lipid mixture, phosphatidylcholine (PC):dioleoylphosphatidylethanolamine (DOPE):Chol:stearylamine (SA), in 4:2:3:1 molar ratio (8.5 mg/mL lipids), for efficient CS entrapment. A lipid phase was prepared by dissolving the lipids in a chloroform:methanol solution (95:5) in a round-bottom flask. A thin dry lipid film resulted after removal of the solvent (continuous rotary evaporation, 150 rpm, at 40°C, for 30 min, under reduced pressure), using a Rotavapor IKA (RV 10 digital V, Ika-Works Inc). The MLVs spontaneously formed by gentle mechanical shaking of the flask upon hydration of the dry thin lipid film with phosphate buffered saline (PBS), pH 7.4 containing CS (10 mg/mL). The dispersion was left undisturbed, at room temperature, for 2–3 h, to allow complete swelling of the lipid film and hence to obtain vesicular suspension. The MLVs were downsized by sonication, at 37°C, in a bath type sonicator (Grant), for 3 × 30 min, followed by probe sonication (Bandelin Sonopuls, Germany), at 70% amplitude, four cycles of 10 s, until a homogeneous population of SUVs was obtained. After sonication, the suspension was allowed to stand at 37°C, for 10 min, rapidly cooled in an ice bath and centrifuged at 10,000 rpm, at 4°C, for 5 min, to remove any titanium fragments. Cycling of the MLV suspension was performed through polycarbonate membrane filters with 200 nm pores (Avanti Mini-Extruder, USA), following the methodology provided (http://www.avantilipids.com). Before use in cell culture experiments, free liposomes and liposomal systems containing CS were filtered under sterile conditions through a 0.22 μm filter (Millipore, Bedford, MA, USA). Vesicular structures thus formed were kept at 4°C, in a N_2 atmosphere to prevent lipid peroxidation.

12.5.3 Liposome structure and characterization

The MLVs have high entrapment ability, but relatively little stability. Multilamellar structures are heterogeneous in size, varying between 1 and 20 μm and, therefore, are little used in cell culture experiments. The morphology of MLVs was observed after Oil Red O staining under an AxioStar Plus microscope (Carl Zeiss, Germany) (Figure 12.1a). This stain is highly soluble in lipids, which appear red under fluorescence microscopy. We have used bath sonication followed by probe sonication and extrusion techniques in order to obtain SUVs. Their morphology was observed under a Zeiss Axio Imager M2 microscope (Figure 12.1b).

The most common method of obtaining SUVs from MLV liposomes is ultrasonic irradiation. This treatment can be applied in a bath type sonicator and in a probe type sonicator that has the rod in direct contact with the liposome suspension. The probe type sonicator develops more power than the standard bath type and it can degrade phospholipids, but its main disadvantage is contamination with metal particles (titanium) from the rod. In turn, the bath type sonicator requires a prolonged operation period. In our study, we developed an optimal method of disrupting the MLV suspension for liposome size reduction using both the bath- (Grant, UK) and probe-type (Bandelin Sonopuls, Germany) sonicators. Sonication was performed at a temperature above the critical temperature (T_c) of lipids for preventing bilayer structural defect occurrence, in an inert atmosphere of N_2. During sonication, the MLV bilayers were broken and were reassigned in unilamellar spherical vesicles, with much smaller diameters. The obtained liposomal systems contained both MLVs and SUVs, and their separation could be achieved by gel filtration (Figure 12.2).

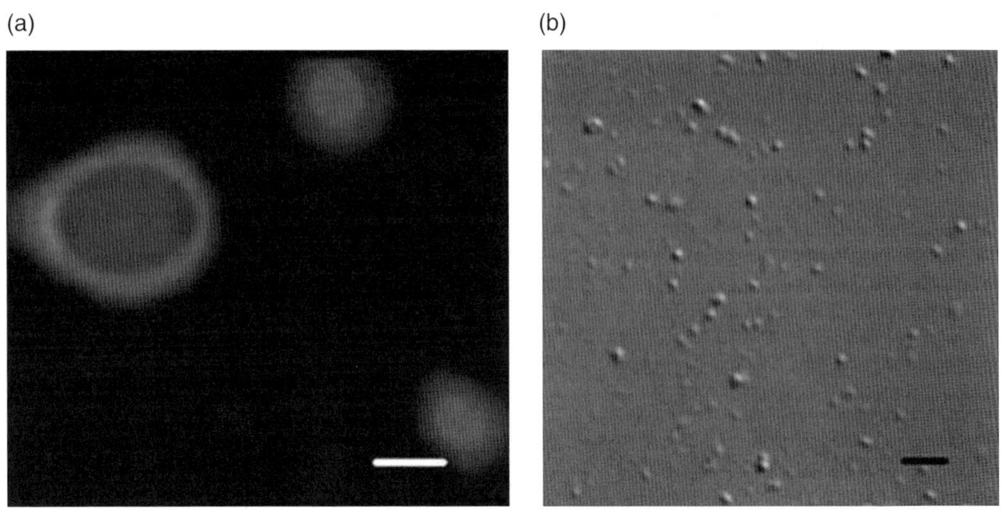

Figure 12.1 (a) Multilamellar lipid vesicles stained with Oil Red O and photographed under an AxioStar Plus microscope (Carl Zeiss, Germany) (bar = 10 μm). (b) Small unilamellar vesicles visualized under a Zeiss upright Axio Imager M2 microscope with ApoTome.2 slider module (bar = 1 μm).

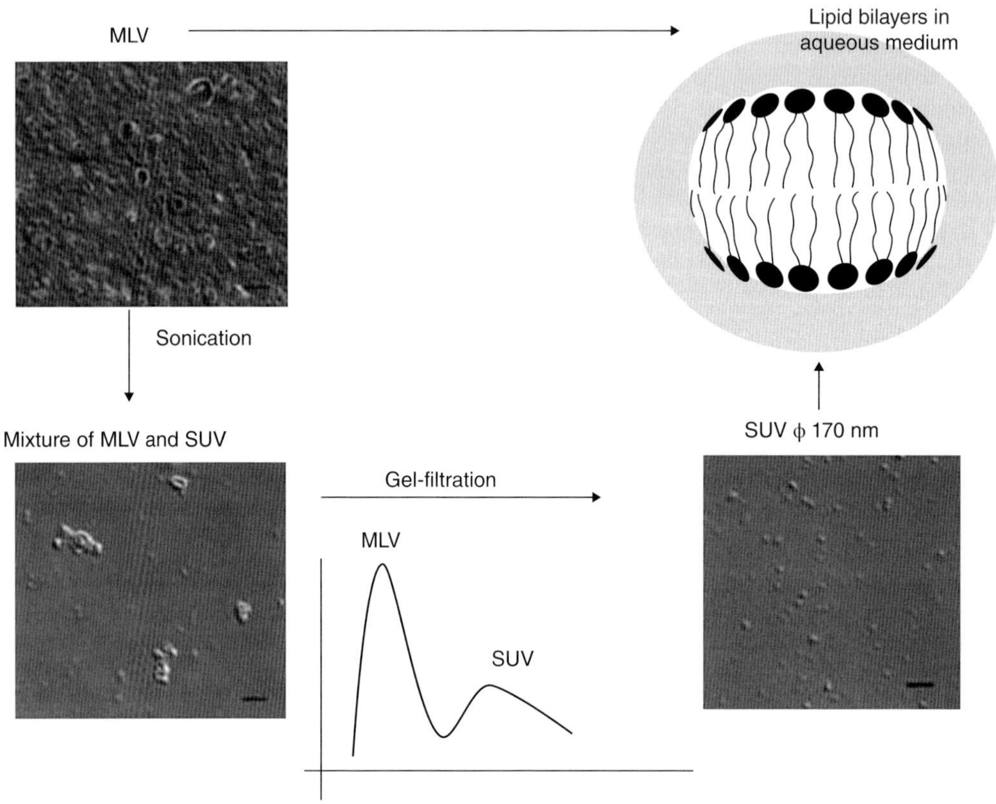

Figure 12.2 Scheme for the preparation of MLV and SUV populations by sonication and their separation by gel filtration.

12.5.4 Size, polydispersity index, and ζ-potential of the liposome system

Mean hydrodynamic diameter of vesicles (liposomes, nanospheres, nanobeads) can be determined with dynamic laser light-scattering instruments, for example, the ZetaPALS instrument (Brookhaven Instruments Corporation, USA). This technique enables determination of z-average mean, polydispersity index (PI) and ζ-potential of the liposomal systems. The z-average mean of vesicles entrapping CS was 250.03 ± 1.2 nm, higher than that of empty liposomes (111.17 ± 3.7 nm) (Figure 12.3).

PI ranged from 0.087 for empty liposomes to 0.335 for liposome-CS vesicles, indicating a good homogeneity. The measurement of ζ-potential indicated that empty liposomes had a positive value (20.65 ± 0.6 mV) (Figure 12.3), due to the presence of the cationic lipid SA in the liposomal bilayer. After CS entrapment in liposomes, the ζ-potential changed to a negative value of −9.44 ± 0.5 mV. These data demonstrated that cationic liposomes, in particular SA lipid, had the ability to interact with the highly negative charged glycosaminoglycan, CS (Trif *et al.*, 2008; Craciunescu et al., 2013).

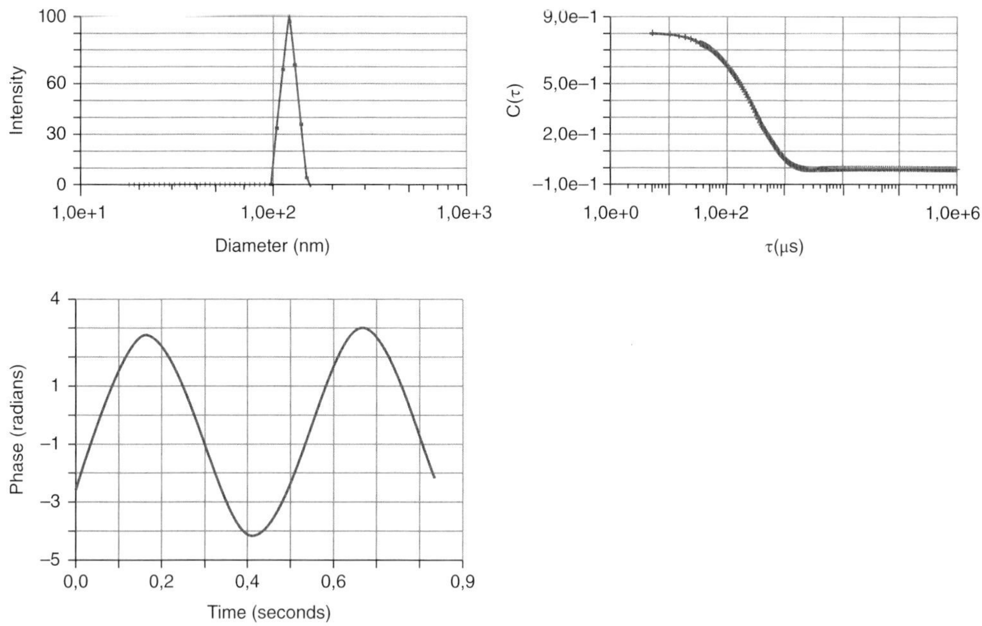

Figure 12.3 Dynamic light scattering diagrams showing the mean diameter, polydispersity, and ζ-potential of the unilamellar liposomes population.

12.5.5 Morphology of the liposome system

Morphology and homogeneity of SUVs obtained by thin-film hydration and sonication were checked by negative staining electron microscopy, as previously reported (Trif *et al.*, 2008). The spherical vesicles were dispersed and presented a well-defined outer lipid bilayer and an endoliposomal aqueous core. The mean diameter, determined by averaging the measurements on transmission electron microscopy (TEM) micrographs, was 168.88 ± 4.6 nm, in agreement with *z*-average mean (111.17 ± 3.7 nm) registered by dynamic light scattering (DLS). These ultrastructural observations highlighted the advantage of using ultrasound treatments that could lead to a homogeneous population of liposomes.

12.5.6 Determination of encapsulation efficiency

It is essential to determine the amount of included material within lipid structures considering the major objective for their use in therapy. To investigate the encapsulation efficiency (EE), comparative experiments were conducted using CS entrapped in SUV structures with different lipid compositions. Liposomes entrapping CS were separated from free-form substances by centrifugation (10,000 rpm, at 4°C, for 1 h). The amount of CS entrapped in liposomes was measured by Blyscan™ glycosaminoglycan assay (Biocolor, UK) (French *et al.*, 2004), after solubilization of liposomal samples in Triton X-100. The EE, defined as the ratio of entrapped CS to the total CS in the system, was calculated using the following equation:

$$\text{EE}\,\% = \text{entrapped CS} / \left(\text{entrapped CS} + \text{nonencapsulated CS}\right) \times 100\% \tag{12.1}$$

The concentration of the encapsulated hydrophilic drug CS in the liposomes varied with phospholipid concentration, the lipid composition, and the method of liposome preparation. The highly negative charge of CS yielded 86% entrapment of CS in cationic liposomes containing PC:DOPE:Chol:SA in a molar ratio of 4:2:3:1 (Craciunescu *et al.*, 2013).

12.5.7 Stability of the liposome system containing CS

Liposome stability is very important for their medical application. Microscopy techniques can provide relevant information on the morphology, size, and stability of liposomal systems. To examine the morphology of liposome-CS stored at 4°C, for 30 days, samples embedded in agarose were processed for TEM. Observations carried out on ultrathin sections showed a heterogeneous population of vesicles, with sizes ranging between 50 and 750 nm. CS appeared as electron-dense particles within the liposomal core, without disrupting the lipid bilayer (Figure 12.4).

Detection of vesicles entrapping CS with diameter size ranging between 500 and 4500 nm could be due to morphological changes caused by both aggregation between liposomal bilayers during storage and electrostatic interactions with CS, as previously observed for MLVs entrapping CS (Trif *et al.*, 2008). Also, the solvents used for sample processing could induce fusion of SUVs in larger vesicles. These observations are in agreement with a previous report on the tendency of liposomes to coalesce during dehydration of samples for TEM (Perrie *et al.*, 2007). The ultrathin-section protocol used in TEM studies appeared to affect SUV structure unlike the negative staining technique.

In order to increase liposomes stability, we included polar lipids in the bilayer. The presence of SA in the liposomal bilayer induced a positive charge and, also, the ability to interact with negatively charged CS. On the other hand, Chol incorporation in the bilayer has been reported to increase membrane stability and encapsulation efficiency of both hydrophilic and lipophilic bioactive molecules. Increasing Chol concentration in the phospholipid bilayer caused a gradual disappearance of the phase transition without affecting the transition temperature (Zhang and Huang, 2012). These data are in agreement with our previous results showing an increased stability of positively charged liposomes,

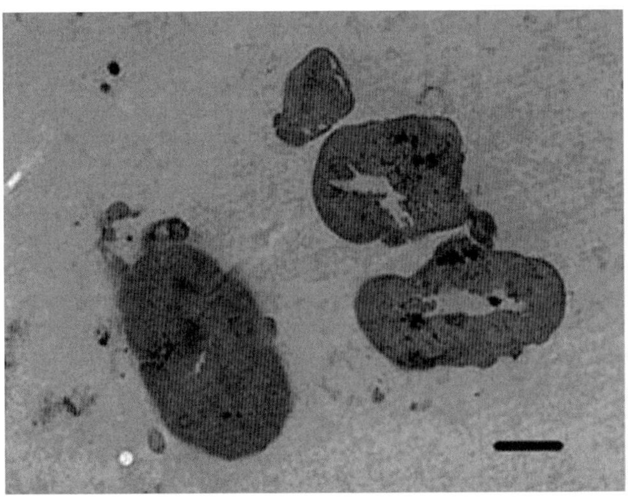

Figure 12.4 Transmission electron microscopy micrograph of a population of lipid vesicles entrapping CS, stored at 4°C for 30 days (bar = 1 μm).

containing SA and Chol, in the presence of human synovial fluid. We also performed comparative experiments using SUVs with different lipid composition and surface charge: negatively charged liposomes made of PC, Chol, phosphatidylserine (PS), pH-sensitive liposomes composed of DOPE, cholesterylhemisuccinate (CHEMS) and positively charged liposomes containing dipalmitoyl-phosphatidylethanolamine (DPPE), Chol, and SA. The latter variant remained intact for at least 5 h and was more stable, with over one-half, than other formulations that tended to disintegrate rapidly in the presence of synovial fluid (Trif *et al.*, 2001). Entrapment of CS into the hydrophilic content of liposomes and its association with positively charged lipid bilayers could also be a benefit for both stability of liposomes and controlled delivery of CS.

12.6 CELL–LIPOSOME INTERACTION

In vitro methodologies are very useful tools for proving DDS efficacy and their biological effect after interaction with cultured cells. It is extremely necessary and important to explore the interaction mechanisms of the liposomal systems with cells, in order to improve their uptake, the targeted release of the encapsulated material to specific organelles, and duration of action, for beneficial *in vivo* effects. Due to its negatively charged groups, CS is able to interact with cationic liposomes, resulting in stable complexes (Trif *et al.*, 2008). Both liposomes encapsulating bioactive molecules (Metselaar and Storm, 2005) and CS alone (Uebelhart *et al.*, 2006) have been reported to interact with several cell types and treatment of inflammatory diseases.

12.6.1 Biocompatibility tests in cell culture

In order to use liposomes and liposomes entrapping CS as nutraceuticals, they must be first tested for *in vitro* cytotoxicity on cell cultures and, subsequently, *in vivo* in preclinical and clinical trials. These tests should serve to establish the optimal concentration and formulation of biocompatible liposomal systems in primary cultures or stabilized lines. According to the International Standard ISO 10993-5, there are several specific ways to analyze the cytotoxicity of medical devices, which involve different aspects of cell function, such as viability and proliferation, cell morphology, loss of membrane integrity, and decrease in cell adhesion.

We performed preliminary cell culture experiments using quantitative determinations of cell viability and qualitative evaluation of cell morphology, in order to establish the concentration threshold of a liposomal formulation of CS on the mouse fibroblast-like NCTC cell line (clone L929), as we previously described (Craciunescu and Moldovan, 2011). Briefly, cells were seeded in 24-well culture plates, at a density of 5×10^4 cells/well, in minimum essential medium supplemented with 10% fetal calf serum (FCS), and plates were incubated in a humidified 5% CO_2 atmosphere, at 37°C, for 24 h, to allow cell adhesion. Then, empty and CS-entrapping liposomes (containing 200 μM lipids final concentration) were added in the culture medium, the plates were incubated at 37°C, for 72 h and 3-[4, 5-dimethyl thiazol-2-y1]-2, 5-diphenyl tetrazolium bromide (MTT) assay (Mossman, 1983), lactate dehydrogenase (LDH) assay (Serrano *et al.*, 2004), and 4',6-diamidino-2-phenylindole (DAPI) staining (Moldovan *et al.*, 2009) were performed.

The results of the MTT test indicated that no cytotoxic effects were induced by empty and CS-entrapping liposomes (Figure 12.5). Cell viability of all tested samples was higher or close to the control culture. No increase in LDH leakage was observed for cells cultivated with liposomal formulations, compared to the untreated cells (Figure 12.5). Cells cultivated with CS-entrapping liposomes maintained their normal morphology, without signs of apoptosis, similar to the control L929 fibroblast cells (Figure 12.5). Quantitative and qualitative data demonstrated that liposomal formulation of CS was highly biocompatible and was able to better interact with L929 fibroblasts and stimulate their cellular activity, compared to CS alone, after 72 h of cultivation.

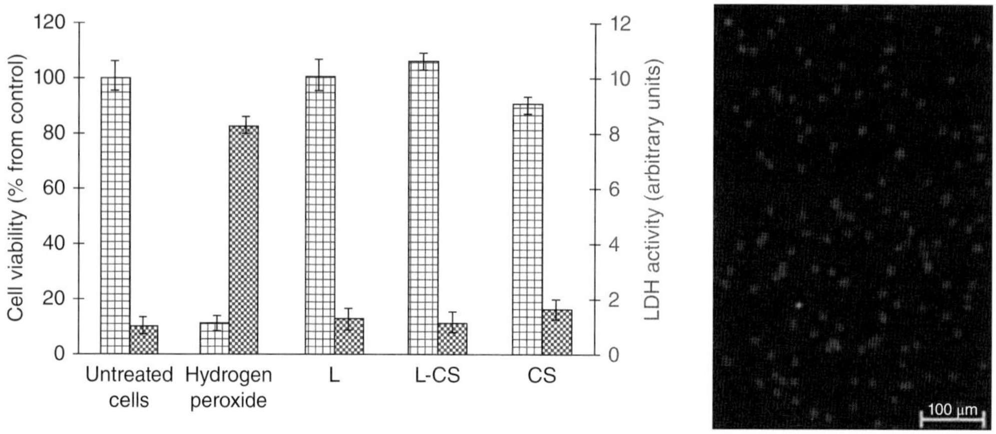

Figure 12.5 Viability and lactate dehydrogenase (LDH) activity of L929 fibroblast cells, after 72 h of cultivation with culture medium (untreated cells), 200 μM hydrogen peroxide (positive control) and samples of liposomes (L), liposomes entrapping CS (L-CS) and CS. The negative control (untreated cells) was considered 100% viable cells and its LDH activity equal to 1. On the right, fluorescence micrograph showing fibroblast cell morphology after cultivation with liposomal variant entrapping CS (DAPI staining, bar = 100 μm).

Previous studies reported a better control of cell growth and metabolic activity of human dermal fibroblasts cultivated with MLV entrapping CS compared to CS alone (Trif *et al.*, 2008). Encapsulation of other dietary agents, such as *N*-acetylcysteine, in liposomes modulated mitochondrial dehydrogenases activity in a culture of A549 epithelial cells, improved cytoprotection, and ameliorated the cellular redox status of stressed cells (Mitsopoulos and Suntres, 2011). CS alone was also tested in cell cultures, in different concentrations, and showed a good biocompatibility in human embryonic cells in the range of 0.2–1 mg/mL (Moldovan *et al.*, 2003). Nanoscale micelles and microscale spheroids containing CS were synthesized by Lim *et al.* (2011) and did not affect the morphology of bovine marrow stromal cells, at moderate concentrations, up to 0.64 mg/mL.

12.6.2 Uptake of liposomes entrapping bioactive molecules by cells

A key research field developed in the past decade is the carrier technology that mediates the intracellular uptake of various bioactive molecules, such as nucleic acids, proteins, peptides, and glycosaminoglycans, because they cannot easily penetrate the cell membrane. A major advantage of using liposomes is the greater quantity of drug that passes through the cell membrane into the cell cytoplasm. In addition, lipophilic drugs could be solubilized by entrapment into the lipid bilayer of the liposomes (Schwendener and Schott, 2010). The main routes of liposomal drug administration are intravenous and topical, but other routes can also be applied. After reaching the systemic circulation or the local area, the interaction between liposomes and the cell takes place by mechanisms of binding and internalization. It is generally believed that the great majority of liposomes undergo adsorption to the cell surface by specific interaction between components, nonspecific hydrophobic interaction, or electrostatic forces and, subsequently, endocytosis by phagocytic cells of the reticuloendothelial system, such as macrophages and neutrophils (Allen *et al.*, 1991). On the other hand, drugs encapsulated in liposomes can be uptaken into the cell cytoplasm after fusion of the plasma cell membrane with the liposomal lipid bilayer. The transfer of liposomal lipids to cellular or subcellular membrane or vice versa without any association of the liposome content has also been reported (Torchilin, 2005; Mansoori *et al.*, 2012). Any of the above methods is operative, either alone or in combination.

It is known that cell–liposome interaction depends on the composition and surface charge of the liposomes. Fluorescence resonance energy transfer and differential scanning calorimetry studies on several lipid formulations were undertaken by Bonicelli *et al.* (2011) in order to evaluate their ability to interact with membrane models. They recommend DMPC cationic liposomes that chemically interact with the cell membrane and act as an efficient intracellular DDS. Subtle modifications in lipid composition and size distribution of DMPC cationic liposomes were also investigated by DSC, DLS, and turbidity measurements and might be highly relevant in controlling organization of lipid bilayers and, hence, their physicochemical and biological features (Barenholz *et al.*, 2011). Additionally, Un *et al.* (2012) showed that Chol, as a liposome component, affected the intracellular trafficking of other PC components to the endoplasmic reticulum and Golgi apparatus. They predicted that the liposomes were intracellulary degraded and the lipid components had different trafficking pathways. In regard to liposome surface charge, Cosimati *et al.* (2013) reported that cationic liposomes produced an alteration of cell-membrane parameters, capacitance, and conductance, evaluated by electrorotation, for a longer period of time than neutral liposomes, as well as a cell radius variation, without affecting cell survival. Other cationic liposomes tested in cell culture showed an electrostatic interaction with the negatively charged cell membranes and cell-surface proteoglycans, facilitating their cellular uptake (Kelly *et al.*, 2011). Our previous experiments investigated the influence of liposome composition and surface charge on their uptake by human synovial fibroblasts (Trif *et al.*, 2001). Negatively charged liposomes (PC:Chol:PS) and pH-sensitive liposomes (DOPE:CHEMS) enhanced drug accumulation in cultured human synovial fibroblasts, unlike the positively charged liposome formulation (DPPE:Chol:SA).

The amount of liposome-entrapped substance that is transferred into cells is an important parameter that affects cell function (Allen and Cullis, 2013). *In vitro* studies established that there is a negative correlation between liposome size and their uptake efficiency by bone-marrow macrophages from mice and rat peritoneal macrophages (Allen *et al.*, 1991). Our previous studies were in agreement with this observation, showing differences in the cellular uptake of liposomes with the same lipid formulation, but with different sizes. The SUVs were internalized more rapidly and more efficiently than MLV liposomes. Concerning the influence of cell line on liposomes recognition and uptake, it was evident that liposomes were uptaken by THP-1 cells in higher quantity than by human synovial fibroblasts (data not published). We have also found that liposomal systems could down-regulate the expression of the bone marrow derived dendritic cells (DCs) surface markers (Trif *et al.*, 2007). A significant decrease was induced in CD40 and CD86 level, suggesting a high interaction between positive liposomes and DCs membrane. Similar experiments using antigen entrapped in cationic liposomes reported the effect of lipid bilayer composition on the interaction with DCs (Foged *et al.*, 2004).

12.6.2.1 Cell uptake of liposomes entrapping CS

Liposome ability to increase intracellular uptake of CS was recently evaluated by our team. Immunohistochemical studies were performed using a monoclonal antibody to CS and secondary antibody coupled with Texas Red for a primary culture of human dermal fibroblasts cultivated with liposomes entrapping CS. After counterstaining with DAPI, cells were observed under a Axiostar Plus microscope (Carl Zeiss, Germany). Micrographs indicated attachment of CS-entrapping liposomes to the cell surface and internalization of CS after 24 h of incubation. When cells were fixed immediately after culture medium removal, the fluorescent spots were visible on the cell surface, while cells that underwent washing with PBS and fixation indicated intracellular immunofluorescent reaction. A high percent of cells incubated with the liposomal formulation of CS were immunostained (Figure 12.6a), while free CS intracellular uptake efficiency was 2–8%, after 48h (Figure 12.6b).

(a) (b)

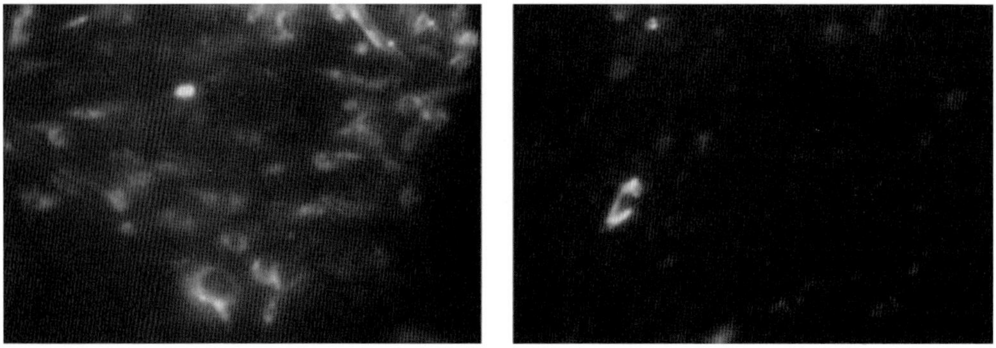

Figure 12.6 Immunohistochemical reaction of CS uptaken by human dermal fibroblast cells, after incubation with (a) liposomes entrapping CS for 24 h and (b) free CS for 48 h.

12.6.3 Ability of liposomes to modify cell response in different models of inflammation

12.6.3.1 Modulation of cytokines level

Downregulation of proinflammatory cytokines, TNF-α, IL-6, and IL-8, as important mediators of inflammation, offers a potential therapeutic strategy to prevent cartilage destruction in inflammatory diseases. Chondroprotective effects of a DDS can be analyzed using *in vitro* experimental models of inflammation and *in vivo* intra-articular administration. Osteoarthritis produces inflammation of the synovial membrane that attracts macrophages and alters synovial fibroblasts and chondrocytes activity. Therefore, two *in vitro* experimental models are mostly used to evaluate the effect of liposomal formulations on the inflammatory process:

1. human monocytic cells (THP-1) differentiated to macrophages and inflamed with lipopolysaccharide (LPS) (Haversen *et al.*, 2002);
2. rabbit chondrocytes inflamed with IL-1β (Aida *et al.*, 2005).

In our studies, human monocytic THP-1 cells were seeded in 24-well culture plates, at a density of 1×10^6 cells/mL and cultivated in RPMI-1640 culture medium, containing 200 ng/mL phorbol myristate acetate (PMA) to differentiate to macrophages. After 48 h, the adhered cells were washed with incomplete RPMI and allowed to settle for 4 h, at 37°C, in RPMI containing 5% FCS and 1% penicillin/streptomycin. Cells were stimulated with 10 ng/mL bacterial LPS (*E. coli* serotype 055:B5, Sigma Chemical Co., St. Louis, MO) for 30 min. Samples of empty liposomes, liposomes entrapping CS, and CS alone were added to each well and the plates were incubated at 37°C, for 18 h. The culture medium from each well was harvested, centrifuged at $400 \times g$ for 10 min, and the supernatants were stored at −20°C. The levels of IL-6, TNF-α, and IL-8 proinflammatory cytokines secreted in the culture media were determined by specific enzyme linked immunosorbent assay (ELISA) sandwich protocols, as recommended by R and D Systems Inc. (Leung *et al.*, 2002).

The results showed that THP-1 macrophage cells stimulated with LPS secreted a very high level of cytokines, in particular TNF-α, in the culture medium. The effect of liposomal formulations and CS alone on cytokines secretion was similar, but to a different extent. Liposomes encapsulating CS inhibited the release of IL-6 and TNF-α in the culture medium in the highest proportion (twofold), followed by CS and empty liposomes (Figure 12.7a). Liposomal formulations were also inhibitors of IL-8 secretion, but to a lesser extent (maximum 1.7-fold in cells cultivated with liposomes entrapping CS).

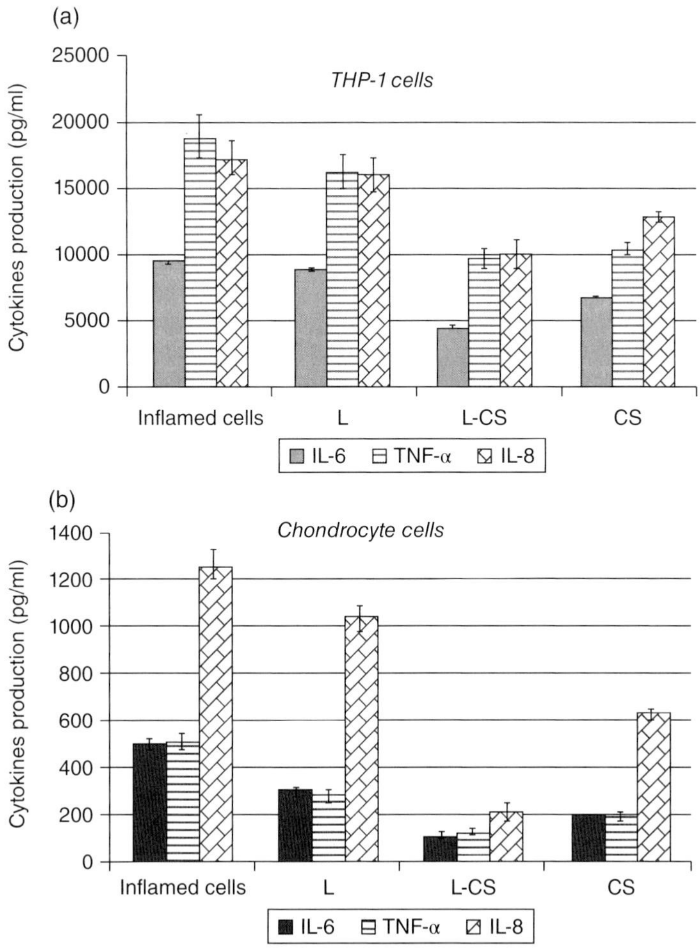

Figure 12.7 The effect of liposomes (L), liposomes entrapping CS (L-CS) and CS on IL-6, TNF-**α**, and IL-8 proinflammatory cytokine production in (a) THP-1 cells and (b) chondrocyte cells.

Overall, liposomal formulation of CS was 1.5 times more efficient than CS alone in reducing LPS-mediated inflammation in THP-1 cells, demonstrating that entrapment of CS into liposomes significantly enhanced the anti-inflammatory capacity of the free compound.

The second experimental model used in our studies was IL-1β-induced inflammation in chondrocytes. Rabbit chondrocytes (1×10^5 cells/well) treated with IL-1β (10 ng/mL) for 24 h were cultivated in the presence of empty liposomes, liposomes entrapping CS, and free CS for 24 h. Conclusive results were also obtained for cytokine production in IL-1β-stimulated rabbit chondrocytes treated with liposomal formulations and were similar to the data obtained for LPS-stimulated macrophage cells. IL-6 and TNF-α were 4.5-fold inhibited in cells treated with liposomal CS compared to 2.6-fold in CS-treated cells (Figure 12.7b). IL-8 secretion presented a greater decline in cells treated with liposomal CS (sixfold decrease) than in CS-treated cells (twofold decrease).The anti-inflammatory activity of liposomal CS could be attributed to the liposome's ability to increase CS accumulation in cells (Craciunescu *et al.*, 2013).

Our findings are in agreement with other *in vitro* studies that reported an enhanced suppresive effect of liposome containing an anti-inflammatory protein (lactoferrin) on LPS-induced TNF-α production in peripheral blood mononuclear leukocytes (Ishikado *et al.*, 2005). Additionally, the anti-inflammatory properties of CS were demonstrated in both *in vitro* and *in vivo* studies by inhibition of several markers of inflammation (Miller and Clegg, 2011). They showed that CS induced a reduction of nuclear factor-kB translocation in IL-1β-treated choncrocyte culture. The authors also showed, on a biochemical basis, that CS treatment presented anti-inflammatory activity in several OA tissues. Campo *et al.* (2009) showed that CS significantly inhibited inflammatory cytokines and inducible nitric oxide synthase in LPS-stimulated mouse chondrocyes by modulating toll-like receptor-4. In a collagen-induced arthritis rat model, treatment with 120 mg CS/kg, intraperitoneally administered once a day (starting at the onset of arthritis) for 10 days, caused a significant reduction in lipid peroxidation (malonaldehyde used as an indicator) and restored the endogenous antioxidants (reduced glutathione and superoxide dismutase), probably by competing for free-radicals scavenging (Campo *et al.*, 2003). As a result, CS could partially reduce several parameters of synovitis, including cell infiltration, fibrosis, and proliferation of synovial lining cells and contributed to preservation of the cartilage integrity.

12.6.3.2 Modulation of matrix metalloproteinase level

Articular cartilage damage is mainly characterized by a progressive degradation of extracellular matrix components, followed by chondrocyte death, tissue fibrillation, and erosion (Martell-Pelletier *et al.*, 2008). Matrix metalloproteinases (MMP) and aggrecanases are the two families that play a crucial role in cartilage matrix breakdown, in particular increasing degradation of collagen and proteogylcans in OA articular cartilage (Tanimoto *et al.*, 2011). Although under physiological conditions the chondrocytes are responsible for the synthesis of cartilage extracellular matrix components (collagens, proteoglycans, glycoproteins), they can be activated to produce enzymes involved in their degradation. This activation is thought to fulfill a critical role in the destruction of cartilage, a characteristic process in OA or RA. *In vitro* experimental models of inflammation in LPS-stimulated macrophages or IL-1β-stimulated chondrocytes were used in MMP–liposomes interaction studies. Activated macrophages produced factors like oxygen radicals and enzymes that activate MMP.

Our study used rabbit chondrocytes isolated from ear cartilage cultured in Dulbecco's modified Eagle's medium containing 10% FCS. Confluent chondrocytes were treated with 50ng/ml IL-1β to induce inflammation. Empty and CS-entrapping liposomes were added in the culture medium of inflamed cells for 48 h and gelatin-zymography was performed, as previously described (Cimpean *et al.*, 2000). Liposomal treatment of inflamed chondrocytes influenced the MMP activity in the culture medium. Due to its ability to modulate the secretion of destructive MMP in inflamed chondrocytes, liposomal formulation of CS could be further investigated as chondroprotective therapeutic agent in arthritis.

From another perspective, upon targeting, the liposomal content release is usually slow. The MMP, in particular high levels of MMP-9, could rapidly release the encapsulated contents from optimized liposomal formulations (Banerjee *et al.*, 2009).

Hofkens *et al.* (2011) have shown that 100-nm-diameter liposomes entrapping an anti-inflammatory molecule induced a dose-dependent suppression of IL-1β production and a significant reduction in mRNA levels of MMP-3, -9, -13, and -14, offering an effective strategy to inhibit the factors that contribute to destruction of cartilage matrix during arthritis.

12.6.3.3 Modulation of reactive oxygen species

Oxidative stress generated by reactive oxygen species (ROS), such as hydrogen peroxide, singlet oxygen, superoxide radical, and hydroxyl radical, plays an important role in the process of OA cartilage degradation and synovial inflammation (Alcaraz *et al.*, 2010). In normal conditions, ROS are involved in activation of some important intracellular signaling cascades (Roman-Blas *et al.*, 2009) but, when oxidative stress occurs, inflammation-type events, cell apoptosis, and degradation of cartilage extracellular matrix take place at the insulted site (Martell-Pelletier *et al.*, 2008). In recent years, it has been found that ROS production is increased in joint diseases, such as OA (Deberg *et al.*, 2005). The antioxidant effect of CS is reviewed by Egea *et al.* (2010) in regard to its therapeutic potential in modulation of local inflammation and local immune response. Development of antioxidant delivery systems could improve oxidant-induced inflammatory processes when administered in sufficient concentrations. Treatment of pathological conditions related to oxidative stress could be investigated in several cell culture experimental models.

To evaluate the cell protective and antioxidant effects of empty and CS-entrapping liposomes, we used a primary culture of rabbit chondrocytes stressed with H_2O_2. Cell suspension was seeded in 24-well culture plates, at a density of 5×10^4 cells/well and the plates were incubated in a humidified 5% CO_2 atmosphere at 37°C to allow cell adhesion, for 24 h. Two treatment groups were designed, as previously described (Craciunescu *et al.*, 2012). Briefly, cells cotreated with liposomal samples (200 μM liposomal lipids final concentration in the culture medium) and 50 μM H_2O_2 and cells pretreated with liposomal samples for 24 h, then H_2O_2-treated for 24 h, were cultivated in standard conditions. Cells from both groups were analyzed for viability by Neutral Red assay and cell morphology observations were carried out by light microscopy after Giemsa staining. Our results showed that liposomal formulations and CS alone could protect cells against H_2O_2-induced stress, but higher values of cell viability were recorded in pretreatment with liposomal formulation of CS than with CS alone, at 24 h after exposure to H_2O_2. There were little difference between the empty liposomes and CS-entrapping liposomes treatments, emphasizing the prevalent cytoprotective activity of lipids over CS. The ability of liposomes to protect chondrocytes against H_2O_2-induced stress could be due to their role in preventing cell membrane alteration, rather than to a direct interaction with H_2O_2. Also, their composition, size, and surface allowed a favorable interaction with chondrocyte cells. Light microscopy observations were in accordance with quantitative cell viability data.

Suntres (2011) has shown that liposomal delivery of antioxidants could improve the therapy with dietary antioxidants that allow direct interaction with ROS involved in several pathological conditions. Systemic administration of pegylated liposomes entrapping superoxide dismutase, an enzyme with an important role in antioxidant defense of cells, augmented liposomes circulation time and the extent of drug localization at arthritic sites (Corvo *et al.*, 1999). Also, encapsulation of antioxidants, such as glutathione, *N*-acetylcysteine, and α-tocopherol, into liposomes could avoid potential drug toxicity and facilitate their delivery to cells (Mitsopoulos and Suntres, 2011; Alipour *et al.*, 2012).

12.7 CONCLUSIONS AND FUTURE PERSPECTIVE

In the nanomedicine area, liposomes have generated great interest, with several formulations being currently approved by the FDA and many others in clinical trials. It is expected that the same advantages can eventually lead to the application of liposomes in foods. Liposomes offer important advantages over other delivery systems, including biocompatibility and control of biological properties via modification of physicochemical parameters (lipid composition, vesicle size, lamellarity, surface charge, lipid membrane fluidity). They are able to encapsulate hydrophilic and lipophilic bioactive molecules, to protect them from degradation in the external environment and to enhance their therapeutic index. Moreover, liposome can increase the solubility of poorly water-soluble molecules and to stabilize a variety of therapeutic agents such as peptides, proteins, and nucleotides. It has also been identified that there are improved phamacokinetic properties and biodistribution with liposomal formulation than with the free form of the therapeutic agent. Liposomes as a DDS can be administrated by many routes (intravenous, intramuscular, intraperitoneal, intra-articular, subcutaneous, transdermal, oral, nasal, ocular) and these can be used for the treatment of various diseases.

Based on all the data presented in this chapter, we conclude that liposomes could represent a valuable controlled delivery system for the treatment of many inflammatory/degenerative disorders. Using a suitable size, lamellarity, lipid combinations, and drug to lipid ratio for a higher acid- and enzyme-resistance in the gastrointestinal tract, liposomes could become a very efficient system for the oral delivery of a large class of anti-inflammatory/nutraceutical compounds. Further studies are needed to identify and support the use of liposomal systems as an efficient therapy for OA management, using intra-articular or oral routes of administration.

Acknowledgments

The authors would like to acknowledge the contribution of Professor Dr Otilia Zarnescu, Faculty of Biology, University of Bucharest and Dr Simona Ghenea, Institute of Biochemistry of the Romanian Academy, to the immunohistochemistry and microscopy studies. We also thank Dr Magdalena Moisei, Dr Lucia Moldovan, and Dr Anca Roseanu for their outstanding collaboration at various stages of the work discussed in this chapter. This work was supported by the Project ESF POSDRU/89/1.5/S/60746 (OC and MT), the Romanian Academy Project No. 3/2012 and LSUAg Center, USA, Subaward 61369 (MT).

References

Aida, Y., Maeno, M., Suzuki, N., Shiratsuchi, H., Motohashi, M. and Matsumura, H. (2005) The effect of IL-1β on the expression of matrix metalloproteinases and tissue inhibitors of matrix metalloproteinases in human chondrocytes. *Life Sciences,* 77, 3210–3221.

Akhatar, A., Prajapati, S.K., Singh, D., Kumar, B. and Shafat, K. (2012) Enhanced bioavailability of drugs via intranasal drug delivery system. *International Research Journal of Pharmacy,* 3, 68–74.

Alcaraz, M.J., Megias, J., Garcia-Arnandis, I., Clerigues, V. and Guillen, M.I. (2010) New molecular targets for the treatment of osteoarthritis. *Biochemical Pharmacology,* 80, 13–21.

Alipour, M., Smith, M.G., Pucaj, K. and Suntres, Z.E. (2012) Acute toxicity study of liposomal antioxidant formulations containing *N*-acetylcysteine, α-tocopherol, and γ-tocopherol in rats. *Journal of Liposome Research,* 22, 158–167.

Allen, T.M. and Cullis, P.R. (2013) Liposomal drug delivery systems: from concept to clinical applications. *Advanced Drug Delivery Reviews*, 65, 36–48.

Allen, T.M., Austin, G.A., Chonn, A., Lin, L. and Lee, K.C. (1991) Uptake of liposomes by cultured mouse bone marrow macrophages: influence of liposome composition and size. *Biochimica et Biophysica Acta,* 1061, 56–64.

Allen, T.M., Cheng, W.W., Hare, J.I. and Laginha, K.M. (2006) Pharmacokinetics and pharmacodynamics of lipidic nano-particles in cancer. *Anti-Cancer Agents in Medicinal Chemistry,* 6, 513–523.

Banerjee, J., Hanson, A.J., Gadam, B., *et al.* (2009) Release of liposomal contents by cell-secreted matrix metalloproteinase-9. *Bioconjugate Chem* 20, 1332–1339.

Barenholz, Y. (2012) Doxil® – the first FDA-approved nano-drug: lessons learned. *Journal of Controlled Release,* 160, 117–134.

Barenholz, Y., Bombelli, C., Bonicelli, M.G., di Profio, P., Giansanti, L., Mancini, G. and Pascale, F. (2011) Influence of lipid composition on the thermotropic behavior and size distribution of mixed cationic liposomes. *Journal of Colloid and Interface Science,* 356, 46–53.

Bonicelli, M.G., Giansanti, L., Ierino, M. and Manicini, G. (2011) Interaction of cationic liposomes with cell membrane models. *Journal of Colloid and Interface Science,* 355, 1–8.

Campo, G.M., Avenoso, A., Campo, S., Ferlazzo, A.M., Altavilla, D. and Calatroni, A. (2003) Efficacy of treatment with glycosaminoglycans on experimental collagen-induced arthritis in rats. *Arthritis Research and Therapy,* 5, R122–R131.

Campo, G.M., Avenoso, A., Campo, S., Traina, P., D'Ascola, A. and Calatroni, A. (2009) Glycosaminoglycans reduced inflammatory response by modulating toll-like receptor-4 in LPS-stimulated chondrocytes. *Archives of Biochemistry and Biophysics,* 491, 7–15.

Canto, G.S., Dalmora, S.L. and Oliveira, A.G. (1999) Piroxicam encapsulated in liposomes: characterization and *in vivo* evaluation of topical anti-inflammatory effect. *Drug Development and Industrial Pharmacy,* 25, 1235–1239.

Chang, H.I. and Yeh, M.-K. (2012) Clinical development of liposome-based drugs: formulation, characterization, and therapeutic efficacy. *International Journal of Nanomedicine,* 7, 49–60.

Chen, H., Torchilin, V. and Langer, R. (1996) Lectin-bearing polymerized liposomes as potential oral vaccine carriers. *Pharmaceutical Research,* 13, 1378–1383.

Chonn, A. and Cullis, P.R. (1998). Recent advances in liposome technologies and their applications for systemic gene delivery. *Advanced Drug Delivery Reviews*, 30, 73–84.

Cimpean, A., Caloianu, M., Alexandru, D., Efimov, N. and Buzgariu, W. (2000) Effect of interleukin-1β on gelatinolytic activity and cell morphology of human osteoarthritic chodrocytes in culture. *Journal of Medical Biochemistry,* 4, 113–129.

Copland, M.J., Baird, M.A., Rades, T., *et* al. (2003) Liposomal delivery of antigen to human dendritic cells. *Vaccine*, 21, 883–890.

Corvo, M.L., Boerman, O.C., Oyen, W.J., Bloois, L.V., Cruz, M.E.M., Crommelin, D.J.A. and Storm, G. (1999) Intravenous administration of superoxide dismutase entrapped in long circulating liposomes. II. *In vivo* fate in a rat model of adjuvant arthritis. *Biochimica et Biophysica Acta,* 1419, 325–334.

Cosimati, R., Milardi, G.L., Bombelli, C., Bonincontro, A., Bordi, F., Manicini, G. and Risuleo, G. (2013) Interactions of DMPC and DMPC/gemini liposomes with the cell membrane investigated by electrorotation. *Biochimica et Biophysica Acta,* 1828, 352–356.

Costin, G.E., Trif, M., Nichita, N., Dwek, R.A. and Petrescu, S.M. (2002) pH-sensitive liposomes are efficient carriers for endoplasmic reticulum-targeted drugs in mouse melanoma cells. *Biochemical and Biophysical Research Communications,* 293, 918–928.

Craciunescu, O. and Moldovan, L. (2011) Designing bio-inspired composite materials for medical applications. In: *Nanocomposites and Polymers with Analytical Methods* (ed. J. Cuppoletti). InTech, Croatia; 309–334.

Craciunescu, O., Constantin, D., Gaspar, A., Toma, L., Utoiu, E. and Moldovan, L. (2012) Evaluation of antioxidant and cytoprotective activities of *Arnica montana* L. and *Artemisia absinthium* L. ethanolic extracts. *Chemistry Central Journal*, 6, 97–107.

Craciunescu, O., Moldovan, L., Moisei, M. and Trif, M. (2013) Liposomal formulation of chondroitin sulphate enhances its antioxidant and anti-inflammatory potential in L929 fibroblast cell line. *Journal of Liposome Research,* 23, 145–153.

Crommelin, D.J. and Storm, G. (2003) Liposomes: from the bench to the bed. *Journal of Liposome Research,* 13, 33–36.

Crommelin, D.J.A. and Zuidam, N.J. (2007) Hydrolysis of phospholipids in liposomes and stability-indicating analytical techniques. In: *Liposome Technology* (ed. G. Gregoriadis). Informa Healthcare, New York; 285–296.

Deberg, M., Labasse, A., Christgau, S., *et al.* (2005) New serum biochemical markers (Coll 2-1 and Coll 2-1 NO_2) for studying oxidative-related type II collagen network degradation in patients with osteoarthritis and rheumatoid arthritis. *Osteoarthritis and Cartilage,* 13, 258–265.

Dezim, Z., Unal, N., Essiz, D. and Abbasoglu, U. (2004) The effect of various liposome formulations on insulin penetration across Caco-2 monolayer. *Life Sciences*, 75(23), 2819–2827

Dua, J.S., Rana, A.C. and Bhandari, A.K. (2012) Liposome: methods of preparation and applications. *International Journal of Pharmaceutical Studies and Research*, III, 14–20.

DuRaine, G., Hu, J. and Athanasiou, K. (2011) Bioengineering in the oral cavity: insights from articular cartilage tissue engineering. *International Journal of Oral and Maxillofacial Surgery,* 26(Suppl), 11–24.

Eddington, N.D., Du, J. and White, N. (2001) Evidence of the oral absorption of chondroitin sulfate as determined by total disaccharide content after oral and intravenous administration to horses. *American Association of Equine Practitioners Proceedings*, 47, 326–328.

Egbaria, K. and Weiner, N. (1990) Liposomes as a topical drug delivery system. *Advanced Drug Delivery Reviews*, 5, 287–300.

Egea, J., Garcia, A.G., Verges, J., Montell, E. and Lopez, M.G. (2010) Antioxidant, antiinflammatory and neuroprotective actions of chondroitin sulfate and proteoglycans. *Osteoarthritis and Cartilage,* 18(Suppl. 1), S24–S28.

Elron-Gross, I., Glucksam, Y. and Margalit, R. (2009) Liposomal dexamethasone-diclofenac combinations for local osteoarthritis treatment. *International Journal of Pharmaceutics,* 376, 84–91.

Florence, A.T. and Jani, P.U. (1993) Particulate delivery: The challenge of the oral route. In: *Pharmaceutical Particulate Carriers* (ed. A Rolland). Marcel Dekker, New York; 31–64.

Foged, C., Arigita, C., Sundblad, A., Jiskoot, W., Atorm, G. and Frokjaer, S. (2004) Interaction of dendritic cells with antigen-containing liposomes: effect of bilayer composition. *Vaccine* 22, 1903–1913.

French, M.M., Rose, S., Canseco, J. and Athanasiou, K.A. (2004) Chondrogenic differentiation of adult dermal fibroblasts. *Annals of Biomedical Engineering,* 32, 50–56.

Gabizon, A., Goren, D., Cohen, R. and Barenholz, Y. (1998) Development of liposomal anthracyclines: from basics to clinical applications. *Journal of Controlled Release,* 53, 275–279.

Gerwin, N., Hops, C. and Lucke, A. (2006) Intraarticular drug delivery in osteoarthritis. *Advanced Drug Delivery Reviews,* 58, 226–242.

Goldberg, R. and Klein, J. (2012) Liposomes as lubricants: beyond drug delivery. *Chem Phys Lipids* 165, 374–381.

Gregoriadis, G. (ed.) (2007) *Liposome Technology*. Informa Healthcare USA, Inc., New York.

Gunasekera, U.A., Pankhurst, Q.A. and Douek, M. (2009) Imaging applications of nanotechnology in cancer. *Targeted Oncology,* 4, 169–181.

Haversen, L., Ohlsson, B.G., Hohn-Zoric, M., Hanson, L.A. and Mattsby-Baltzer, I. (2002) Lactoferrin down-regulates the LPS-induced cytokine production in monocytic cells via NF-kappa B. *Cellular Immunology,* 220, 83–95.

Henrotin, Y., Mathy, M., Sanchez, C. and Lambert, C. (2010) Chondroitin sulfate in the treatment of osteoarthritis : from *in vitro* studies to clinical recommendations. *Therapeutic Advances in Musculoskeletal Disease,* 2, 335–348.

Hincha, D.K. (2008) Effects of alpha-tocopherol (vitamin E) on the stability and lipid dynamics of model membranes mimicking the lipid composition of plant chloroplast membranes. *FEBS Letters,* 582, 3687–3692.

Hofkens, W., Storm, G., van den Berg, W.B. and van Lent, P.L. (2011) Liposomal targeting of glucocorticoids to the inflamed synovium inhibits cartilage matrix destruction during murine antigen-induced arthritis. *International Journal of Pharmaceutics,* 416, 486–492.

Iovu, M., Dumais, G. and du Souich, P. (2008) Anti-inflammatory activity of chondroitin sulfate. *Osteoarthritis and Cartilage,* 16, S14–S18.

Ishikado, A., Imanaka, H., Takeuchi, T., Harada, E. and Makino, T. (2005) Liposomalization of lactoferrin enhanced its anti-inflammatory effects via oral administration. *Biological and Pharmaceutical Bulletin,* 28, 1717–1721.

Iwanaga, K., Ono, S., Narioka, K., *et al.* (1999) Application of surface-coated liposomes for oral delivery of peptide: effects of coating the liposome's surface on the GI transport of insulin. *Journal of Pharmaceutical Sciences,* 88, 248–252.

Jain, S.K., Jain, A., Gulbake, A., Shilpi, S. and Khare, P. (2010) Steroid-coupled liposomes for targeted delivery to tumor. *Therapeutic Delivery,* 1, 345–357.

Jithan, A.V. and Swathi, M. (2010) Development of topical diclofenac sodium liposomal gel for better antiinflamatory activity. *International Journal of Pharmaceutical Sciences and Nanotechnology*, 3, 986–993.

Katre, N.V. (2004) Liposome-based depot injection technologies: how versatile are they? *American Journal of Drug Delivery,* 2, 213–227.

Kelly, C., Jefferies, C. and Cryan, S.-A. (2011) Targeted liposomal drug delivery to monocytes and macrophages. *Journal of Drug Delivery,* 2011, 727241.

Kobo, M., Ando, K., Mimura, T., Matsusue, Y. and Mori, K. (2009) Chondroitin sulfate for the treatment of hip and knee osteoarthritis: current status and future trends. *Life Sciences,* 85, 477–483.

Kotrotsiou, O., Kotti, K., Dini, E., Kammona, O. and Kiparissides, C. (2005) Nanostructured materials for selective recognition and targeted drug delivery. *Journal of Physics: Conference Series,* 10, 281–284.

Lasic, D.D. and Papahadjopoulos, D. (eds) (1998) *Medical applications of liposomes*. Elsevier, Amsterdam.

Lasic, D.D., Joannic, R., Keller, B.C., Frederik, P.W. and Auvray, L. (2001) Spontaneous vesiculation. *Advances in Colloid and Interface Science,* 89–90, 337–349.

Lee W. C. and Tsai T. H. (2010) Preparation and characterization of liposomal coenzyme Q10 for *in vivo* topical application, *International Journal of Pharmaceutics,* 395, 78–83.

Lehtinen, J., Raki, M., Bergstrom, K.A., *et al.* (2012) Pre-targeting and direct immunotargeting of liposomal drug carriers to ovarian carcinoma. *PLoS One* 7, e41410.

Leung, B.P., Conacher, M., Hunter, D., McInnes, I.B., Liew, F.Y. and Brewer, J.M. (2002) A novel dendritic cell-induced model of erosive inflammatory arthritis: distinct roles for dendritic cells in T cell activation and induction of local inflammation. *J Immunol* 169, 7071–7077.

Li, H., Song, J. H., Park, J. S. and Han, K. (2003) Polyethylene glycol-coated liposomes for oral delivery of recombinant human epidermal growth factor. *International Journal of Pharmaceutics,* 258, 11–19.

Lim, J.J., Hammoudi, T.M., Bratt-Leal, A.M., *et al.* (2011) Development of nano- and micro-scale chondroitin sulfate particles for controlled growth factor delivery. *Acta Biomaterialia,* 7, 986–995.

Maestrelli, F., Gonzalez-Rodriguez, M.L., Rabasco, A.M. and Mura, P. (2005) Preparation and characterisation of liposomes encapsulating ketoprofen-cyclodextrin complexes for transdermal drug delivery, *International Journal of Pharmaceutics,* 298, 55–67.

Mansoori, M.A., Agrawal, S., Jawade, S. and Khan, M.I. (2012) A review on liposome. *International Journal of Pharmaceutical Studies and Research,* 1, 453–464.

Martell-Pelletier J., Boileau C., Pelletier J.P. and Roughley P.J. (2008). Cartilage in normal and osteoarthritis conditions. *Best Practice and Research Clinical Rheumatology,* 22, 351–384.

Martin, F. and Lasic, D. (eds) (1995) *Stealth Liposomes.* CRC Press, Boca Raton, FL.

Maurer, N., David, B., Fenske, D.B. and Cullis, P.R. (2001) Developments in liposomal drug delivery systems. *Expert Opinion on Biological Therapy,* 1, 1–25.

McNary, S.M., Athanasiou, K.A. and Reddi, A.H. (2012) Engineering lubrication in articular cartilage. *Tissue Engineering B,* 18, 88–100.

Metselaar, J.M. and Storm, G. (2005) Liposomes in the treatment of inflammatory disorders. . *Expert Opinion on Drug Delivery,* 2, 465–476.

Miller, C.R., Bondurant, B., McLean, S.D., McGovern, K.A. and O'Brien, D.F. (1998) Liposome-cell interactions *in vitro*: effect of liposome surface charge on the binding and endocytosis of conventional and sterically stabilized liposomes. *Biochemistry,* 37, 12875–12883.

Miller, K.L. and Clegg, D.O. (2011) Glucosamine and chondroitin sulfate. *Rheumatic Disease Clinics of North America,* 37, 103–118.

Mitsopoulos, P. and Suntres, Z.E. (2011) Protective effects of liposomal N-acetylcysteine against paraquat-induced cytotoxicity and gene expression. *Journal of Toxicology,* 2011, 808967.

Moghimi, S.M. and Agrawal, A. (2005) Lipid-based nanosystems and complexes in experimental and clinical therapeutics. *Current Drug Delivery,* 2, 295.

Moghimi, S.M. and Szebeni, J. (2003) Stealth liposomes and long circulating nanoparticles: critical issues in pharmacokinetics, opsonization and protein-binding properties. *Progress in Lipid Research,* 42, 463–478.

Moldovan, L., Buzgariu, W., Zarnescu, O., Flamanzeanu, M. and Craciunescu, O. (2003) *In vitro* biocompatibility evaluation of some corneal biopolymers. *Romanian Biological Science,* 1, 23–34.

Moldovan, L., Craciunescu, O., Oprita, E.I., Balan, M. and Zarnescu, O. (2009) Collagen-chondroitin sulfate-hydroxyapatite porous composites: preparation, characterization and *in vitro* biocompatibility testing. *Romanian Biotechnology Letters,* 14, 4459–4466.

Morelli, V., Naquin, C. and Weaver, V. (2003) Alternative therapies for traditional disease states: osteoarthritis. *American Family Physician,* 67, 339–344.

Mossman, T. (1983) Rapid colorimetric assay for cellular growth and survival: application to proliferation and cytotoxic assays. *Journal of Immunological Methods,* 65, 55–63.

Mozafari, M.R. (2005) Liposomes: an overview of manufacturing techniques. *Cellular and Molecular Biology Letters,* 10, 711– 719.

Mozafari, M.R., Johnson, C., Hatziantoniou, S. and Demetzos, C. (2008) Nanoliposomes and their applications in food nanotechnology. *Journal of Liposome Research,* 18, 309–327.

Muthu, M.S. and Feng, S.-S. (2010) Nanopharmacology of liposomes developed for cancer therapy. *Nanomedicine (London),* 5, 1017–1019.

Muthu, M.S. and Singh, S. (2009) Targeted nanomedicines: effective treatment modalities for cancer, AIDS and brain disorders. *Nanomedicine (London),* 4, 105–118.

Niu, M., Lu, Y, Hovgaard, L. and Wu, W. (2011) Liposome containing glycocolate as potential oral insulin delivery system: preparation, in vitro characterization and improved protection against enzymatic degradation. *International Journal of Nanomedicine,* 6, 1155–1166).

Park, J.W. (2002) Liposome-based drug delivery in breast cancer treatment. *Breast Cancer Research,* 4, 95–99.

Perrie, Y., Mohammed, A.U.R., Vangala, A. and McNeil, S.E. (2007) Environmental scanning electron microscopy offers real-time morphological analysis of liposomes and niosomes. *Journal of Liposome Research,* 17, 27–37.

Pierre, M.B.R. and dos Santos Miranda Costa, I. (2011) Liposomal systems as drug delivery vehicles for dermal and transdermal applications. *Archives of Dermatological Research,* 303, 607–621.

Pinheiro, M., Lucio, M., Lima, J.L.F.C. and Reis, S. (2011) Liposomes as drug delivery systems for the treatment of TB. *Nanomedicine,* 6, 1413–1428.

Rodgers, M.R. (2006) Effects of oral glucosamine and chondroitin sulfates supplementation on frequency of intra-articular therapy of the horse tarsus. *International Journal of Applied Research in Veterinary Medicine,* 4, 155–162.

Rogers, J.A. and Anderson, K.E. (1998) The potential of liposomes in oral drug delivery. *Critical Reviews in Therapeutic Drug Carrier Systems,* 15, 421–480.

Roman-Blas, J.A., Contreras-Blasco, M.A., Largo, R., Alvarez-Soria, M.A., Castaneda, S. and Herrero-Beaumont, G. (2009) Differential effects of the antioxidant n-acetylcysteine on the production of catabolic mediators in IL-1β-stimulated human osteoarthritic synoviocytes and chondrocytes. *European Journal of Pharmacology*, 623, 125–131.

Samad, A., Sultana, Y. and Aqil, M. (2007) Liposomal drug delivery systems: an update review. *Current Drug Delivery*, 4, 297–305.

Schwendener, R.A. and Schott, H. (2010) Liposome formulations of hydrophobic drugs. *Methods in Molecular Biology*, 605, 129–138.

Serrano, M.C., Pagani, R., Vallet-Regi, M., Pena, J., Ramila, A., Izquierdo, I. and Portoles, M.T. (2004) *In vitro* biocompatibility assessment of poly(ε-caprolactone) films using L929 mouse fibroblasts. *Biomaterials*, 25, 5603–5611.

Shivhare, U.D., Ambulkar, D.U., Mathur, V.B., Bhusari, K.P. and Godbole, M.D. (2009) Formulation and evaluation of pentoxifylline liposome formulation. *Digest Journal of Nanomaterials and Biostructures*, 4, 631–637.

Sihorkar, V. and Vyas, S.P. (2001) Potential of polysaccharide anchored liposomes in drug delivery, targeting and immunization. *Journal of Pharmacy and Pharmaceutical Sciences*, 4, 138–158.

Sivan, S., Schroeder, A., Verberne, G., *et al.* (2010) Liposomes act as effective biolubricants for friction reduction in human synovial joints. *Langmuir*, 26, 1107–1116.

Song, J., Shi, F., Zhang, Z., *et al.* (2011) Formulation and evaluation of celastrol-loaded liposomes. *Molecules*, 16, 7880–7892.

Storm, G. and Crommelin, D.J.A. (1998) Liposomes: quo vadis? *Pharmaceutical Sciences and Technology*, 1, 19–31.

Suntres, Z.E. (2011) Liposomal antioxidants for protection against oxidant-induced damage. *Journal of Toxicology*, 2011, 152474.

Taira, M.C., Chiaramoni, N.S., Pecuch, K.M. and Alonso-Romanowski, S. (2004) Stability of liposomal formulations in physiological conditions for oral drug delivery. *Drug Delivery*, 11, 123–128.

Tang, J., Wei, H., Liu, H., Ji, H., Dong, D., Zhu, D. and Wu, L. (2010) Pharmacokinetics and biodistribution of itraconazole in rats and mice following intravenous administration in a novel liposome formulation. *Drug Delivery*, 17, 223–230.

Tanimoto, K., Iwabuchi, Y., Tanne, Y., *et al.* (2011) Interleukin-1 beta affects cyclooxygenase-2 expression and cartilage metabolism in mandibular condyle. *Archives of Oral Biology*, 56, 1412–1418.

Tomoyasu, Y., Tatsuji, M., Maeda, S., Higuchi, H. and Miyawaki, T. (2011) Liposome-encapsulated midazolam for oral administration. *Journal of Liposome Research*, 21, 166–172.

Torchilin, V.P. (2005) Recent advances with liposomes as pharmaceutical carriers. *Nature Reviews Drug Discovery*, 4, 145–160.

Touitou, E., Junginger, H.E., Weiner, N.D., Nagai, T. and Mezei, M. (1994) Liposomes as carriers for topical and transdermal delivery. *Journal of Pharmaceutical Sciences*, 83, 1189–1203.

Trif, M., Guillen, C., Vaughan, D.M., Telfer, J.M., Brewer, J.M., Roseanu, A. and Brock, J.H. (2001) Liposomes as possible carriers for lactoferrin in the local treatment of inflammatory diseases. *Experimental Biology and Medicine*, 226, 559–564.

Trif, M., Roseanu, A., Brock, J.H. and Brewer, J.M. (2007) Designing lipid nanostructures for local delivery of biologically active macromolecules. *Journal of Liposome Research*, 17, 237–248.

Trif, M., Moldovan, L., Moisei, M., Craciunescu, O. and Zarnescu, O. (2008) Liposomes-entrapped chondroitin sulphate: ultrastructural characterization and *in vitro* biocompatibility. *Micron*, 39, 1042–1045.

Tsamaloukas, A, Szadkowska, H. and Heerklotz, H. (2006) Thermodynamic comparison of the interactions of cholesterol with unsaturated phospholipid and sphingomyelins. *Biophysics Journal*, 90, 4479–4487.

Uebelhart, D., Knols, R., de Bruin, E.D. and Verbruggen, G. (2006) Treatment of knee osteoarthritis with oral chondroitin sulfate. *Advances in Pharmacology*, 53, 523–539.

Uhumwangho, M.U. and Okor, R.S. (2005) Current trends in the production and biomedical applications of liposomes: a review. *Journal of Medicine and Biomedical Research*, 4, 9–21.

Un, K., Sakai-Kato, K., Oshima, Y., Kawanishi, T. and Okuda, H. (2012) Intracellular trafficking mechanism, from intracellular uptake to extracellular efflux, for phospholipid/cholesterol liposomes. *Biomaterials*, 33, 8131–8141.

Van den Hoven, J.M., van Tomme, S.R., Metselaar, J.M., Nuijen, B., Beijnen, J.H. and Storm, G. (2011) Liposomal drug formulations in the treatment of rheumatoid arthritis. *Molecular Pharmacology* 8, 1002–1015.

Vanniasinghe, A.S., Bender, V. and Manolios, N. (2009) The potential of liposomal drug delivery for the treatment of inflammatory arthritis. *Seminars in Arthritis and Rheumatism*, 39, 182–196.

Verberne, G., Schroeder, A., Halperin, G., Barenholz, Y. and Etsion, I. (2010) Liposomes as potential biolubricant additives for wear reduction in human synovial joints. *Wear* 268, 1037–1042.

Vorauer-Uhl K., Furnschlief, Wagner E.A., Ferko B. and Katinger H.. (2001) Topically applied liposome encapsulated superoxide dismutase reduces postburn wound size and edema formation, *European Journal of Pharmaceutical Sciences,* 14, 63–67.

Wang, S. Zhang, J. Jiang T., *et al.* (2011) Protective effect of Coenzyme Q10 against oxidative damage in human lens epithelial cells by novel ocular drug carriers, *International Journal of Pharmaceutics,* 403, 219–229.

Williams, A., Topley, N., Dojcinov, S., Richards, P. and Williams, B. (2001) Amelioration of rat antigen-induced arthritis by liposomally conjugated methotrexate is accompanied by down-regulation of cytokine mRNA expression. *Rheumatology (Oxford)*, 40, 375–383.

Woitiski, C.B., Carvalho, R.A., Ribeiro, A.J., Neufeld, R.J. and Veiga, F. (2008) Strategies toward the improved oral delivery of insulin nanoparticles via gastrointestinal uptake and translocation. *Drug Delivery,* 22, 223–237

Wong, A.W., Ormsby, E., Zhang, H., Seo, J.W., Mahakian, L.M., Caskey, C.F. and Ferrara, KW. (2013) A comparison of image contrast with (64)Cu-labeled long circulating liposomes and (18)F-FDG in a murine model of mammary carcinoma. *American Journal of Nuclear Medicine and Molecular Imaging,* 3, 32–43.

Wu, Z.H., Ping, Q.N., Wei, Y. and Lai, J.M. (2004) Hypoglycemic efficacy of chitosan-coated insulin liposomes after oral administration in mice. *Acta Pharmacologica Sinica,* 25, 966–972.

Xu, H., Deng, Y.H., Wang, K.Q. and Chen, D.W. 2012, Preparation and characterization of stable pH-sensitive vesicles composed of alfa-tocopherol hemisuccinate. *AAPS PharmSciTech*, 13, 1377–1385.

Yang, C., Liu, H.Z., Fu, Z.X. and Lu, W.D. (2011) Oxaliplatin long-circulating liposomes improved therapeutic index of colorectal carcinoma. *BMC Biotechnology,* 11, 21.

Zhang, L., Gu, F.X., Chan, J.M., Wang, A.Z., Langer, R.S. and Farokhzad, O.C. (2008) Nanoparticles in medicine: therapeutic applications and developments. *Clinical Pharmacology and Therapeutics,* 83, 761–769.

Zhang, Z. and Huang, G. (2012) Micro- and nano-carrier mediated intra-articular drug delivery systems for the treatment of osteoarthritis. *Journal of Nanotechnology,* 2012, 748909.

13 Solid lipid nanoparticles and applications

Maria Fernanda San Martin-Gonzalez

Food Science Department, Purdue University, West Lafayette, Indiana, USA

13.1 INTRODUCTION

The use of solid lipid nanoparticles (SLNs) for the encapsulation and delivery of bioactive components, primarily drugs, has been of interest in the pharmaceutical industry for the past 20 years, and, more recently, in the field of food science. As knowledge about the beneficial properties of specific food components on health evolves, approaches to selectively deliver those beneficial components are now considered during development of new functional food products. Several bioactive compounds, drugs included, typically exhibit extremely low water solubility, thus, their incorporation into colloidal carrier systems is under extensive research. Conventional carrier systems used in drug delivery, and potentially in the manufacture of functional foods, are divided into two groups: polymeric or lipidic, depending on the carrier material used in their manufacture (Bunjes and Siekmann, 2006). Solid lipid nanoparticles are a type of disperse system that differs from emulsion systems by the fact that the lipids in the dispersed phase exist in solid state, as opposed to the liquid state of droplets in an emulsion.

The interest in novel colloidal carriers systems seeks to address some of the limitations frequently encountered such as: (i) low *in vivo* concentration of drug or bioactive component due to poor bioavailability, rapid absorption, and elimination; (ii) extremely low drug or bioactive component solubility; (iii) fluctuation in plasma levels due to interactions with foods during ingestion (Mehnert and Mader, 2001). Existing delivery systems partially address some of those limitations, but no single system fully overcomes them. Examples of delivery systems that are currently under investigation include polymeric nanoparticles, nanoemulsions, nanosuspensions, micelles, and liposomes.

The objective of this chapter is to present an overview of SLNs as colloidal delivery systems, their advantages and disadvantages, currently available manufacturing methods, and common ingredients used in the production of SLN, as well as their current and potential applications in the development of functional foods.

13.2 SOLID LIPID NANOPARTICLES AS DELIVERY SYSTEMS

Lipidic carrier systems are classified into two main groups: emulsions or SLNs. Both these types of disperse systems consist of a lipid phase dispersed and stabilized in a continuous water phase using surfactants. The bioactive compound is carried in the lipid phase. The main difference between a

Nanotechnology and Functional Foods: Effective Delivery of Bioactive Ingredients, First Edition.
Edited by Cristina M. Sabliov, Hongda Chen, and Rickey Y. Yada.

nanoemulsion and a SLN dispersion is that in the former, lipids are in a liquid state, whereas in SLNs, as its name implies, lipids are in a solid state. Nanoemulsions can also exist as water droplets dispersed into a continuous lipid phase, but those are less common. Throughout this chapter when comparisons between SLN and nanoemulsions are made, we assume that the lipid is the dispersed phase. The use of nanoemulsions as a delivery system is extensively discussed in Chapter 11.

The use of lipid nanoemulsions in parenteral nutrition dates back to the 1950s, where vegetable oils were emulsified using phospholipids to increase the caloric content of the formulas (Wretlind, 1981). Later efforts focused on using the nanoemulsions as delivery vehicles, however, the availability of nanoemulsion-based systems in the current pharmaceutical market is limited to very few drugs, due to limitations of the carrier system to stabilize the drug and to release the compound in a controlled manner (Bunjes and Siekmann, 2006). In the food industry, nanoemulsions are widely used for the encapsulation of flavors and essential oils included as ingredients for beverages and other products, where the controlled release of a flavor compounds upon consumption is not a priority, but other properties such as transparency of emulsions are. The limited stability of nanoemulsions has been in part attributed to high mobility of the drug within the liquid oil, which allows the compounds to diffuse out to the interface, disrupting the interfacial layer or altering the zeta potential (surface charge) of the droplets, ultimately affecting the stability of the entire formulation (Washington and Evans, 1995; Washington, 1996; Bujes and Siekmann, 2006). One potential way to decrease the mobility of bioactive compounds is by substituting oils by solid lipids. In addition to reducing the mobility, the use of solid lipids in the dispersed phase allows for controlled release of the encapsulated compounds, as the degradation of lipid structure gradually releases the active compound. Thus, back in the early 1990s SLN emerged as an alternative to polymeric nanoparticles, liposomes, and emulsion-based delivery systems (Müller *et al.*, 2000; Pardeike *et al.*, 2009).

The advantages of SLN as compared to other delivery vehicles include (Mehnert and Mader, 2001; Müller *et al.*, 2002a, b; Bunjes and Siekmann, 2006; Teeranachaideekul *et al.*, 2007, 2008; Muchow *et al.*, 2008; Zhuang *et al.*, 2010):

- availability of technology for large-scale production;
- enhanced physical and chemical stability of matrix and bioactive due to reduced mobility;
- potential for controlled and targeted release of bioactive compounds;
- potential for surface modification due to static interface;
- reduced interaction between encapsulated bioactive compounds and surfactant;
- ability to encapsulate hydrophilic or lipophilic components;
- biodegradability and decreased toxicity due to availability of ingredients generally recognized as safe (GRAS) for their manufacture;
- no residues from organic solvents as they are not necessarily required in the manufacture of SLN.

Even though SLNs show promising advantages as delivery systems, some drawbacks have been identified, including (Müller *et al.*, 2000a, b; Radtke and Müller, 2001; Okuda *et al.*, 2005; Tikekar and Nitin, 2011):

- limited loading capacity due to drug or bioactive compound's solubility in lipid;
- drug expulsion due to reduced solubility upon cooling or to polymorphic transitions during storage;
- existence of other colloidal structures such as surfactant micelles, vesicles, drug nanocrystals;
- inhibited crystallization and formation of supercooled melts;
- limited maximum particle concentration to lower than 30%.

In order to overcome the drug expulsion problem, a new concept for the production of the next generation of lipid delivery systems, called nanostructured lipid carriers (NLC), was proposed and patented by Müller *et al.* (2000a, b) in Germany (Radtke and Müller, 2001). In SLN the drug or bioactive compound is located between fatty acid chains, which during transition to more stable polymorphic forms, from α to β' to β, can physically displace the molecules out of the crystalline lattice. The NLCs, on the other hand, are made using mixtures of liquid and solid lipids. The inherent differences in fatty acid composition between oil and fat triglycerides prevent the formation of perfect crystalline structures that delay or inhibit the expulsion of the bioactive component. Several types of NLC with decreased drug expulsion have been produced, for example, lipid structures containing crystalline regions with embedded oil nanocompartments, amorphous lipid matrices containing the dispersed drug, and imperfect crystalline structures capable of better retaining drug molecules within the imperfection zones.

13.3 MANUFACTURING SLN DISPERSIONS

The manufacturing method used in the production of SLNs and NLCs, as well as the type of ingredients used, will largely dictate the particle size of the dispersions, which range in sizes from 40 to 1000 nm (Lucks and Müller, 1991). There are three main ingredients used in the manufacture of SLN or NLC dispersions. The dispersed phase is typically made from biodegradable triglycerides, fatty acids, steroids, or waxes, in which the drug or bioactive component is dissolved prior to subsequent processing. The use of edible triglycerides is preferred for food applications in order to decrease toxicity risks (Mehnert and Mader, 2001; Shidaye *et al.*, 2008).

The second key ingredient in the manufacture of SLNs and NLCs is a surfactant to stabilize the interface. In food applications, phospholipids such as soy or egg lecithins are commonly used, in combination with smaller molecule surfactants such as polysorbates (20, 40, 60, 65, 80 or 85) to stabilize SLN dispersions by homogenization methods. The selection and concentration of a surfactant is critical in determining the stability of the dispersion as well as other features, such as nanoparticle size; furthermore, combinations of surfactants are more efficient at increasing the stability of the dispersions than single surfactants.

Various studies have demonstrated that the type of surfactant used influences the crystallization process and the rate of polymorphic transitions during storage as well as the stability of the encapsulated bioactive compound. Qian *et al.* (2012) studied the impact of surfactant and lipid composition on the crystallization behavior of tripalmitin SLN. They reported that the stability in aggregation behavior was dependent on storage temperature. At storage temperatures < 20°C modified starch yielded more stable SLN towards gelation and aggregation than polysorbate 20, whereas at higher temperatures, SLN stabilized with polysorbate 20 showed better long-term storage stability than those stabilized with modified starch. Qian *et al.* (2012) also noted that tripalmitin mixed with medium chain triglycerides or with orange oil at 70:30 ratio decreased the crystallization temperature of the lipid phase and substantially improved the stability towards gel formation for polysorbate 20 stabilized SLN. Similarly, Teeranachaideekul *et al.* (2007) reported that the stability of ascorbyl-palmitate encapsulated in NLC of identical lipid composition but different emulsifiers depended on the type of emulsifier used due to variations in lipid polymorphism. Long-chain surfactants are thought to act as heterogeneous nucleation sites and initiate lipid crystallization in the surface, which then proceeds centripetally and delays drug expulsion (Bunjes and Koch, 2005; Bunjes *et al.*, 2007; Helgason *et al.*, 2009).

Current available methods for manufacturing SLNs or NLCs may be classified according to the energy required in the process. High-energy methods include hot (or melt) homogenization, cold homogenization, and precipitation from solvent-in-water emulsions; whereas those requiring low energy include precipitation from warm microemulsions and precipitation from organic solutions (Bunjes and Siekmann, 2006). In general, homogenization methods are preferred due to their ability to incorporate higher amounts of bioactive compounds and the availability of large scale production units.

13.3.1 Hot homogenization

High-pressure homogenization is the method most commonly used in the production of SLNs. The homogenization process can be performed at high (above the melting point of the lipids used) or low (below the melting point of the lipid) temperatures. Hot homogenization involves the production of an emulsion that is subsequently cooled to induce solidification of the melted fats. In this method, the fats are heated to temperatures above their melting points, which depending on the type of fat used may be as high as 60°C or above. The drug or bioactive compound is then mixed and dissolved within the molten fat. The mixture of fat and bioactive component is then added to an aqueous phase, typically a buffered solution that contains the emulsifier or emulsifier mixture that has been preheated to the same temperature as the lipid mixture to prevent solidification upon mixing. The lipid phase is then dispersed into the aqueous phase at high speeds using a high shear mixer or an ultrasound horn. At this point, a coarse emulsion is formed and ready to be homogenized using a high-pressure homogenizer. Depending on the type of equipment used and the level of pressure applied the hot emulsion may be passed through the homogenizer more than once. An increase in the number of cycles tends to decrease the polydispersity of the particle size distribution. During high-pressure homogenization, the large droplets in the coarse emulsion are forced through a very narrow gap at the homogenizing valve or nozzle, experiencing enormous shear and friction forces that deform and break the droplets into smaller sizes, ranging between 50 and 200 nm (McClements, 2005; Huang *et al.*, 2010). The final step in the hot homogenization process involves reducing the temperature of the newly formed nanoemulsion to induce crystallization of the lipids. Cooling rate is a key component in determining nanoparticle structure and shape that is frequently overlooked (Weiss *et al.*, 2008). Some studies have reported that polymorphic transitions from the least stable α to the more stable β' and β polymorphs result in changes of particle shape from spherical to platelet or needle-like configurations that render the dispersions prone to particle aggregation and gel formation (Awad *et al.*, 2008; Helgason *et al.*, 2008). The formation of true SLNs needs to be verified experimentally, typically by differential scanning calorimetry (DSC), since depending on the composition and cooling conditions applied, lipids may remain as supercooled melts. In the supercooled state lipids remain in liquid form for extended periods of time, thus, while stable during storage, do not provide the advantages attributed to solid lipid forms.

13.3.2 Cold homogenization

The production of SLNs by cold homogenization involves the dissolution of the bioactive component into a melted fat. While still above the melting point, the bioactive component is added and dissolved within the liquid lipid. Once the bioactive compound has been thoroughly incorporated the mixture is cooled to temperatures well below the melting point of the fat. When the mixture has solidified, it is ground in the presence of a cryogen such as liquid nitrogen or dry ice to produce a fine lipid powder. The powder is then added to the aqueous phase that contains the emulsifier or emulsifier mixture. This nanosuspension is processed using a high-pressure homogenizer. Since the objective of the homogenization step is to physically break the large powder fat particles into nanoparticles, processing conditions such as homogenization pressure and number of cycles are usually higher than those used in hot homogenization. Cold homogenization is preferred for processing heat labile bioactive components. Furthermore, since the lipid is solidified prior to homogenization, the partition of the bioactive compound into the aqueous phase is limited. Solid lipid nanoparticle dispersions produced by cold homogenization methods exhibit larger particle size and polydispersity index than those obtained by hot homogenization.

13.3.3 Solvent emulsification/evaporation

This method allows the production of nanoparticles smaller than those obtained by hot homogenization methods. The steps involved in solvent emulsification are very similar to hot homogenization with the difference that the lipid and bioactive compounds are dissolved in an organic solvent. The solvent is then mixed with the aqueous phase that contains the emulsifiers and dispersed using a high-shear mixer to prepare a coarse emulsion. The coarse emulsion is then processed in a high-pressure homogenizer to obtain a solvent nanoemulsion. The final step in the production of SLN by this method requires removal of the solvent from the droplets. Solvent removal is performed under vacuum at temperatures below the melting point of the lipid and, as solvent removal progresses, SLNs eventually precipitate into the aqueous phase. Limitations noted for this method include very low lipid concentration in the final dispersion along with even lower bioactive component concentration. Furthermore, the droplet volume dependence on the third power of the particle's radius requires that large volumes of solvent be used to attain significant reduction in particle size.

13.3.4 Precipitation from warm microemulsions

This preparation method for SLNs belongs to the low-energy category. Microemulsions, as opposed to nanoemulsions obtained from high-pressure homogenization, are thermodynamically stable systems that form spontaneously under specific conditions. The method was first presented by Gasco and Morel (1990; Bunjes and Siekmann, 2006). An aqueous microemulsion is prepared using the lipid matrix, bioactive component, surfactant, and co-surfactants at elevated temperature (~70°C). Then the microemulsion is dispersed into a cold aqueous phase (~2–3°C) under stirring to promote solidification of the lipids. Common dilution rates for microemulsion:cold-aqueous-phase range between 1:5 and 1:10 or greater (Weiss *et al.*, 2008). Once the SLNs have been solidified, the dispersions are washed to remove water-soluble components such as surfactants that were used to formulate the microemulsion.

13.3.5 Precipitation from organic solutions

The precipitation from organic solutions method was adapted from the production of polymeric nanoparticles. Organic solvents that are miscible with water are used as the dispersion medium for the lipids, bioactive components, and emulsifier (e.g. phospholipids). Then the solution is injected into an aqueous phase containing a co-surfactant, where the SLNs precipitate upon mixing. Solvents commonly used in this production method include acetone and ethanol and their mixtures (Bunjes and Siekmann, 2006).

13.4 APPLICATIONS OF SLN AS CARRIERS FOR FOOD BIOACTIVE COMPONENTS

Solid lipid nanoparticles are a promising system for the delivery of bioactive components in functional foods. Successful developments in the pharmaceutical industry for delivery of lipophilic drugs could be applied for foods, since the water solubility of several bioactive components in foods is extremely low. While several studies report promising results for SLNs and NLCs as controlled release and targeted delivery vehicles in the pharmaceutical field, the number of studies on SLNs or NLCs as specific carriers for bioactive components in food applications is just beginning to increase. Examples of applications of interest in food science are described next.

13.4.1 Nanostructured lipid carriers and SLNs for encapsulation of β-carotene

The suitability of NLC for encapsulation of β-carotene was investigated by Hentschel *et al.* (2008). Nanostructured lipid carriers were prepared using a commercial propylene glycol monostearate to which a commercial dispersion of β-carotene in sunflower oil (10% w/w) was added (β-carotene 0.35–0.45% (w/w) of lipid). The aqueous phase was prepared by mixing Tween 80 (0.5–2.5% (w/w) in dispersion) and deionized water heated to 80°C. The lipid and aqueous phase were combined and dispersed using a high-shear mixer to obtain a coarse emulsion that was further homogenized at 50 MPa for three cycles. Initial particle size was ~200–400 nm but stability towards aggregation depended greatly on storage temperature. Low temperatures (4–8°C) resulted in particles ~400 nm for up to 30 weeks, whereas storage at 20°C promoted rapid particle aggregation. Degradation of β-carotene occurred rapidly when the NLC suspensions were diluted in water from the original 10% lipid content to 0.3% lipid particles. After 15 days of storage, ~50% of the initial β–carotene had been degraded in the NLC formulation, whereas > 90% degraded β-carotene was observed in the diluted system.

The impact of surfactant-type on SLNs for encapsulation of β-carotene was recently investigated by Helgason *et al.* (2009). Four types of surfactants were evaluated: two types of lecithin, high- or low-melting at 2.4% (w/w), and Tween 60 or Tween 80 at 1.4% (w/w), all of them combined with taurodeoxycholate (0.6%) as co-surfactant. Solid lipid nanoparticles were prepared by hot homogenization. Tripalmitin was melted and β-carotene (1 mg/g of lipid) was added and dissolved. Separately, each surfactant with co-surfactant was added to a phosphate buffer solution (pH 7) and kept at 70–80°C. The lipid and aqueous phases were mixed and dispersed using a hand=held mixer to obtain a coarse emulsion that was subsequently homogenized at high pressure for five cycles. The emulsions were then held at 20°C and allowed to crystallize. Cooling rate was reported to be about 1–2°C/min. Particle diameter ranged from 150 to 173 nm depending on surfactant used. The authors proposed that high-melting surfactants (HM-lecithin and Tween 60) may promote surface-heterogeneous nucleation as indicated by higher crystallization temperatures observed in DSC thermograms, as compared to low-melting surfactants. Furthermore, thermograms obtained after 1 day storage showed that a fraction of lipids in SLNs prepared with high-melting surfactants had α-crystals, whereas lipids in SLNs from low-melting surfactants had only β' and β-crystals (the more stable forms). The authors also noticed that even though the enthalphy of α-crystals decreased over storage, after 21 days, SLNs from high-melting surfactants still retained a fraction of lipids in α-form. However, after 21 days the SLN stabilized by HM-lecithin were not stable towards aggregation and particle size increased to the micrometer range. Results from β-carotene degradation showed that SLN from HM-lecithin were most effective at retarding degradation, with only 11 ± 4% of the initial β-carotene degraded after 21 days, whereas degradation of β-carotene in SLNs from low-melting surfactants was greater than 90% after 21 days. The authors proposed that when low-melting surfactants (LM-lecithin and Tween 80) are used, tripalmitin recrystallizes rapidly into the more stable β-polymorph and that this process is accompanied by the expulsion of β-carotene from the lipid matrix, which would explain the rapid degradation during storage. On the other hand, the use of high-melting surfactants (HM-lecithin and Tween 60) causes tripalmitin to crystallize into less stable polymorphs (α or β') that better retain β-carotene molecules within the lipid matrix. Even though recrystallization occurs, the process is comparably slower and results in decreased degradation of the bioactive compound. It is important to note that in this study, the authors compared side-by-side tripalmitin (SLN) and medium chain triglycerides (nanoemulsion) as β-carotene carriers, and noted that, in spite of high mobility within the lipid phase, nanoemulsions may perform better at protecting encapsulated compounds than SLN if recrystallization behavior is not considered during formulation.

13.4.2 Solid lipid nanoparticles for the encapsulation of ergocalciferol (D_2)

Particle size and crystallization behavior of ergocalciferol-loaded SLN were characterized by Patel and San Martin (2012). Tripalmitin SLNs were prepared by dissolving ergocalciferol (vitamin D_2) in tripalmitin melted at 80°C. The lipid phase accounted for 5% of the total dispersion, with ergocalciferol concentrations in the lipid phase ranging between 0 and 20% (w/w). Separately, an aqueous phase was prepared by dissolving Tween 20 in deionized water to a concentration of 5% in the final dispersion. The lipid and aqueous phases were combined and dispersed using a high-shear mixer, and subsequently homogenized at 138 MPa at 80°C for 10 cycles. Particle hydrodynamic radius as estimated from photon correlation spectroscopy (PCS) ranged from 70 to 120 nm, for 20% ergocalciferol and 0% ergocalciferol, respectively. The DSC thermograms show the increasing ergocalciferol concentration in the lipid phase leads to a higher content of the least stable α-polymorph during the melting cycle. The authors noted that vitamin D_2 may be successfully encapsulated using SLNs and further research to investigate the stability of encapsulated ergocalciferol is needed.

13.4.3 Solid lipid nanoparticles for enhanced bioavailability of γ-tocotrienol

Encapsulation of lipophilic vitamins in SLNs has been studied. Abuasal *et al.* (2012) investigated the ability of SLNs to increase the bioavailability of γ-tocotrienol. Solid lipid nanoparticles were prepared by a hot oil-in-water microemulsion method. Cetyl palmitate, γ-tocotrienol (0.2% w/w), and glyceryl behenate were melted at 80°C and used as the lipid phase. Separately, 0.2% (w/w) poloxamer 188 was dissolved in deionized water preheated to 80°C. Both phases were then mixed and dispersed using a high-shear mixer. The hot pre-emulsion was ultrasonicated for 10 min and nanoparticles allowed to precipitate by overnight storage at 4°C. The performance of SLNs was compared to a mixed micelles (MMs) system prepared by mixing diluted γ-tocotrienol in methanol, phospatidylcholine diluted in methanol, taurocholate diluted in 96% ethanol, and mono- and diglycerides evaporated to dryness under nitrogen. A transport buffer was added to the dried micelles for further cell culture and *in vivo* studies. Peak concentrations and time of peak concentrations were 786 ± 63 ng/mL and 3 ± 0.9 h and 275 ± 11 ng/mL and 5 ± 0.9 h for SLNs and MMs respectively. The authors demonstrated that enhanced intestinal absorption of γ-tocotrienol in SLNs was not a consequence of enhanced endocytosis, but attributed instead to enhanced passive permeability of γ-tocotrienol-loaded SLN through the intestinall wall and reduced contribution of the γ-tocotrienol transporter NPC1L1.

13.4.4 Solid lipid nanoparticles for delivery of coenzyme Q10

Hot homogenization was used by Liu *et al.* (2012) to encapsulate coenzyme Q10 (CoQ10) in SLNs. The lipid phase was prepared by mixing glycerol monolaurate, acetylated mono- and diglycerides and octyl and decyl glycerate at a ratio of 3:2:3. The lipids were melted at 55°C and CoQ10 added at 10% (w/w). The aqueous phase was prepared by dispersing three surfactants: polyglyceryl-10 laurate, gum arabic and powdered soybean lecithin at a 10:1:7 ratio in water heated to 55°C. The concentration of the surfactant mixture was 6% (w/w) of the dispersion, whereas the lipid phase accounted for 10% (w/w). The lipid and aqueous phases were combined and dispersed using a high-shear mixer. The coarse emulsion was then homogenized at 70 MPa for three cycles. The average particle size increased from 160 to 220 nm over 3 months storage at ~20°C. The concentration of CoQ10 decreased to about 70% of the initial concentration after 3 months. The authors also demonstrated through *in vitro* studies that peak plasma concentrations and time to peak concentrations were two times greater for the SLN system than when CoQ10 was administered as a suspension prepared by dissolving CoQ10 in octyl and decyl glycerate. While the authors mentioned that different absorption routes may occur, they were not tested.

13.4.5 Solid lipid nanoparticles to enhance bioavailability of curcumin

The health benefits of *Curcuma longa* Linn have been recognized for a long time, however, one of the major challenges to fully exploit its potential health benefits is to increase the extremely low oral bioavailability of curcumin. Kakkar *et al.* (2011) studied the oral bioavailability of curcuminoids encapsulated in a commercial lipid, Compritol 888 ATO (glyceril behenate), stabilized with a mixture of soy lecithin and Tween 80. In this study, a microemulsion was prepared by dissolving the surfactants in water to which curcumin was added. Then the aqueous phase containing emulsifiers and curcumin was slowly added to the melted lipid (82–85°C) under continuous stirring to produce a hot, clear microemulsion. The microemulsion was added to cold water (~21°C) whereby SLNs crystallized.

Entrapment efficiency ranged between 92.3 ± 1.6% and 81.9 ± 2.9% and particle size determined by laser diffraction was 134.6 ± 15.4 nm, although transmission electron microscopy micrographs showed particles within the 40–120 nm range. Results from *in vitro* release into a 50% ethanol (v/v) solution indicate that curcumin was released rapidly during the first 2 days with 47 ± 6% of curcumin released, but that the rate slowed down and was extended until day 7. *In vivo* pharmacokinetic study showed that at an equivalent dose (50 mg/kg) maximum plasma concentration of curcumin was greatly increased when administered by SLNs (14.3 mg/mL) as compared to free curcumin (0.30 mg/mL). The authors proposed that curcumin-containing SLNs have potential to be exploited as therapeutical agents rather than as a dietary supplement in nonencapsulated forms.

13.4.6 Solid lipid nanoparticles for enhanced bioavailability of quercetin

The bioavailability of quercetin encapsulated in SLNs was investigated by Li *et al.* (2009). Solid lipid nanoparticles were prepared using a hot emulsification method. The lipid phase was prepared using glyceryl monostearate, quercetin, and soy lecithin dissolved in a warm chloroform:acetone (1:1 (v/v)) mixture. The lipid phase was sonicated to obtain a clear phase. The aqueous phase was prepared by dissolving the surfactants Tween 80 and PEG 400 in water heated to 80°C. Both lipid and aqueous phases were combined under mechanical agitation and solvents were then evaporated by holding the mixture at 80°C. The emulsion was then held at 0°C for 1 h to crystallize the particles. In vitro release of quercetin was measured using dyalisis bags, whereas in vivo gastrointestinal absorption and pharmacokinetic studies were conducted in rats. Nanoparticle size ranged from 20 to 500nm, indicating the coexistence of other colloidal structures. Entrapment efficiency was 91 ± 0.56% with average quercetin loading of 13.20 ± 0.18%. Kinetic results indicated that SLN released quercetin at lower rate than quercetin dissolved in a poly(ethylene glycol) (PEG) solution, and that SLNs exhibited sustained release for up to 48 h. Pharmacokinetic results after a single dose (50 mg/kg) showed that peak plasma concentrations and time to reach the peak concentrations were 5.90 ± 1.24 ng/mL and 5.00 ± 1.18 h for free quercetin and 12.22 ± 2.15 ng/mL and 8.00 ± 1.38 h for quercetin-loaded SLNs. The major absorption site for quercetin SLNs was the ileum and colon.

13.4.7 SLN for antimicrobial delivery

In a recent study, Prombutara *et al.* (2012) investigated the feasibility of using SLN for sustained antimicrobial activity through controlled release of nisin. They used a commercial lipid, Imwitor 900 (>40–50% 2,3-dihydroxyprpyl octadecanoate), stabilized with poloxamer 188 and sodium deoxycholate. Solid lipid nanoparticles were prepared by hot homogenization at pressures between 50–150MPa and three cycles, followed by cooling to room temperature. The smallest particle size was obtained for the highest pressure, with hydrodynamic diameter ranging from 150nm to 240nm after 28 days of storage a 30°C. Entrapment efficiency of nisin into the SLN was about 73.6% ± 0.04% for 2.0% (w/v) nisin concentration. In vitro released studies showed that nisin was rapidly released during the first day, and that the release rate was strongly dependent on the pH of the medium, with pH 2 releasing about 2.5 times the nisin released at pH 7. The antimicrobial activity against *Listeria monocytogenes* DMST and *Lactobacillus plantarum* TISTR 850 of nisin-loaded SLN was tested over 25 days and compared to the activity of free nisin. While free nisin rapidly decreased microbial populations during the first day, bacterial concentration increased rapidly after the third day and reached levels comparable to initial concentration after 7 days. On the other hand, bacterial populations treated with SLN increased about 1.5 log after the first day, but microbial concentration gradually decreased over 20 days. After 15 days, bacterial concentration was comparable to that obtained by free nisin after 3 days. Thus, SLN are a promising delivery system for sustained release of antimicrobial peptides.

13.4.8 Solid lipid nanoparticles for encapsulation of resveratrol

The antioxidant activity of resveratrol-loaded SLNs and NLCs for transdermal application was studied by Gokce *et al.* (2012). Results showed that while free resveratrol was not able to permeate skin tissue, both SLN and NLC systems allowed the penetration and accumulation of resveratrol into the skin.

13.5 FINAL REMARKS

Both SLN and NLC are promising colloidal carriers for delivery and controlled release of bioactive components that could potentially be used in the formulation of functional foods. One of the challenges in the production of SLNs and NLCs for food applications is that the number of emulsifiers approved for use in foods is limited, thus formulations need to be optimized to develop successful delivery systems. Another aspect to consider is that if SLN or NLC encapsulates are to be used in the formulation of more complex products, the stability of the systems under subsequent processing should be studied. For example, if SLNs are to be added to fortify a beverage, a pasteurization treatment is likely to be applied, thus the stability of the SLN and the encapsulated compound should be assessed. Finally, let us keep in mind that since SLNs can substantially increase the bioavailability of bioactive components, there may be a potential risk for reaching toxic concentrations. Therefore the application of bioactive compounds encapsulated in SLN or NLC should be carefully envisioned. Toxicological studies should be conducted to assess the safety of the new ingredients.

References

Abuasal, B.S., Lucas, C., Peyton, B., Alayoubi, A., Nazzal, S., Sylvester, P.W. and Kaddoumi, A. (2012) Enhancement of intestinal permeabiity utilizing solid lipid nanoparticles increases γ-tocotrienol oral bioavailability. *Lipids*, 47, 461–469.

Awad, T.S., Helgason, T., Kristbergsson, K., Decker, E.A., Weiss, J. and McClements, D.J. (2008) Effect of cooling and heating rates on polymorphic transformations and gelation of tripalmitin solid lipid nanoparticle (SLN) suspensions. *Food Biophysics*, 3, 155–162.

Bunjes, H. and Koch, M.H.J. (2005) Saturated phospholipids promote crystallization but slow down polymorphic transitions in triglyceride nanoparticles. *Journal of Controlled Release*, 107(2), 229–243.

Bunjes, H. and Siekmann, B. (2006). Manufacture, characterization and applications of solid lipid nanoparticles as drug delivery systems. In, *Microencapsulation - Methods and Industrial Applications*, 2nd edn (ed. S. Benita). Taylor and Francis, New York; 213–268.

Bunjes, H., Steiniger, F. and Richter, W. (2007) Visualizing the structure of triglyceride nanoparticles in different crystal modifications. *Langmuir*, 23(7), 4005–4011.

Gasco, M.R. and Morel, S. (1990) Lipospheres from microemulsions. *Il Farmaco*, 45, 1127–1128.

Gokce, E.H., Korkmaz, E., Dellera, E., Sandri, G., Bonferoni, M.C. and Ozer, O. (2012) Resveratrol-loaded solid lipid nanoparticles versus nanostructured lipid carriers, evaluation of antioxidant potential for dermal applications. *International Journal of Nanomedicine*, 7, 1841–1850.

Helgason, T., Awad, T., Kristbergsson, K., McClements, D.J. and Weiss, J. (2008) Influence of polymorphic transformations on gelation of tripalmitin solid lipid nanoparticle suspensions. *Journal of the American Oil Chemists' Society*, 85(6), 501–511.

Helgason, T., Awad, T.S., Kristbergsson, K., Decker, E.A., McClements, D.J. and Weiss, J. (2009) Impact of surfactant properties on oxidative stability of carotene encapsulated within solid lipid nanoparticles. *Journal of Agricultural and Food Chemistry*, 57, 8033–8040.

Hentschel, A., Gramdorf, S., Müller, R.H. and Kurz, T. (2008) β-Carotene-loaded nanostructured lipid carriers. *Journal of Food Science*, 73(2), N1–N6.

Huang, Q., Yu, H and Ru, Q. (2010) Bioavailability and delivery of nutraceuticals using nanotechnology. *Journal of Food Science*, 75, R50–R57.

Kakkar, V., Singh, S., Singla, D. and Kaur, I.P. (2011) Exploring solid lipid nanoparticles to enhance the oral bioavailability of curcumin. *Molecular Nutrition and Food Research*, 55(3), 495–503.

Li, H., Zhao, X.B., Ma, Y.K., Zhai, G.X., Li, L.B. and Lou, H.X. (2009) Enhancement of gastrointstinal absorption of quercetin by solid lipid nanoparticles. *Journal of Controlled Release*, 133, 238–244.

Liu, G.Y., Wang, J.M. and Xia, Q. (2012) Application of nanostructured lipid carrier in food for the improved bioavailability. *European Food Research and Technology*, 234, 391–398.

Lucks, J.S. and Müller, R.H. (1991) Medication vehicles made of solid lipid particles (solid lipid nanospheres SLN). EP0000605497.

McClements, D.J. (2005) *Food Emulsions – Principles, Practices and Techniques*, 2nd edn. CRC Press, Abingdon, UK; 609 pp.

Mehnert, W. and Mader, K. (2012) Solid lipid nanoparticles: production, characterization and applications. *Advanced Drug Delivery Reviews*, 47, 165–196.

Muchow, M., Maincent, P. and Müller, R.H. (2008) Lipid nanoparticles with a solid matrix (SLN, NLC, LDC) for oral drug delivery. *Drug Development and Industrial Pharmacy*, 34, 1394–1405.

Müller R.H., Mader, K. and Gohla, S. (2000) Solid lipid nanoparticles (SLN) for controlled drug delivery – a review of the state of the art. *European Journal of Pharmaceutics and Biopharmaceutics*, 50, 161–177.

Müller, R.H., Radtke, M. and Wissing, S.A. (2002a) Nanostructured lipid matrices for improved microencapsulation of drugs. *International Journal of Pharmaceutics*, 242, 121–128.

Müller, R.H., Radtke, M. and Wissing, S.A. (2002b) Solid lipid nanoparticles (SLN) and nanostructured lipid carriers (NLC) in cosmetic and dermatological preparations. *Advanced Drug Delivery Reviews*, 54, S131–S155.

Okuda, S., McClements, D.J. and Decker, E.A. (2005) Impact of lipid physical state on the oxidation of methyl linolenate in oil-in-water emulsions. *Journal of Agricultural and Food Chemistry*, 53, 9624–9628.

Pardeike, J., Hommos, A. and Müller, R.H. (2009) Lipid nanoparticles (SLN, NLC) in cosmetic and pharmaceutical dermal products. *International Journal of Pharmaceutics*, 366(1–2), 170–184.

Patel, M.R. and San Martin-Gonzalez, M.F. (2012) Characterization of ergocalciferol loaded solid lipid nanoparticles. *Journal of Food Science*, 77(1), N8–N13.

Prombutara, P., Kulwatthanasal, Y., Supaka, N., Sramala, I. and Chareonpornwattana, S. (2012) Production of nisin-loaded solid lipid nanoparticles for sustained antimicrobial activity. *Food Control*, 24, 184–190.

Qian, C., Decker, E.A., Xiao, H. and McClements, D.J. (2012) Solid lipid nanoparticles: effect of carrier oil and emulsifier type on phase behavior and physical stability. *Journal of the American Oil Chemists' Society*, 89(1), 17–28.

Radtke, M. and Müller, R.H. (2001) Nanostructured lipid drug carriers. *New Drugs*, 2, 48–52.

Shidhaye, S.S., Vaidya, R., Sutar, S., Patwardhan, A. and Kadam, V.J. (2008) Solid lipid nanoparticles and nanostructured lipid carriers – innovative generations of solid lipid carriers. *Current Drug Delivery*, 5, 324–331.

Teeranachaideekul, V., Müller, R.H. and Junyaprasert, V.B. (2007) Encapsulation of ascorbyl palmitate in nanostructured lipid carriers (NLC) – effects of formulation parameters on physicochemical stability. *International Journal of Pharmaceutics*, 340, 198–206.

Teeranachaideekul, V., Souto, E.B., Müller, R.H. and Junyaprasert, V.B. (2008) Physicochemical characterization and *in vitro* release studies of ascorbyl palmitate-loaded semi-solid nanostructured lipid carriers (NLC) gels. *Journal of Microencapsulation*, 25(2), 111–120.

Tikekar, R.V. and Nitin, N. (2011) Effect of physical state (solid vs liquid) of lipid core on the rate of transport of oxygen and free radicals in solid lipid nanoparticles and emulsions. *Soft Matter*, 7, 8149–8157.

Washington, C. (1996) Stability of lipid emulsions for drug delivery. *Advanced Drug Delivery Reviews*, 20(2–3), 131–145.

Washington, C. and Evans, K. (1995) Release rate measurements of model hydrophobic solutes from submicron triglyceride emulsions. *Journal of Controlled Release*, 33(3), 383–390.

Weiss, J., Decker, E. A., McClements, D. J., Kristbergsson, K., Helgason, T. and Awad, T. (2008) Solid lipid nanoparticles as delivery systems for bioactive food components. *Food Biophysics*, 3, 146–154.

Wretlind, A. (1981) Development of fat emulsions. *Journal of Parenteral and Enteral Nutrition*, 5(3), 230–235.

Zhuang, C.Y., Li, N., Wang, M., *et al.* (2010) Preparation and characterization of vinpocetine loaded nanostructured lipid carriers (NLC) for improved oral bioavailability. *International Journal of Pharmaceutics*, 394(1–2), 179–185.

14 Protein–polysaccharide complexes for effective delivery of bioactive functional food ingredients

Yunqi Li and Qingrong Huang

Department of Food Science, Rutgers University, New Brunswick, New Jersey, USA

14.1 INTRODUCTION

Increasing concerns in food safety and food nutrition are calling for novel functional food ingredients to replace conventionally used potent health-threatening additives and to improve the bioavailability of health-beneficial compounds. The major goal of novel functional food ingredients is to devise delivery vehicles that can entrap, preserve, deliver, and release various molecules relating to aroma, color, taste, and texture in food products. Natural biomacromolecules already present in our diet are ideal sources to construct such delivery vehicles. Among them, proteins and polysaccharides are the two most widely distributed and easily achievable biomacromolecules in nature. Although protein and polysaccharide are most likely to be present in natural foods, it is necessary to re-assemble them with different combinations and processes in order to devise delivery vehicles with the desired properties. Proteins used in functional foods may come from multiple convenient sources such as plants, including legumes (soy protein, globulin, and lactoglobulin), cereal (globulin, gluten, prolamin, etc.), nuts (globulin), animals, including milk (casein, lactoglobulin, serum albumin, etc.), meat (gelatin, actin, myosin, titin, etc.), and eggs (ovalbumin, lysozyme). Polysaccharides occur in (agars, alginate, carrageenans), plant cell walls (cellulose, pectin), plant seeds, tubers, and roots (starch, guar gum, locust bean gum), plant exudates (gum arabic, gum tragacanth), and microorganisms (xanthan gum and gellan gum), as well as in derivatives (modified starch, carboxymethyl cellulose, propylene glycol alginate, etc.)

Both protein and polysaccharide are widespread natural polymers in foods. They have complicated structures and complex aggregation behaviors with distinguishable differences in multiple aspects (Dickinson, 1998, 2003, 2008, 2009). One of the most common features may be the formation of complexes with a mixture of protein and polysaccharides. The term "complex coacervate" coined by Tiebackx (1911), with the further classification of "simple coacervate" added by Bungenberg de Jong and Kruyt (1929), was introduced for such aggregation behavior around 100 years ago. In addition to the complex coacervation behavior, many other types of behavior in protein–polysaccharide mixtures have also attracted numerous studies and comprehensive reviews

Nanotechnology and Functional Foods: Effective Delivery of Bioactive Ingredients, First Edition.
Edited by Cristina M. Sabliov, Hongda Chen, and Rickey Y. Yada.

(Schmitt *et al.*, 1998; de Kruif and Tuinier, 2001; Tolstoguzov, 2003; Turgeon *et al.*, 2003; Schmitt and Turgeon, 2011; Veis, 2011; Kayitmazer *et al.*, 2013). The major feature of complex coacervation that distinguishes itself from other phase transitions, like gelation, precipitation, crystallization and vitrification, is that it is an associative liquid–liquid phase separation, and the coacervate is a viscous rather than solid-like substance. At the molecular level, there are a variety of interactions contributing to the association of proteins and polysaccharides, such as electrostatic, hydrophobic, and entropic (e.g., contour ion condensation, depletion, and release of solvent molecules). Among these, electrostatic interaction is the most widely accepted dominant force and the major interaction considered to tune complex coacervation. Beyond the molecular interaction level, microstructures and mechanisms associated with protein–polysaccharide complex (PPC) become divergent and system dependant.

Therefore, in this chapter, we will first update the understanding of PPCs and then introduce various applications of the complex as a functional food ingredient for the desired purpose. Finally, a possible breakthrough of PPCs for functional foods applications will also be reviewed.

14.2 UNDERSTANDING PROTEIN–POLYSACCHARIDE COMPLEXES

14.2.1 Overview of protein–polysaccharide complexes

In the ingestion and digestion processes of food in body, the number of proteins and polysaccharides are greatly reduced. Generally, the proteins and polysaccharides present in the food consumed are almost all at the generally recognized as safe (GRAS) level. However, to address some physical problems associated with the formation of protein–polysaccharide complexes and the phase behavior that results from large-scale aggregation of such complexes, some PPCs have been thoroughly investigated. Although the distinction between polysaccharides and polyelectrolytes has raised many concerns (Cooper *et al.*, 2005), protein–polysaccharide and protein–polyelectrolyte complexes share much in common. Comprehensive reviews of different proteins–polysaccharide and protein–polyelectrolyte complexes have been presented by Schmitt *et al.* (1998), Turgeon *et al.* (2003, 2007), Schmitt and Turgeon (2011), and Grinberg and Tolstoguzov (1997). Therefore, this overview of protein–polysaccharide complexes will include not only proteins and polysaccharides from food sources such as plants, meat, and milk, but also other similar biopolymers like polypeptides and polyelectrolytes in circumstances where clear clues for the understanding of such complex systems can be provided.

Food protein (Tolstoguzov, 1991) species can be of high purity, such as bovine serum albumin (BSA), β-lactoglobulin (BLG), casein, gelatin, lysozyme, and zein, or some can be in combined form, such as whey protein, oat protein, and pea protein, or some can be in the raw state, such as whey protein isolates and oat protein isolates, which are widely used in the fundamental studies of PPCs. Polysaccharides also have many variants in chemical composition, sources, molecular weight, rigidity, and side chains. Seeking to propose general features of PPCs greatly enriches our knowledge and also brings great challenge. Fortunately, the variation of the functional group in polysaccharides is limited in number. For example, chitosan is the only cationic polysaccharide with functional group of $-NH_3^+$; pectin, gum Arabic, sodium alginate, hyaluronic acid, carboxymethyl cellulose and carboxymethyl dextran, with $-COO^-$, as well as carrageenan and dextran sulphate, with $-SO_3^-$, are anionic polysaccharides; and dextran, amylopectin, and Ficoll are neutral polysaccharides.

Intensive studies on the above systems have made three very important contributions to science. First, although the structure–property relationship is complicated in such complex systems, it is not as complicated as understanding the origin of life (Oparin, 1953; Weinbreck, 2004). Second, elucidating the relationship among interaction, structure, phase behavior, and properties of such complex fluids greatly facilitates our understanding of a number of biological processes (Lipfert and Doniach, 2007). Third, a clear understanding of the mechanism for aggregation and phase behavior of PPCs enables their application as multiple-purpose delivery systems (George and Abraham, 2006; Dickinson, 2008; Jones and McClements, 2011). Beyond these findings, the association of protein and polysaccharide together with the self-assembly of individual biopolymers have also demonstrated there is much room for improvement in the preparation of structured materials. For example, Girod *et al.* (2004) found that the charge neutralization of ι-carrageenan by poly(L-lysine) can largely promote helix formation of ι-carrageenan chains; Seyrek *et al.* (2003) introduced insulin and heparin into a PPC and found that low salt concentration can promote complex formation, while high salt concentration inhibits complexation due to counterion condensation, and the salt impact was found to be consistent in different combinations of proteins and polysaccharides. Weinbreck *et al.* (2003a, b, 2004a–d) systematically studied complexes of whey protein with various ionic polysaccharides, and found that these systems exhibited common dependence on extrinsic conditions, with small difference related to the specificity of each system (see also de Vreis *et al.*, 2003; de Kruif *et al.*, 2004). Overall, the behaviors of protein and polysaccharide associations have several common features, but obvious distinctions in their responses to extrinsic stimuli are still widely reported.

From a geometric view, almost all PPCs can be regarded as coexisting mixtures of particles (of different size, surface topology, surface charge, and penetrability) and polymers (of variant size, architecture, stiffness, and charge). The stable coexistence status of such mixture may be determined by either the kinetic or thermodynamic contribution, depending on the processing and the storage time for functional food products. Since functional foods require a long shelf life, thus thermodynamic stability is a prerequisite: detailed discussions in several recent reviews are available (Ulrich *et al.*, 2006a; Gucht *et al.*, 2011; Schmitt and Turgeon, 2011). Here we will provide a summary of the experimental approaches and physical properties of the complexes, in order to explore interaction, structure, association, and phase behaviors in PPCs.

14.2.2 Experimental approaches for complex characterization

Benefited from advances in science and technology, experimental approaches for characterization of complexes can surpass the traditional macroscopic scale using absorption spectroscopy, thermal detection, and mechanical characterization, to monitor *in situ* the interaction and structure at molecular level: the size of PPCs spans from subnanometer for interactions to tens of micrometers for phase domains. Most of modern scientific techniques cover this size region, thus a large set of various instruments can be applied to study PPCs, and those generally used are listed in Table 14.1. According to the characteristics being measured, these experimental approaches can be catogorized into thermal (such as isothermal titration calorimetry), electrical (ζ potential), mechanical (rheology), spectral (IR, UV-Vis, turbidimetric titration and circular dichroism), scattering (small-angle neutron (SANS), small-angle X-ray (SAXS), and dynamic-light), probe tracing (microrheology, fluorescence), and molecular detection (morphology capture, quartz crystal microbalance with dissipation (QCMD), surface plasmon resonance (SPR), nuclear magnetic resonance (NMR)). Although high-resolution and high-sensitivity techniques are always in demand to provide more information on PPCs, hybrid approaches using complementary techniques to address challenging problems are now widely reported in the prominent scientific literature.

Representative reports on these various techniques ranged from microscopic scale studies of molecular interactions to macroscopic studies of phase behavior and properties of complexes. Nuclear magnetic resonance spectroscopy enhanced by nuclear Overhauser measurements allows the detection

Table 14.1 Experimental methods to study protein–polysaccharide complexes.

Method	Focus	Advantages	Limitation
Turbidimetric titration	Mesoscopic aggregation and macroscopic phase behavior	Simple and direct index for phase separation; efficient to screen physicochemical parameters and intrinsic properties on the aggregation and eventual phase separation; turbidity reveals the rate of aggregation and the size of aggregates in solution; can be used for fractal dimension determination for large aggregates; easy to control experiments and to collect highly reproducible data	Less information other than phase diagram; cannot use for qualification; not capable of distinguishing phase domain through coacervation and precipitation
Isothermal titration calorimeter (ITC)	Thermodynamic parameters associated with molecular aggregation	Direct and powerful to quantitatively collect thermodynamic parameters including binding constant, stoichiometry of binding sites, free energy (ΔG), entropy (ΔH), and enthalpy changes (ΔS); different phase transitions coupling with aggregation of molecules will not affect the measurement	Low sensitivity if molecular aggregation has low exothermal/endothermal effect or the binding is slow (diffusion controlled); high purity and narrow distribution of sample is preferred; special attention should be paid in sample preparation
Rheology	Bulk viscosity and modules.	Dedicatedly characterizes viscoelastic properties; coupling with suitable physical models can provide stringent structural–mechanical properties relationship	Reproducibility is low; very sensitive to artificial settings, environmental conditions, and sample purity; most of the complex exhibits non-Newtonian behavior; needs careful interpretation
Morphology capture: optical microscopy, confocal scanning laser microscopy, scanning/ transmission electron microscopy and atomic force microscopy	Morphology and aggregate status	Instinctive view of the complex aggregates; orientation of aggregates or molecules; phase coexistence status; strong proof for small phase domain or depletion, etc.	Small portions of material in capture may not be representative of the whole sample; contrast of compounds need special consideration to select suitable capture instrument; sample preparation may greatly change the observed morphology
UV-Vis and Fourier transform infrared (FT-IR) spectroscopy	New peaks and the shift of peaks to view interaction; IR can also detect secondary structure	Traditional, easy in sample preparation and measurements, fast and cheap	Need other complementary measurements to make conclusive statement; impurity with strong adsorption at given wavelength should be strictly controlled; avoid complex systems with low transmission (but not limited for reflective FT-IR)

(conutined)

Table 14.1 *(continued)*

Method	Focus	Advantages	Limitation
Small angle neutron/X-ray scattering	Interaction, conformation, and sizes of molecules or small aggregates	Gives information on internal structure of the aggregates and domain organization; can be used for structure/property relationship; distribution and organization of different species within sample can be obtained by contrast matching	Limited access possibilities; not very sensitive; high sample concentration
Microrheology (particle tracking microrheology, optical tweezers)	Internal structure, local viscosity	Cutting-edge technique provides extensive new information; flexible in selecting probe to detect the internal structure and interaction	Experimental conditions should be strictly controlled; interpretation of massive or active diffusion needs professional knowledge
Surface plasmon resonance and quartz crystal microbalance with dissipation monitoring	Interactions at molecule level	Sensitive to molecular binding; label-free to determine complex at molecule level	Immobilization of one molecule is needed; insensitive to weak binding; upper limit of loading is low
Nuclear magnetic resonance	Interaction and binding sites identification	Precisely identifies binding sites and their distribution; reveals interaction types in short range	Expensive, high purity sample and needs expertise for data interpretation
Zeta potential (electrophoretic mobility)	Particle surface charge	It is straightforward to provide surface charge of nanoparticles, identify isoelectric point of particles, and reveal the relationship between net charge and the stability of dispersions	Salt and pH have significant impact on surface charge; definite influence is still unclear; monodispersed particles are preferred to deliver reliable measurements
Dynamic light scattering	Hydrodynamic radius of dispersed particles	Readily obtains the size and size distribution of dispersed particles	Dust particles and high turbidity of sample may have significant impact on measurements
Fluorescence, including fluorescence recovery after photobleaching and fluorescence resonance energy transfer	Kinetics of tagged molecules in complex	Very sensitive for detecting binding; both the local structure of media and the dynamic properties of tagged molecules can be easily revealed	Hard to repeat; needs dye, but introduction of dye molecules may affect the structure and interaction in the complex
Circular dichroism	Protein secondary structure	Monitoring conformational changes of proteins originated from complex	Aromatic group polysaccharide in complex may affect qualification; overall average of secondary structures at protein rather than residue level

of intra- and intermolecular hydrogen bonding and van der Waals interaction. It is extremely useful for detecting structure and interactions and achieving spatial resolution at the chemical group level for protein complexes with oligosaccharides (Bush *et al.*, 1999), or in reverse order, study of polysaccharide-based vaccines (Clement *et al.*, 2006). Nuclear magnetic resonance can recognize the exact binding sites, which can greatly facilitate rational design at the molecular level. The drawback of this technique is that experts are needed to deliver reliable interpretation and to clarify binding behavior. Moreover, high-purity samples are necessary in order to provide good results. Alternatively, QCMD (Kim *et al.*, 2007; Noel *et al.*, 2007; Wang *et al.*, 2007a) and SPR (Clememt *et al.*, 2002; Demolliens *et al.*, 2008; Ning *et al.*, 2011) measure the binding between protein and polysaccharide through immobilization of the target protein, and then by exposing the chip with the immobilized protein to polysaccharide buffers, a signal with resolution at the molecular level can be recorded following individual intermolecular binding. Both of these methods have very similar mechanisms and can be applied to various systems.

In addition to investigating interaction at the subnanometer scale, scattering methods can provide information on interaction, aggregation, and structure at the nanometer scale. Ideally, as illustrated in Figure 14.1 (adapted from the study of lysozyme/pectin complexes using SANS by Schmidt *et al.* (2009)), the scattering intensity profile can cover the corresponding spatial scale of the complex

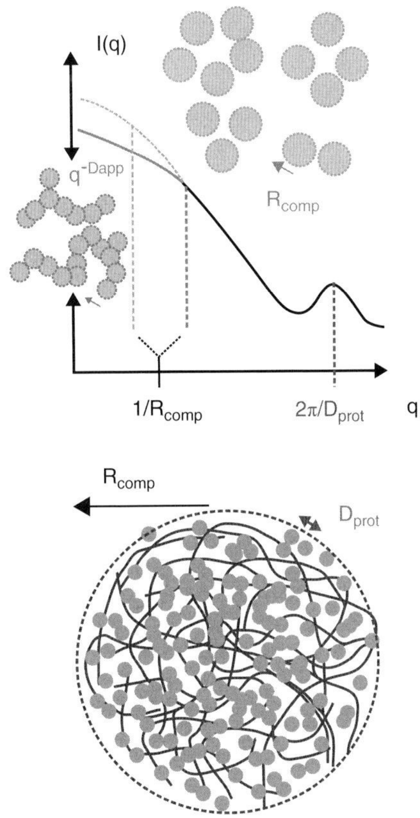

Figure 14.1 A schematic diagram of the ideal scattering-intensity profile, which provides information on the individual proteins and structure of a protein–polysaccharide complex. (Schmidt *et al.*, 2009. Reproduced with permission from American Chemical Society.)

structure, as well as providing individual conformation of both protein and polysaccharide. The scattering intensity profile also reveals the interaction contributed to the stabilization of the complex and the response of the distribution of molecules in the complex to changing physicochemical conditions. The power of small-angle scattering makes it an ideal tool to correlate the information of chemical composition with macro phase and properties. Renard *et al.* (1997) used SANS to study subtle structure changes in response to shear stress in complexes of globular proteins and various polysaccharides. Weinbreck *et al.* (2004d) used SAXS to study whey protein–gum arabic complex coacervates and found that the coacervate is dense and structured at the pH conditions when electrostatic interaction was maximized in the complex. Boral and Bohidar (2010) used SANS to study selective binding behavior in gelatin–agar complex systems, and verified the asymmetric distribution of charge patches in the protein surface. Alternatively, light scattering has also been widely applied for characterizing the hydrodynamic radius of complex nanoparticles, or to provide conformation of macromolecules. Microrheology and the various scattering methods plays a central role in the accumulation of direct information on interaction and structure at the nanometer scale, and recent developments and applications indicate microrheology as a promising technique to greatly facilitate the study of PPCs at the nanometer scale (Mason and Weitz, 1995; Mason *et al.*, 1997; Xu *et al.*, 1998; Waigh, 2005; Larsen and Furst, 2008; Khair and Squires, 2010; Lopez-Diaz and Castillo, 2011).

In addition to obtaining information at the nanometer scale, phase behavior in complex systems can also be experimentally determinated. The most widely used technique to detect phase boundaries based on the size and the distribution of complex aggregates is turbidimetric titration. Since PPCs can eventually lead to liquid–liquid phase separation, which is known as complex coacervation, particles sharply increase in size and population in association with the phase transition, thus greatly enhancing Debye scattering and the turbidity of dispersions. Critical pH values shown in the turbidity–pH profile indicate well-defined phase boundaries for the formation of soluble complexes and insoluble complex coacervates (Weinbreck *et al.*, 2003a; Wang *et al.*, 2007b). An important finding based on turbidity titration profiles is that salt and temperature play a similar role in suppressing the complex formation (de Kruif *et al.*, 2004). Intergration of turbidity titration profiles with molecular simulation and theoretical calculation is also a solid way to reveal thermodynamic mechanisms associated with PPCs (Li *et al.*, 2012, 2014; Li and Huang 2013). Other applications of titration to trace protein–polysaccharide complex in various systems are also available (Laos *et al.*, 2007; McCullough *et al.*, 2010).

Beyond characterizing the interaction, structure and phase behavior, thermodynamics, and mechanical properties of PPCs using the latest techniques, it is also critical to correlate findings with traditional knowledge in order to provide rational directions towards the manufacture of desired food products. At earlier times, differential scanning calorimetry (DSC) was widely used to acquire thermal parameters during phase transition in protein complexes (Imeson *et al.*, 1977). Later, the development in sensitivity and data analysis flow from DSC to isothermal titration calorimetry (ITC) for tracing molecular binding during complex formation became feasible (Freyer and Lewis, 2008). Bastos *et al.* (2010) systematically studied the electrostatic interactions driving complexes of whey-protein isolate and chitosan using ITC and rheology. They found that the viscous and viscoelastic behaviors of the complex coacervates were mainly contributed from chitosan chains. Alternatively, in the study of PPCs (Hu *et al.*, 2011), fluorescence with quenching implemented also can provide information on thermodynamic parameters, including binding constant, stoichiometry of binding sites, free energy (ΔG), entropy (ΔH), and enthalpy changes (ΔS), for qualification of the binding of protein and polysaccharide when combined with suitable models, such as the modified Stern–Volmer equation (Xia *et al.*, 1997; Hu *et al.*, 2011).

Mechanical properties are also important for the preparation and application of PPCs. These complexes are normally viscoelastic and measurement of rheology is therefore capable of providing comprehensive characterization. Sanchez *et al.* (1997) found that a whey-protein isolate (WPI) and xanthan complex behaved like heat-set gel with typical shear-thinning flow behavior. They found this to be a result of segregative phase separation with high self-aggregation proneness of whey proteins in the presence of incompatible xanthan. A rheology study of BLG and κ-carrageenan complexes revealed that increasing κ-carrageenan concentration promotes complex aggregation and eventually leads to gelation, but the subtle structure of the complex remains unchanged in both complex and complex coacervate gels (Croguennoc *et al.*, 2001). Andrade *et al.* (2010) studied the thermorheological properties of a pea-protein isolate and κ-carrageenan complex, and found that the mechanical properties of the complex were dominated by conformational change of κ-carrageenan in the gelation process. These studies illustrate the complicated properties that such complexes can display, and also that they can be versatile functional-food ingredients by satisfying various requirements and strongly responsible to the environmental stimuli.

Other experimental approaches using direct or indirect methods for characterizing morphology, structure, and interaction have also been widely applied. Affinity capillary electrophoresis was developed to determine the binding properties between an anticoagulant polysaccharide and an antithrombin protein (Varenne *et al.*, 2010). Chourpa *et al.* (2006) applied Raman spectroscopy and found that the optimal complex coacervation pH of globulin and gliadin with gum arabic is produced by tight coupling with secondary structural changes in proteins. Schmitt *et al.* (2001) using confocal scanning laser microscopy (CSLM) and diffusing wave spectroscopy (DWS) found that self-aggregation of BLG played an important role in a BLG and acacia gum complex, this effect was confirmed through Monte Carlo simulation to be contributed from the large negative charge patch in BLG (Li and Huang, 2013). Additionally, there has been a rapid growth of research interest in microrheology and its application in the study of PPCs, which has further enriched our understanding of such systems (Waigh, 2005; Sato and Breedveld, 2006; Pesce *et al.*, 2009; Khair and Squires, 2010). Recent success of a study on salt superposition over time based on rheological data from an electrostatic complex (Spruijt *et al.*, 2010) indicated that parallel comparison studies on the behavior of a complex with other similar systems will provide new insights to understand the properties of a complex. Further development of methods and techniques to detect subtle structures and distribution of molecules in complex systems is still ongoing (Huber and Scheler, 2012).

14.2.3 Physical view of protein–polysaccharide complexes

As PPCs comprise soft condensed matter, theoretical and simulation approaches have been widely explored to understand the interaction, structure, phase behavior, thermodynamics, and mechanical properties. These approaches may stem from classic concepts such as free energy, enthalpy, and entropy, but empirical interpretations based on specific experimental approaches have also played an important role in helping our understanding of the dependence of the complex properties on its intrinsic characters and changing extrinsic conditions. Additionally, since heterogeneity is ubiquitous in PPCs, using mean field approximations presents a great challenge to accurately describe such systems: a widely accepted physical model remains elusive.

Protein–polysaccharide complexes normally are sustained by several weak physical (nonbonding) interactions. Such weak and cooperative interactions bestow on the complexes strong environmental implications. A brief summary of the qualitative trend and dominant contribution of individual interactions can be found in Dickinson's (1998) review. An overall view of how interactions interplay in PPC formation and its stability is presented here: (i) under the conditions where electrostatic interaction is promoted, such as low salt and suitable pH range to allow the protein and polysaccharide to have charges of opposite sign and with higher charge density in the ionic polysaccharide, the electrostatic interaction dominates the complex formation and further aggregation; (ii) under the condition where electrostatic interaction is suppressed, such as a protein complexed with nonionic polysaccharide, a pH range where protein and polysaccharide bring charges of the same sign or the disassociation/protonation of ionizable group polysaccharide in polysaccharide is inhabited, a high salt content where short-range electrostatic interaction is screened, and a high excess of one components, other interactions may take charge of the complex stabilization. Under these two conditions, the change of PPC in response to extrinsic stimuli can be easily understood. For example, β-lactoglobulin interacting with acacia gum shows typical changes in association behaviors according to the polysaccharide ratio and pH (Schmitt *et al.*, 1999). Since the electrostatic interaction is sensitive to pH (changing the charge status of biomacromolecules) and salt (screening electrostatic interaction and causing counterion release entropy change), adding ionic polysaccharide can prevent target proteins from unwanted denaturation and aggregation in high pressure and heat treatments (Galazka *et al.*, 1999a, b). It also has been reported that the maximum protein–electrostatic affinity is at ionic strengths corresponding to the Debye radii of the same order as the protein radius (Seyrek *et al.*, 2003). Jones *et al.* (2010a) reported that the strong electrostatic interaction between carrageenan and BLG fiber allows the polysaccharide to greatly change the fibrillation. Through analogy with colloid suspension and colloidal gel systems (Stradner *et al.*, 2004), several reports show that the stabilization of a complex is a balance of short-range attraction to form the complex and long-range repulsion to stabilize to almost charge-neutral complex particles (Chen *et al.*, 2011). Although implementation of the discussion about molecular interactions is different depending on the system studied, theoretical explanations associating with all these experimental measurements still follow the two conditions stated above.

Alternatively, the explicit contribution from each interaction can be quantified through computer simulations. At one extreme point where the polysaccharide is much smaller than the protein, it becomes a binding problem that has been intensively studied (Bush *et al.*, 1999) based on some classic force fields such as CHARMM (Chemistry at HARvard Macromolecular Mechanics; MacKerell *et al.*, 1998; Brooks *et al.*, 2009), AMBER (Assisted Model Building and Energy Refinement; Duan *et al.*, 2003) and OPLS (optimized potential for liquid simulation; Jorgensen *et al.*, 1996). Other than these systems, a coarse grain is necessary to handle a system with more than 10^5 atoms. Linse's group has systematically explored the impact of various factors on PPCs through coarse-grained Monte Carlo (MC) simulations (Wallin and Linse, 1996; Carlsson *et al.*, 2001, 2003; Jonsson and Linse, 2001a; Akinchina and Linse, 2002, 2003; Skepo and Linse, 2003). Meanwhile, Stoll's group, with additional emphasis on the presence of nanoparticles, also presented some detailed information on PPCs (Brynda *et al.*, 2002; Carnal *et al.*, 2004; Ulrich *et al.*, 2005, 2006a, b, 2011; Cooper *et al.*, 2006; Seijo *et al.*, 2006, 2008). Our recent development of coarse-grained MC simulations also revealed new features in PPCs (Li *et al.*, 2012, 2014; Li and Huang, 2013). Additionally, using an adaptive Poisson–Boltzmann solver (APBS) to calculate the distribution of charges and conterions on protein surfaces, Dubin's group also provided good qualitative description of the complicated behaviors in several complex systems (Grymonpre *et al.*, 2001; Cooper *et al.*, 2006; Chen *et al.*, 2011; Xu *et al.*, 2011). Other propositions such as charge regulation (da Silve *et al.*, 2006; da Silva and Jönsson, 2009) and contourion condensation (Seijo *et al.*, 2008, 2009; Ulrich *et al.*, 2011; Huber and Scheler, 2012) play roles in the stabilization of PPCs under certain conditions. Although both theoretical and simulation approaches have revealed many interesting details in PPCs, their complicated systems still require more accurate modelling to account for the more elaborate features and to provide robust predictions.

Beyond molecular interactions, the structure and the associative phase behavior in protein–polysaccharide mixtures are extremely important. Since many of the PPCs are far from thermodynamic equilibrium and the diffusion of biopolymers is quite slow, kinetic arrest has significant impact on the final structure and the apparent phase behavior. Various physicochemical factors including the size and weight ratios, the total biomacromolecular concentration, the charge density in proteins and polysaccharides, pH, and salt concentration of the solutions, as well as temperature and pressure treatment, may dramatically change the structure and phase behavior. A tentative overview to show the major influence of these factors on the structure of PPCs is shown in Figure 14.2, and detailed descriptions

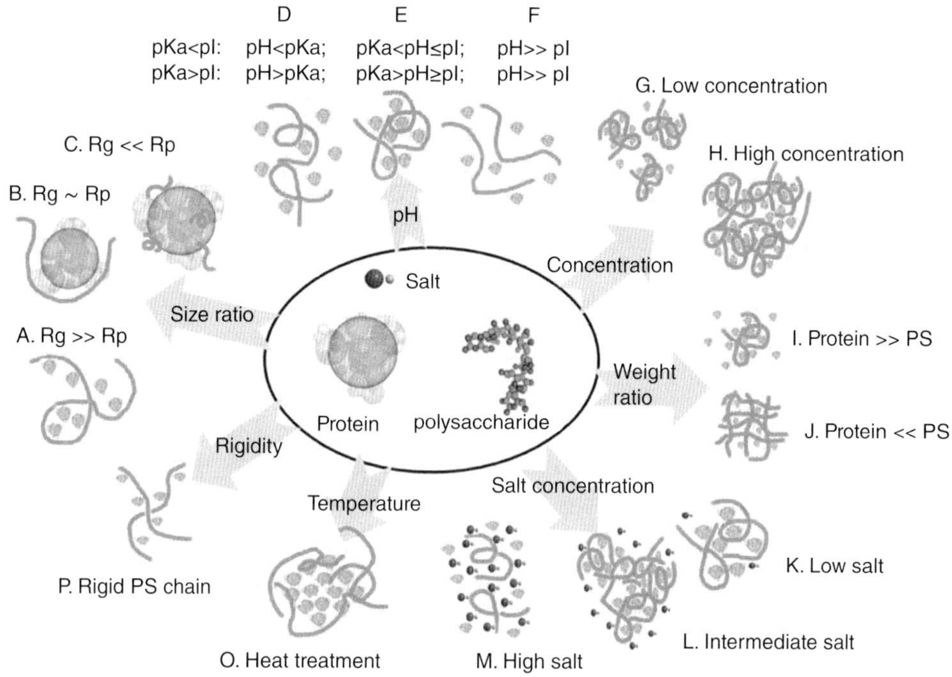

Figure 14.2 Schematic diagram of the structures of a protein–polysaccharide complex as a function of the size ratio, pH, total biomacromolecule concentration, the protein to polysaccharide weight ratio, salt concentration, temperature, and the rigidity of polysaccharide chains.

of each structural status are presented in Table 14.2. Protein–ionic polysaccharide complexes normally are electrostatic in nature, and the physicochemical conditions affect the interaction, structures, and properties of the complex, and follow several dominant rules: such as changing pH towards neutral net charge of the complex will promote complex formation; very low salt concentration promotes complex formation and prohibits further aggregation of the complex due to nonscreened electrostatic interactions; increasing the salt concentration facilitates complex formation through screening of long-range electrostatic repulsion; and high salt content can break down complex association and reduce complex yield when short-range electrostatic attraction is effectively screened out. It is a central idea that the competition of short-range attraction and long-range repulsion attributed to salt concentration always plays a key role in the protein–ionic polysaccharide complex formation and further aggregation. Other rules are also ubiquitous such as elevating temperature normally increases thermal fluctuations, and then reduces the stability of a complex. It is worthy to note that in reality, as protein–polysaccharide mixtures have various extrinsic controls and intrinsic properties of biopolymers, the general picture shown here represents only the major trends based on published reports. There are, however, a number of reports that observed different consequence upon changing conditions in different systems.

In addition to the structure characterization in PPCs, the stability of the complex can be largely categorized by four mechanisms: charge neutralization (Weinbreck *et al.*, 2003b; Mekhloufi *et al.*, 2005; Gummel *et al.*, 2008; De Santis *et al.*, 2010; Xu *et al.*, 2011), charge regulation (da Silva and Jönsson, 2009; Boral and Bohidar, 2010), minimization of electrostatic interaction (Weinbreck *et al.*, 2004d), and minimization of free energy (enthalpy and entropy contributions, or energy functions engaged in computer simulations) (Jonsson and Linse, 2001b; Girard *et al.*, 2003; Biesheuvel and Cohen Stuart, 2004; Pawar and Bohidar, 2010; Veis, 2011). Our recent work found that these four mechanisms can be applied in different protein–polysaccharide systems, some of their predictive

Table 14.2 Description of different microstructures in response to physicochemical condition changes shown in Figure 14.2.

Schema	Variable	Description	References
A	$R_g >> R_p$	Each polysaccharide chain can bind multiple proteins to neutralize the charge	Bush *et al.*, 1999; Schmitt *et al.*, 2000; Sanchez *et al.*, 2002; Clement *et al.*, 2006; Wang *et al.*, 2007c; Newhouse *et al.*, 2009
B	$R_g \sim R_p$	Complex contains protein and polysaccharide according to the stoichiometric composition	
C	$R_g << R_p$	When polysaccharide is much smaller than protein, it acts as ligand binding to protein	
D	pH > pKa > pI, or pH < pKa < pI	Polysaccharide chains almost are neutral at such pH conditions; protein and polysaccharide have weak binding force majorly contributed by van der Waals interaction; there is equilibrium between isolated biomacromolecules and soluble complex	Schmitt *et al.*, 1999; Schmitt *et al.*, 2000; Sanchez *et al.*, 2002; Sanchez and Renard, 2002; Galazka *et al.*, 1999b; Weinbreck *et al.*, 2004d; Laos *et al.*, 2007; Ru *et al.*, 2007
E	pH is in the range between pKa and pI	Polysaccharide and protein have opposite charge; complex widely formed and phase separation can be observed, with broad aggregation as aneutralized complex; there is equilibrium between complex coacervate/precipitate with suspended complex particles	
F	pH >> pI > pKa, or pH << pI < pKa	Polysaccharide and protein have same sign of charge; polysaccharide chains become stiff because of the charge; there is equilibrium between disassociated and associated complexes	
G	Low concentration	Protein–polysaccharide complex is suspended in solution; soluble complex formed; stability of complex is similar to colloid suspensions	Sanchez *et al.*, 1997; Schmitt *et al.*, 1999; Schmitt *et al.*, 2001; Ru *et al.*, 2007; Kaloti and Bohidar, 2010
H	High concentration	Large aggregation of complex occurs, complex coacervate may have gel-like network structure or the packing of colloid, depending on the interaction among neutralized complex particles.	
I	Protein >> polysaccharide	Polysaccharide chains were saturated or over charge neutralized; excess proteins can prohibit further aggregation of complex	Schmitt *et al.*, 1999; Schmitt *et al.*, 2001; Sanchez *et al.*, 2002; Wang *et al.*, 2007c; Kaloti and Bohidar, 2010; Ru *et al.*, 2012
J	Protein << polysaccharide	Proteins were entirely wrapped by polysaccharide chains; excess polysaccharide chains with unneutralized charges suppress complex coacervation	
K	Low salt	Both short-range attraction and long-range repulsion are not screened; complex aggregate of limit scale	Galazka *et al.*, 1999b; Seyrek *et al.*, 2003; Weinbreck *et al.*, 2004b; Wang *et al.*, 2007c; Boral and Bohidar, 2010; Ru *et al.*, 2012
L	Intermediate salt	The screening length is close to the size of complex particle; intercomplex repulsion is screened while attraction holding complex kept unchanged; large-scale aggregation of complex occurs	
M	High salt	Both short-range attraction and long-range repulsion are screened, which leads to protein and polysaccharide tending to stay in isolated form	
O	Heat treatment	Increasing temperature can partially denature protein and induce protein aggregate; polysaccharide chains then enwrap the protein aggregates through strong interactions	Imeson *et al.*, 1977; Galazka *et al.*, 1999b; Croguennoc *et al.*, 2001; Boral and Bohidar, 2010; Jones *et al.*, 2010a, b
P	Rigid polysaccharide chain	Proteins bound to polysaccharide chains decrease with rigid chain due to additional penalty in polysaccharide conformational entropy; the rigidity majorly contributed to from charges in polysaccharide chain through either density of charge group polysaccharide change or the pH dependant dissociation/protonation of charge group polysaccharide	Ulrich *et al.*, 2005; Burova *et al.*, 2007; Schmidt *et al.*, 2009

ability from that work overlaps with experimental observation, but some of the predictive ability is exclusive to quantifying the phase boundary for a given system (Li *et al.*, 2014). In certain systems, kinetics factors may also take charge of the stability of a complex, where complex coacervation may experience either a spinodal decomposition or a nucleation and growth mode (Sanchez *et al.*, 2002; Turgeon *et al.*, 2003; Veis, 2011). There are also some systems where electrostatic interaction does not absolutely dominate the complex formation, which raises an extraordinary challenge for interpretation. For example, Chourpa *et al.* (2006) found that stablization of the complex of α-gliadin or globulin with gum Arabic is contributed to by electrostatic and non-Coulombic interactions (e.g., intermolecular hydrogen bonds), which is closely related to the overall structure and the charge density profile of these two proteins. Overall, the stability of a complex can be qualitatively predicted, accounting for the detail from the interaction, the structure, and the aggregation behavior in the complex systems.

In addition to the stability of PPCs, the desired thermodynamics and mechanical properties are also critical for their application in functional foods. The viscoelastic nature of PPCs means that they can be typically used in emulsion, suspension, or hydrogel, both in the support matrix and as nanoparticle carriers. Andrade *et al.* (2010) studied the impact of temperature and concentration on the rheological behavior of mixtures of pea-protein isolate and κ-carrageenan, and they found that the viscosity and the stability of food products were easily adjusted by temperature with the complex. Picone and da Cunha (2010) found that tri-polymeric systems of whey protein, sodium caseinate, and gellan gum showed high compatibility and inhibition of complex coacervate with lower mechanical properties than in the bi-polymeric gels without complexes (Carvalho *et al.*, 2006; Li and McClements, 2014; Thongkaew *et al.*, 2014; Zhang and Vardhanabhuti, 2014).

14.3 PROTEIN–POLYSACCHARIDE COMPLEX IN FUNCTIONAL FOODS

14.3.1 Major roles and conventional applications

Functional foods have the potential to provide a positive impact to human health. Because some of the beneficial health components in conventional foods are difficult to ingest, its versatile compatibility, natural source, and excellent bioactivity retention ability (Xia *et al.*, 1997) means that the PPC is an ideal candidate for functional food ingredients. Among the best known functional food compounds, for the supply of probiotics and prebiotics, natural antioxidants and anti-microbials are the most important ones (Grajek *et al.*, 2005). Their general purposes, representative compounds, and reports/reviews are presented in Table 14.3. Each focus also inherent hot topics, such as: antioxidant active components may also have in anticancer, anti-aging, anti-amyloid, and anti-inflammatory effects; antimicrobial includes antibiotics, antivirals, antifungals, and antiparasitics. Most of these bioactive compounds raise challenges for delivery systems to enable them to be properly ingested by humans. Most of these bioactive compounds have adverse properties, such as low water solubility, chemical instability, crystallinity, low bioavailability, and poor compatibility with delivery systems. Fortunately, many well-designed PPCs are capable of encapolysaccharideulating these functional compounds, and their great response to physiological stress through structural change has proved them to be an excellent delivery system for targeted delivery and controlled release.

Protein–polysaccharide complexes can encapsulate other bioactive compounds or they can be bioactive for novel beneficial functions as a functional food ingredient. First of all, many proteins are health beneficial but overcoming various physical instabilities, such as denaturing, surface adsorption, aggregation, and precipitation (Manning *et al.*, 2010), is necessary for efficient intake. As a delivery system, PPCs have long been used to microencapsulate various oils (Weinbreck *et al.*, 2004a) and proteins (Burgess and Ponsart, 1998) with specific bioactivities. The can complex systems also use

Table 14.3 List of several current focuses to reduce health-threat risks in functional foods.

Focus	Major function	Bioactive components	Reference
Antioxidant	Inhibits chemical and physical reactions relating to irregular oxidization, or overreaction to harmful stress, or deleterious aggregation in the body	Vitamins (C, E), polyphenols (catechins, EGCG, penta-galloyl glucose, tea phenols, etc.), carotenoids (lycopene, carotenes, lutein, curcumin ,etc.), unsaturated fats (omega-3 fatty acids, etc.), protein/peptide (bromelain, glutathione etc.)	Gibbs *et al.*, 1999; Zhang *et al.*, 2011
Antimicrobial	Kills or inhibits the growth of deleterious microorganisms	Polypeptides (poly(L-lysine), subtiloson); organic acids (lactic acid, citric acid, acetic acid, and their salts); essential oils (cinnamon, garlic, lemon, mint oil); polysaccharide (chitosan)	Foschiatti *et al.*, 2009; McCullough *et al.*, 2010; Hu *et al.*, 2011
Supply of prebiotics and probiotics	Improves the activity of health beneficial microorganisms and suppresses the health-threat microorganisms in the gastrointestinal tract	Prebiotics: soluble fiber from whole grains, beans, seeds, vegetables and nuts, inulin, oligosaccharide, etc. Probiotics: lactic acid bacteria, bifidobacteria, and yeast	Roberfroid, 2000; Grajek *et al.*, 2005; Macfarlane *et al.*, 2006; Livney, 2010; Figueroa-González *et al.*, 2011
Anti-obesity	Reduces appetite, alters gastrointestinal metabolism and absorption of calories	Orlistat, sibutramine, rimonabant, lipoic acid, raspberry ketone, etc.	Roberfroid, 2000; Yang *et al.*, 2005; Sanz *et al.*, 2008; Kumar *et al.*, 2009

bioactive protein or polysaccharide such as poly(L-lysine) (Girod *et al.*, 2004) and chitosan to enhance beneficial functions in food products. Whey-protein complexes can be used in yogurt-like products to improve the stability and quality, so to enhance the intake of probiotics (Weinbreck *et al.*, 2003b). Extensive studies undertaken to incorporate protein–polysaccharide into functional foods for each of the major focus areas are listed in Table 14.3. Encapsulation of other natural products, such as cumin oil, phenols extracted from tea and traditional flavors, which has been reported with multiple beneficial in anti-inflammatory, anti-allergic anti-oxidant, anticancer, antiviral, and general well-being effects, require further studies on PPCs in order to enhance their efficiency as versatile delivery systems in functional foods.

The biological nature of protein and polysaccharide allows their complexes to have low allergenic risks and environmental hazards, and combined with their broad and flexible adjustable compatibilities and stimuli responses makes PPCs a versatile and ideal functional foods ingredient. Protein–polysaccharide complexes are a convenient way to control the texture and stability of food products, and to tune the rheological properties in food processing (Imeson *et al.*, 1977; Dickinson, 1998). They are also readily used for microencapsulation in order to reduce the off-flavours produced by some vitamins and minerals, to control the release of nutrients, to reduce reactivity of bioactive molecules with other ingredients, and to enhance the stability of nutraceuticals for longer shelf life (Dziezak, 1988). Other applications of PPCs include a milk-protein–xanthan gum complex that replaces fat and a soy protein with alginate complex that produces a meat analogue through a series of treatments (Schmitt *et al.*, 1998), and different processing of milk proteins also makes complexes that can be used in functional foods with multiple purposes (Livney, 2010). With the development and understanding of functional foods, PPCs received more and more attentions, and their widely adjustable compatibilities and readily tunable association and disassociation have specially made them attractive as material for encapsulation.

14.3.2 Controlled assembly of PPC nanoparticles

In order to improve the stability, appearance, and nutrition in functional foods using PPCs, rational design for controlled assembly of given systems under target physicochemical and physiological conditions is integrant. Controlled-assembly complexes should have the desired shape, such as particle, fiber, and vesicle, a stable dispersion status, such as suspension, emulsion, gel, and film, and the desired thermodynamics and mechanical properties. The assembly of PPCs can be maintained through short-range electrostatic attraction, hydrophobic interaction, depletion=induced aggregation, etc., and stabilized through long-range electrostatic repulsion or hydration of the exterior layer. The assembly can be through self-assembly or by additional force (such as centrifugation, or high-pressure and high-speed homogenization) under kinetic arrest status. The versatility of PPCs can be judged from the following examples applied for various purposes. First, the size of stable sodium caseinate and gum Arabic complex nanoparticles can be easily tuned through change in the composition (Ye *et al.*, 2006). Introduction of either protein- or polysaccharide-sensitive salt is another convenient way to control the desired structure and association of the complex (Kunz, 2010). The designed complex nanoparticles can be easily used for the delivery of target proteins (Gaudana *et al.*, 2011). Ru *et al.* (2010) reported that ι-carrageenan and β-lactoglobulin complexes can efficiently enhance the stability of emulsion, which can help the delivery of epigallocatechin-3-gallate (EGCG), and stabilization can be applied to the emulsion with a tunable droplet size, which is controlled by processing. Adjusting the composition of BLG–acacia gum dispersions can easily control the structure as complex particles or vesicles (Sanchez *et al.*, 2002). A delivery system of BSA–pectin and 2-octanone is readily compatible for flavor encapsulation and release (Burova *et al.*, 1999). Controlled-assembly complex nanoparticles also can be used as submicroscopic reservoirs for ocular delivery (Motwani *et al.*, 2008). Since PPC nanoparticles have colloidal stability, where long-range repulsion can inhabit further aggregation of the nanoparticles, layer-by-layer deposition to form a complex has promising interest. A newly proposed complex coacervate core material (C3M) using such technique has been developed for delivery systems as functional food ingredients (Hofs *et al.*, 2008; Brzozowska *et al.*, 2010a, b). Cai *et al.* (2005) used a chitosan–gelatin complex coated on titanium colloids to improve its biocompatibility using a layer-by-layer electrostatic interaction dominated association. The tunable association affinity of polyelectrolytes also allows the electrostatic complex to be an ideal building block for hydrogels (Hunt *et al.*, 2011). Increasing interests involved the complex of functional proteins to latex colloids for novel biosensor are very active research focuses (Caruso and Mohwald, 1999; Cooper, 2003; Bouvree *et al.*, 2009).

14.3.3 Complex coacervate as supporting matrix

When PPCs are broadly aggregated, this leads to associative phase separation (Fang *et al.*, 2006), known as complex coacervation. The coacervate normally has a gel-like structure (Wang *et al.*, 2007c), which enables the complex to be a good supporting matrix. Analogous to the colloidal stability of complex nanoparticles, the electrostatic interaction in protein–polysaccharide complex also can be used to enhance the short-term and long-term stabilities of emulsions. The stabilization shares many similarities, with surfactant, colloids and pure proteins as emulsifiers, but has distinct mechanisms and against Ostwald ripening still can be clearly seen (Dickinson, 2009). In comparison to homogeneous hydrogel, the complexes have the great merits in their broad compatibilities and natural sources. The complex coacervate of sodium-montmorillonite nanoplatelets and fish gelatin behaves like a highly recoverable gel, where electrostatic interaction and the conformation of gelatin are both in charge of the stability and the reversibility (Qazvini *et al.*, 2012). It has been reported that whey-protein isolate and low methoxyl pectin complexes can be used as a matrix for hydrosoluble food ingredient entrapment in acidic foods (Bedie *et al.*, 208). Other than this, the complex also can act as a surfactant to stabilize emulsions. It may perform better than a monocomponent of surfactant because both electrostatic interaction and steric repulsion of PPCs may synergistically enhance the stability of emulsions.[18] For example, whey protein/gum Arabic complex can be used to encapsulate sunflower oil, lemon, and orange oil flavor and make chewable components used in cheese (Weinbreck *et al.*, 2004a). A gelatin–gum Arabic complex matrix was used to entrap long-chain polyunsaturated fat acid (Liu *et al.*, 2010). Complexes formed by β-lactoglobulin and Acacia gum electrostatically can be used to stabilize foams (Schmitt *et al.*, 2005). Additionally, the development of biosensors required stable and small colloidal dispersions, and the long-range electrostatic interactions in PPCs should be stabilized easily. Such layer-by-layer deposition of PPCs on designed template particles has been reported to greatly facilitate the development of versatile biosensors (Caruso and Mohwald, 1999; Sanchez *et al.*, 2002).

14.3.4 Polypeptide–polysaccharide complex

Polypeptides from food proteins such as casein and β-lactoglobulin can inherit the emulsifying capability and the binding affinity of polysaccharides. Their complexes bring novel aspects to tailor delivery systems. The polypeptides can be generated from proteins by simple hydrolysis or degradation with the help of enzymes. The caseinophosphopeptides can enhance the lipopolysaccharide and phytohemagglutinin-induced proliferation and it has been reported that eventually these peptides could be used as a humoral immunostimulator in cell cultures (Hata *et al.*, 1998). Detailed study of self-association β-lactoglobulin polypeptides is beneficial for understanding the aggregation of milk proteins (Kosters *et al.*, 2010). Furthermore, Priftis and Tirrell (2012) found that a complex of polylysine and polyglutamic acid can be easily adjusted by polypeptide mixing ratio (stoichiometry), ionic strength (salt concentration), total polymer concentration, pH, and temperature, which can be used to form either precipitate or coacervate. Complexes formed by caseinophosphopeptides (CPP) and chitosan have been systematically studied by Hu *et al.* in recent years (Hu *et al.*, 2011, 2012a, b, 2013). They used the resultant nanocomplexes to encapsulate EGCG in an oral formula, and found the bioaviluability of the bioactive compound was significantly enhanced. The CPP–chitosan complex structure and size is tunable with a change in ratio, which can be understood in terms of charge neutralization and binding saturation (Hu *et al.*, 2011). They also found than the CPP–chitosan nanoparticle with a ζ-potential of 32.2 mV can significantly promote EGCG permeability in Caco-2 model, indicating bioavailability enhancement (Hu *et al.*, 2012b). Overall, the polypeptide–polysaccharide complexes have exhibit promising behaviors, and although the purification of polypeptides is challenging, their application in functional foods is still a cheap and safe way.

14.4 FUTURE DIRECTIONS

Increasing knowledge of PPCs by considering the features of both of the individual biopolymers and the complex, and using theoretical and computer simulation approaches, is still required in order to provide predictive rules. One possible break point may emerge from comparisons with other similar systems, such as polymer systems (Rubinstein and Colby, 2003), colloid systems (Russel *et al.*, 1989; Hiemenz *et al.*, 1997; Birdi, 2009), and their mixtures (Likos, 2001; Doublier and Cuvelier, 2006), which already have well-developed theoretical models for characterizing interaction, conformation, and other properties. The challenge is that all complex systems respond synergistically to both intrinsic and extrinsic variables: determining decoupling rules at broad temporal and spatial scales is difficult. However, interdisciplinary approaches to consideration of the numerous reports on PPCs may shed new light in this field.

For the development and application of PPCs, the systems outlined above have exhibited amazing behaviors and properties, which will attract more attention in the future. There are, however, only a limited number of natural proteins and polysaccharides available, and extensive modification of their chemical structures may be required to allay safety concerns and meet various market demands. Most recently, complexes of polysaccharide and protein fibrils (Jones *et al.*, 2010a, b, 2011), protein hydrolyzates (Hu *et al.*, 2011; Li and Huang, 2012), and protein-assembled nanoparticles (Jones and McClements, 2011; Luo *et al.*, 2011; Mahmoudi *et al.*, 2011; Mukhidinov *et al.*, 2011; Ulrich *et al.*, 2011; Larsen-Smith *et al.*, 2012), as well as complexes assembled with polymer nanoparticles and colloids (Kaloti and Bohidar, 2010; Livney, 2010; Qi *et al.*, 2010; Chen *et al.*, 2011; Gaudana *et al.*, 2011), have attracted increasing interest. In a very similar system, protein–polysaccharide conjugates also exhibited excellent performance in reducing allergic reaction and enhancing emulsifying ability (Hattori *et al.*, 2000; Kobayashi *et al.*, 2001, 2003; Xia *et al.*, 2015), as well as against the denaturation, precipitation, and β-fiberification of proteins. External processing methods, including heat treatment combined with high-speed and high-pressure homogenization and electrospinning techniques will greatly facilitate the final dispersion status of PPCs. Applications of these complexes also can be greatly broadened to fields not only in functional foods, but also in cosmetics, nutraceuticals and pharmaceuticals.

References

Akinchina, A. and Linse, P. (2002) Monte Carlo simulations of polyion-macroion complexes. 1. Equal absolute polyion and macroion charges. *Macromolecules,* 35(13), 5183–5193.

Akinchina, A. and Linse, P. (2003) Monte Carlo simulations of polyion-macroion complexes. 2. Polyion length and charge density dependence. *Journal of Physics and Chemistry B,* 107(32), 8011–8021.

Andrade, R.J., Azevedo, A.G., Musampa, R.M. and Maia, J.M. (2010) Thermo-rheological behavior of model protein-polysaccharide mixtures. *Rheologica Acta,* 49(4), 401–410.

Bastos, D.S., Barreto, B.N., Souza, H.K.S., Bastos, M., Rocha-Leao, M.H.M., Andrade, C.T. and Goncalves, M.P. (2010) Characterization of a chitosan sample extracted from Brazilian shrimps and its application to obtain insoluble complexes with a commercial whey protein isolate. *Food Hydrocolloids,* 24(8), 709–718.

Bedie, G.K., Turgeon, S.L. and Makhlouf, J. (2008) Formation of native whey protein isolate-low methoxyl pectin complexes as a matrix for hydro-soluble food ingredient entrapment in acidic foods. *Food Hydrocolloids,* 22(5), 836–844.

Biesheuvel, P.M. and Cohen Stuart, M.A. (2004) Electrostatic free energy of weakly charged macromolecules in solution and intermacromolecular complexes consisting of oppositely charged polymers. *Langmuir,* 20(7), 2785–2791.

Birdi, K.S. (2009) *Handbook of Surface and Colloid Chemistry,* 3rd edn. CRC Press/Taylor & Francis, Boca Raton, FL.

Boral, S. and Bohidar, H. B. (2010) Effect of Ionic Strength on Surface-Selective Patch Binding-Induced Phase Separation and Coacervation in Similarly Charged Gelatin-Agar Molecular Systems, *Journal of Physics and Chemistry B,* 114(37), 12027–12035.

Bouvree, A., Feller, J.F., Castro, M., Grohens, Y. and Rinaudo, M. (2009) Conductive polymer nano-biocomposites (CPC): Chitosan–carbon nanoparticle a good candidate to design polar vapour sensors. *Sensors and Actuators B – Chemical,* 138(1), 138–147.

Brooks, B.R., Brooks, C.L., Mackerell, A.D., *et al.* (2009) CHARMM: the biomolecular simulation program. *Journal of Computational Chemistry,* 30(10), 1545–1614.

Brynda, M., Chodanowski, P. and Stoll, S. (2002) Polyelectrolyte-particle complex formation. Polyelectrolyte linear charge density and ionic concentration effects. Monte Carlo simulations. *Colloid and Polymer Science,* 280(9), 789–797.

Brzozowska, A.M., de Keizer, A., Norde, W., Detrembleur, C. and Stuart, M.A.C. (2010a) Grafted block complex coacervate core micelles and their effect on protein adsorption on silica and polystyrene. *Colloid and Polymer Science,* 288(10–11), 1081–1095.

Brzozowska, A.M., Zhang, Q., de Keizer, A., Norde, W. and Stuart, M.A.C. (2010b) Reduction of protein adsorption on silica and polysulfone surfaces coated with complex coacervate core micelles with poly(vinyl alcohol) as a neutral brush forming block. *Colloids and Surfaces A,* 368(1–3), 96–104.

Bungenberg de Jong, H. G. and Kruyt, H. R. (1929) Coaservation. *Proc. K. Ned. Akad. Wet.*, 32, 849–956.

Burgess, D.J. and Ponsart, S. (1998) beta-Glucuronidase activity following complex coacervation and spray drying microencapsulation. *J. Microencapsulation*, 15(5), 569–579.

Burova, T.V., Grinberg, N.V., Golubeva, I.A., Mashkevich, A.Y., Grinberg, V.Y. and Tolstoguzov, V.B. (1999) Flavour release in model bovine serum albumin/pectin/2-octanone systems. *Food Hydrocolloids*, 13(1), 7–14.

Burova, T.V., Grinberg, N.V., Grinberg, V.Y., Usov, A.I., Tolstoguzov, V.B. and de Kruif, C. G. (2007) Conformational changes in iota- and kappa-carrageenans induced by complex formation with bovine beta-casein. *Biomacromolecules*, 8(2), 368–375.

Bush, C.A., Martin-Pastor, M. and Imberty, A. (1999) Structure and conformation of complex carbohydrates of glycoproteins, glycolipids, and bacterial polysaccharides. *Annual Review of Biophysics and Biomolecular Structure*, 28, 269–93.

Cai, K., Rechtenbach, A., Hao, J., Bossert, J. and Jandt, K.D. (2005) Polysaccharide-protein surface modification of titanium via a layer-by-layer technique: Characterization and cell behaviour aspects. *Biomaterials*, 26(30), 5960–5971.

Carlsson, F., Linse, P. and Malmsten, M. (2001) Monte Carlo simulations of polyelectrolyte-protein complexation. *Journal of Physics and Chemistry B*, 105(38), 9040–9049.

Carlsson, F., Malmsten, M. and Linse, P. (2003) Protein-polyelectrolyte cluster formation and redissolution: A Monte Carlo study. *Journal of the American Chemical Society*, 125(10), 3140–3149.

Carnal, F., Laguecir, A., Stoll, S. and Laguecir, A. (2004) Simulations and scattering functions of polyelectrolyte-macroion complexes. *Colloid and Polymer Science*, 283(3), 317–328.

Caruso, F. and Mohwald, H. (1999) Protein multilayer formation on colloids through a stepwise self-assembly technique. *Journal of the American Chemical Society*, 121(25), 6039–6046.

Carvalho, E., Póvoas, M.J., Mateus, N. and de Freitas, V. (2006) Application of flow nephelometry to the analysis of the influence of carbohydrates on protein–tannin interactions. *Journal of the Science of Food and Agriculture*, 86(6), 891–896.

Chen, K.M., Xu, Y.S., Rana, S., *et al.* (2011) Electrostatic selectivity in protein–nanoparticle interactions. *Biomacromolecules*, 12(7), 2552–2561.

Chourpa, I., Ducel, V., Richard, J., Dubois, P. and Boury, F. (2006) Conformational modifications of alpha gliadin and globulin proteins upon complex coacervates formation with gum Arabic as studied by Raman microspectroscopy. *Biomacromolecules*, 7(9), 2616–2623.

Clement, G., Boquet, D., Frobert, Y., *et al.* (2002) Epitopic characterization of native bovine β-lactoglobulin. *Journal of Immunological Methods*, 266(1–2), 67–78.

Clement, M.J., Fortune, A., Phalipon, A., *et al.* (2006) Toward a better understanding of the basis of the molecular mimicry of polysaccharide antigens by peptides – the example of Shigella flexneri 5A. *J. Biol. Chem.*, 281(4), 2317–2332.

Cooper, C.L., Dubin, P.L., Kayitmazer, A.B. and Turksen, S. (2005) Polyelectrolyte-protein complexes *Current Opinion in Colloid and Interface Science*, 10(1–2), 52–78.

Cooper, C.L., Goulding, A., Kayitmazer, A.B., *et al.* (2006) Effects of polyelectrolyte chain stiffness, charge mobility, and charge sequences on binding to proteins and micelles. *Biomacromolecules*, 7(4), 1025–1035.

Cooper, M.A. (2003) Label-free screening of bio-molecular interactions. *Analytical and Bioanalytical Chemistry*, 377(5), 834–842.

Croguennoc, P., Nicolai, T., Durand, D. and Clark, A. (2001) Phase separation and association of globular protein aggregates in the presence of polysaccharides: 2. Heated mixtures of native beta-Lactoglobulin and k-Carrageenan. *Langmuir*, 17(14), 4380–4385.

Da Silva, F.L.B. and Jönsson, B. (2009) Polyelectrolyte–protein complexation driven by charge regulation. *Soft Matter*, 5, 2862–2868.

Da Silva, F.L.B., Lund, M., Jönsson, B. and Åkesson, T. (2006) On the complexation of proteins and polyelectrolytes. *Journal of Physical Chemistry B*, 110(9), 4459–4464.

De Kruif, C.G. and Tuinier, R. (2001) Polysaccharide protein interactions. *Food Hydrocolloids*, 15(4–6), 555–563.

De Kruif, C. G., Weinbreck, F. and de Vries, R. (2004) Complex coacervation of proteins and anionic polysaccharides. *Current Opinion in Colloid and Interface Science*, 9(5), 340–349.

De Santis, S., Ladogana, R.D., Diociaiuti, M. and Masci, G. (2010) Pegylated and thermosensitive polyion complex micelles by self-assembly of two oppositely and permanently charged diblock copolymers. *Macromolecules*, 43(4), 1992–2001.

De Vries, R., Weinbreck, F. and de Kruif, C.G. (2003) Theory of polyelectrolyte adsorption on heterogeneously charged surfaces applied to soluble protein–polyelectrolyte complexes. *Journal of Chemical Physics*, 118(10), 4649–4659.

Demolliens, A., Boucher, C., Durocher, Y., Jolicoeur, M., Buschmann, M. D. and De Crescenzo, G. (2008) Tyrosinase-catalyzed synthesis of a universal coil-chitosan bioconjugate for protein immobilization, *Bioconjugate Chemistry,* 19(9), 1849–54.

Dickinson, E. (1998) Stability and rheological implications of electrostatic milk protein-polysaccharide interactions. *Trends in Food Science and Technology,* 9(10), 347–354.

Dickinson, E. (2003) Hydrocolloids at interfaces and the influence on the properties of dispersed systems. *Food Hydrocolloids,* 17(1), 25–39.

Dickinson, E. (2008) Interfacial structure and stability of food emulsions as affected by protein–polysaccharide interactions. *Soft Matter,* 4, 932–942.

Dickinson, E. (2009) Hydrocolloids as emulsifiers and emulsion stabilizers. *Food Hydrocolloids,* 23(6), 1473–1482.

Doublier, J.-L. and Cuvelier, G. (2006) Gums and hydrocolloids: functional aspects. In: *Carbohydrates in Food,* 2nd edn (ed. A.-C. Eliasson). Marcel Dekker Inc., New York, NY; 233–272.

Duan, Y., Wu, C., Chowdhury, S., *et al.* (2003) A point-charge force field for molecular mechanics simulations of proteins based on condensed-phase quantum mechanical calculations. *Journal of Computational Chemistry,* 24(16), 1999–2012.

Dziezak, J.D. (1988) Microencapsulation and encapsulated ingredients. *Food Technology,* 4, 136–157.

Fang, Y., Li, L., Inoue, C., Lundin, L. and Appelqvist, I. (2006) Associative and segregative phase separations of gelatin/κ-carrageenan aqueous mixtures. *Langmuir,* 22(23), 9532–9537.

Figueroa-González, I., Quijano, G., Ramírez, G. and Cruz-Guerrero, A. (2011) Probiotics and prebiotics—perspectives and challenges. *Journal of the Science of Food and Agriculture,* 91(8), 1341–1348.

Foschiatti, M., Cescutti, P., Tossi, A. and Rizzo, R. (2009) Inhibition of cathelicidin activity by bacterial exopolysaccharides. *Molecular Microbiology,* 72(5), 1137–1146.

Freyer, M.W. and Lewis, E.A. (2008) Isothermal titration calorimetry: Experimental design, data analysis, and probing macromolecule/ligand binding and kinetic interactions. *Methods in Cell Biology,* 84, 79–113.

Galazka, V., Smith, D., Ledward, D. and Dickinson, E. (1999a) Interactions of ovalbumin with sulphated polysaccharides: effects of pH, ionic strength, heat and high pressure treatment. *Food Hydrocolloids,* 13(2), 81–88.

Galazka, V.B., Smith, D., Ledward, D.A. and Dickinson, E. (1999b) Complexes of bovine serum albumin with sulphated polysaccharides: effects of pH, ionic strength and high pressure treatment. *Food Chemistry,* 64(3), 303–310.

Gaudana, R., Khurana, V., Parenky, A. and Mitra, A.K. (2011) Encapsulation of protein-polysaccharide HIP complex in polymeric nanoparticles. *Journal of Drug Delivery,* 2011, 458128.

George, M. and Abraham, T.E. (2006) Polyionic hydrocolloids for the intestinal delivery of protein drugs: Alginate and chitosan – a review. *Journal of Controlled Release,* 114(1), 1–14.

Gibbs, B.F., Kermasha, S., Alli, I. and Mulligan, C.N. (1999) Encapsulation in the food industry: a review. *International Journal of Food Sciences and Nutrition,* 50(3), 213–24.

Girard, M., Turgeon, S. L. and Gauthier, S. F. (2003) Thermodynamic Parameters of β-Lactoglobulin–Pectin Complexes Assessed by Isothermal Titration Calorimetry, *Journal of Agricultural and Food Chemistry,* 51(15), 4450–4455.

Girod, S., Boissière, M., Longchambon, K., Begu, S., Tourne-Pétheil, C. and Devoisselle, J. M. (2004) Polyelectrolyte complex formation between iota-carrageenan and poly(L-lysine) in dilute aqueous solutions: a spectroscopic and conformational study, *Carbohydrate Polymers,* 55(1), 37–45.

Grajek, W., Olejnik, A. and Sip, A. (2005) Probiotics, prebiotics and antioxidants as functional foods, *Acta Biochimica Polonica,* 52(3), 665–671.

Grinberg, V.Y. and Tolstoguzov, V.B. (1997) Thermodynamic incompatibility of proteins and polysaccharides in solutions. *Food Hydrocolloids,* 11(2), 145–158.

Grymonpre, K.R., Staggemeier, B.A., Dubin, P.L. and Mattison, K.W. (2001) Identification by integrated computer modeling and light scattering studies of an electrostatic serum albumin-hyaluronic acid binding site. *Biomacromolecules,* 2(2), 422–429.

Gucht, J. van der, Spruijt, E., Lemmers, M. and Cohen Stuart, M.A. (2011) Polyelectrolyte complexes: Bulk phases and colloidal systems. *Journal of Colloid and Interface Science,* 361(2), 407–422.

Gummel, J., Boue, F., Clemens, D. and Cousin, F. (2008) Finite size and inner structure controlled by electrostatic screening in globular complexes of proteins and polyelectrolytes. *Soft Matter,* 4(8), 1653–1664.

Hata, I., Higashiyama, S. and Otani, H. (1998) Identification of a phosphopeptide in bovine alpha(s1)-casein digest as a factor influencing proliferation and immunoglobulin production in lymphocyte cultures. *Journal of Dairy Research,* 65(4), 569–578.

Hattori, M., Nagasawa, K., Ohgata, K., *et al.* (2000) Reduced immunogenicity of beta-lactoglobulin by conjugation with carboxymethyl dextran. *Bioconjugate Chemistry,* 11(1), 84–93.

Hiemenz, P. C., Hiemenz, H. and Rajagopalan, R. (1997) *Principles of Colloid and Surface Chemistry*, 3rd edn. Marcel Dekker, New York.

Hofs, B., de Keizer, A., van der Burgh, S., *et al.* (2008) Complex coacervate core micro-emulsions. *Soft Matter,* 4(7), 1473–1482.

Hu, B., Wang, S.S., Li, J., Zeng, X.X. and Huang, Q.R. (2011) Assembly of bioactive peptide-chitosan nanocomplexes. *Journal of Physics and Chemistry B,* 115(23), 7515–7523.

Hu, B., Ting, Y.W., Yang, X.Q., Tang, W.P., Zeng, X.X. and Huang, Q.R. (2012a) Nanochemoprevention by encapsulation of (−)-epigallocatechin-3-gallate with bioactive peptides/chitosan nanoparticles for enhancement of its bioavailability. *Chemical Communications,* 48(18), 2421–2423.

Hu, B., Ting, Y.W., Zeng, X.X. and Huang, Q.R. (2012b) Cellular uptake and cytotoxicity of chitosan-caseinophosphopeptides nanocomplexes loaded with epigallocatechin gallate. *Carbohydrate Polymers,* 89(2), 362–370.

Hu, B., Ting, Y.W., Zeng, X.X. and Huang, Q.R. (2013) bioactive peptides/chitosan nanoparticles enhance cellular antioxidant activity of (−)-epigallocatechin-3-gallate. *Journal of Agricultural and Food Chemistry,* 61(4), 875–881.

Huber, K. and Scheler, U. (2012) New experiments for the quantification of counterion condensation. *Current Opinion in Colloid and Interface Science,* 17(2), 64–73.

Hunt, J.N., Feldman, K.E., Lynd, N.A., *et al.* (2011) Tunable, high modulus hydrogels driven by ionic coacervation. *Advanced Materials,* 23(20), 2327-+.

Imeson, A.P., Ledward, D.A. and Mitchell, J.R. (1977) On the nature of the interaction between some anionic polysaccharides and proteins, *Journal of the Science of Food and Agriculture,* 28(8), 661–668.

Jones, O. G. and McClements, D. J. (2011) Recent progress in biopolymer nanoparticle and microparticle formation by heat-treating electrostatic protein–polysaccharide complexes. *Advances in Colloid and Interface Science,* 167(1–2), 49–62.

Jones, O.G., Adamcik, J., Handschin, S., Bolisetty, S. and Mezzenga, R. (2010a) Fibrillation of beta-lactoglobulin at low pH in the presence of a complexing anionic polysaccharide. *Langmuir,* 26(22), 17449–17458.

Jones, O.G., Decker, E.A. and McClements, D.J. (2010b) Comparison of protein–polysaccharide nanoparticle fabrication methods: impact of biopolymer complexation before or after particle formation. *Journal of Colloid and Interface Science,* 344(1), 21–29.

Jones, O. G., Handschin, S., Adamcik, J., Harnau, L., Bolisetty, S. and Mezzenga, R. (2011) Complexation of β-lactoglobulin fibrils and sulfated polysaccharides. *Biomacromolecules,* 12(8), 3056–3065.

Jonsson, M. and Linse, P. (2001a) Polyelectrolyte-macroion complexation. I. Effect of linear charge density, chain length, and macroion charge. *Journal of Chemical Physics,* 115(7), 3406–3418.

Jonsson, M. and Linse, P. (2001b) Polyelectrolyte-macroion complexation. II. Effect of chain flexibility. *Journal of Chemical Physics,* 115(23), 10975–10985.

Jorgensen, W.L., Maxwell, D.S. and TiradoRives, J. (1996) Development and testing of the OPLS all-atom force field on conformational energetics and properties of organic liquids. *Journal of the American Chemical Society,* 118(45), 11225–11236.

Kaloti, M. and Bohidar, H.B. (2010) Kinetics of coacervation transition versus nanoparticle formation in chitosan-sodium tripolyphosphate solutions. *Colloids and Surfaces B,* 81(1), 165–173.

Kayitmazer, A.B., Seeman, D., Minsky, B.B., Dubin, P.L. and Xu, Y.S. (2013) Protein–polyelectrolyte interactions. *Soft Matter,* 9(9), 2553–2583.

Khair, A.S. and Squires, T.M. (2010) Active microrheology: a proposed technique to measure normal stress coefficients of complex fluids. *Physical Review Letters,* 105(15).

Kim, J., Weber, N., Shin, G., Huang, Q. and Liu, S. (2007) The study of β-lactoglobulin adsorption on polyethersulfone thin film surface using QCM-D and AFM. *Journal of Food Science,* 72(4), E214-E221.

Kobayashi, K., Hirano, A., Ohta, A., Yoshida, T., Takahashi, K. and Hattori, M. (2001) Reduced immunogenicity of beta-lactoglobulin by conjugation with carboxymethyl dextran differing in molecular weight. *Journal of Agricultural and Food Chemistry,* 49(2), 823–831.

Kobayashi, K., Yoshida, T., Takahashi, K. and Hattori, M. (2003) Modulation of the T cell response to beta-lactoglobulin by conjugation with carboxymethyl dextran, *Bioconjugate Chemistry,* 14(1), 168–176.

Kosters, H.A., Wierenga, P.A. and Gruppen, H. (2010) SELDI-TOF-MS as a rapid tool to study food related protein-peptide interactions. *Food Hydrocolloids,* 24(6–7), 667–673.

Kumar, S.G., Rahman, M.A., Lee, S.H., Hwang, H.S., Kim, H.A. and Yun, J.W. (2009) Plasma proteome analysis for anti-obesity and anti-diabetic potentials of chitosan oligosaccharides in *ob/ob* mice. *Proteomics,* 9(8), 2149–2162.

Kunz, W. (2010) Specific ion effects in colloidal and biological systems. *Current Opinion in Colloid and Interface Science,* 15(1–2), 34–39.

Laos, K., Brownsey, G. J. and Ring, S. G. (2007) Interactions between furcellaran and the globular proteins bovine serum albumin and β-lactoglobulin, *Carbohydrate Polymers,* 67(1), 116–123.

Larsen, T.H. and Furst, E.M. (2008) Microrheology of the liquid-solid transition during gelation. *Physical Review Letters,* 100(14).

Larson-Smith, K., Jackson, A. and Pozzo, D.C. (2012) SANS and SAXS analysis of charged nanoparticle adsorption at oil–water interfaces. *Langmuir,* 28(5), 2493–2501.

Li, J. and Huang, Q. (2012) Rheological properties of chitosan–tripolyphosphate complexes: From suspensions to microgels. *Carbohydrate Polymers,* 87(2), 1670–1677.

Li, Y. and Huang, Q. (2013) Influence of protein self-association on complex coacervation with polysaccharide: a Monte Carlo study. *Journal of Physics and Chemistry B,* 117(9), 2615–2624.

Li, Y. and McClements, D.J. (2014) Modulating lipid droplet intestinal lipolysis by electrostatic complexation with anionic polysaccharides: Influence of cosurfactants. *Food Hydrocolloids,* 35, 367–374.

Li, Y., Shi, T., An, L. and Huang, Q. (2012) Monte Carlo Simulation on Complex Formation of Proteins and Polysaccharides. *Journal of Physics and Chemistry B,* 116(10), 3045–3053.

Li, Y., Zhao, Q. and Huang, Q. (2014) Understanding complex coacervation in serum albumin and pectin mixtures using a combination of the Boltzmann equation and Monte Carlo simulation. *Carbohydrate Polymers,* 101(0), 544–553.

Likos, C.N. (2001) Effective interactions in soft condensed matter physics. *Physics Reports – Review Section of Physics Letters,* 348(4–5), 267–439.

Lipfert, J. and Doniach, S. (2007) Small-angle X-ray scattering from RNA, proteins, and protein complexes. *Annual Review of Biophysics and Biomolecular Structure,* 36, 307–327.

Liu, S., Low, N. H. and Nickerson, M.T. (2010) Entrapment of flaxseed oil within gelatin-gum Arabic capsules. *Journal of the American Oil Chemists' Society,* 87(7), 809–815.

Livney, Y.D. (2010) Milk proteins as vehicles for bioactives. *Current Opinion in Colloid and Interface Science,* 15(1–2), 73–83.

Lopez-Diaz, D. and Castillo, R. (2011) Microrheology of solutions embedded with thread-like supramolecular structures. *Soft Matter,* 7(13), 5926–5937.

Luo, Y.C., Zhang, B.C., Whent, M., Yu, L.L. and Wang, Q. (2011) Preparation and characterization of zein/chitosan complex for encapsulation of alpha-tocopherol, and its in vitro controlled release study. *Colloids and Surfaces B,* 85(2), 145–152.

Macfarlane, S., Macfarlane, G.T. and Cummings, J.H. (2006) Review article: prebiotics in the gastrointestinal tract. *Alimentary Pharmacology and Therapeutics,* 24(5), 701–714.

MacKerell, A.D., Bashford, D., Bellott, M., *et al.* (1998) All-atom empirical potential for molecular modeling and dynamics studies of proteins. *Journal of Physics and Chemistry B,* 102(18), 3586–3616.

Mahmoudi, M., Lynch, I., Ejtehadi, M.R., Monopoli, M.P., Bombelli, F.B. and Laurent, S. (2011) Protein–nanoparticle interactions: opportunities and challenges. *Chemical Reviews,* 111(9), 5610–5637.

Manning, M.C., Chou, D.K., Murphy, B.M., Payne, R.W. and Katayama, D.S. (2010) Stability of Protein Pharmaceuticals: An Update. *Pharmaceutical Research,* 27(4), 544–575.

Mason, T.G. and Weitz, D.A. (1995) Optical measurements of frequency-dependent linear viscoelastic moduli of complex fluids. *Physical Review Letters,* 74(7), 1250–1253.

Mason, T.G., Ganesan, K., vanZanten, J.H., Wirtz, D. and Kuo, S.C. (1997) Particle tracking microrheology of complex fluids. *Physical Review Letters,* 79(17), 3282–3285.

McCullough, B. J., Kalapothakis, J. M., Chin, W., *et al.* (2010) Binding a heparin derived disaccharide to defensin inspired peptides: insights to antimicrobial inhibition from gas-phase measurements. *Physical Chemistry Chemical Physics,* 12(14), 3589–3596.

Mekhloufi, G., Sanchez, C., Renard, D., Guillemin, S. and Hardy, J. (2005) pH-Induced structural transitions during complexation and coacervation of beta-lactoglobulin and acacia gum. *Langmuir,* 21(1), 386–394.

Motwani, S.K., Chopra, S., Talegaonkar, S., Kohl, K., Ahmad, F.J. and Khar, R.K. (2008) Chitosan–sodium alginate nanoparticles as submicroscopic reservoirs for ocular delivery: Formulation, optimisation and *in vitro* characterisation. *European Journal of Pharmaceutics and Biopharmaceutics,* 68(3), 513–525.

Mukhidinov, Z.K., Kasimova, G.F., Bobokalonov, D.T., *et al.* (2011) Pectin–zein microspheres as drug delivery systems. *Pharmaceutical Chemistry Journal,* 44(10), 564–567.

Newhouse, E.I., Xu, D., Markwick, P.R., *et al.* (2009) Mechanism of glycan receptor recognition and specificity switch for avian, swine, and human adapted influenza virus hemagglutinins: a molecular dynamics perspective. *Journal of the American Chemical Society,* 131(47), 17430–17442.

Ning, S.H., Huang, Q.Y., Sun, X.Y., *et al.* (2011) Carboxymethyl dextran-coated liposomes: Toward a robust drug delivery platform. *Soft Matter,* 7(19), 9394–9401.

Noel, T.R., Krzeminski, A., Moffat, J., Parker, R., Wellner, N. and Ring, S.G. (2007) The deposition and stability of pectin/protein and pectin/poly-L-lysine/protein multilayers. *Carbohydrate Polymers,* 70(4), 393–405.

Oparin, A. I. (1953) *The Origin of Life*. Dover Publications, Inc., New York.

Pawar, N. and Bohidar, H.B. (2010) Statistical thermodynamics of liquid-liquid phase separation in ternary systems during complex coacervation. *Physical Review E,* 82(3), 036107.

Pesce, G., De Luca, A.C., Rusciano, G., Netti, P.A., Fusco, S. and Sasso, A. (2009) Microrheology of complex fluids using optical tweezers: a comparison with macrorheological measurements. *Journal of Optics A – Pure and Applied Optics,* 11(3), 034016.

Picone, C.S.F. and da Cunha, R.L. (2010) Interactions between milk proteins and gellan gum in acidified gels. *Food Hydrocolloids,* 24(5), 502–511.

Priftis, D. and Tirrell, M. (2012) Phase behaviour and complex coacervation of aqueous polypeptide solutions. *Soft Matter*, 8, 9396–9405.

Qazvini, N.T., Bolisetty, S., Adamcik, J. and Mezzenga, R. (2012) Self-healing fish gelatin/sodium montmorillonite biohybrid coacervates: structural and rheological characterization. *Biomacromolecules,* 13(7), 2136–2147.

Qi, J.N., Yao, P., He, F., Yu, C.L. and Huang, C.O. (2010) Nanoparticles with dextran/chitosan shell and BSA/chitosan core-Doxorubicin loading and delivery. *International Journal of Pharmaceutics,* 393(1–2), 176–184.

Renard, D., Boue, F. and Lefebvre, J. (1997) Protein-polysaccharide mixtures structure of the systems and the effect of shear studied by SANS. *Physica B,* 234, 289–291.

Roberfroid, M.B. (2000) Prebiotics and probiotics: are they functional foods? *The American Journal of Clinical Nutrition,* 71(6), 1682s–1687s.

Ru, Q., Cho, Y. and Huang, Q.R. (2007) Biopolymer-stabilized emulsions on the basis of interactions between β-lactoglobulin and ι-carrageenan. *Frontiers of Chemical Science and Engineering,* 3, 399–406.

Ru, Q.M., Yu, H.L. and Huang, Q.R. (2010) Encapsulation of epigallocatechin-3-gallate (EGCG) using oil-in-water (O/W) submicrometer emulsions stabilized by iota-carrageenan and beta-lactoglobulin. *Journal of Agricultural and Food Chemistry,* 58(19), 10373–10381.

Ru, Q., Wang, Y., Lee, J., Ding, Y. and Huang, Q. (2012) Turbidity and rheological properties of bovine serum albumin/pectin coacervates: Effect of salt concentration and initial protein/polysaccharide ratio. *Carbohydrate Polymers,* 88(3), 838–846.

Rubinstein, M. and Colby, R.H. (2003) *Polymer Physics*. Oxford University Press.

Russel, W. B., Saville, D. A. and Schowalter, W. R. (1989) *Colloidal Dispersions*. Cambridge University Press, Cambridge, UK.

Sanchez, C. and Renard, D. (2002) Stability and structure of protein-polysaccharide coacervates in the presence of protein aggregates. *International Journal of Pharmaceutics,* 242(1–2), 319–324.

Sanchez, C., Schmitt, C., Babak, V.G. and Hardy, J. (1997) Rheology of whey protein isolate-xanthan mixed solutions and gels. Effect of pH and xanthan concentration. *Nahrung-Food,* 41(6), 336–343.

Sanchez, C., Mekhloufi, G., Schmitt, C., *et al.* (2002) Self-assembly of beta-lactoglobulin and acacia gum in aqueous solvent: Structure and phase-ordering kinetics. *Langmuir,* 18(26), 10323–10333.

Sanz, Y., Santacruz, A. and De Palma, G. (2008) Insights into the roles of gut microbes in obesity. *Interdisciplinary Perspectives on Infectious Diseases,* 2008, ID 829101. http://dx.doi.org/10.1155/2008/829101

Sato, J. and Breedveld, V. (2006) Transient rheology of solvent-responsive complex fluids by integrating microrheology and microfluidics. *Journal of Rheology,* 50(1), 1–19.

Schmidt, I., Cousin, F., Huchon, C., Boue, F. and Axelos, M. A. (2009) Spatial structure and composition of polysaccharide-protein complexes from small angle neutron scattering. *Biomacromolecules,* 10(6), 1346–1357.

Schmitt, C. and Turgeon, S. L. (2011) Protein/polysaccharide complexes and coacervates in food systems. *Current Opinion in Colloid and Interface Science,* 167(1–2), 63–70.

Schmitt, C., Sanchez, C., Desobry-Banon, S. and Hardy, J. (1998) Structure and technofunctional properties of protein-polysaccharide complexes: A review. *Critical Reviews in Food Science and Nutrition,* 38(8), 689–753.

Schmitt, C., Sanchez, C., Thomas, F. and Hardy, J. (1999) Complex coacervation between beta-lactoglobulin and acacia gum in aqueous medium, *Food Hydrocolloids,* 13(6), 483–496.

Schmitt, C., Sanchez, C., Despond, S., Renard, D., Thomas, F. and Hardy, J. (2000) Effect of protein aggregates on the complex coacervation between beta-lactoglobulin and acacia gum at pH 4.2. *Food Hydrocolloids,* 14(4), 403–413.

Schmitt, C., Sanchez, C., Lamprecht, A., *et al.* (2001) Study of beta-lactoglobulin/acacia gum complex coacervation by diffusing-wave spectroscopy and confocal scanning laser microscopy. *Colloids and Surfaces B,* 20(3), 267–280.

Schmitt, C., Palma da Silva, T., Bovay, C., *et al.* (2005) effect of time on the interfacial and foaming properties of β-lactoglobulin/acacia gum electrostatic complexes and coacervates at pH 4.2, *Langmuir,* 21(17), 7786–7795.

Seijo, M., Ulrich, S., Filella, M., Buffle, J. and Stoll, S. (2006) Effects of surface site distribution and dielectric discontinuity on the charging behavior of nanoparticles. A grand canonical Monte Carlo study. *Physical Chemistry Chemical Physics,* 8(48), 5679–5688.

Seijo, M., Ulrich, S., Filella, M., Buffle, J. and Stoll, S. (2008) Modeling the surface charge evolution of spherical nanoparticles by considering dielectric discontinuity effects at the solid/electrolyte solution interface. *Journal of Colloid and Interface Science,* 322(2), 660–668.

Seijo, M., Ulrich, S., Filella, M., Buffle, J. and Stoll, S. (2009) Modeling the adsorption and coagulation of fulvic acids on colloids by Brownian dynamics simulations, *Environmental Science and Technology,* 43(19), 7265–7269.

Seyrek, E., Dubin, P.L., Tribet, C. and Gamble, E. A. (2003) Ionic strength dependence of protein–polyelectrolyte interactions. *Biomacromolecules,* 4(2), 273–282.

Skepo, M. and Linse, P. (2003) Complexation, phase separation, and redissolution in polyelectrolyte-macroion solutions. *Macromolecules,* 36(2), 508–519.

Spruijt, E., Sprakel, J., Lemmers, M., Stuart, M.A.C. and van der Gucht, J. (2010) Relaxation dynamics at different time scales in electrostatic complexes: time-salt superposition. *Physical Review Letters,* 105(20).

Stradner, A., Sedgwick, H., Cardinaux, F., Poon, W.C.K., Egelhaaf, S.U. and Schurtenberger, P. (2004) Equilibrium cluster formation in concentrated protein solutions and colloids. *Nature,* 432(7016), 492–495.

Thongkaew, C., Gibis, M., Hinrichs, J. and Weiss, J. (2014) Polyphenol interactions with whey protein isolate and whey protein isolate-pectin coacervates. *Food Hydrocolloids,* 41, 103–112.

Tiebackx, F. W. Z. (1911) Gleichzeitige Ausflockung zweier Kolloide. *Chem. Ind. Kolloide,* 8, 198–201.

Tolstoguzov, V. (2003) Some thermodynamic considerations in food formulation *Food Hydrocolloids,* 17(1), 1–23.

Tolstoguzov, V.B. (1991) Functional properties of food proteins and role of protein-polysaccharide interaction. *Food Hydrocolloids,* 4(6), 429–468.

Turgeon, S.L., Beaulieu, M., Schmitt, C. and Sanchez, C. (2003) Protein-polysaccharide interactions: phase-ordering kinetics, thermodynamic and structural aspects. *Current Opinion in Colloid and Interface Science,* 8(4–5), 401–414.

Turgeon, S.L., Schmitt, C. and Sanchez, C. (2007) Protein-polysaccharide complexes and coacervates. *Current Opinion in Colloid and Interface Science,* 12(4–5), 166–178.

Ulrich, S., Laguecir, A. and Stoll, S. (2005) Complexation of a weak polyelectrolyte with a charged nanoparticle. Solution properties and polyelectrolyte stiffness influences. *Macromolecules,* 38(21), 8939–8949.

Ulrich, S., Seijo, M., Laguecir, A. and Stoll, S. (2006a) Nanoparticle adsorption on a weak polyelectrolyte. Stiffness, pH, charge mobility, and ionic concentration effects investigated by Monte Carlo Simulations. *Journal of Physics and Chemistry B,* 110(42), 20954–20964.

Ulrich, S., Seijo, M. and Stoll, S. (2006b) The many facets of polyelectrolytes and oppositely charged macroions complex formation. *Current Opinion in Colloid and Interface Science,* 11(5), 268–272.

Ulrich, S., Seijo, M., Carnal, F. and Stoll, S. (2011) Formation of complexes between nanoparticles and weak polyampholyte chains. Monte Carlo simulations. *Macromolecules,* 44(6), 1661–1670.

Varenne, A., Gareil, P., Colliec-Jouault, S. and Daniel, R. (2003) Capillary electrophoresis determination of the binding affinity of bioactive sulfated polysaccharides to proteins: study of the binding properties of fucoidan to antithrombin. *Analytical Biochemistry,* 315(2), 152–159.

Veis, A. (2011) A review of the early development of the thermodynamics of the complex coacervation phase separation. *Advances in Colloid and Interface Science,* 167(1–2), 2–11.

Waigh, T.A. (2005) Microrheology of complex fluids. *Reports on Progress in Physics,* 68(3), 685–742.

Wallin, T. and Linse, P. (1996) Monte Carlo simulations of polyelectrolytes at charged micelles .1. Effects of chain flexibility. *Langmuir,* 12(2), 305–314.

Wang, X., Ruengruglikit, C., Wang, Y.-W. and Huang, Q. (2007a) Interfacial Interactions of Pectin with Bovine Serum Albumin Studied by Quartz Crystal Microbalance with Dissipation Monitoring: Effect of Ionic Strength, *Journal of Agricultural and Food Chemistry,* 55(25), 10425–10431.

Wang, X., Wang, Y.-W., Ruengruglikit, C. and Huang, Q. (2007b) Effects of salt concentration on formation and dissociation of β-lactoglobulin/pectin complexes. *Journal of Agricultural and Food Chemistry,* 55(25), 10432–10436.

Wang, X.Y., Lee, J.Y., Wang, Y.W. and Huang, Q.R. (2007c) Composition and rheological properties of beta-lactoglobulin/pectin coacervates: Effects of salt concentration and initial protein/polysaccharide ratio. *Biomacromolecules,* 8(3), 992–997.

Weinbreck, F. (2004) *Whey protein / polysaccharide coacervates: structure and dynamics*. Utrecht University, The Netherlands.

Weinbreck, F., de Vries, R., Schrooyen, P. and de Kruif, C.G. (2003a) Complex coacervation of whey proteins and gum Arabic *Biomacromolecules,* 4(2), 293–303.

Weinbreck, F., Nieuwenhuijse, H., Robijn, G.W. and de Kruif, C.G. (2003b) Complex formation of whey proteins: Exocellular polysaccharide EPS B40. *Langmuir,* 19(22), 9404–9410.

Weinbreck, F., Minor, M. and De Kruif, C.G. (2004a) Microencapsulation of oils using whey protein/gum Arabic coacervates. *Journal of Microencapsulation,* 21(6), 667–679.

Weinbreck, F., Nieuwenhuijse, H., Robijn, G.W. and de Kruif, C.G. (2004b) Complexation of whey proteins with carrageenan. *Journal of Agricultural and Food Chemistry,* 52(11), 3550–3555.

Weinbreck, F., Rollema, H.S., Tromp, R.H. and de Kruif, C.G. (2004c) Diffusivity of whey protein and gum Arabic in their coacervates. *Langmuir,* 20(15), 6389–95.

Weinbreck, F., Tromp, R.H. and de Kruif, C.G. (2004d) Composition and structure of whey protein/gum Arabic coacervates. *Biomacromolecules,* 5(4), 1437–45.

Xia, J.L., Mattison, K., Romano, V., Dubin, P.L. and Muhoberac, B.B. (1997) Complexation of trypsin and alcohol dehydrogenase with poly(diallyldimethylammonium chloride). *Biopolymers,* 41(4), 359–365.

Xia, S., Li, Y., Xia, Q., Zhang, X. and Huang, Q. (2015) Glycosylation of bovine serum albumin via Maillard reaction prevents epigallocatechin-3-gallate-induced protein aggregation. *Food Hydrocolloids,* 43, 228–235.

Xu, J.Y., Viasnoff, V. and Wirtz, D. (1998) Compliance of actin filament networks measured by particle-tracking microrheology and diffusing wave spectroscopy. *Rheologica Acta,* 37(4), 387–398.

Xu, Y., Mazzawi, M., Chen, K., Sun, L. and Dubin, P.L. (2011) Protein purification by polyelectrolyte coacervation: influence of protein charge anisotropy on selectivity. *Biomacromolecules,* 12(5), 1512–1522.

Yang, Q., Graham, T.E., Mody, N., *et al.* (2005) Serum retinol binding protein 4 contributes to insulin resistance in obesity and type 2 diabetes. *Nature,* 436(7049), 356–362.

Ye, A.Q., Flanagan, J. and Singh, H. (2006) Formation of stable nanoparticles via electrostatic complexation between sodium caseinate and gum Arabic. *Biopolymers,* 82(2), 121–133.

Zhang, B., Luo, Y. and Wang, Q. (2011) Effect of acid and base treatments on structural, rheological, and antioxidant properties of [alpha]-zein. *Food Chemistry,* 124(1), 210–220.

Zhang, S. and Vardhanabhuti, B. (2014) Acid-induced gelation properties of heated whey protein-pectin soluble complex (Part II): Effect of charge density of pectin. *Food Hydrocolloids,* 39, 95–103.

15 Bicontinuous delivery systems

Graciela Padua

Department of Food Science and Human Nutrition, University of Illinois, Urbana, Illinois, USA

15.1 INTRODUCTION

Success for the food products of the future will depend on their capacity to deliver valuable nutritional/bioactives cargo. Thus, new technologies are needed to develop safe and effective carriers for the delivery of such compounds to human, animal, and plant cells, and nanostructuring in this context has good potential. This chapter describes the structure and preparation of bicontinuous systems in this respect, and highlights their application as delivery vehicles, with emphasis placed on self-assembled nanostructures comprising distinctive internal compartmentalization.

Surfactant molecules (or amphiphiles) self-assemble in aqueous solutions into a rich variety of liquid-crystalline phases. Self-aggregation of surfactants in water is a thermodynamically favorable process, which results in orientation of the hydrophilic regions at the aqueous interface and concurrent association of the hydrophobic groups, to minimize their exposure to bulk water. This dewetting process drives the self-assembly of amphiphiles in water. Depending on the prevailing conditions, a wide array of structures with different degrees of complexity may arise. Aggregates can have two-dimensional order, such as classic monolayer and bilayer structures (lamellar phase), or complex three-dimensional architectures, including micellar (L), hexagonal (H), and bicontinuous cubic phases (Q). Liquid-crystalline phases exhibit long-range order but have no short-range order, that is, the hydrophobic chains are disordered. The formation of liquid-crystalline phases may be lyotropic or thermotropic. Among the various lyotropic liquid-crystalline phases, the lamellar phase with stacks of lipid bilayers, see Figure 15.1 (http://education.mrsec.wisc.edu/nanoquest/self_assembly/index.html), is the best known in biological systems (Fraser *et al.*, 2009).

Bicontinuous cubic phases were first reported by Luzzati *et al.* (1968) in the 1960s, although their detailed study did not begin until the 1980s. Since then, bicontinuous phases have been found in various lyotropic systems, including lipid–water systems and block copolymers. A cubic-phase liquid-crystalline gel material is a temperature stable, optically isotropic, viscous gel with a high surface area (400 m^2/g of cubic gel). It may solubilize hydrophilic, hydrophobic, as well as amphiphilic compounds (Yang *et al.*, 2004). The cubic phase is important to cellular processes, including membrane fusion and protein function.

The most popular application of the cubic phase is as a delivery vehicle for materials that are solubilized in the cubic gel but diffuse out in a controlled-release manner. Cubic gels in equilibrium with excess water can be dispersed into nanoparticles. In analogy with liposomes, which are formed from lamellar liquid-crystalline materials, these nanoparticles are termed cubosomes (Spicer *et al.*, 2002).

Nanotechnology and Functional Foods: Effective Delivery of Bioactive Ingredients, First Edition.
Edited by Cristina M. Sabliov, Hongda Chen, and Rickey Y. Yada.

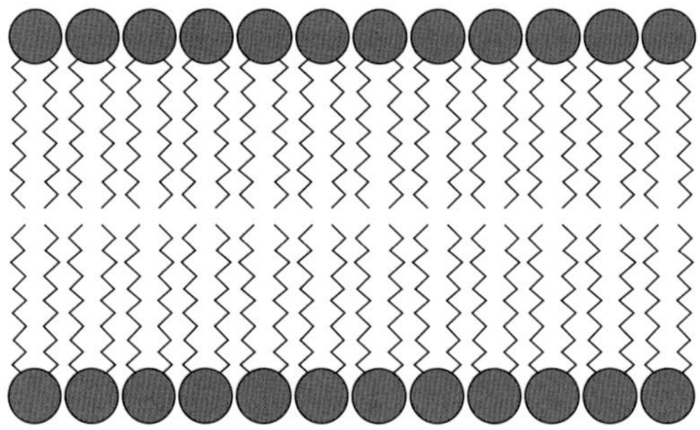

Figure 15.1 Structure of a lipid bilayer.

It has been demonstrated that protein molecules (such as lysozyme) can be entrapped within the water channel system of the cubic structure. Cubosomes have also recently attracted great interest for new potential applications in the area of pharmaceutical technology. These nanostructured fluids show potential as drug nanocarriers since they are composed of naturally occurring biocompatible lipids tailored for solubilizing high amounts of amphiphilic, hydrophobic, and hydrophilic drugs in their highly ordered self-assembled interiors (Yang *et al.*, 2004).

15.2 STRUCTURE OF THE CUBIC MESOPHASE

The most distinctive characteristic of surfactants is the capacity of their water mixtures to exist as liquid crystal phases. Liquid crystals behave as fluids, even though they are more structured than ordinary liquids. They exhibit long-range periodicity in two and three dimensions. They remain very soft phases and their structure and morphology have been studied by small-angle X-ray scattering (SAXS), small-angle neutron scattering (SANS), cryo-transmission electron microscopy (cryo-TEM), atomic force microscopy (AFM), nuclear magnetic resonance spectroscopy, transmission and freeze–fracture electron microscopy (Lindblom and Rilfors, 1989). Other techniques applied to study cubic lipid–water systems are time-resolved fluorescence, Fourier transform infrared spectroscopy (FTIR), differential scanning calorimetry, and resonance Raman spectroscopy (Yang *et al.*, 2004; Sagalowicz *et al.*, 2006).

Scriven (1976) first suggested that bicontinuous cubic phases have interface structures based on periodic minimal surfaces, a view that is now generally accepted. A cubic phase consists of a single continuous curved lipid bilayer that forms a complex, symmetrical three-dimensional cubic matrix separating two continuous, nonintersecting water channels. The internal topography of a bicontinuous phase is commonly described as an infinite periodical minimal surface (IPMS) with zero mean curvature (Hyde *et al.*, 1984). In a cubic liquid-crystalline lipid–water phase, the lipid bilayers are arranged in periodic three-dimensional cubic-lattice structures. By bending the bilayers into infinite periodic minimal surfaces, the mean curvature at any point on the surface is zero. That is, all points on the bilayer surface are as concave as they are convex. The minimal surface is defined as the middle of the bilayer where the terminal methyl groups of the hydrophobic tails intersect (Larsson and Tiberg, 2005).

The complementary three-dimensional labyrinth network is formed by two distinct water channels. The diameter of the aqueous channels, a function of lipid composition and phase type, can range from 4 to 20 nm (Benedicto and O'Brien, 1997). Bicontinuous cubic phases show molecular orientational order and may be characterized by their structural symmetry. The cubic-phase structure exhibits crystallographic periodicity, with three types of IPMS structures, the gyroid (G), diamond (D), and primitive (P) surfaces, corresponding to the crystallographic space groups Ia3d, Pn3m, and Im3m, respectively (Figure 15.2) (Seddon *et al.*, 2000). These structures have been verified using SAXS (Fraser *et al.*, 2009; Angelova *et al.*, 2011), in addition to mathematical modeling (Chung and Caffrey, 1995; Squires *et al.*, 2005; Templer, 1998).

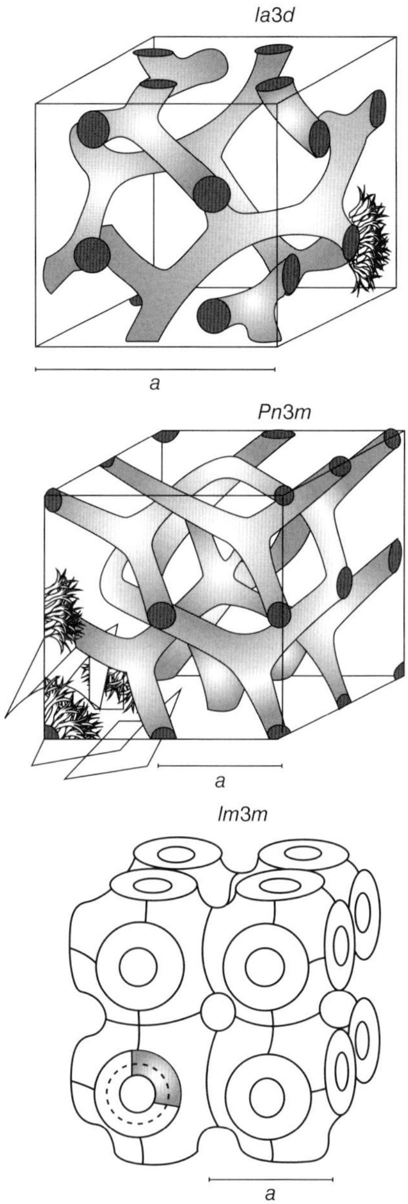

Figure 15.2 Structure of bicontinuous mesophase showing crystallographic space groups. (Adapted from Seddon *et al.*, 2000. Reproduced with permission from the PCCP Owner Societies.)

Bicontinuous cubic phases may be considered to be normal (type I) or inverse (type II) phases depending on the relative spontaneous curvature of the hydrophobic phase at the aqueous interface (oil-in-water or water-in-oil). The conventional notation for normal and inverse cubic phases is QI and QII, respectively. The propensity toward type-II phase behavior is strongly influenced by the geometry of individual surfactant molecules (Sagalowicz *et al.*, 2006). Interestingly, the interconversion of cubic phases readily occurs. Squires *et al.* (2002) reported that inverse cubic phases formed by lauric acid–diauroyl

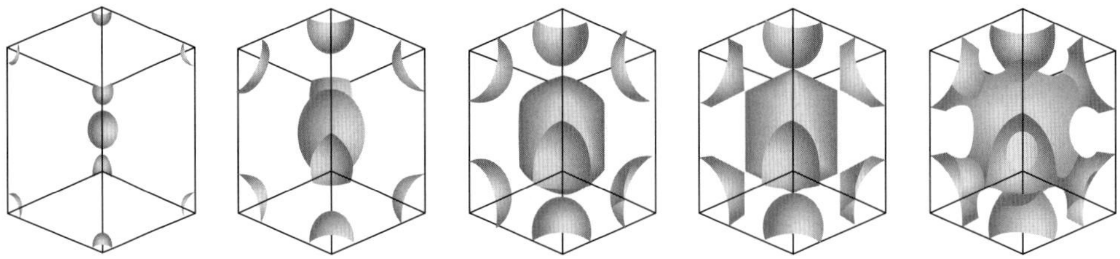

Figure 15.3 Simulation of the transition from soft spheres in a body-centered cubic lattice to a bicontinuous structure. (Shin *et al.*, 2009. Reproduced with permission from The Royal Society of Chemistry.)

phosphatidyl choline–water easily undergo phase transitions from the QII D to the QII G, that is, cubic-phase transition from diamond (D)-type minimal surface to one with a gyroid (G) minimal surface.

The amphiphile most commonly used to study inverse cubic-phase behavior is monoacylglycerol (monoolein or glyceryl monooleate) (Ericsson *et al.*, 1983). The cis-double bond in the acyl chain increases the effective volume occupied by the tail region, thereby favoring the formation and stability of inverse phases. In terms of phase behavior, an increase in water content at a fixed temperature initially results in a phase transition from the lamellar phase (La) to the QII G phase and a further increase in water content favors an intercubic phase transition from the QII G phase to QII D phase (Qiu and Caffrey, 2000). The modification of cubic-phase structure may be desirable where a 'swelled' structure or a structure of greater porosity is required for the entrapment of large substrates (e.g., proteins). A QII phase with contracted aqueous channels or reduced porosity may be preferential for the controlled transport of substrates in applications such as drug delivery. The addition of a lipid to the cubic phase may be useful to control phase behaviour and, therefore, the structure of the system. In glyceryl monooleate–water binary systems, the addition of a lipid has a pronounced effect on the QII phase behaviour and structure for each binary mixture (Fraser *et al.*, 2009).

The bicontinuous cubic phase is the intermediate phase between other liquid crystal phases. The structure may be visualized as consisting of spheres along the three axes x, y, and z, which touch and then fuse at the contact points (Figure 15.3). If the spheres of this network can be deformed and strive to minimize the total surface area like a soap film, increasing the outer pressure will shrink the surface and when outer and inner space is equal we have the primitive (P)-surface and the surface minimal area. Minimal surfaces are best described by analogy to soap films, their most readily observed natural form (Lovett and Tilley, 1994). The surface formed by a soap film between two rings is a catenoid, a simple form of minimal surface in which two principal curvatures are equal but opposite in sign at every point, resulting in an average curvature of zero and a negative Gaussian curvature. The P-surface and the D-surface (or diamond surface) were discovered mathematically in 1890, whereas the G-surface (or gyroid surface) was not discovered until the 1960s. In cubic phases, the minimal surface is formed by the self-assembled bilayer that occurs as the hydrophobic or hydrophilic portions of the surfactant molecules line up to minimize their interaction with their opposites. The three structures are bicontinuous (i.e., they divide space into two continuous but nonintersecting regions). In the case of cubic phases, two separate regions of hydrophilic material (water channels) form. The monoolein–water system forms the D-surface at high water levels and the G-surface at lower levels. The P-surface is formed in the monoolein–water system, but only when a third component, such as casein or amphiphilic block copolymers are added.

The rationale for the minimal surface approach to cubic phase characterization is its mathematical basis. A significant advance came in the discovery that nodal surfaces are closely related to minimal surfaces and offer a simple equation that can be used to plot and study cubic phase structures (von Schnering and Nesper, 1991). Extensive reviews and mathematical descriptions of bicontinuous phases are given in Anderson *et al.* (1990) and Hyde *et al.* (1997). A more recent review (Schwarz and Gompper, 2002) discusses bicontinuous geometry in amphiphilic systems. Nodal surfaces were used

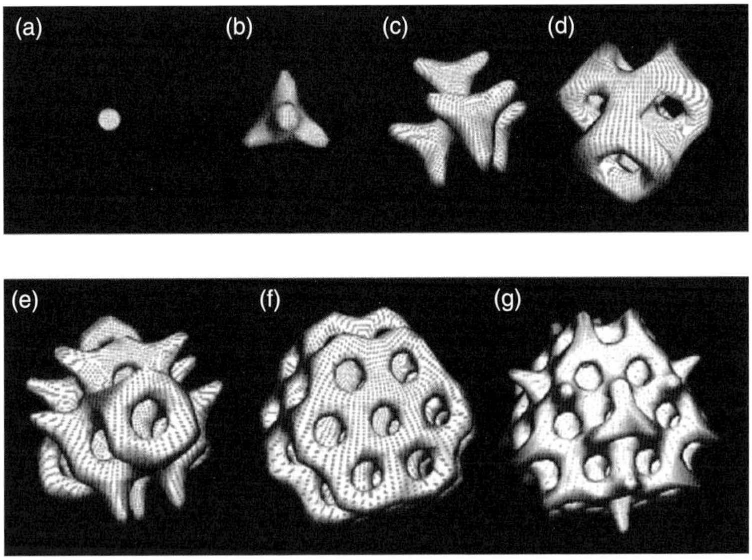

Figure 15.4 Mechanism for three-dimensional growth of cubosome nanoparticles (a–g) represent N = 1, 5, 17, 29, 71, 147, and 191. N is the number of nodes (repeat volumes) in the structure. (Angelov *et al.*, 2006. Reproduced with permission from American Chemical Society.)

to simulate the growth of a diamond type cubic network of aqueous channels starting from a curved lipid bilayer nucleus (nanovesicle) (Angelov *et al.*, 2006). The diameter of the curved lipid bilayer unit corresponds to the sum of the thickness of two lipid bilayers and the diameter of one aqueous nanochannel. The simulated nanovesicle has a diameter of 8.8 nm. Diverse cubosomic nanostructures were generated as a function of the number of repeated elementary unit volumes or nodes (N) in the growing nanochannel network. The three-dimensional cubosomic structures were modeled in stages starting from a single nanovesicle and building up to a cubosomic assembly formed by hundreds of nodes: $N = 29$ yielded the first stable lipid bilayer configuration described by a diamond-type periodic surface (Figure 15.4).

Dispersed particles of bicontinuous cubic phases were first observed during studies of fat digestion that simulated stomach contents by combining oil with lipase and bile salts (Patton and Carey, 1979). Olive oil droplets, mainly triolein, in contact with lipase formed small particles of cubic phases as the enzyme broke the triglyceride down into monoolein, which then hydrated. Lindstrom *et al.* (1981) studied the phenomenon further, observing that sodium cholate bile salts provided dispersion and colloidal stability of the cubic-phase particles formed by monoolein and water, and that such particles formed in other ternary systems as well. The term "cubosome" first appeared in a review by Larsson and Tiberg (2005) of the implications of bicontinuous cubic phases to biological membranes. Later research on cubosomes focused on the ternary phase behavior of the Poloxamer 407 (a PEO99–PPO67–PEO99 block copolymer)–monoolein–water system because of the polymer's ability to provide colloidal stability to cubosomes against recoalescence to a bulk cubic phase. The PPO region of the block copolymer exists either at the surface of the cubic phase particles, or within the bilayer structure, whereas the PEO chains remain in the bulk water phase. Landh (1994) observed cubosome formation in a three-phase region including disordered bicontinuous (L3 or sponge) phase, cubic phase, and water.

Although small amounts of cubosomes are apparently produced by simple addition of dispersing agents such as sodium cholate to a bulk cubic phase, mechanical dispersion of the bulk cubic phase is the most studied and most direct cubosome formation technique. Gustafsson *et al.* (1996, 1997) formed cubosomes by high-shear microfluidization and ultrasonication of the bulk cubic phase in contact with aqueous Poloxamer 407. The cubosomes showed a P-surface structure, based on X-ray diffraction and, interestingly, resemble cubes despite their nonsolid nature. The molecular-scale cubic symmetry of the lipid bilayer comprising cubosomes is apparent in the nanometer scale of cubosome particles, although their appearance is often orientation dependent. The particles range in size from about 10 to 500 nm in diameter. Cubosomes are immediately identifiable as square particles with a well-defined cubic lattice of dots. Each dot corresponds to the presence of a pore-containing aqueous phase. Along with well-formed cubosomes, more disordered cubosomes that have a more spherical shape are also visible. Gustafsson *et al.* (1996) point out that the absence of grain boundaries in the cubosomes indicates they are single-crystal structures.

Ultrasonication of the bulk cubic phase produces mostly vesicles (Gustafsson *et al.*, 1996) that, over the course of several weeks, transform into cubosomes, likely via membrane fusion. Such metastability is characteristic of cubosome systems because of the slow transport processes involved in forming the high-viscosity crystalline structure and the high-energy input required to fragment the bulk cubic phase. Distinct from the dispersed vesicles, the surfaces of some of the cubosome particles have hemispherical vesicle structures that are thought to block the exposure of the hydrophobic bilayer at the termination of the cubosome edge structure, but may also aid in colloidal stabilization of the cubosomes.

A cubosome dissection along a fracture plane may serve to elucidate the mechanism of protein nanoencapsulation and confinement in the triply periodic networks of fluid membrane interfaces, and eventually give the respective location of the biomacromolecules with regard to the aqueous nanopores. The introduction of guest molecules causes changes in the local curvature of the lipid bilayer, which results in a pattern with lipid bilayer distortions. In the presence of guest biomolecules, the cubic membrane architecture is no longer perfect as compared to the single-crystal systems of pure monoolein. The observed undulations of the three-dimensional lipid membrane are typical for fluid-phase architectures. Such fluctuations suggest a dynamic structure where aqueous nanochannels could open or close at cubosome interfaces, thus affecting the exchange of molecules between the cubosome interior and the excess aqueous reservoir. The internal cubosome nanostructure was investigated for the case of entrapment of water-soluble fragments of the protein immunoglobulin in fully hydrated monoolein cubic assemblies. In this case, the macromolecule length (~7 nm) is larger than the cubosome aqueous nanochannel diameter (D_w) of 3.6 nm. This steric incompatibility may induce local perturbations of the overall cubic supramolecular organization. The entrapped protein macromolecules are believed to locate along the contact interfaces of the adjacent cubosome nanodroplets. The protein locations could be attributed to a defect network that tessellates the cubic lattice into an array of nanocubosomes.

The location of a protein in the three-dimensional self-assembled cubic architecture will depend on its hydrophobic–hydrophilic balance and its affinity to interact with the lipid bilayer. Small proteins of amphiphilic nature may be able to adsorb or penetrate into the lipid bilayer and, thus, alter the mean interfacial curvature. Above a critical concentration, the guest protein could destabilize the host cubic lipid matrix and cause its transformation into new phases. The protein cytochrome c was embedded in monoolein cubic mesophases (Angelov *et al.*, 2006). In contrast, the protein transferrin induced swelling of the monoolein cubic lattice without changing the space group symmetry. Lipid-based lyotropic phases are particularly attractive because they can be nanostructurally customized, for instance, to closely resemble cellular components, or to formulate delivery vehicles for biomolecules and drugs. In the presence of water or aqueous buffers, lipid molecules can self-assemble into a wide range of nanostructures. The nanostructure can be tuned by varying the composition of the hydrophobic phase (lipid and oil), which is quantified by the mass fraction of the lipid phase (δ). The value of δ can be varied from 100% down to 50% to obtain corresponding nanostructures of bicontinuous-cubic (Pn3m), hexagonal (H2), micellar-cubic (Fd3m), and fluid-isotropic (L2) phases.

15.3 PREPARATION

The glycerol monooleate–water system, one of the most studied lipid–water systems, exhibits two types of bicontinuous cubic phases at water contents between 20 and 40 wt%, at room temperature (Hyde *et al.*, 1984). An increase in the water content results in a transition from the lamellar phase to the gyroid (G) cubic phase then to the double-diamond (D) cubic phase. Phase diagrams of the glyceryl monooleate (GMO)–water system are available (Qiu and Caffrey, 2000). Other lipids such as phosphatidyl-cholines, phosphatidylethanolamines, polyethylene glycol modified phospholipids, and various monoglycerides have also been shown to form bicontinuous cubic phases.

Cubic gels have interesting and potentially useful properties; however, they may be difficult to handle and process. Their dispersion into smaller structures (cubosomes) that retain at least some of the properties of the bulk material would facilitate their use in various applications. At its most basic, the production of cubosomes entails the formation of nanoparticles. This may be accomplished following top-down or bottom-up techniques. Most cubosome research has focused on top-down techniques, whereby the bulk cubic phase is first produced and then dispersed by high-energy processing into cubosome nanoparticles (Salentinig *et al.*, 2008). Cubic particles from monoolein–water systems are usually produced by mixing monoolein, dispersing agent, and water, followed by incubation for 24 h. The cubic-gel product is then dispersed into nanometer- to micron-scale particles by the application of mechanical or ultrasonic energy. Homogenization, often employed for the dispersion process, requires high pressure and numerous cycles before homogeneous cubic-particle dispersions are produced. The size can be reduced further in a high-pressure emulsifier. One disadvantage of this process is the large amount of mechanical energy required for homogenization, which increases the risk of degrading the cubic lipid–water assembly. Furthermore, the viscous nature of the cubic gel makes scaling-up difficult. Several low-energy input processes have been developed to replace the viscous gel and high-energy-input dispersion methods.

In the bottom-up approach the key factor is the use of a hydrotrope. Spicer *et al.* (2001) reported the development of a hydrotrope process for producing cubic particles spontaneously by dilution of a monoolein–ethanol–water system. In this case, a hydrotrope is defined as a molecule with both hydrophobic and hydrophilic character but incapable of displaying surfactant behavior. Hydrotropes are capable of disrupting liquid-crystalline phases, as well as functioning as facilitators of liquid-crystalline particle dispersions, although they do not form liquid crystals by themselves. However, when mixed in substantial quantities, they do help increase the solubility of water-insoluble lipids such as monoolein. Hydrotropes can dissolve water-insoluble lipids to create liquid precursors and prevent the formation of viscous liquid crystals at high concentration.

Ethanol can be used as a hydrotrope in combination with monoolein to create a liquid precursor that forms cubic nanoparticles spontaneously when diluted with water. Dilution of monoolein–water–hydrotrope systems can produce particles as small as 50 nm or as large as 20 μm. The dilution process was found to produce smaller cubic nanoparticles more consistently than conventional methods. In principle, any lipid and hydrotrope combination that forms cubic liquid-crystalline phases can be dispersed by this method. Colloidally stabilized, controlled-release dispersions can be produced by this process by the direct inclusion of active ingredients and colloidal stabilizers into either of the two solutions. This process could also be used in applications that require the formation of cubic nanoparticles *in situ*. Although the hydrotrope process provides the advantage of low-energy input and easier handling, for many applications it would be desirable to have dry powders or granular materials rather than liquid products (Spicer *et al.*, 2001).

Compared with the top-down approach, this dilution-based approach can produce cubosomes without the need for high-energy input fragmentation. Moreover, this approach is more efficient at generating small particles. The dilution-based approach can be regarded as a process of small particles forming large particles through aggregation, whereas the top-down approach is analogous to the attrition of large particles. In addition, cubosomes prepared through dilution show long-term stability, which might be attributed to the homogeneous dispersion of stabilizers on the surface of cubosomes. The use of a hydrotrope can simplify the preparation process and produce cubosomes possessing properties similar to those fabricated by the top-down approach. It should be noted that this approach requires precise control of the process and knowledge of the ternary phase diagram lipid–water–hydrotrope.

Freeze-drying is a well-known method for the production of liposomes for drug delivery applications. Powdered precursors of drug-containing vesicles have been prepared by freeze-drying for the delivery of DNA and proteins by inhalation. Kim *et al.* (2000) treated powdered nanoparticle precursors by freeze-drying. However, the cubic nature of the particles was not confirmed. Another method for producing powdered precursors for lipid particles is spray-drying. Freitas and Muller (1998) reported the formation of lipid nanoparticle powders from nonliquid crystalline lipids by spray-drying. Spicer *et al.* (2002) applied this method in the production of two types of dry-powder cubic nanoparticle precursors, a starch-encapsulated monoolein precursor from a dispersion of cubic liquid-crystalline particles in an aqueous starch solution, and a dextran encapsulated monoolein by spray-drying an emulsion from ethanol-dextran-monoolein-water system. The encapsulants were introduced to decrease powder cohesion during spray-drying and to function as soluble colloidal stabilizers upon hydration of the powder precursors. Both powdered precursors formed 600-nm colloidally stable cubic nanoparticles upon hydration of the powder. The dextran-stabilized system gave a narrower powder-size distribution as a result of the relative ease of dispersing and spraying an emulsion in contrast to a dispersion. The powder-precursor process widens the range and ease of applications for cubic nanoparticles. However, this process is limited by the large proportion of polymer required for encapsulation, that is, 75 wt% for starch and 60 wt% for dextran. The smaller proportion of lipids encapsulated means that only a limited amount of active ingredients can be incorporated for subsequent delivery. Therefore, the dry-powder-precursor process is only applicable for high-value-added materials such as pharmaceuticals and enzymes.

It is well known that lamellar phases in coexisteance with water can be easily dispersed into stable vesicles termed liposomes. Bicontinuous cubic phases do not form stable aqueous dispersions, as exposure of the hydrophobic domains to the aqueous medium surrounding the particles effectively disrupts the crystallinity of the inner structure. That is, however, if the unfavorable interaction between the hydrophobic domain and the aqueous media can be avoided without breaking the inner cubic structure. By the addition of amphiphilic dispersion agents, it is possible to form nanoparticles with a cubic liquid-crystalline inner structure. The correct choice of dispersing agent is the key, as the dispersing agent will participate in the lipid–water assembly, and needs to do so without disrupting the cubic liquid-crystalline structure.

Poloxamer 407 (also called pluronic F127), a nonionic triblock copolymer composed of polyethylene oxide (PEO) and polypropylene oxide (PPO), PEO98PPO67PEO98, is a widely used dispersing agent (Yang *et al.*, 2004). While the hydrophobic PPO block is adsorbed onto or incorporated in the surface of the stabilized particle, the hydrophilic PEO blocks extend to cover the surface, providing steric shielding and stabilizing the colloid particles. The ternary system glycerol monooleate (89.5 wt%)–poloxamer 407–water exhibits an extended inverted cubic-phase field in which four different cubic structures were detected by X-ray diffraction (Landh, 1994). A similar behavior was observed when oleic acid, a fatty acid that has the same acyl chain as monoolein and a negative spontaneous curvature, was mixed in the lipid region of the cubic particles from a monoolein–water mixture. The X-ray diffraction pattern revealed a transformation from a bicontinuous cubic to an inverted hexagonal phase with increasing amounts of oleic acid. The inner microstructure of these dispersed particles was also examined by ^{13}C-NMR relaxation measurements. A comparison of the ^{13}C-NMR relaxation rates measured for monoolein in the dispersed particles with those measured for the corresponding nondispersed cubic samples indicated that the lipid organization and dynamic properties of the original cubic assemblies were retained. These dispersed samples were reported to be stable for up to 6 months.

Dispersions of monoolein-rich monoglycerides (MO), with or without phospholipids (PL), may be prepared by equilibration of a MO–PL–water cubic phase, subsequent fragmentation with a poloxamer 407 solution, sonication, and homogenization. This method requires only moderate energy input independently of the homogenization equipment. Between two and five homogenization cycles at 300–400 bar were sufficient to obtain submicron-sized particles.

Whereas the cubic bulk phases are always homogenous, clear and isotropic, dispersions of cubic phases vary considerably in their macroscopic appearance after sonication or homogenization, from almost transparent dispersions, slightly turbid systems, opaque dispersions to milky emulsion-like systems depending on preparation parameters. The mean average particle diameter ranges from 80 to 350 nm. Homogenized dispersions which differ in their macroscopic appearance also show differences in particle size. The transparent, slightly turbid systems with a macroscopically homogeneous appearance typically show mean average diameters in the range of 80–130 nm. The mean average diameters of the opalescent and opaque dispersions are typically between 120 and 200 nm. Emulsion-like systems are in the same size range or above (up to more than 350 nm). The milky dispersions typically have broad size distributions. The macroscopic appearance of monoolein dispersions stored at room temperature tends to change with time. Dispersions that are almost clear and transparent directly after preparation may turn increasingly turbid with storage time. A number of dispersions and macroscopically visible particles with sizes in the micrometer to millimeter range form during storage within days or weeks after preparation. Storage of homogenized dispersions at refrigeration temperatures result in the formation of white, semi-solid, ointment-like gels.

Considerable particle growth was detected in most systems during storage at room temperature. Storage at 5°C resulted in the formation of ointment-like gels, which may be attributed to the crystallization of MO. Synchrotron radiation X-ray diffraction indicated that high-energy input during disintegration of the cubic phase leads to very complex systems in which particles with a cubic structure and MO/PL vesicles may coexist. The characteristic reflections of cubic systems were absent in the diffraction patterns of almost transparent or slightly turbid dispersions. The results indicate a strong dependence of ultrastructure of the dispersions on the preparation parameters (Garti *et al.*, 2012).

Dispersions of lyotropic phases that contain dispersed particles with internal nanostructure similar to the original nanostructure of the lyotropic phase, are called internally self-assembled somes (isasomes). Isasomes present advantages with respect to bulk lyotropic phases due to their tunable viscosity, by varying the amounts of the dispersed phase and hydrogelator. Isasomes (mainly cubosomes and hexosomes) are well-known for their potential applications in food, cosmetic, pharmaceutical, medical, and agricultural production. They have been studied extensively as vehicles for the delivery of functional molecules (especially drugs) to biological targets. Isasomes, are submicrometer-sized lipid particles nanostructured in a form of various liquid-crystalline or microemulsion phases, which facilitate their loading with hydrophilic, hydrophobic, and amphiphilic molecules. Isasomes can be incorporated into polymer hydrogel networks. Such entrapment of isasomes, driven by the increase in viscosity of the aqueous medium, can be controlled by changing the amount of hydrogelator in the mixture. The thermoreversible properties of hydrogels enable entrapment and release of the isasomes both at low and high temperatures. Although the isasomes are confined in the hydrogel network, their immobilization in a film would further increase the possibilities of application of these systems.

Hydrogelling agents, including κ-carrageenan and methyl cellulose, are primarily distributed in the aqueous continuous phase. They form a hydrogel network that increases the viscosity of the system, thus enabling the entrapment of isasomes. Drying of such systems leads to the formation of solid films that immobilize the isasomes. The isasomes lose their typical internal nanostructures during the drying process through loss of their internal water. The subsequent rehydration of the films gives rise to water-continuous media as they were before drying. The isasomes stay essentially intact even after the dried solid films are resolubilized and they imbibe in water and regain their original nanostructure. κ-Carrageenan and methyl cellulose hydrogels and films are good candidates for incorporation, drying, and recovering the isasomes after their rehydration. Many other polysaccharides such as modified starch, dextran, agarose, and chitosan, have also been used for similar applications. Such polysaccharides have a negligible effect on the lipid phases and they are used for encapsulating lipid phases or emulsions. Even after drying and subsequent rehydration of the loaded films, the internal nanostructure of isasomes is retained, which apparently means that they also preserve their ability to solubilize various types of molecules (hydrophilic, hydrophobic, or amphiphilic) in such hydrogel systems with hierarchically organized structure. This opens up new avenues in biosciences, as these systems can be handled in dry and solid form. Systems could therefore be potentially designed for long-term storage of isasomes, as well as for chemical and/or biosensor applications (Kulkarni *et al.*, 2011).

Another amphiphile that readily forms cubic phases is the aliphatic alcohol 3,7,11,15,-tetramethyl-1,2,3,-hexadecanetriol, or 'phytantriol' (PHYT). This amphiphile differs greatly in structure from the GMO. Phytantriol is comprised of an isoprenoid hydrophobic tail rather than an oleoyl group. The large volume of the tail is derived from the curvature of the isoprenoid group, and the inverse cubic phase appears at ambient temperatures (20–40°C) when the water content exceeds 15% (Barauskas and Landh, 2003). Also, the attachment of the triol head group to the hydrophobic tail is through a C–C bond rather than an ester linkage, which confers greater chemical stability relative to the GMO. The PHYT in excess water can form an inverse bicontinuous cubic phase at ambient temperatures, an inverse hexagonal phase and an inverse micellar phase at higher temperatures. Some glycolipids also exhibit the cubic-water miscibility gap required for cubosome formation.

Although Pluronic F127, a commercially available triblock copolymer, is often used to impart steric stabilization to these particles, the need for alternative stabilizers has emerged in the field of oral drug and nutrient delivery. The potential of using β-casein to disperse two lipids, GMO and PHYT, into cubosome dispersions was studied by Zhai *et al.* (2011). β-Casein is highly amphiphilic and has a flexible, linear structure, which has made it an excellent stabilizer in other dispersion systems such as oil-in-water emulsions. Studies using synchrotron radiation circular dichroism (SRCD) spectroscopy and neutron reflectivity have shown that β-casein adsorbs to the oil–water interface of emulsions and adopts a characteristic interfacial structure to provide steric stabilization against oil particle aggregation. In dispersed cubosomes systems, there is little understanding of the interaction between protein and the lipid–water interfaces. It was found that β-casein alone could provide sufficient stabilization to disperse the GMO in excess water, although not for the PHYT-based system.

15.4 APPLICATIONS

Lyotropic liquid-crystal systems, such as reversed bicontinuous cubic and hexagonal mesophases, are attracting more and more attention because of their unique microstructures and physicochemical properties (Guo *et al.*, 2010). Inverse cubic and hexagonal mesophases are often formed by polar lipids in an aqueous environment. The structure-forming lipids can absorb a certain amount of water and then spontaneously form gel-like phases with unique internal structures, into which functional compounds can be incorporated. Bioactive molecules such as nutrients, drugs, peptides, or proteins may be solubilized in either the aqueous or oil phase and thus be protected from hydrolysis or oxidation.

The cubic phase is attractive for controlled release because of its small pore size (5–10 nm), its ability to solubilize hydrophobic, hydrophilic, and amphiphilic molecules, and its biodegradability by enzymatic action. The tortuous structure of the cubic phase lends itself well to slowing diffusive release of solubilized actives. The cubic phase is strongly bioadhesive and is thought to be a skin-penetration enhancer, suggesting excellent compatibility with topical and mucosal deposition and delivery of active ingredients. Recent studies have emphasized similarities between the bicontinuous structures formed in human skin layers and those comprising cubic phases, offering the promise of better skin-drug transport and treatment.

Cubosome dispersions possess a nanometer-scale structure identical to the bulk cubic phase, but the dispersion itself has a much lower, water-like viscosity. Compared to liposomes or vesicles, cubosomes possess much higher bilayer area-to-particle volume ratios as well as higher viscous resistance to rupture. Although the bulk cubic phase has sufficient length scale to allow controlled release of solutes, cubosomes are too small and have too high a surface area for such performance, exhibiting instead burst release. Turning this concept around, cubosomes should be quite useful for uptake instead of release because they can rapidly absorb pollutants (for water treatment or cosmetic skin protection) and retain an amount determined by the solute partition coefficient. From a materials science perspective, the creation of ordered structures with nanoscale pore geometries is of great interest to numerous fields including electronics, photonics, catalysis, and medicine.

Loading of water-soluble actives in cubic-phase liquid-crystal dispersions (cubosomes) is difficult, given the large fraction of bulk water present. The inherent problem reflects the preference of the water-soluble actives to associate with water rather than with the liquid crystals. Ideally, the properties of the liquid crystal can be tailored to enhance the association of the liquid crystal with the active, thereby increasing loading. It is found that the inclusion of surfactants into the liquid crystal can provide this function. For example, the loading of negatively charged, water-soluble ketoprofen was enhanced by the inclusion of positively charged surfactants into the liquid crystal. Ketoprofen is an analgesic used to reduce inflammation of the joints and muscle tissue. Loading differences resulting from the inclusion of dioctadecyl dimethyl ammonium chloride (DODMAC) and dioctadecyl ammonium chloride (DOAC) into the liquid crystal demonstrate that the magnitude of the enhancement is dependent on the surfactant concentration and the steric nature of its head group. The inclusion of di(canola ethyl ester) dimethyl ammonium chloride (DEEDAC) yielded formulas with loads higher than 20 wt%, an order of magnitude higher than previous reports.

The liquid cube (Q_L) mesophase (primitive cubic symmetry) is transparent and has a low viscosity, making it a good system for solubilization of drugs and nutraceuticals and for the formation of loaded soft nanoparticles. In an experiment by Efrat *et al.* (2009), four biomolecules, carbamazepine (CBZ), coenzyme Q10 (CoQ10), and two sterols, cholesterol (Chl) and phytosterols (PSs, plant sterols), were solubilized in the discontinuous QL mesophase of monoolein. Analysis of their effects on the enthalpy changes in the presence of even small quantities of guest molecules permitted their sorting into two types and provided insight into their localization in the bilayer and their influence on phase behavior at high loads. Type I molecules (CBZ and Chl) had their hydrophilic moieties arranged in the vicinity of the lipid headgroups, while type II molecules (PS and CoQ10) appeared to be located mainly within the lipid tails, relatively far from the hydrophilic lipid moiety. The effect of guest molecules on the formation and structure of nanoparticles upon dispersion of the drug-loaded mesophase is consistent with their influence on the bulk mesophase. Type I solubilizates affect the lipid curvature, thus yielding soft nanoparticles with no or low internal order. Type II molecules do not change the packing, thus they allow the formation of cubosomes. It was considered that the nanoparticles are reconstructed from the solution by a bottom-up process through fusion events between lipid elements. The growth starts by fusion between two small vesicles to form intermediate structures. As the process proceeds through additional fusions, more complex intermediates form with larger ordered domains, up to the formation of a cubic array. The type of guest compound dictates the final particulate structure and determines the degree of internal order, the smaller the effect on Gaussian curvature and elastic modulus of lipid monolayers and on lipid–lipid interactions, the more developed the order inside nanoparticles. It was interesting to note that the capacity of the monoolein discontinuous QL mesophase for solubilizing guest molecules was highly dependent on the molecular characteristics of guest molecules and independent of their localization within the lipid bilayer. Thus, much larger loads of the flat Chl and PS molecules could be incorporated in the lipid compared with the quantities of the bulky CBZ and long CoQ10.

The three-dimensional organization of a lipid–water assembly of a bicontinuous cubic double-diamond (D-type) architecture in the crystallographic space group Pn3m reveals two cubic networks of nanochannel compartments that are separated by lipid bilayer membrane compartments. The inherent multicompartment organization provides improved encapsulation efficiency for biomolecules (related to the high surface-to-volume ratio) and favors slow release processes. Small proteins could be accommodated and protected either in the aqueous channel compartments (water-soluble proteins) or in the lipid bilayer compartments (membrane proteins). Investigations using SAXS established the mechanism of nanoconfinement of proteins in multicompartment lipid carriers, which is essential for slow release processes. Multicompartment phospholipid containers comprising a large outer vesicle membrane (0.1–1.0 μm) with a liquid reservoir encapsulating a dispersion of small vesicles (20–500 nm). Phospholipid nanoparticles entrapping therapeutic proteins in their aqueous reservoirs (liposomal nanomedicines) represent one of the most advanced classes of drug-delivery systems (Angelova *et al.*, 2011). Thermostability of the inverse bicontinuous cubic phase is governed by the curvature energy of the lipid monolayers, the stretching energy of the hydrocarbon chains, and the thermal fluctuations of the lipid bilayer. These energies are functions of the curvature, thickness, and rigidity of the lipid membrane (Angelova *et al.*, 2011).

Studies of hydration and swelling of nonlamellar lipid phases, formed by glycerol monooleate, monolinolein, or phytantriol, and in particular of their bicontinuous cubic assemblies constitute a fundamental step in the development of nanocarriers with inner multicompartment organizations. The maximum water uptake is sensitive to temperature and determines the three-dimensional surface-to-volume ratio that controls bioactives loading and release rate of protein and peptides. Pioneering studies on protein-containing MO–water systems have been conducted at low hydration levels corresponding to about 40 wt% aqueous phase and 60 wt% lipid. The protein nanoconfinement interfered with the structural impact of the macromolecules on the self-assembly organization of the lipid, which competes with the proteins for hydration water. Under these conditions, bicontinuous cubic gyroid (Ia3d) or primitive cubic (Im3m) phases have formed in some of the partially hydrated systems. The diameter, D_w, of the aqueous channels in a double diamond (Pn3m) lipid cubic lattice of a normal type (D Normal) is usually in the range of 2.5–3.5 nm. Despite the fact that the nanochannel size in the cubic lattice formed by the hydrated MO does not exceed the dimensions of medium and large proteins, like apo-ferittin (474 kDa), transferrin (75 kDa), immunoglobulin (145 kDa), and fibrinogen (420 kDa), these proteins have been entrapped in self-assembled lipid cubic phases. The SAXS studies revealed that transferrin, a medium-size protein, changes the nanochannel hydration and swelling (the D_w size) without modification of the cubic crystallographic space group. The swollen cubic-phase states, with channels larger than the protein size, are stable at body temperature. A small quantity of octyl glucoside (OG) (5 mol%) appears to be efficient for enlargement of the aqueous compartments in self-assembled MO/OG cubic nanostructures through a smooth curvature change and increased hydration of the lipid lattice ($D_w > 6$ nm).

Considering that the hydrodynamic diameter of globular proteins is approaching or exceeds the size of the aqueous nanochannel diameter in the cubosome carrier, a question is raised about the location of the entrapped proteins. The SAXS results revealed that the structural organization of proteocubosome assemblies, characterized by a swollen D-type cubic lattice, remains in the same crystallographic space group upon encapsulation of a protein. Because the protein macromolecules are not able to locate inside the aqueous channels of the normal diamond cubic lattice ($D_w \approx 3$ nm), they might be encapsulated in cubosomal pockets of the supramolecular architecture. Geometrically, the cubic bilayer membrane can fold like an infinite periodic minimal surface. However, the lipid bilayer cannot be infinite in real soft-matter nanoparticles obtained by fragmentation of self-assembled amphiphilic cubic phases. The presence of nanopockets, which are randomly distributed in the template of cubic symmetry, should facilitate the entrapment of proteins in the cubosome particle. The encapsulation of biomolecules is sensitive to the structural asymmetry. Indeed, the diamond-type cubic particle involves two networks of aqueous channels. Upon contact with the medium containing dissolved biomolecules, one of the nanochannel networks is open, while the other is closed. Such an asymmetry may have a role in the protein entrapment through the pockets in the cubic structure. Protein-loaded cubosomes have been obtained upon spontaneous lipid–protein assembly with proteins of amphiphilic character. Freeze-fracture electron microscopy (FF-EM) investigations established the hierarchical structural organizations of the proteocubosomes. The lipid–protein nanoassemblies appear to be built-up by cubosomal nanodroplets: they lack a perfect single crystal organization. The entrapped proteins might be confined in the nanopocket defects created. The FF-EM imaging confirms the concept of "nanopockets" for inclusion of biomolecules. The proteins may locate at the interfaces between the nanodroplet fragmented cubosome particles or associate with the surface of the interconnected cubosomal entities. Functionalization by 1 mol% colipid (MO-PEG2000) does not modify the cubic symmetry, established by X-ray diffraction (XRD), but it has an effect on the local nanodomain topology involving states of lower crystalline order ("sponge-like" local disorder).

Lyotropic liquid crystals (LLC) have been shown to provide sustained release of drug molecules with a range of physicochemical properties. The cubic phase was shown to host a range of water-soluble biomacromolecules, ranging from a single amino acid (tryptophan) to complex proteins and DNA, for use in controlled release applications. It was found that for a given cubic phase, the rate of diffusion depends on the molecular size of the guest molecule. The rate of transport for a given diffusant was adjusted when the aqueous channel size of the carrier was changed by using lipids with different acyl chains. Additional control over release was demonstrated when the strength of the electrostatic interaction between the diffusant and the walls of the channels were fine tuned. Differences in liquid-crystal nanostructures have previously been shown to change drug diffusion and hence release. The release was found to be diffusion controlled, with the generally larger aqueous channels of the cubic phase providing a faster release rate than the inverted hexagonal (HII) mesophase. Specifically, it was shown that the release of hydrophilic drugs from the phytantriol-based Q2 structure is significantly faster than from the HII mesophase (prepared by addition of 10% vitamin E acetate to phytantriol). Based on this idea, phytantriol and monoolein-based cubic (Q2) and (HII) mesophases were developed to allow change to the nanostructure in response to external change in temperature, to control drug release rates *in vivo* (Garti *et al.*, 2012).

Encapsulation of DNA in lipid medium has been intensively investigated toward preparation of nonviral carriers of genetic drugs. Studies using XRD have revealed that self-assembled complexes of DNA with charged lipids could form inverted hexagonal or cubic liquid crystal (LC) phases. Owing to the rigidity and the large size of the DNA strains, it is impossible to accommodate DNA macromolecules (molecular weight (MW) $\approx$ 6000 KDa) in the nanochannels of a lipid cubic structure. As a result, the encapsulated DNA adopts a confined state. The transport properties of the cubic phase, being dependent on the nanochannel diameter, do not permit diffusion of entrapped large DNA molecules along the cubic network architecture.

Lyotropic mesophases can serve as delivery vehicles for many possible food additives including flavor compounds. The LLC were used as microreactors for the synthesis of flavor compounds. The release of some aroma compounds from cubic and micellar mesophases demonstrated the advantages of these mesophases over emulsions and microemulsions. Phenylacetaldehyde, melonal (2,6-dimethyl-5-heptenal), linalool, and trans-4-decenal (t-4-d) are four hydrophobic aroma molecules with major oxidation and degradation sensitivity once exposed to air. Flavor and aroma compounds are usually stored at low temperatures, in the dark, and under nitrogen. These four molecules (MW < 200 g/mol/L) have moderate volatility, are soluble in vegetable oils, and belong to the flowery-fruity aroma group. Medium chain lipophilic moieties were successfully packed within the tails of the GMO molecules and modified the critical packing parameter (CPP) of the GMO. Up to 10 wt% of phenylacetaldehyde was solubilized at room temperature in the GMO–water system but only lamellar and cubic structures were formed, similar to what was obtained in the binary mixture of GMO–water alone. Three different structures were formed depending on the water content and the GMO–water ratio. The dominant structure is the lamellar, next (with more water added) is a mixture of lamellar and cubic mesophases, and in the presence of additional water a third region of cubic mesophase was formed. Increasing temperature induces thermal motions of GMO tails, which caused a decrease in the lattice parameter, but at certain temperatures the HII mesophase disintegrated, transforming to an isotropic fluid phase (reverse micellar) that will cause an aroma release. The aroma that has the strongest curvature effect will resist the temperature increase more, and as a result will be more protected and released at higher temperatures. Thus, lipophilic aromas can be solubilized in lyotropic mesophases, transitioning lamellar and cubic mesophases to reverse hexagonal mesophase. This may offer new options of using sensitive aromas both in long-term storage and in final use of food products.

Protein and peptide self-assembly is emerging as a new route to produce novel materials and to modify the properties of existing materials (ceramics, metals, alloys, synthetic polymers, and composite materials) (Zhang *et al.*, 2002). Peptides and proteins are now used as building blocks to produce a wide range of biomaterials for diverse applications (Holmes *et al.*, 2000; Petka *et al.*, 1998; Aggeli *et al.*, 2001; Hartgerink *et al.*, 2001) including scaffolds for tissue repair, drug delivery, and biological surface engineering. Proteins may assemble into a vast array of complex mesophases, which include bicontinuous structures with multiple symmetries and quasi-crystalline packings of block copolymers. This rich behavior offers the opportunity for designing materials with complex target structures at length scales long enough to be used as biomaterials. However, the fundamentals of this process, including interactions at the molecular level and the thermodynamics of long-range assemblies, are still not well understood. At sufficient densities, these interactions result in stable, finite size clusters with lamellar, cylindrical, or spherical geometries.

It is generally known that small polymers, which contain within them a hydrophobic region and a separate hydrophilic region, will self-assemble in aqueous solution to form distinct structures such as micelles, vesicles and tubules. This is largely due to a hydrophobic effect, which drives nonpolar regions of each polymer molecule away from water and towards one another. Protein amphiphiles with the familiar alternating polar–nonpolar regions, with each polypeptide consisting of ~200 amino acids arranged as a di-block copolymer, have been reported to self-assemble into strong and fast-recovering hydrogels that respond to pH changes (Petka *et al.*, 1998) or are heat resistant to 90°C (Nowak *et al.*, 2002). Also, upon addition of monovalent alkaline cations or introduction of the self-assembling peptide solutions into physiological media, these peptides spontaneously assemble to form macroscopic structures that can be fabricated into various geometric shapes (Holmes *et al.*, 2000; Kisiday *et al.*, 2002). Scanning electron microscopy (SEM) and atomic force microscopy reveal that the matrices are made of interwoven nanofibers with a diameter of 10–20 nm and pores about 50–200 nm (Holmes *et al.*, 2000; Marini *et al.*, 2002). Wang *et al.* (2008) reported SEM images of the formation of water-channel networks in zein systems (Figure 15.5).

It is important in food and pharmaceutical research to provide a family of potentially safe and highly effective nanoparticulate dispersions as novel nanocarriers for delivering bioactive materials (nutrients, flavors, peptides, proteins, and drugs). In this respect, it is important to emphasize that the effective utilization of self-assembled nanostructured aqueous dispersions requires a detailed understanding of their physicochemical properties and their stability under various conditions (solutes, salts, temperature, and pH changes). Future achievements in controlling the nanostructures confined

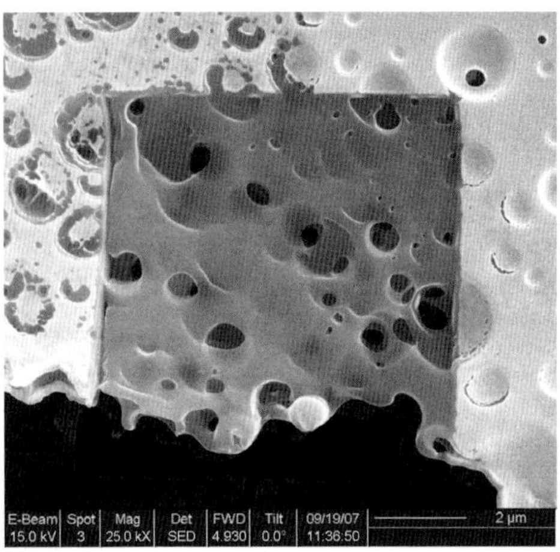

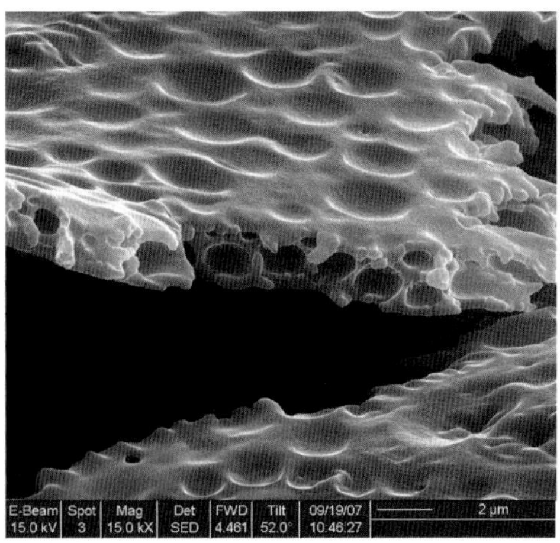

Figure 15.5 Zein bicontinuous structures.

in the dispersed particles could also open the door for further applications within material science and medicine. The field of nanostructured aqueous dispersions remains an area where further advances are still very much needed, especially the effective loading and the controlled delivery of bioactive materials, particularly proteins, and poorly water-soluble drugs and foods (Yaghmur and Glatter, 2009).

References

Aggeli, A., Nyrkova, I.A., Bell, M., *et al.* (2001) Hierarchical self-assembly of chiral rod-like molecules as a model for peptide beta-sheet tapes, ribbons, fibrils, and fibers. *Proceedings of the National Academy of Sciences of the United States of America*, 98, 11857–11862.

Anderson, D.M., Davis, H.T., Scriven, L.E. and Nitsche, J.C.C. (1990) Periodic surfaces of prescribed mean curvature. *Advances in Chemical Physics*, 77, 337–396.

Angelov, B., Angelova, A., Papahadjopoulos-Sternberg, B., *et al.* (2006) Detailed structure of diamond-type lipid cubic nanoparticles. *Journal of the American Chemical Society*, 128, 5813–5817.

Angelova, A., Angelov, B., Mutafchieva, R., Lesieur, S. and Couvreur, P. (2011) Self-assembled multicompartment liquid crystalline lipid carriers for protein, peptide, and nucleic acid drug delivery. *Accounts of Chemical Research*, 44, 147–156.

Barauskas, J. and Landh, T. (2003) Phase behavior of the phytantriol/water system. *Langmuir*, 19, 9562–9565.

Benedicto, A. and OBrien, D. (1997) Bicontinuous cubic morphologies in block copolymers and amphiphile/water systems: Mathematical description through the minimal surfaces. *Macromolecules*, 30, 3395–3402.

Chung, H. and Caffrey, M. (1995) Polymorphism, mesomorphism, and metastability of monoelaidin in excess water. *Biophysical Journal*, 69, 1951–1963.

Efrat, R., Kesselman, E., Aserin, A., Garti, N. and Danino, D. (2009) Solubilization of hydrophobic guest molecules in the monoolein discontinuous QL cubic mesophase and its soft nanoparticles. *Langmuir*, 25, 1316–1326.

Ericsson, B., Larsson, K. and Fontell, K. (1983) A cubic protein–monoolein–water phase. *Biochimica et Biophysica Acta*, 729, 23–27.

Fraser, S., Separovic, F. and Polyzos, A. (2009) Cubic phases of ternary amphiphile–water systems. *European Biophysics Journal with Biophysics Letters*, 39, 83–90.

Freitas, C. and Muller, R. (1998) Spray-drying of solid lipid nanoparticles (SLNTM). *European Journal of Pharmaceutics and Biopharmaceutics*, 46, 145–151.

Garti, N., Libster, D. and Aserin, A. (2012) Lipid polymorphism in lyotropic liquid crystals for triggered release of bioactives. *Food and Function*, 3, 700–713.

Guo, C., Wang, J., Cao, F., Lee, R.J. and Zhai, G. (2010) Lyotropic liquid crystal systems in drug delivery. *Drug Discovery Today*, 15, 1032–1040.

Gustafsson, J., Ljusberg-Wahren, H., Almgren, M. and Larsson, K. (1996) Cubic lipid-water phase dispersed into submicron particles. *Langmuir* 12, 4611–4613.

Gustafsson, J., Ljusberg-Wahren, H., Almgren, M. and Larsson, K. (1997) Submicron particles of reversed lipid phases in water stabilized by a nonionic amphiphilic polymer. *Langmuir*, 13, 6964–6971.

Hartgerink, J.D., Beniash, E. and Stupp, S.I. (2001) Self-assembly and mineralization of peptide-amphiphile nanofibers. *Science*, 294, 1684–1688.

Holmes, T.C., de Lacalle, S., Su, X., Liu, G.S., Rich, A. and Zhang, S.G. (2000) Extensive neurite outgrowth and active synapse formation on self-assembling peptide scaffolds. *Proceedings of the National Academy of Sciences of the United States of America*, 97, 6728–6733.

Hyde, S.T., Andersson, S., Ericsson, B. and Larsson, K. (1984) A cubic structure consisting of a lipid bilayer forming an infinite periodic minimum surface of the gyroid type in the glycerolmonooleate-water system. *Zeitschrift fur Kristallographie*, 168, 213–219.

Hyde, S., Andersson, S., Larsson, K., *et al.* (1997) *The Language of Shape: The Role of Curvature in Condensed Matter: Physics, Chemistry and Biology*. Elsevier, Amsterdam.

Kim, J.S., Kim, H.K., Chung, H., Sohn, Y.Y., Kwon, I.C. and Jeong, S.Y. (2000) Drug formulations that form a dispersed cubic phase when mixed with water. *Proceedings of the International Symposium on Controlled Release of Bioactive Materials*, 27, 1118–1119.

Kisiday, J., Jin, M., Kurz, B., *et al.* (2002) Self-assembling peptide hydrogel fosters chondrocyte extracellular matrix production and cell division: Implications for cartilage tissue repair. *Proceedings of the National Academy of Sciences of the United States of America*, 99, 9996–10001.

Kulkarni, C.V., Tomsic, M. and Glatter, O. (2011) Immobilization of nanostructured lipid particles in polysaccharide films. *Langmuir*, 27, 9541–9550.

Landh, T. (1994) Phase-behavior in the system pine oil monoglycerides-poloxamer 407-water at 20°C. *Journal of Physical Chemistry*, 98, 8453–8467.

Larsson, K. and Tiberg, F. (2005) Periodic minimal surface structures in bicontinuous lipid-water phases and nanoparticles. *Current Opinion in Colloid and Interface Science*, 9, 365–369.

Lindblom, G. and Rilfors, L. (1989) Cubic phases and isotropic structures formed by membrane lipids – possible biological relevance. *Biochimica et Biophysica Acta*, 988, 221–256.

Lindstrom, M., Ljusbergwahren, H., Larsson, K. and Borgstrom, B. (1981) Aqueous lipid phases of relevance to intestinal fat digestion and absorption. *Lipids*, 16, 749–754.

Lovett, D.R. and Tilley, J. (1994) *Demonstrating Science With Soap Films*. Institute of Physics Publishers, Briston, Philadelphia.

Luzzati, V., Tardieu, A., Gulik-Krzywicki, T., Rivas, E. and Reiss-Husson, F. (1968) Structure of the cubic phases of lipid–water systems. *Nature*, 220, 485–488.

Marini, D., Hwang, W., Lauffenburger, D., Zhang, S. and Kamm, R. (2002) Left-handed helical ribbon intermediates in the self-assembly of a beta-sheet peptide. *Nano Letters*, 2, 295–299.

Nowak, A., Breedveld, V., Pakstis, L., *et al.* (2002) Rapidly recovering hydrogel scaffolds from self-assembling diblock copolypeptide amphiphiles. *Nature*, 417, 424–428.

Patton, J.S. and Carey, M.C. (1979) Watching fat digestion. *Science*, 204, 145–148.

Petka, W.A., Harden, J.L., McGrath, K.P., Wirtz, D. and Tirrell, D.A. (1998) Reversible hydrogels from self-assembling artificial proteins. *Science*, 281, 389–392.

Qiu, H. and Caffrey, M. (2000) The phase diagram of the monoolein/water system: metastability and equilibrium aspects. *Biomaterials*, 21, 223–234.

Sagalowicz, L., Mezzenga, R. and Leser, M.E. (2006) Investigating reversed liquid crystalline mesophases. *Current Opinion in Colloid and Interface Science*, 11, 224–229.

Salentinig, S., Yaghmur, A., Guillot, S. and Glatter, O. (2008) Preparation of highly concentrated nanostructured dispersions of controlled size. *Journal of Colloid and Interface Science*, 326, 211–220.

Schwarz, U. and Gompper, G. (2002) Bicontinuous surfaces in self-assembling amphiphilic systems. *Morphology of Condensed Matter*, 107–151.

Scriven, L. (1976) Equilibrium bicontinuous structure. *Nature*, 263, 123–125.

Seddon, J.M., Robins, J., Gulik-Krzywicki, T. and Delacroix, H. (2000) Inverse micellar phases of phospholipids and glycolipids. *Physical Chemistry Chemical Physics*, 2, 4485–4493.

Shin, H., Grason, G.M. and Santangelo, C.D. (2009) Mesophases of soft-sphere aggregates. *Soft Matter*, 5, 3629–3638.

Spicer, P., Hayden, K., Lynch, M., Ofori-Boateng, A. and Burns, J. (2001) Novel process for producing cubic liquid crystalline nanoparticles (cubosomes). *Langmuir*, 17, 5748–5756.

Spicer, P., Small, W., Lynch, M. and Burns, J. (2002) Dry powder precursors of cubic liquid crystalline nanoparticles (cubosomes). *Journal of Nanoparticle Research*, 4, 297–311.

Squires, A.M., Templer, R.H., Seddon, J.M., *et al.* (2002) Kinetics and mechanism of the lamellar to gyroid inverse bicontinuous cubic phase transition. *Langmuir*, 18, 7384–7392.

Squires, A.M., Templer, R.H., Seddon, J.M., *et al.* (2005) Kinetics and mechanism of the interconversion of inverse bicontinuous cubic mesophases. *Physical Review E*, 72, 011502.

Templer, R.H. (1998) Thermodynamic and theoretical aspects of cubic mesophases in nature and biological amphiphiles. *Current Opinion in Colloid and Interface Science*, 3, 255–263.

Von Schnering, H.G. and Nesper, R. (1991) Nodal surfaces of Fourier series: fundamental invariants of structured matter. *Zeitschrift Fur Physik B-Condensed Matter*, 83, 407–412.

Wang, Q., Yin, L. and Padua, G.W. (2008) Effect of hydrophilic and lipophilic compounds on zein microstructures. *Food Biophysics*, 3, 174–181.

Yaghmur, A. and Glatter, O. (2009) Characterization and potential applications of nanostructured aqueous dispersions. *Advances in Colloid and Interface Science*, 147–48, 333–342.

Yang, D., Armitage, B. and Marder, S. (2004) Cubic liquid-crystalline nanoparticles. *Angewandte Chemie-International Edition*, 43, 4402–4409.

Zhai, J., Waddington, L., Wooster, T.J., Aguilar, M. and Boyd, B.J. (2011) Revisiting beta-casein as a stabilizer for lipid liquid crystalline nanostructured particles. *Langmuir*, 27, 14757–14766.

Zhang, S., Marini, D., Hwang, W. and Santoso, S. (2002) Design of nanostructured biological materials through self-assembly of peptides and proteins. *Current Opinion in Chemical Biology*, 6, 865–871.

16 Self-assembly of amylose, protein, and lipid as a nanoparticle carrier of hydrophobic small molecules

Genyi Zhang, Deepak Bhopatkar, Bruce R. Hamaker, and Osvaldo H. Campanella

Whistler Center for Carbohydrate Research, Department of Food Science, Purdue University, West Lafayette, Indiana, USA

16.1 INTRODUCTION

Starch, proteins, and lipids are biomaterials present in nature in abundant quantities. Chemically, starch is a homopolymer with only one building block (glucose) linked through α-1,4 and α-1,6 glycosidic bonds to form linear and branched molecules. Proteins are heterogeneous polymers formed by combinations of different amino acids. Fatty acids are the major building blocks for lipids, are nonpolymer molecules smaller than starch and proteins, and structurally consist of a hydrophobic aliphatic tail and a hydrophilic carboxylate head group.

Amylose, one of the two starch molecules, is essentially linear and naturally exists in conformational forms as a random coil, single helix, or double helix, which in the latter case can align to form crystalline structures. Interactions leading to aggregation depend on amylose concentration and the solvent conditions. Proteins exhibit secondary structures such as α-helices and β-sheets, and tertiary or quaternary structures. Lipid molecules also may exhibit different structural forms such as a spherical liposomes, micelles, and lipid bilayers that are formed through self-assembling mechanisms. All three of these common food macronutrients have self-assembly characteristics with like molecules that can generate nanoparticles and have a good potential in the area of nanotechnology because they are generally safe, abundant, and inexpensive, and they have proved to form different nanoscale structures with novel properties and functions.

Starch has a semi-crystalline structure, and starch nanocrystals produced by acid hydrolysis in combination with ultrasonication (Bel Haaj *et al.*, 2013a; Kim *et al.*, 2012, 2013a, b) and other methods have been used as fillers to improve the mechanical and barrier properties of biocomposites (Le Corre *et al.*, 2010) and to stabilize emulsions (Bel Haaj *et al.*, 2013b). Proteins have a variety of functional groups, and nanoparticles prepared from them exhibit unique properties as binding ligands of valuable molecules that allow them to act as drug delivery vehicles (Wang and Uludag, 2008). For example, nanoparticles prepared using the human protein albumin have been clinically used as a drug delivery system (Hawkins *et al.*, 2008) with good targeted delivery properties. Milk protein has been used to prepare nanoparticles that are used to deliver bioactive compounds through hydrophobic micelles,

Nanotechnology and Functional Foods: Effective Delivery of Bioactive Ingredients, First Edition.
Edited by Cristina M. Sabliov, Hongda Chen, and Rickey Y. Yada.

nanotubes, and core–shell nanoparticles (Livney, 2010). Nanoscale delivery systems based on lipids such as liposomes, emulsions, and solid lipid nanoparticles are being used (Müller *et al.*, 2000).

Since starch, proteins, and lipids are inexpensive and ubiquitous in foods, their self-assembly potential to form nanoscale structures may allow for the development of novel foods with enhanced nutritional properties notably with improved bioavailability of certain bioactive components. Nanoparticles prepared from amylose–lipid inclusion complexes are natural carriers for lipophilic compounds (Feng *et al.*, 2011); proteins and lipids can also form complexes through hydrophobic interactions in which the lipophilic/hydrophobic regions of the proteins are the binding site of the lipids. Lipoproteins that can solubilize lipids are a good example of self-assembled nanocarriers for delivery of lipids. Concerning interaction between starch and proteins, although thermodynamically incompatible, they can form conjugates, which are produced under appropriate conditions so that the physiochemical properties of the starch and protein components can be enhanced.

16.2 DISCOVERY OF THE TERNARY COMPLEX OF STARCH, PROTEIN, AND FREE FATTY ACIDS

The complex formed by interaction among starch, proteins, and fatty acids through self-assembly was discovered in 1997 by our group while studying the pasting properties of sorghum flour. The pasting characteristics of cereal flours mainly depend on the physicochemical properties of the starch present in the flour. Pasting properties of flours are commonly determined by using the rapid visco analyzer (RVA). In the RVA test, a dispersion of the flour with a given solid content is heated while it is continuously agitated with a rotating impeller. Temperature profiles include heating, holding, and cooling steps using various heating and cooling rates and temperature holding periods. During the test, viscosity changes of the mixed and heated/cooled suspension are monitored. Changes in the suspension viscosity upon the heating/holding/cooling steps in the presence of mixing provide a profile that is characteristic of the flour system tested. For most flours tested in the RVA instrument, a typical profile shows an increase of viscosity as the temperature increases, followed by a decrease of viscosity as the paste structure composed mostly of swollen starch granules begins to break down. Upon cooling most flours exhibit a second more noted increase of viscosity that is caused by amylose retrogradation (a recrystallization phenomenon) (Figure 16.1, curve 1). During RVA testing of sorghum flours, a second high-viscosity peak in the cooling stage was observed. A later investigation of Zhang and Hamaker (2005) showed that free fatty acids (FFA) liberated by lipase action during flour storage were associated with the high cooling stage viscosity (Figure 16.1, curves 2 and 3 and associated chromatography).

A pure system containing starch, whey protein, and FFA in a weight ratio of 20:1:2 showed a three-component interaction that appeared to explain the presence of the higher cooling stage viscosity peak (Zhang and Hamaker, 2003). The thin layer chromatograph shown in Figure 16.1 illustrates that the incremental increase in the free fatty acid content over 3 months of storage correlated with the formation of the three-component complex that caused the increase of viscosity in the cooked flour.

Using high-pressure size-exclusion chromatography (HPSEC), a distinct peak was found, eluting between the amylopectin and amylose peaks (Figure 16.2) with a molecular weight of ~10^6–10^7 Da that corresponded to the three-component complex (Zhang *et al.*, 2003; Shah *et al.*, 2011). Figure 16.2 also shows that the two-component system combinations of amylose–FFA and protein–FFA did not show the peak (Shah *et al.*, 2011).

16.3 MECHANISM OF FORMATION OF THE NANOCOMPLEX

Amylose, whey protein, and FFA are the three components that self-assemble to form the nanoscale complex. The previously demonstrated self-assembled inclusion complex between amylose and FFA through hydrophobic interaction is one structural component of the complex (Zhang and Hamaker,

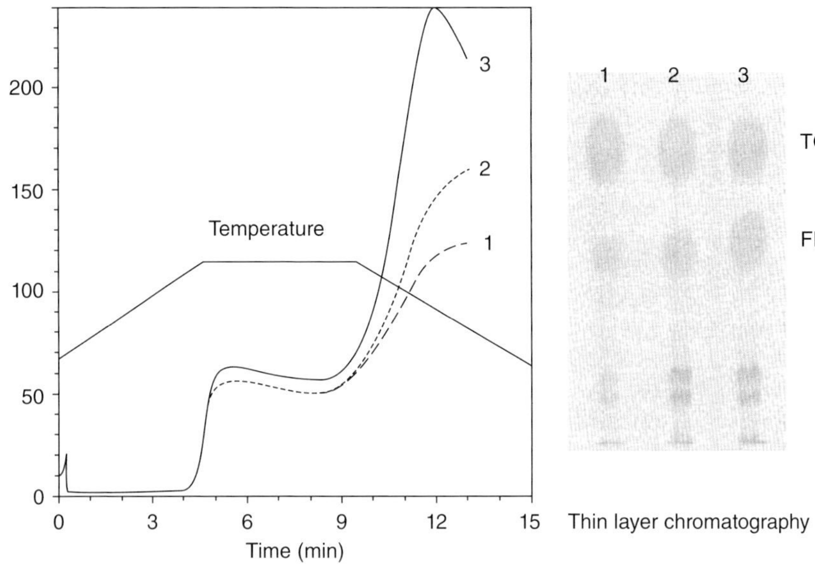

Figure 16.1 The RVA profiles and the liberated fatty acids of whole grain sorghum flour stored for different time periods: 1, 2, and 3 months. (Zhang and Hamaker, 2005. Reproduced with permission from the AACC.)

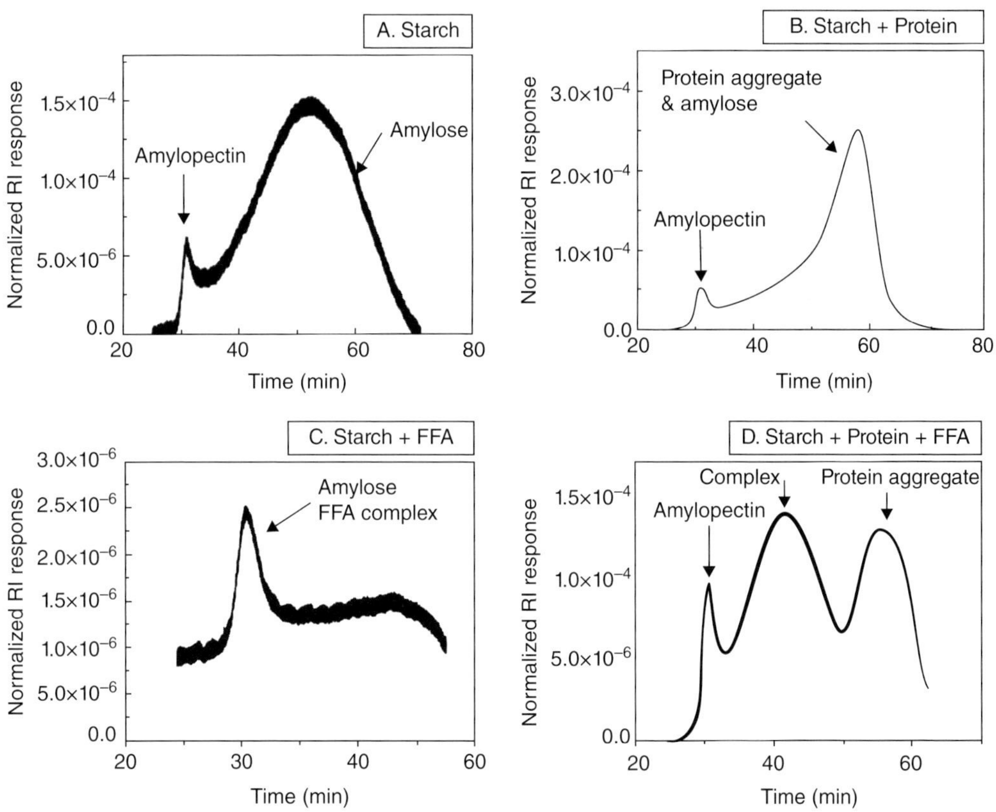

Figure 16.2 The three-component complex formed among amylose, whey protein, and linoleic acid. (Shah *et al.*, 2011. Reproduced with permission from Elsevier.)

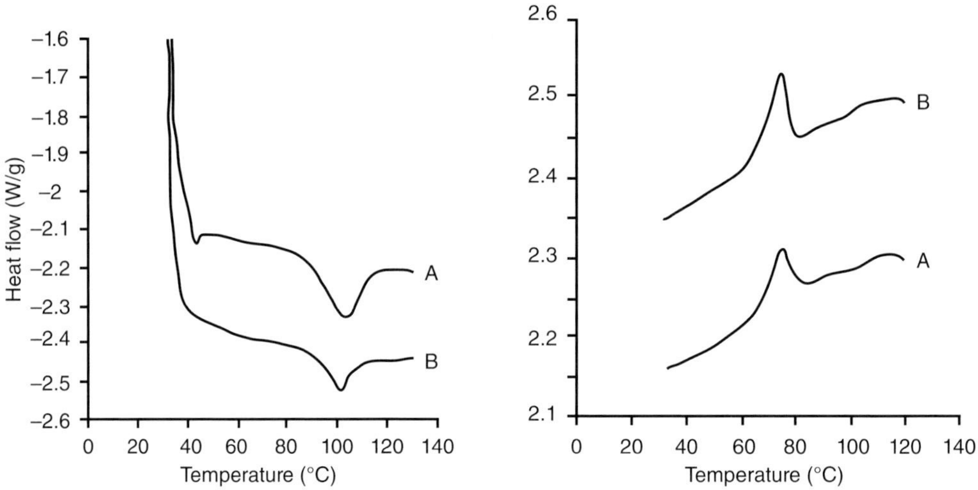

Figure 16.3 DSC thermograms of the melting (100°C)–left and formation (75°C)–right of an amylose–free fatty acid complex (A) and a three component complex (B), indicating the reversibility of formation. (Zhang and Hamaker, 2004. Reproduced with permission from Elsevier.)

2004), which can be observed by differential scanning calorimetry (DSC; Figure 16.3) and X-ray powder diffraction showing a crystalline V-type pattern. The presence of whey protein, when the other two components are present, has a significant influence on the self-assemblage of amylose–FFA complex by decreasing the degree of its complexation (as measured by DSC) but increasing the crystalline order of the V-type starch–FFA complex, which were observed as clear peaks in the complex X-ray diffraction pattern.

Rheological studies using a pasting cell attached to a rheometer also confirmed the formation of the three-component complex. Protein content, shear rate, and temperature all had a significant effect on complex formation and its resulting properties (Shah *et al.*, 2011). Like common FFA, the bioactive conjugated linoleic acid (CLA) also participated in the formation of the three-component complex, indicating that the self-assembled three-component complex can be used as a delivery system for functional lipids such as CLA.

During the self-assembly process of the three-component complex the presence of salt was found to have a significant effect on the complex properties. It was demonstrated that the addition of different types of salt with ions included in the Hofmeister series can alter the shape and size of the nanoparticles formed from the interaction. This, and other data, indicated that electrostatic interaction is involved in the complex formation, and that the carboxylate head of the fatty acid is the electrostatic bridge between the amylose–FFA complex and the basic amino acids of the protein (Zhang *et al.*, 2010).

A study of the complex formation showed that the molecular features of amylose and the fatty acid have a significant impact on properties of the nanocomplex. The influence of the FFA structure on the complexation and thus the nanocomplex formed is mainly through the inclusion complex in which the conformation of the FFA (saturated or unsaturated, number of unsaturated double bonds) has an important role on the physicochemical properties of the amylose–FFA complex. Consequently, different FFAs also affected the size of the three-component nanocomplex (Zhang *et al.*, 2010; Shah *et al.*, 2011). The impact of the amylose structure on the formation of the nanocomplex is also of importance, as shown in Table 16.1. Using amylose from normal corn starch and high amylose corn starch (50% amylose) with different molar mass and polydispersity index, the resulting nanoparticles had different, though incongruent, differences in molar mass, polydispersity index, and radius of gyration (R_g). It appears that the polydispersity index of amylose is related to the hydrodynamic size of the

Table 16.1 Average molar mass, polydispersity index, and radius of gyration of different types corn starch and nanocomplexes formed using the different amyloses. (Source: Bhopatkar, 2012. PhD Thesis.)

Type	Average Molar Mass (g/mol)	Polydispersity index	Radius of gyration (nm)
Normal corn starch (NCS)	1.09E+07	3.61 ± 0.48	NA
High amylose corn starch (HAC)	5.71E+06	1.44 ± 0.27	NA
NCS–nanoparticle	2.73E+07	2.27 ± 0.38	70.4 ± 8.1
HAC–nanoparticle	3.47E+07	1.36 ± 0.14	48.3 ± 2.7

Table 16.2 Thermodynamic parameters during the formation of the nanoparticles. (Source: Bhopatkar, 2012. PhD Thesis.)

Samples	T_F (K)	Cp (J/g/K)	ΔS (J/K)	ΔG (J/g)
Pure protein	330.1	3.52	−52.959	0.161
Protein–fatty acid	335.6	2.47	0.018	−5.946
Amylose–fatty acid	336.8	2.70	0.014	−1.376
Nanoparticle 2.5 mg protein	341.6	3.30	0.036	−13.723
Nanoparticle 5 mg protein	346.9	3.84	0.168	−62.049
Nanoparticle 10 mg protein	348.9	3.64	0.198	−72.486
Nanoparticle 15 mg protein	349.9	3.96	0.157	−61.288

nanoparticles, but the similar molar mass and decreased polydispersity index of the nanocomplex is likely to be determined by the presence of the whey protein, which plays a structural role in the final nanocomplex formation.

The self-assembly of the nanoparticles was also studied from a thermodynamic viewpoint by using a calorimetric method (Bhopatkar, 2012). Results showed that the formation of the nanoparticles by the self-assembling mechanism described above was accompanied by a negative change in Gibbs free energy (ΔG). As the formation of the nanoparticle requires a cooling stage, the thermodynamic change associated with the assembly was calculated from a heating–cooling process on formulated three-component mixtures in a DSC pan, where all the associated thermal events were recorded. Results obtained for the different formulations are reported in Table 16.2 (Bhopatkar, 2012). The negative values of ΔG determined indicate that the formation of nanoparticle is a favorable process; although negative ΔG values were obtained for the amylose–FFA and protein–FFA interactions indicating that both these interactions are thermodynamically driven processes, the formation of the three-component complex appears to be an even more favorable event given the lower calculated free-energy change. Results show that whey protein is crucial to the formation of the three-component nanoparticle over the formation of binary complexes. Table 16.2 also shows that there appears to be an optimum amount of protein to form a more stable nanoparticle.

16.4 PROPERTY AND STRUCTURE OF THE TERNARY FOOD-BASED NANOPARTICLE

The freeze-dried self-assembled three-component nanoparticle is a white, hygroscopic, crystalline powder that has a discrete regular structure (Figure 16.4), which is consistent with the low polydispersity as viewed from the HPSEC chromatograph.

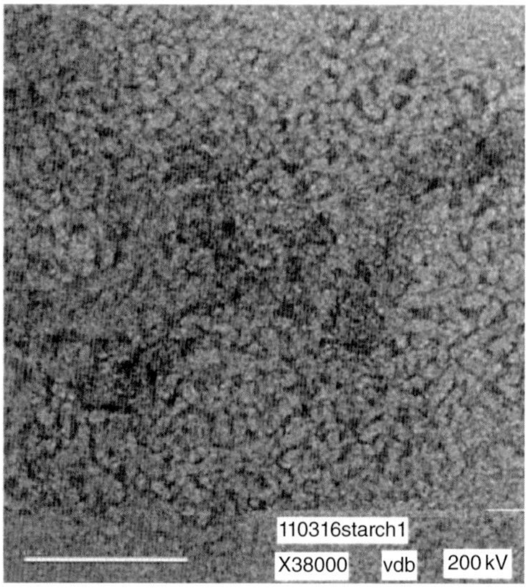

Figure 16.4 Cryo-transmission electron microscopy image of the purified three-component nanoparticles indicating the discrete, regular, and rod-shaped structure. The scale bar is 100 nm. (Bhopatkar, 2012. PhD Thesis.)

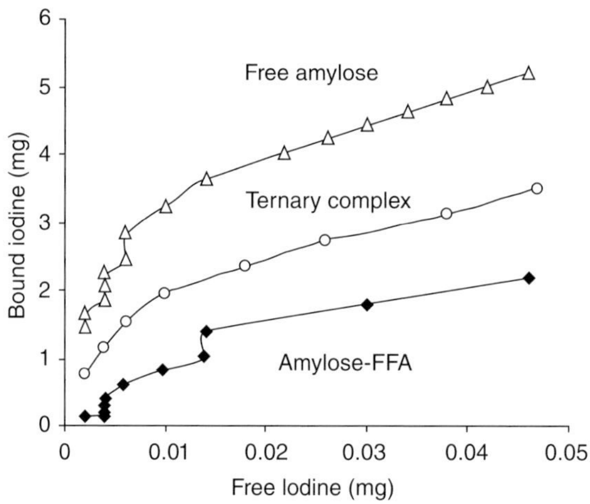

Figure 16.5 The iodine binding property of the nanoparticles. (Liu *et al.*, 2007. Reproduced with permission from Elsevier.)

The iodine binding property, which is used to measure empty space in the amylose lumen, of a dispersion of nanoparticles was studied and it was found that more iodine was bound to the nanoparticles in comparison to the amylose–FFA complex (Figure 16.5) (Liu *et al.*, 2009). In addition, the λ_{max} of the sample also shifted to a lower wavelength (Zhang *et al.*, 2010). These results indicate the amylose–FFA interaction does not form a typical inclusion complex that occupies virtually the whole of the hydrophobic lumen of amylose, but instead forms partially occupied amylose helices that are present in the complex, thereby providing some space for the inclusion other hydrophobic components. This unique structural property of the nanoparticle indicates that it can be used as a nano-carrier for water-insoluble small molecules, such as are found in some nutraceuticals and drugs.

Thermodynamically, the nanoparticle formed by the interaction of the three components is a stable system. However, it is dissociated at around 100°C, which represents the melting the crystalline structure of the amylose–FFA complex in the nanoparticle. Once the temperature is decreased the nanoparticle reforms, so it appears that this self-assembling–dissociation process is reversible. This reversibility is also a unique property of the nanoparticle, meaning that any modification for specific purposes can be carried out without ultimate loss of its structure at room, or somewhat elevated, temperature, for example the thermal-reversibility can facilitate the loading of drugs to the nanoparticle. Another important property of the nanoparticle is its slow enzyme digestion of amylose compared to pure amylose (Zhang *et al.*, 2010). This slow digestion likely means, for instance, that a drug would be slowly delivered.

16.5 CASE STUDY: USE OF NANOPARTICLES AS A NANOCARRIER OF 5-FLUOROURACIL

The available space in the hydrophobic lumen of amylose of the three-component nanoparticle might be utilized to incorporate, disperse, and protect invaluable hydrophobic small molecules. In a case study, 5-fluorouracil (5-FU), a drug used for cancer treatments was used to test the carrying property of the nanoparticles. Potentiometric titration demonstrated that the 5-FU is effectively incorporated in the hydrophobic interior of the amylose lumen of the nanoparticle, and its incorporation induces a conformational change of the nanoparticle (Figure 16.6). A Caco-2 BBe MTT cytotoxicity study showed that 5-FU release can be attained (Figure 16.7), demonstrating that this dispersable soft nanoparticle made from simple food components has the capacity to carry sparingly soluble small molecules through their incorporation inside the hydrophobic lumen of the amylose helix, with ultimate applications ranging from flavor carrying or flavor masking to drug delivery.

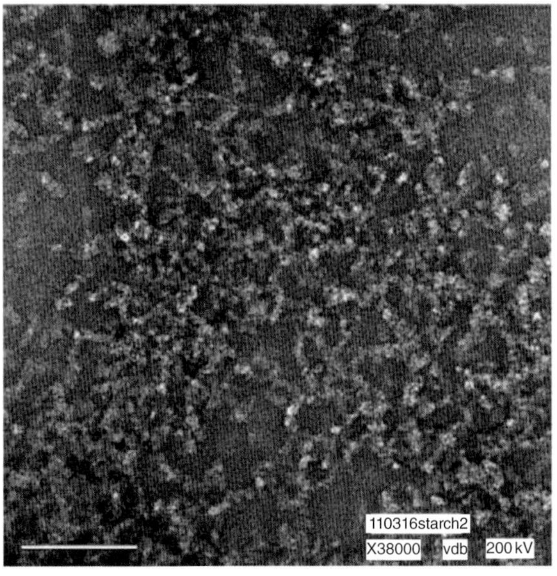

Figure 16.6 Cryo-transmission electron microscopy image of nanoparticles loaded with 5-FU, indicating the nanostructure with minor changes compared to an unloaded nanoparticle. (Bhopatkar, 2012. PhD Thesis.)

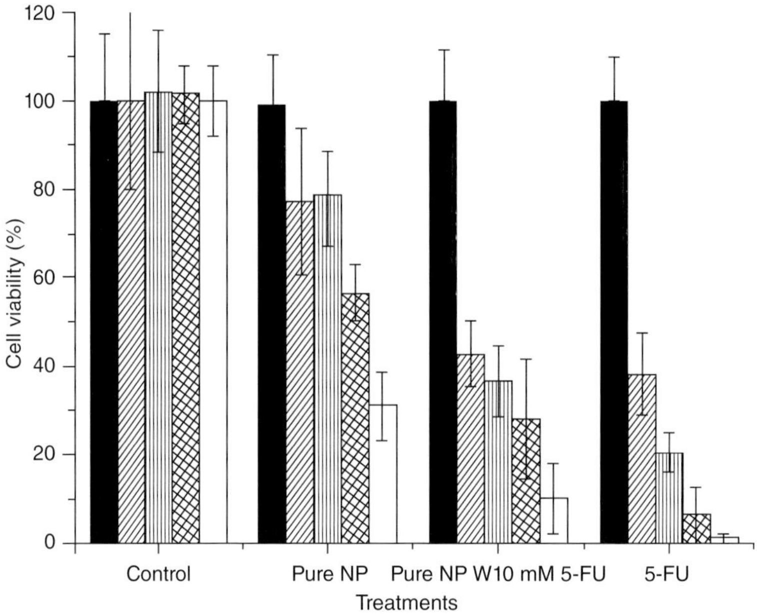

Figure 16.7 Cytotoxicity test of the nanoparticle to Caco2 cells. NP: nanoparticle without 5-FU. (Bhopatkar, 2012. PhD Thesis.)

16.6 PERSPECTIVE

A food-constituent-composed nanoparticle has been identified and its physiochemical property and carrying capability are discussed. It might have the potential for widespread applications, but one important aspect that needs to be taken into consideration is its scale-up production, as well as a better understanding of other properties it may have that can be utilized, especially if the nanoparticle is used as a carrier of valuable food and non-food hydrophobic molecules.

Acknowledgment

We would like to thank the USDA AFRI program (grant #07-35603-17741) for partial support of this work.

References

Bel Haaj, S., Magnin, A., Pétrier, C. and Boufi, S. (2013a) Starch nanoparticles formation via high power ultrasonication. *Carbohydrate Polymers*, 92(2), 1625–1632.

Bel Haaj, S., Ben Mabrouk, A., Thielemans, W. and Boufi, S. (2013b) A one-step miniemulsion polymerization route towards the synthesis of nanocrystal reinforced acrylic nanocomposites. *Soft Matter*, 9(6), 1975–1984.

Bhopatkar, D. (2012) *Food biopolymer-based soft nanoparticles for solubilization of sparingly soluble small molecules*. Purdue University, West Lafayette.

Feng, T., Zhuang, H.N., Xiao, Z.B. and Tian, H.X. (2011) Recent patents on amylose-flavor inclusion complex nano particles preparation and their application. *Recent Patents on Food, Nutrition and Agriculture*, 3(3), 179–186.

Hawkins, M.J., Soon-Shiong, P. and Desai, N. (2008) Protein nanoparticles as drug carriers in clinical medicine. *Advanced Drug Delivery Reviews*, 60(8), 876–885.

Kim, H.-Y., Lee, J.H., Kim, J.-Y., Lim, W.-J. and Lim, S.-T. (2012) Characterization of nanoparticles prepared by acid hydrolysis of various starches. *Starch – Stärke*, 64(5), 367–373.

Kim, H.-Y., Han, J.-A., Kweon, D.-K., Park, J.-D. and Lim, S.-T. (2013a) Effect of ultrasonic treatments on nanoparticle preparation of acid-hydrolyzed waxy maize starch. *Carbohydrate Polymers*, 93(2), 582–588.

Kim, H.-Y., Park, D.J., Kim, J.-Y. and Lim, S.-T. (2013b) Preparation of crystalline starch nanoparticles using cold acid hydrolysis and ultrasonication. *Carbohydrate Polymers*, 98(1), 295–301.

Le Corre, D., Bras, J. and Dufresne, A. (2010) Starch nanoparticles: a review. *Biomacromolecules*, 11(5), 1139–1153.

Liu, J., Fei, L., Maladen, M., Hamaker, B.R. and Zhang, G. (2009) Iodine binding property of a ternary complex consisting of starch, protein, and free fatty acids. *Carbohydrate Polymers*, 75(2), 351–355.

Livney, Y.D. (2010) Milk proteins as vehicles for bioactives. *Current Opinion in Colloid and Interface Science*, 15(1), 73–83.

Müller, R.H., Mäder, K. and Gohla, S. (2000) Solid lipid nanoparticles (SLN) for controlled drug delivery, a review of the state of the art. *European Journal of Pharmaceutics and Biopharmaceutics*, 50(1), 161–177.

Shah, A., Zhang, G., Hamaker, B.R. and Campanella, O.H. (2011) Rheological properties of a soluble self-assembled complex from starch, protein and free fatty acids. *Journal of Food Engineering*, 105(3), 444–452.

Wang, G. and Uludag, H. (2008) Recent developments in nanoparticle-based drug delivery and targeting systems with emphasis on protein-based nanoparticles. *Expert Opinion on Drug Delivery*, 5(5), 499–515.

Zhang, G. and Hamaker, B.R. (2003) A three component interaction among starch, protein, and free fatty acids revealed by pasting profiles. *Journal of Agricultural and Food Chemistry*, 51(9), 2797–2800.

Zhang, G. and Hamaker, B.R. (2004) Starch-free fatty acid complexation in the presence of whey protein. *Carbohydrate Polymers*, 55(4), 419–424.

Zhang, G. and Hamaker, B.R. (2005) Sorghum (*Sorghum bicolor* L. Moench) flour pasting properties influenced by free fatty acids and protein 1. *Cereal Chemistry Journal*, 82(5), 534–540.

Zhang, G., Maladen, M.D. and Hamaker, B.R. (2003) Detection of a novel three component complex consisting of starch, protein, and free fatty acids. *Journal of Agricultural and Food Chemistry*, 51(9), 2801–2805.

Zhang, G., Maladen, M., Campanella, O.H. and Hamaker, B.R. (2010) Free fatty acids electronically bridge the self-assembly of a three-component nanocomplex consisting of amylose, protein, and free fatty acids. *Journal of Agricultural and Food Chemistry*, 58(16), 9164–9170.

17 Polymeric nanoparticles for food applications

Cristina M. Sabliov and Carlos E. Astete

Department of Biological and Agricultural Engineering, Louisiana State University and LSU Agricultural Center, Baton Rouge, Louisiana, USA

17.1 INTRODUCTION

Solubilization of hydrophobic natural pigments in water, protection of peptides and proteins against oxidation or degradation, controlled and stable vitamin release, and increase of bio-absorption of poorly absorbed minerals or vitamins are ways in which polymeric nanoparticle can enhance bioactive functionality. Interactive polymeric nanodelivery systems with sensing ability are developed as second-generation polymeric nanoparticle with applications in food to further improve functionality of the delivered bioactive.

Polymeric nanoparticles have been developed for delivery of active components since the late 1970s and early 1980s, when several research groups started to show improved drug efficacy for doxorubicin (used for cancer treatment) when entrapped in polymers such as polyalkylcyanoacrylate. The research groups of R. Langer, J. Kreuter, and P. Couvreur were first to identify the advantages and challenges of using polymers as drug delivery systems (Couvreur *et al.*, 1977, 1979a, b; Langer, 1980; Langer *et al.*, 1980a, b; Kreuter, 1983a–d; Kreuter *et al.*, 1983). Polymeric nanoparticles offered several advantages such as drug protection, reduced side effects, enhanced activity, and controlled release, sufficient to confirm the importance of polymeric nanoparticles in disease treatment. The same advantages proven in the medical world for polymeric nanoparticles as drug delivery systems also have the potential to revolutionize the food industry: polymeric particles can be designed as potential bioactive delivery systems for development of new food products, of better quality, designed to prevent disease, and to enhance health.

Polymeric nanoparticles with a typical size range of 20 to 1000 nm consist of a polymeric core with the active component entrapped in the polymeric matrix, usually surrounded by a surfactant layer that stabilizes the system. The limitation of in the use of polymeric nanoparticles in the food industry stems from concerns that particles may pose health issues to the host exposed repeatedly to nanoengineered particles for long periods of time. In response to this concern, only food-grade components are selected for particle synthesis designed for food use in the form of natural polymers, and biodegradable and biocompatible surfactants. Polysaccharides such as alginic acid, carrageenan, xanthan gum, chitosan, and pullulan, and proteins such as casein, zein, and β-lactoglobulin are examples of

Nanotechnology and Functional Foods: Effective Delivery of Bioactive Ingredients, First Edition.
Edited by Cristina M. Sabliov, Hongda Chen, and Rickey Y. Yada.

currently used natural polymers for nanoparticle synthesis, which are replacing synthetic polymers such ξ-caprolactone, poly(lactic-co-glycolic acid) (PLGA) and poly(lactic acid) (PLA), mainly used in the biopharmaceutical area.

Polymeric nanoparticles synthesized from generally recognized as safe (GRAS) components can be designed to provide protection of entrapped active components, to control the release of the bioactive over time, to promote desired interactions with the food matrix for improved flavor, texture, and other sensorial properties, to act on demand based on external conditions for improved bioavailability of the entrapped bioactive, with the goal of improving the quality of the food in which they are entrapped and the health of the consumer, without eliciting toxic responses from the host on the basis of formation of toxic by-products during degradation and biodistribution. In this chapter, advantages of polymeric nanoparticles are considered, the materials available for the nanoparticle synthesis described, the available methods used to make particles of controlled properties detailed with a focus on nanoparticle size control, and a few applications of polymeric nanoparticles in foods are provided as case studies. The chapter closes with conclusions and future directions.

17.2 COMPONENTS OF POLYMERIC NANOPARTICLES

Polymeric nanoparticles are complex systems formed of a polymer or mixture of polymers (i.e., chitosan, alginic acid, albumin) and surfactants (i.e., lecithin, Tween, Span), with bioactive components entrapped in the polymeric matrix. The active component can be entrapped in the core of the matrix, or at the interface between the polymer and surfactant, depending on the properties of the bioactive and polymer in which they are entrapped.

Several options are available for materials used in nanoparticle synthesis, including synthetic polymers and natural polymers (e.g., proteins and hydrocolloids), each posing challenges and offering advantages for specific bioactive delivery applications in foods. The need to replace synthetic polymers with natural and biofriendly components suitable for food applications led to development of polymeric particles made out of polysaccharides and proteins of various surface charge, hydrophobic/hydrophilic ratio, solubility, targeting properties, release profiles, entrapment efficiencies of active components, and degradation profiles.

17.2.1 Polymers with potential applications in synthesis of polymeric nanoparticles for food use

17.2.1.1 Synthetic polymers

Among the synthetic polymers, the polyesters family (i.e., PLA, poly(e-caprolactone) (PCL), poly(glycolic acid) (PGA)) are of interest in the biomedical area because of their biocompatibility and biodegradability properties. Poly (D,L-lactide-co-glycolide) (PLGA) (Figure 17.1) is an example of a common synthetic polymer used in biological applications approved by the Food and Drugs Agency (FDA) as GRAS for pharmaceutical applications (Jalil and Nixon, 1990; Jain, 2008). Its degradation is thought to occur mostly through hydrolysis, where it breaks down into its monomers, lactic acid, and glycolic acids. The monomers, lactic acid, and glycolic acids, are metabolized through the Kreb's cycle and eliminated as harmless waste products, carbon dioxide, and water (Brady *et al.*, 1973; Tice and Cowsar, 1984). The degradation profiles of these components affect the drug release profile. Changing the copolymer ratio, the PLGA copolymer can be more or less hydrophilic (50:50, 75:25, and 85:15 PLGA molar ratios are the most common copolymer ratios available commercially), which affects the polymer degradation profile. The PLGA copolymers formed from D,L-PLA and PGA are amorphous in nature, whereas L-PLA and PGA are crystalline (Beletsi *et al.*, 1999; Bala *et al.*, 2004). The copolymer ratio 75:25 showed faster drug release as compared with 85:15: after 15 h 95 wt% was

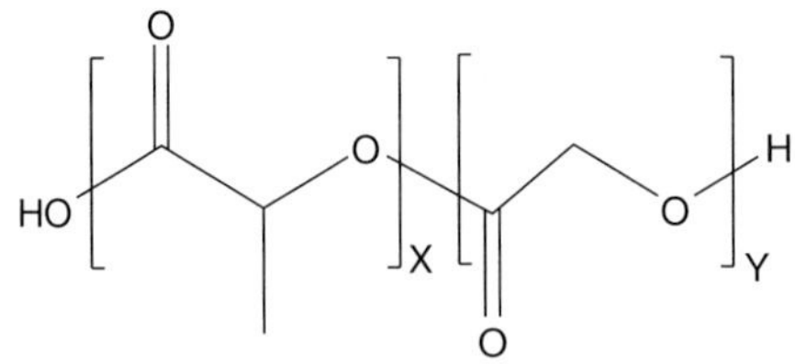

Figure 17.1 Structure of poly(lactic-co-glycolic acid).

released from 75:25, but only 80 wt% from 85:15 (Jeon *et al.*, 2000a). Nanoparticles with a higher amount of lactide monomers showed a slower release rate for hydrophobic drugs. The more hydrophobic the polymer, the more intense hydrophobic interaction between the polymer and drug entrapped, and thus a slower release profile (Lamprecht *et al.*, 2001; Fonseca *et al.*, 2002; Lee *et al.*, 2002; Saxena *et al.*, 2004; Mainardes and Evangelista, 2005; Win and Feng, 2005; Zhang and Zhuo, 2005).

The molecular weight of the polymer impacts the degradation process and the release profile, also mainly during the degradation step. Polymer molecular weight ranges from a few thousand Daltons to over 100,000 Daltons. Polymers with higher molecular weight present a slower degradation associated with a prolonged constant drug release over time. Diffusion processes are affected by the polymer molecular weight if the drug is mainly entrapped in the core of the nanoparticle (Anderson and Shive, 1997; Lamprecht *et al.*, 2001; Avgoustakis *et al.*, 2002; Lee *et al.*, 2002; Astete and Sabliov, 2006; Zigoneanu *et al.*, 2008; Astete *et al.*, 2011; Murugeshu *et al.*, 2011).

17.2.1.2 Natural polymers

Nanoparticles designed for food use are synthesized from food-grade polymers, hydrocolloids, and proteins, in an attempt to minimize potential toxic effects associated with ingestion of the particles. Figure 17.2 shows the structure of natural polymers commonly used for synthesis of nanoparticles designed for food use. Several examples of biopolymers are described in more details in the following sections.

Pectin

Pectin is a hydrophilic polysaccharide with at least 65% galacturonic acid units. The acid group may be free, naturally esterified with methanol, or an acid amide. Some sugar residues and other impurities are found in commercial products.

Pectin polymeric nanoparticle in the size range of 200–800 nm can be synthesized, but usually in the presence of $CaCl_2$. The addition of Ca^{++} ions promotes ionic interaction with the $-COO^-$ group of pectin and cross-linking of the pectin chains. Some common pectins used for particle synthesis are beet pectin, low methoxyl pectin, and high methoxyl pectin. Pectins will gel only in the presence of sugars or other co-solutes at low pH (<4.0) because the acid groups are not completely ionized and available for the hydrogen bonding required to stabilize the gel formed (Oakenfull, 1991). It is difficult to control cross-linking, and hence the sample polydispersity is high. Pectin nanoparticles are degraded by a β-elimination mechanism at ambient temperature, or above neutral or alkaline pH, by the ester groups' hydrolyzation under alkaline or acidic conditions, or by pectin esterases, or a combination of mechanisms.

Chitosan

N-acetyl-D-glucosamine and D-glucosamine are the main components of chitosan (β-(1–4)-2-amino-2-deoxy-D-glucose), which is produced by removing an acetate group from chitin. The naturally occurring linear biodegradable polysaccharide presents several advantages, such as mucoadhesion,

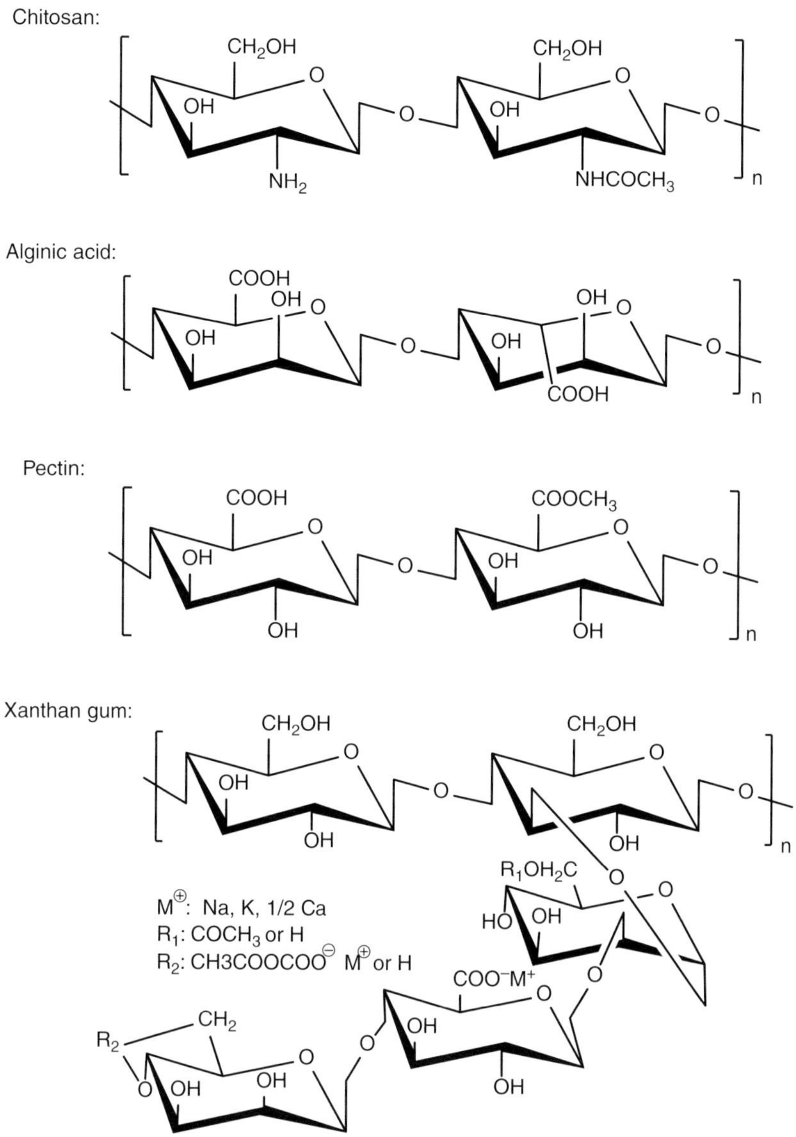

Figure 17.2 Chemical structure of some common hydrocolloids.

antibacterial property, minimal toxicity, low production cost, biocompatibility, and biodegradability, which justify its potential for use in delivery systems. The cationic polymer can be dissolved in water under acidic conditions that promote amine protonation. The pH usually is 3.5–4 to ensure complete solubilization.

Chitosan nanoparticles are usually synthesized by ionotropic gelation by addition of pentasodium tripolyphosphate (TPP) based on the interaction between the negative groups of the TPP and the positive charges of chitosan amine groups at pH lower than the pK 6.3 (Berger *et al.*, 2004a, b). At neutral or basic pH, chitosan nanoparticles form large aggregates, varying in size from 150 nm to over 1 μm depending on the concentration, molecular weight, purity of the polymer, and surfactant used. Zeta potential of the chitosan nanoparticles is positive and can rise to 60 mV depending on the surfactants and other additives used during synthesis, and the conditions under which the measurement is performed. Chitosan is degraded *in vivo* by an enzymatically driven process. Chitosan with a high degree of deacetylation has a higher rate of degradation compared with chitosan with a lower degree of deacetylation, which affects the release profile of the entrapped active components (Paul and Sharma, 2000; Borges *et al.*, 2006; Zhang *et al.*, 2008; Honarkar and Barikani, 2009; Nagpal *et al.*, 2010; Murugeshu *et al.*, 2011; Singh *et al.*, 2011; Wang *et al.*, 2011; Zhao *et al.*, 2011).

Alginic acid

Alginic acid is a hydrophilic colloidal polysaccharide obtained from various species of brown seaweed (*Phaeophyceae*). It is a linear copolymer consisting mainly of residues of D-mannuronic acid and L-guluronic acid, often arranged in homogeneous blocks (random or alternating units of mannuronic acid and guluronic acid). Hydration of alginic acid increases the viscosity of the aqueous solution by intermolecular binding. Monovalent metal ions form soluble salts, and divalent or multivalent cations (except Mg^{++}) form gels or precipitate alginic acid by cross-linking the negatively charged chains of alginic acid (Tonnesen and Karlsen, 2002; Augst *et al.*, 2006). The pK values of mannuronic acid and guluronic acid of alginate chains are pK 3.4 and 3.6, respectively (Sarmento *et al.*, 2006, 2007).

Particles synthesized with alginic acid are typically in the size range of 100 nm to over 1 μm, and the size distribution is usually broader compared to that of particles made out of synthetic polymers. Reported that the ζ potential of alginic acid nanoparticles is under −20 mV. The addition of $CaCl_2$ in different amounts promotes different degrees of cross-linking of alginic acid and can be used to modify the release of the active component over time (Rastogi *et al.*, 2007; Lertsutthiwong *et al.*, 2008; Zohri *et al.*, 2010). The alginic acid degradation occurs by hydrolysis (proton-catalyzed) depending on time, pH, and temperature (Augst *et al.*, 2006; Martins *et al.*, 2007; Nesamony *et al.*, 2012).

Xanthan gum

Xanthan gum is an extracellular anionic high molecular weight polysaccharide secreted by *Xanthomonas campestris*. It has a linear (1–4) linked β-D-glucose backbone with D-glucosyl, D-mannosyl, and D-glucuronyl acid residues in a 2:2:1 molar ratio and variable proportions of *O*-acetyl and pyruvyl residues. The trisaccharide side chains on every other glucose at C3 have a glucuronic acid residue linked (1–4) to a terminal mannose unit and (1–2) to a second mannose residue that connects to the backbone. Acetyl groups are often present as 6-O substituents on the internal mannose residues. Under low ionic strength or high temperature, xanthan chains appear as disordered and flexible structures, whereas at low temperature or high ionic strength they present ordered structures (Dumitriu and Chornet, 1997; Chellat *et al.*, 2000). The *O*-acetyl and pyruvyl groups are deprotonated at pH > 4.5, and Ca^{++} ions can physically cross-link the polymer along the side chains (Dario *et al.*, 2011). Xanthan gum is soluble in cold water and resistant to enzymatic degradation (Xue and Sethi, 2012). Typically, xanthan gum is used as a coating agent on nanoparticles made out of chitosan or proteins as a core material. The ionic interactions between xanthan gum carboxylic groups and amine groups of the core material allow deposition of the xanthan gum over the nanoparticle surface for a variety of reasons (Chellat *et al.*, 2000; Argin-Soysala *et al.*, 2009).

Zein

Zein is a corn protein rich in prolamines, which has been under extensive study due to its hydrophobicity, biocompatibility, and biodegradability properties. The main fractions of zein are α-zein with (21 to 25 kDa), β-zein (17 to 18 kDa), and γ-zein (27 kDa). α-Zein is the most commonly used due to its relative abundance, which is around 80% of the total zein, and around 50% of the 225 amino acid residues are hydrophobic (i.e., proline, leucine, and alanine). The presence of glutamine promotes the polarity regions and protonation of some side chains. The hydrophobic/hydrophilic behavior allows using zein as a carrier for active components, including lutein, vitamin D, omega-3 fatty acids, curcumin, and others (Wang and Padua, 2007; Luo *et al.*, 2010; Patel *et al.*, 2010, 2012; Luo *et al.*, 2011; Mehta *et al.*, 2011; Dhandayuthapani *et al.*, 2012; Hu *et al.*, 2012; Regier *et al.*, 2012; Wu *et al.*, 2012; Zou *et al.*, 2012). Nanoparticles synthesized with zein measured 100 to 700 nm, with a polydispersity index of 0.15 to 0.4, as a function of the pH of the solution, presence of surfactant or stabilizers, active component entrapped, synthesis method, purification process, and presence of cryoprotectants for powder form (Wu *et al.*, 2012). Zein concentration is highly correlated to the final particle size. As shown by Podaralla and Perumal (2010), zein concentration of 0.09% w/v and 0.5% w/v formed nanoparticles with sizes of 230 nm to 400 nm, respectively. In general, to obtain zein nanoparticles under 200 nm, zein concentration should be lower that 0.1% w/v (Xu *et al.*, 2011). The isoelectric point (pI) for zein nanoparticles formed with pluronic and lecithin as surfactants loaded with 6,7-dihydroxycoumarin as a model drug is, at pH 5.6, smaller than the pure zein pI of 6.8, showing the effect of the active component entrapped (Cabra *et al.*, 2005; Luo *et al.*, 2010, 2011, 2012). The release of active components is affected by molecular diffusion from the core and degradation of zein. Zein hydrolysis occurs very slowly but can be accelerated by enzymes, such as trypsin or collagense (Sun *et al.*, 2005; Hurtado-Lopez and Murdan, 2006; Xu *et al.*, 2011; Luo *et al.*, 2012).

Albumin

There are three main types of albumin protein currently used in delivery systems: ovalbumin, bovine serum albumin, and human serum albumin. Ovalbumin is a monomeric phosphoglycoprotein with 385 amino acid residues (one internal disulfide bond and four free sulphydryl groups). The molecular weight is 47,000 g/mol and the pI is 4.8 (Elzoghby *et al.*, 2012a). The bovine serum albumin (BSA) has a molecular weight of 69,323 g/mol with a pI 4.7, (Tantra *et al.*, 2010) and it is widely used in biopharmaceutics for delivery purposes (Zimmer *et al.*, 1994, 1995; Frassineti *et al.*, 2002; Ibrahim *et al.*, 2002a; K. N. Lee *et al.*, 2011; Shen *et al.*, 2011; Wang *et al.*, 2012). To avoid immunologic responses *in vivo*, human serum albumin (HSA) is used for pharmaceutical purposes. Human serum albumin is a soluble globular protein with 585 amino acid residues and a molecular weight of 66,500 g/mol. It is stable in pH 4–9 and a temperature of 60°C for up to 10 h (Hu *et al.*, 2006; Tantra *et al.*, 2010; Elzoghby *et al.*, 2012a). Biodegradability, biocompatibility, low immunoreactivity, versatility, and cost are the main advantages of albumin as a biomaterial for polymeric nanoparticle synthesis. Albumin nanoparticles loaded with paclitaxel are currently used for drug delivery purposes for treatment of cancer in the form of Abraxame® (Ibrahim *et al.*, 2002b; Elzoghby *et al.*, 2012a). There are several techniques currently used for albumin nanoparticles synthesis, with desolvation, emulsification and thermal treatment being the most frequently used.

Usually, the particle size can range from 100 nm to 600 nm and can entrap a variety of active components. Albumin nanoparticle suspension can be stable for over 6 months when stored at 4°C, with no apparent change in size (Yu *et al.*, 2006; Dreis *et al.*, 2007). The addition of cryoprotectants such us trehalose, sucrose, or glucose minimizes aggregation and particle-size increase during freeze-drying (Anhorn *et al.*, 2008). Zeta potential ranges from −20 mV at pH 9 to +15 mV at pH 2 for albumin nanoparticles without surfactant (in the presence of calcium citrate tetrahydrate added during the synthesis) (Jun *et al.*, 2011). Zeta potential of albumin nanoparticles resuspended at pH 7 in phosphate buffer saline (PBS) when glutaraldehyde is used as the cross-linker is around −20 to −25 mV (Bansal *et al.*, 2011). Cross-linking results in a lower pI for albumin nanoparticles compared to pure protein (nanoparticle pI 4 and albumin pI 4.7) (Elzoghby *et al.*, 2012a). The stabilization procedure not only affects size and ζ potential, it also affects the enzymatic degradability of HSA nanoparticles, with a significant impact on the release of the embedded drug.

Milk proteins: casein

Casein is a protein that makes up over 90% of the milk proteins. There are four caseins in bovine milk: α-s1-casein (23.6 kDa, isoionic pH 4.94), α-s2-casein (25.3 kDa, isoionic pH 5.23–5.45), β-casein (24 kDa, isoionic pH 5.14), and κ-casein (19 kDa, isoionic pH 5.61) (Semo *et al.*, 2007). Casein is self-assembled in micelles that are stabilized by calcium phosphate and the surface is covered with kapa-casein. Casein can be cross-linked by the enzyme transglutaminase (TGase) (Horne, 2009). β-Casein is an unstructured amphiphilic that self-assembles in nanostructures of different morphologies at different pH values, with a disk shape at low pH and as non-uniform oblate micelles at neutral pH (Horne, 2002; Elzoghby *et al.*, 2011). Casein can be transformed to cationic salt when the pH of a solution is raised (pH > 5.5) by addition of a monovalent cation base. At pH < 3.5 casein is soluble but with higher solution viscosity (Reddy and Kinsella, 1990; Semo *et al.*, 2007; Livney, 2010; Shapira *et al.*, 2010a; Zimet *et al.*, 2011; Haham *et al.*, 2012). β-Casein has a critical micelle concentration (CMC) that ranges from 0.05% to 0.2% w/v as a function of the temperature, pH, solvent, and ionic strength (Portnaya *et al.*, 2006). Nanoparticles made out of casein can range from 110 nm to 270 nm when a nano-spray dryer is used, and ζ potential is around −35 mV for particles formed with genipin as a cross-linker agent. The casein protein is stable at high temperature, so the spray dryer is suitable to synthesize casein nanoparticles (Elzoghby *et al.*, 2013). The casein concentration affects the final particle size of casein nanoparticles, as shown by Shapira *et al.* (2010b), with particles increasing in size from 200 nm to 600 nm when β-casein concentration changed from 1 mg/mL to 3 mg/mL.

Whey proteins (β-lactoglobulin and α-lactalbumin)

β-Lactoglobulin is a small globular protein with a molecular weight of 18,400 g/mol with 162 amino acid residues, two disulphide bonds, and one free thiol group. β-Lactoglobulin is the major whey protein of ruminant species, and it has a pI of 4.6 (Chanasattru *et al.*, 2009; Ron *et al.*, 2010; Abd El-Salam and El-Shibiny, 2012). β-Lactoglobulin belongs to the protein family lipocalins, known for their ligand-binding properties (Brownlow *et al.*, 1997; Kontopidis *et al.*, 2002, 2004). β-Lactoglobulin presents hydrophobic pockets that promote interactions with more hydrophobic molecules, such as retinol, fatty acids, vitamin D, and polyphenols. The protein is stable at low pH and resistant to proteolitic degradation in the stomach (Reddy *et al.*, 1988), and nanoparticle synthesis usually involves the use of solvents, buffers with more basic pH (pH >7.5), heat, and cross-linking. Desolvation is a common technique used for β-lactoglobulin nanoparticle formation, and acetone or alcohol is usually used (desolvation agents) to promote particle aggregation. Ko and Gunasekaran (2006) obtained small β-lactoglobulin nanoparticles of 59 ± 6 nm when the protein solution was heated at 60°C for 30 min before the desolvation step. It is important to mention that the particle size without the heating step was around 130 nm. Zeta potential fluctuated from +25 mV to −40 mv at pH ranges of 3–11. The β-lactoglobulin nanoparticles were more stable under basic conditions (PBS buffer at pH 7.4) compared to acidic conditions (PBS buffer at pH 2), but the presence of enzymes (pepsin and trypsin) speeded up degradation from 20 h without enzymes to 8 to 14 h with enzymes under both acidic and basic environments (Ko and Gunasekaran, 2006).

α-Lactalbumin is a protein (123 amino acid residues) present in the milk of all mammals. Its molecular weight is 14,000 g/mol and the pI is between 4.2 and 4.5 (Mehravar *et al.*, 2009). The protein has 123 amino acid residues with a high proportion of essential amino acids, such as tryptophan, lysine, and cysteine (Kamau *et al.*, 2010; Rodiles-Lopez *et al.*, 2010). The nanoparticle synthesis out of α-lactalbumin is based on desolvation with additional heat treatment and cross-linking steps, at the following parameters: protein concentration ranges between 1 and 2% w/v, the solvent used is acetone or ethanol at 4:1 ratio to water phase, the heating is applied for 30 min at 60°C, and the cross-linking is performed with 8% glutaraldehyde solution. The particle size synthesized by this method ranged from 150 to 300 nm, with ζ potential of +10 mV to −40 mV at pH 3.1 to 5.5, respectively. Proteolytic enzymes easily degraded the α-lactalbumin nanoparticles after 1 h of mixing at 37°C with pectin and pancreatin simulated gastric solution (Esmaeilzadeh *et al.*, 2011; Arroyo-Maya *et al.*, 2012).

17.2.2 Surfactant

If top-down chemical techniques are used for synthesis of polymeric nanoparticles, many require the addition of surfactants during synthesis to stabilize the polymeric nanoparticles. Surfactant selection is critical, considering that presence of surfactant on the surface of the particles affects the surface characteristics of the particle, and in turn nanoparticle interaction with food and biological systems. Examples of surfactants available for food applications are listed in Tables 17.1 and 17.2 with their classification numbers in the United States and European Union.

Table 17.1 Generally recognized as safe food surfactants by US Food and Drug Agency (FDA) and European Union (EEC) regulations. (Source: Hasenhuettl, 2008.)

Emulsifier	FDA (21CFR)	EEC
Diacetyltartaric esters of monoglycerides (DATEM)	184.1101	E472e
Lecithin	184.1400	E322
Mono- and diglycerides	184.1505	E471
Monosodium phosphate derivatives of mono- and diglycerides	184.1521	–

Table 17.2 Surfactants used as direct food additives and accordance with US Food and Drug Agency (FDA) and European Union (EEC) regulations. (Source: Hasenhuettl, 2008.)

Emulsifier	FDA (21CFR)	EEC
Acetylated mono- and diglycerides	172.828	E472a
Calcium stearoyl lactylate	172.844	E482
Citric acid esters of mono- and diglycerides	172.832	E472c
Ethoxylated mono- and diglycerides	172.834	–
Lactic acid esters of mono- and diglycerides	172.850	E472b
Magnesium salts of fatty acids	172.863	E470b
Polyglycerol polyricinoleate	–	E476
Polysorbate 60	172.836	–
Polysorbate 65	172.838	–
Polysorbate 80	172.840	–
Propylene glycol esters of fatty acids	172.856	E477
Salts of fatty acids	172.863	E470a
Sodium stearoyl lactylate	172.846	E481
Sodium stearoyl fumarate	172.826	–
Sorbitan monolaurate	–	E493
Sorbitan monooleate	–	E494
Sorbitan monopalmitate	–	E495
Sorbitan monostearate	172.842	E491
Sorbitan tristearate	–	E492
Stearyl tartrate	–	E483
Succinylated mono- and diglycerides	172.830	–
Sucrose acetate isobutyrate (SAIB)	172.833	–
Sucrose esters of fatty acids	172.859	E473
Tartaric acid esters of mono- and diglycerides	–	E472d

17.3 SYNTHESIS METHODS WITH EMPHASIS ON CONTROL OF NANOPARTICLE PROPERTIES

Common techniques used in polymeric nanoparticle synthesis are emulsion evaporation, emulsion diffusion, solvent displacement, or salting out. New synthetic approaches have resulted in response to the need to form polymeric nanoparticles out of materials suitable for food use. The new methods employ ionic interaction, hydrophobic/hydrophilic forces, stearic repulsion, and pH and temperature manipulations to promote polymer aggregation and particle stabilization. The characteristics of these methods are presented in Table 17.3.

17.3.1 pH-cycling treatment

Proteins are sensitive to pH variations and temperature changes, leading to precipitation, unfolding, and denaturation in the more acidic or basic conditions and high temperatures. On this basis, various processes can be developed for formation of nanoparticles of various properties (Britten and Giroux, 2001; Gunasekaran *et al.*, 2007; Giroux *et al.*, 2010; Livney, 2010). The first step in the process of making protein nanoparticles by the pH-cycling treatment is to suspend the protein at pH 7 and to warm up the dispersion to 70–80°C for 15–20 min. Next, the suspension is cooled to room temperature and the concentration of the protein adjusted to no more than 3% w/v. Following a pH adjustment step to around pH 5–6, a further protein precipitation step can be applied by adding $CaCl_2$ salts. Following a resting time of several hours at 4°C that allows formation of disulfide bonds between protein polymers, the pH is again adjusted to pH 7 in order to reduce other interactions and to minimize aggregate formation. Finally, for further reduction in aggregation, a homogenization step is applied. Type of protein, protein concentration and purity, type of active component, resting time, homogenization, and other parameters had been studied to optimize the size of particles, entrapment efficiencies, and shape of the nanoparticles. Particles made by this method ranged from 150 to 500 nm following entrapment of an active component; presence of clusters was observed, with a negative impact on the polydispersity of the protein nanoparticles (Britten and Giroux, 2001; Giroux *et al.*, 2010; Giroux and Britten, 2011). The active component, research shows, should be added in the first step, for a higher entrapment efficiency and product recovery (Giroux *et al.*, 2010). The type of active component entrapped affects the shape, size, polydispersity, and ζ potential.

17.3.2 Thermal treatment for proteins

The heat treatment denatures the protein allowing exposure of more hydrophobic moieties of the protein. Livney's group (Shpigelman *et al.*, 2010, 2012) entrapped active components using the heating process. In this approach, the initial step was to dissolve the protein (β-lactoglobulin) in PBS buffer at pH 7 with further warming up to 75–85°C for 20 min. The active component (EGCG) was dissolved in PBS at pH 2.5 and added to the protein suspension. The mixture was cooled at room temperature and mixed to reduce aggregates, and the final pH was adjusted to 7 (Shpigelman *et al.*, 2010, 2012): size of the protein aggregate ranged from 20 nm (after heating) to 400 nm. An important factor to consider when this process is followed is the sensitivity of the active component to heat; if the active component is thermolabile, an alternative method should be followed for its entrapment in a polymeric matrix. In a different approach, albumin particles can be formed in the presence of an adjuvant such as lysozyme (Yu *et al.*, 2006), dextran, and chitosan (Qi *et al.*, 2010). The role of the adjuvant is to aid particle condensation by interacting with the unfolded protein through hydrogen bonding, electrostatic or hydrophobic interactions, and disulfide–sulfhydryl interchange reaction (Bronich *et al.*, 2005; Yu *et al.*, 2006; Qi *et al.*, 2010).

Table 17.3 Important characteristics of top-down techniques.

Important characteristics	Methods						
	pH cycling	**Heat treatment**	**Spraying technique**	**Solvent displacement**	**CO_2 SC**	**Salting out**	**Emulsion evaporation**
Set-up time	High	High	High	High	High	Medium	Low
Process time	Medium	Medium–high	Medium–high	Medium–low	Medium–high	Medium	Medium–low
Solvents	Mainly buffers	Mainly buffers	Buffer or alcohols	Low toxicity (benzyl alcohol)	Supercritical fluids	Low toxicity	Low toxicity (ethyl acetate)
Energy consumption	Low	Medium	Medium	Low	Medium	Low	Medium
Water requirements	High	High	Medium	High	Medium	Medium	Low
Polymer concentration	Medium–high	Medium	Medium	Linear relation with Np size	Medium	Medium–high	Medium–high without increase in NP size
Hydrophobic components entrapment	Low-medium	Low-medium	Medium	High entrapment	Medium–High	High	High
Hydrophilic components entrapment	High	Medium–high	Medium–high	Medium	Medium–high	Low entrapment	Medium entrapment (by double emulsion)
Surfactant	Not important	Not Important	Not important	Recommended	Recommended	Recommended	Recommended
Process time	Medium	Medium	Medium	Medium–high	Medium–high	Medium	Medium
Purification	Low	Low	Low	Medium	low	High	high
Versatility	Medium	Low	Medium	Medium	Low	Medium	Medium
Observations	Good for proteins	Not for temperature labile bioactives	Proteins and carbohydrates	Solvent limitation	High pressures	Presence of salts	For hydrophobic and hydrophilic bioactives

NP, nanoparticle.

McClements's research group studied the addition of polysaccharides during heat treatment of the globular protein β-lactoglobulin to form nanostructures (Jones *et al.*, 2009, 2010a, b). Solutions of β-lactoglobulin were heated up to 75–80°C for 15 min at pH 4.75 in the presence of polysaccharides such as pectin or carrageenan. The aggregates formed were 200–800 nm in size at a concentration of 0.1% β-lactoglobulin and 0.05% polysaccharides for pectin and carrageenan, respectively. The differences in size were explained by the difference in electrostatic interaction between the different polysaccharides and protein. The particles changed size when solution pH was changed from pH 7 to 3, from 200 nm to over 800 nm for low methoxylated pectin, and from 180 nm to over 1000 nm for carrageenan. At lower pH carrageenan dissociated and formed larger aggregates due to the loss in charge, whereas pectins interacted stronger with the protein of increased positive charges at these pH conditions. In general, the net negative charge of the protein–polysaccharide decreased at lower pH and larger aggregates formed. The highly methoxylated pectin nanoparticles kept a constant size (around 250 nm) in the pH range tested, mainly explained by the hydrophobic interactions between the protein–polysaccharide and lower ionic net charge. Zeta potential for high methoxylated nanoparticles increased from −22 mV to −8 mV at pH 5 and 3, respectively. The low methoxylated pectin nanoparticles had much lower ζ potential in the same pH range, ranging from −42 mV to −11 mV. In comparison, carrageenan nanoparticles had a ζ potential of −60 mV to −49 mV in the same pH range (Jones *et al.*, 2009, 2010a).

17.3.3 Desolvation with polar solvents

This technique uses the principle of decreasing polarity, which favors hydrogen bonding and reduces hydrophobic interactions. In general, polymer and active component are dissolved in a polar, water-miscible solvent such as acetone, acetonitrile, ethanol, or methanol. The solution is then poured in a controlled manner (i.e., drop-wise addition) into an aqueous solution with surfactant. Nanoparticles are formed instantaneously by rapid solvent diffusion. Finally, the solvent is removed under reduced pressure by evaporation. Important parameters to be considered in controlling the nanoparticle size, among other physical characteristics, are: polymer/surfactant ratio, polymer concentration, surfactant nature and concentration, solvent nature, viscosity, additives, active component, and phase injection rate (Niwa *et al.*, 1993; De Jaeghere *et al.*, 1999; Govender *et al.*, 1999; Kwon *et al.*, 2001; Panagi *et al.*, 2001; Ahlin *et al.*, 2002; Konan *et al.*, 2003; Prakobvaitayakit and Nimmannit, 2003; Ameller *et al.*, 2004; Csaba *et al.*, 2004, 2005; Kumar *et al.*, 2004; Oster *et al.*, 2004; Ravi Kumar *et al.*, 2004; Saxena *et al.*, 2004; Lee *et al.*, 2005).

To form whey protein isolate aggregates, as an example, the first step is to prepare a solution of whey protein pH-adjusted to 9 with NaOH (Ko and Gunasekaran, 2006; Gunasekaran *et al.*, 2007; Gulseren *et al.*, 2012b). The solution can be filtered to eliminate impurities and stirred over night at 4°C. Next, ethanol is added drop-wise at 1 mL/min in order to reach the desired final ration between water and ethanol. Finally, the protein aggregate suspension is diluted (1:100) with buffer at pH 3: at neutral pH the aggregates are not stable (Gulseren *et al.*, 2012a). An optional sonication step can be added to minimize formation of large aggregates in the suspension. Also, temperature increase with homogenization can be used for shape and size control of protein aggregates. The protein aggregates formed by desolvation with polar solvent ranged in size from 20 to 100 nm, with size affected by the amount of ethanol used in the synthesis. The more ethanol used in desolvation, the larger the protein aggregates. An increase in temperature to over 70°C at pH 3 promotes an increase in aggregate size up to 1000 nm. Gulseren *et al.* (2012a) showed that with the addition of a homogenization step after heating, the size distribution can be narrowed and better control of size distribution is obtained.

Similarly, albumin nanoparticles can be synthesized by drop-wise addition of ethanol to an aqueous solution of albumin. The coacervates are hardened by addition of a cross-linker such as glutaraldehyde. The particle size usually ranges between 150 and 300 nm, with the smaller particles formed at an aqueous pH of 5. It is common to add lysine to cap the free aldehydic groups of glutaraldehyde (Weber *et al.*, 2000a–c; Shen *et al.*, 2011).

17.3.4 Spraying techniques

Common examples of spraying techniques include electrohydrodynamic atomization and spray drying. The electrohydrodynamic atomization technique is based on small and uniform droplets formation by applying an electric field. The liquid passes through a metal tube such as a nozzle or needle, and the liquid meniscus at the tip of the tube is electrically stressed by the potential difference between the tube and the counter electrode. Important parameters that affect the output are strength of the electric field, flow rate, and liquid properties. The powder is collected in a plate fixed below the needle tip and connected to the grounded counter electrode of the power supply. The electrospray droplets are dried during the fly time, on the way to the surface of the collector plate covered with aluminum foil (Gomez-Estaca *et al.*, 2012). Hernandez-Munoz's research group studied the effect of polymer concentration (zein) in aqueous ethanol solution, the flow rate, and the applied voltage on the polymeric structure's morphology and size. The particle size ranged from 400 to 900 nm for zein concentrations under 15%; at concentrations exceeding 1% zein fibers were formed instead. The particle shape was greatly affected by the polymer concentration, with irregular shaped systems forming at concentrations over 5%. The entrapment efficiency of curcumin, the active component used as a model, ranged from 85% to 90% (Gomez-Estaca *et al.*, 2012).

Albumin nanoparticles were synthesized by using a nanospray dryer (Buchi®) that has an electrostatic particle collector. The particles measured 460 nm in diameter and the yield was 72% by using a 4 µm nozzle mesh. The following parameters were used: albumin concentration 0.1% w/v, drying flow rate 150 L/min and inlet temperature 120°C (S. H. Lee *et al.*, 2011). In general, the lower the solution concentration and the nozzle mesh, the smaller the particles formed.

17.3.5 Supercritical fluids

Supercritical antisolvents, such as carbon dioxide (CO_2), have been used extensively in different pharmacy and food industry applications. The main advantages of supercritical processing are the mild conditions of operation, formation of solvent-free end products, and small droplet size formation. The main drawbacks are the requirement of high-pressure equipment, high initial cost, and specialized maintenance. Adequate nozzle selection to avoid clogs and solution preparation are important to avoid impurities that can obstruct the nozzle in a batch mode operation. A typical set-up involves the supercritical fluid and the organic solution (zein-lutein plus organic solvents) separately feeding into a nozzle. The mixture formed during jet formation in the high-pressure chamber is continuously discharged from the bottom. The final step is the slow depressurization of the precipitation vessel to collect the sample (Hu *et al.*, 2012).

The process was applied by Hu *et al.* (2012) to form zein nanoparticles with entrapped lutein, ranging from 200 to 350 nm and spherical in shape. The entrapment efficiency ranged from 30 to 85% based on the type of solvent, pressure (10–15 mPa), flow rate (0.5–1.25 mL/min), and temperature (32–45°C). The organic solvent mixture used was acetone and dimethylsulfoxide (7:3 v/v) (Hu *et al.*, 2012).

17.3.6 Salting-out method

In this method, the polymer is dissolved in the organic phase, which should be water-miscible (e.g., acetone, tetrahydrofuran, or ethanol). The organic phase is added to an aqueous phase, which contains the emulsifier and a high concentration of salts that are not soluble in the organic phase, under strong mechanical shear stress. There is no diffusion of the solvent in the water phase due to the presence of salts. However, upon fast addition of pure water to the oil in water emulsion, under mild stirring, the ionic strength is reduced and it leads to the migration of the water-soluble organic solvent into the aqueous phase, inducing polymer aggregation (Nakache *et al.*, 2000). The final step is purification by cross-flow filtration or centrifugation required to remove the salting-out agent. Common salting-out agents are electrolytes (sodium chloride, magnesium acetate, or magnesium chloride) or nonelectrolytes, such as sucrose (Konan *et al.*, 2002; Avgoustakis, 2004; Eley *et al.*, 2004; Zweers *et al.*, 2004). The polymeric nanoparticles formed by salting out range in size from 100 nm to 600 nm, with a narrow size distribution. The main parameters such as polymer concentration, molecular weight, stirring rate, type of solvent, type of surfactant, and concentration are required in the set-up in order to obtain nanoparticles in the right size distribution.

17.3.7 Emulsion evaporation method

The method is based on the emulsification of an organic solution of the polymer in an aqueous phase followed by the evaporation of the organic solvent. Normal emulsions of oil in water (o/w) or water in oil (w/o) and double emulsions (w/o/w) can be used to accommodate the entrapment of active components with different properties. The single o/w emulsion method is used to entrap hydrophobic active components, whereas the w/o/w double emulsion method is used for the entrapment of hydrophilic active components. Usually, surfactants are added for emulsion and nanoparticle stabilization, but this approach adds cost and processing time due to the subsequent purification step (Alex and Bodmeier, 1990; Zhao and Goveas, 2001; Lopez-Quintela, 2003; Mohammed and Hassan, 2003; Obeidat and Price, 2003; Schmitt *et al.*, 2003; Takada *et al.*, 2003; Bouchemal *et al.*, 2004).

17.3.7.1 Single emulsion method (o/w or w/o)

In summary, the method is based on the emulsification of an organic solution that contains the polymer and the active component in an aqueous phase, followed by the evaporation of the organic solvent. Different surfactants, such as poly(vinyl alcohol), sodium dodecyl sulfate, Span 80, didodecyldimethylammonium bromide (DMAB), and pluronic F68, can be dissolved in the aqueous phase. The size reduction of the emulsion droplet is accomplished by sonication or microfluidization. The evaporation step is required to eliminate the organic solvent present in the organic phase. This leads to the precipitation of the polymer as nanoparticles with a diameter in the nanometer range. A final purification step to reduce the amount of surfactant is required. The nanoparticle size ranges from 50 nm to 500 nm, with a narrow size distribution, and the main parameters which affect the nanoparticle size are: polymer molecular weight and concentration, copolymer ratio and end groups, surfactant nature, phase ratio, solvent nature, evaporation rate, drug entrapment, additives, shear stress, and sterilization (Julienne *et al.*, 1992; Song *et al.*, 1997; Zhao and Goveas, 2001; Chung *et al.*, 2002; Pietzonka *et al.*, 2002; Panyam *et al.*, 2004).

Albumin nanoparticles can be made by emulsion evaporation. In one study, the aqueous phase with dispersed albumin was added to an organic phase (e.g., cotton seed oil) to obtain a w/o emulsion. The emulsion was homogenized with a high-speed homogenizer and the albumin was stabilized by heating at 175°C for 10 min (Yang *et al.*, 2007; Jahanshahi and Babaei, 2008), or by denaturation in the presence of a cross-linker such as formaldehyde or glutaraldehyde (Jahanshahi and Babaei, 2008; Crisante *et al.*, 2009). Centrifugation was used to recover the albumin nanoparticles, which showed a size range from 500 to 900 nm as a function of the synthesis parameters (i.e., temperature, protein concentration, emulsification speed) (Yang *et al.*, 2007).

17.3.7.2 Double emulsion method (w/o/w)

The two-step emulsification involves the formation of w/o emulsion, where the aqueous solution contains the hydrophilic active component and the organic phase contains the polymer and a suitable surfactant (Span 80, pluronic F 68, or others) of a low hydrophilic–lipophilic balance. The emulsion is formed under strong shear stress (i.e., sonication, microfluidization, high-speed homogenization). Next, the w/o is poured into a second water phase that has a suitable surfactant such as poly(vinyl alcohol), forming the w/o/w. The emulsion is sonicated or homogenized for droplet-size reduction, and the synthesis parameters such as time, temperature, and others must be rigorously controlled in order to minimize the hydrophilic active-component diffusion to the external aqueous phase. Evaporation is used to remove the organic solvent, and usually it is done under vacuum. The main drawback of the double emulsion method is the large size of the nanoparticles formed and the leakage of the hydrophilic active component, responsible for low entrapment efficiencies. The leakage effect can be reduced by using a high polymer concentration and a high polymer molecular weight, accompanied by an increase in the viscosity of the inner water phase, and an increase in the surfactant molecular weight (Domb *et al.*, 1996). The nanoparticle size ranges from 120 nm to 600 nm, and the parameters that affect nanoparticle size are polymer/surfactant ratio, polymer concentration, surfactant nature, viscosity, solvent nature, shear stress, evaporation, and first/second phase volume ratios (Jeon *et al.*, 2000b; Sahoo *et al.*, 2002; Vandervoort and Ludwig, 2002; Aukunuru *et al.*, 2003; Mu and Feng, 2003; Dillen *et al.*, 2004a, b; Prabha and Labhasetwar, 2004; Vandervoort *et al.*, 2004).

17.4 APPLICATION OF POLYMERIC NANOPARTICLES

Polymeric nanoparticles can address several limitations currently faced by the food industry in processing, food formulations, and health claims. The potential to increase nutritional quality of food matrices, limit nutritional losses during processing, add new flavors or mask bad taste, add new texture, offer release of aroma for longer periods of time, or to improve stability, absorption and functionality of active components are some of the important applications in which polymeric nanoparticles are foreseen to contribute in the future. Food design will be driven by the individualized needs of the population, whether genetic predispositions, or perceived health risks. Products that offer not only proven health claims, but also new and improved flavor, taste, texture, and efficiency will be enabled by this technology. The potential applications of polymeric nanoparticles in foods will require a thorough study that will enable understanding of polymeric nanoparticle interactions in the body, most importantly their safety. The information provided will allow better particle designs that will interact in a known manner with the food matrix and with the mouth, stomach, small intestine, colon, and lymph components for the most efficient and safe delivery of bioactives. Some successful applications of polymeric nanoparticles in food reported in the literature are detailed in the following sections.

17.4.1 Delivery of natural pigments as food colorants

The use of natural pigments (e.g. carotenes, curcumin, zeaxanthins, and lutein) in water-based food applications is limited due to their poor solubility in water. In addition, lack of stability of these natural bioactives is a drawback. It would be of interest to the food industry to package these pigments in polymeric nanoparticles that would enable their dispersion in water and prevent their degradation. Such nanodelivery systems would enable their use in water-based foods in order to provide additional health benefits, including antioxidant and anti-inflammatory effects (Nickols-Richardson and Piehowski, 2008; Bouwmeester *et al.*, 2009; Garcia *et al.*, 2010; Huang *et al.*, 2010; Forster *et al.*, 2011; Neethirajan and Jayas, 2011; Rashidi and Khosravi-Darani, 2011; Elzoghby *et al.*, 2012b; Kuan *et al.*, 2012; Pahuja *et al.*, 2012). Astete *et al.* (2009) developed a system designed for β-carotene entrapment, formed by alginic acid, lecithin, and β-carotene. The nanoparticle size ranged from 120 nm to 200 nm with a narrow size distribution. Stability studies showed that addition of $CaCl_2$ during synthesis improved the β-carotene stability. β-Carotene polymeric nanoparticle aqueous suspension changed color upon dilution from red (1:1 v/v water to β-carotene nanoparticle suspension), orange (1:40 v/v water to sample), and yellow (1:100 water to sample) (Astete *et al.*, 2009).

Another example of polymeric nanoparticle developed for colorant delivery is the curcumin–zein nanoparticles proposed by Gomez-Estaca *et al.* (2012). The entrapment of curcumin was performed by electro-atomization of zein and curcumin in ethanol. The nanoparticle size ranged from 200 to 800 nm with spherical shape for zein concentrations of less than 5%. The entrapment efficiency ranged from 85 to 95%. The sample stored at 23°C and 43% relative humidity was stable for 3 months. The colorant properties were studied with milk as a model food and indicated that a significant difference in lightness, chroma, and hue angle was found between the curcurmin–zein polymeric nanoparticle and the controls, suggesting interesting colorant potential of curcumin delivered with polymeric nanoparticles for food applications (Gomez-Estaca *et al.*, 2012).

17.4.2 Vitamin E

α-Tocopherol, a vitamin E vitamer has been researched extensively in order to understand the entrapment of hydrophobic active components in polymeric nanoparticles (Zigoneanu *et al.*, 2008; Somchue *et al.*, 2009; Luo *et al.*, 2011; Murugeshu *et al.*, 2011). Luo *et al.* (2011) studied the entrapment of α-tocopherol in zein–chitosan nanoparticles. The particles ranged from 200 to 800 nm with a polydispersity index of 0.2 to 0.4 (monodisperse size distribution < 0.1, high polydisperse > 0.7). The entrapment efficiencies ranged from 76% to 87% for the different ratios of polymers tested. The cumulative release was reduced when chitosan was increased in the formulations. For 5:1 zein:chitosan 40% of α-tocopherol was released, whereas the 20:1 formulation released 70%, at 6 h. Interestingly, when the α-tocopherol loading was increased to 30%, a slower release profile was obtained compared with lower loadings (10% and 20%). The data suggested that hydrophobic interactions between zein and α-tocopherol were predominant at higher loadings. The particle stability was analyzed in a simulated gastric medium with pepsin and pancreatin enzymes (30 min with pepsin (pH 2) and 6 hwith pancreatin at pH 7). The lower the ratio of zein to chitosan, the faster the release of α-tocopherol, suggesting that enzymatic degradation was the main mechanism of release due to zein degradation. The addition of chitosan retarded the release of α-tocopherol, suggesting that polysaccharides act as a barrier against enzymes diffusion to the zein matrix. The particles released more α-tocopherol in the simulated gastric fluid compared with the simulated intestinal fluid for all conditions tested (Somchue *et al.*, 2009; Luo *et al.*, 2011).

17.4.3 Vitamin D_3

Vitamin D_3 was entrapped in zein–carboxymethyl chitosan (CMCS) nanoparticles in the presence of calcium. The nanoparticle size ranged from 90 nm to 200 nm with a narrow polydispersity index and a ζ potential from −15 to −28 mV for different polymer ratios and calcium solution concentration. The entrapment efficiency ranged from 52% to 87% for the different formulations. The calcium addition to zein–CMCS nanoparticles modified the release profiles of vitamin D_3 entrapped. The calcium system released 45% of vitamin D_3 in 6 h, whereas zein–CMCS without calcium released 80% and zein alone released 85%. The photochemical stability was tested in the three systems (zein, zein–CMCS, and zein–CMCS–calcium) against vitamin D_3 in solution. After 9.5 h, around 70% of free vitamin D_3 was degraded in the control sample. In contrast, the nanoparticle systems tested were able to protect vitamin D_3 against UV-light-induced degradation, and less than 30% of vitamin D_3 was degraded during the same time frame (Luo *et al.*, 2012).

17.4.4 Epigallocatechin-3-gallate

The entrapment of EGCG by using β-lactoglobulin was studied by Shpigelman *et al.* (2010, 2012). The polymeric nanoparticles formed by thermal treatment measured 12–20 nm for different ratios of β-lactoglobulin to EGCG (1:2 to 1:8), with a relatively high polydispersity index. The entrapment efficiency was around 70%. The EGCG protection against oxidation was studied by heating the solutions at 85°C for 20 min, and the data showed improved protection of EGCG entrapped, as indicated by absorbance readings around four to five times less than that of the controls (solution of EGCG, mixtures of EGCG, and β-lactoglobulin). The EGCG-β-lactoglobulin can be freeze-dried and the nanoparticle reconstitution in water at pH 6.6 showed no significant difference in size from the fresh nanoparticles. The release kinetics were evaluated under gastric conditions (pH 2, 37°C, and pepsin), and showed that the EGCG was not released in the first 90 min. Around 75% of the EGCG remained in the nanoparticles/aggregates as a result of EGCG attachment to small peptides, which are products of β-latoglobulin degradation by pepsin. A sensory test was performed by a professional panel (six persons) that tasted the EGCG–polymer nanoparticles resuspended in phosphate buffer with 5 g of sugar/100 mL. The panel found that the system of β-lactoglobulin–EGCG showed significantly less bitterness and astringency compared to the same concentration of non-entrapped EGCG (EGCG nanoparticles: 1.4 g β-lactoglobulin/100 mL; control: 154 mg EGCG/100 mL in phosphate buffer at pH 7). The remaining bitterness and astringency was attributed to the free EGCG present after synthesis (entrapment efficiency of 70% and no purification step was applied) (Shpigelman *et al.*, 2010, 2012).

17.4.5 Lutein

Lutein is a natural pigment that is a derivative of α-carotene. When entrapped in zein by a supercritical fluids method, the particle size ranged from 200 nm to 300 nm and the entrapment efficiency varied from 35% to 85% as a function of the processing parameters, such as temperature, pressure, concentration, and polymer to active components ratio. The release profile followed a near-zero order release. The mechanism for lutein release can be controlled by swelling and erosion of the zein matrix: water molecules penetrate into the zein matrix, which promotes swelling and widening of the network allows lutein to diffuse out. About 90% of the lutein was released after 120 min. The release media was PBS with pH 7.4 and 0.2% sodium dodecyl sulfate (Hu *et al.*, 2012).

17.4.6 Omega 3 fatty acids

Omega-3 polyunsaturated fatty acids (PUFAs), associated with major health benefits such as reducing the risk of cardiovascular disease, metabolic syndrome, and others, are very hydrophobic and unstable. An interesting approach was proposed by Zimet *et al.* (2011), which consisted of entrapping omega-3 fatty acids in casein micelles polymeric nanoparticles and studying the stability and protection of docosahexanoic acid (DHA) in this system. The entrapment efficiency in particles measuring 290 nm was around 63%, which suggested some lossof PUFAs or degradation during synthesis by the solvent displacement method. The polymeric nanoparticle system provided protection to the entrapped DHA against oxidative processes as compared with nonentrapped DHA (dissolved in ethanol solution). The concentration of entrapped DHA remained constant for 37 days at 4°C. In comparison, a decrease in stability was observed for DHA dissolved in ethanol–water. The particle size remained stable for the shelf-stability study with a size of 285 nm after 37 days at 4°C.

17.4.7 Aroma and flavor compounds

Aroma and flavor are important components in food formulations. Processing, packaging, and storage conditions will affect the aroma intensity, with subsequent change in the food quality associated with its aroma. Technologies that allow control and modulation of aroma release are promising in food applications. Delivery systems can provide a means to protect the aroma component from undesirable interactions with food, to prevent oxidation, control its release profile, and increase aroma shelf life. The research of Giroux and Britten (2011) showed that entrapment of ethyl hexanoate in whey protein to form 200 nm nanoparticles was affected by pH conditions and $CaCl_2$ addition. The release of ethyl hexanoate entrapped in polymeric nanoparticles was lowest compared with the control (38% improvement in the 0.01% samples) (Giroux and Britten, 2011).

17.4.8 Minerals

The entrapment of selenite was accomplished by using chitosan cross-linked with tripolyphosphate (TPP). The selenite was mixed with chitosan before addition of TPP. The nanoparticle size ranged from 130 nm to 300 nm as a function of the ratio of chitosan and selenite added during the synthesis step, and entrapment efficiency was 65% on average. The 95% selenite was released in 120 min under simulated gastric medium. In an attempt to improve release profile, a zein layer was added to the delivery system by mixing the chitosan nanoparticle with zein. The particle size increased to 250–400 nm for formulations with 1:1 and 1:3 w/w of chitosan:zein, respectively. By the addition of zein, the release was decreased from 95% to 25% in the first 120 min, following a more linear release profile (Luo *et al.*, 2010).

The entrapment of zinc was accomplished by mixing a solution of whey protein isolates with $ZnCl_2$ following a solvent displacement technique by Gulseren *et al.* (2012b). The whey polymeric nanoparticles size was 200 nm with a zinc entrapment efficiency of over 90%; an increase in zinc concentration from 1 to 10 mM, promoted aggregation and larger size. The polymeric nanoparticles remained stable for 30 days at 22°C, with the nanoparticle size slightly increasing from 200 nm to 240 nm. The amount of zinc incorporated was within the range of daily zinc requirements for healthy adults, around 2–3 mg daily intake (Gulseren *et al.*, 2012b). Another approach used pectin mixed with $Zn(NO_3)_{2*}6H_2O$, to which, after full dissolution, was added 0.125 M NaOH. After 24 h of stirring at 20°C, aggregates of 150 nm were confirmed by transmission electron microscopy. The samples presented a strong aggregation, but ZnO was linked to pectin (Shi and Gunasekaran, 2008).

17.5 CONCLUDING REMARKS

New materials, systems, processes, and formulations are constantly developed to solve old and new problems faced by the food industry and to provide new sources for innovation in food. Significant progress has been made recently in the area of polymeric nanoparticles as bioactive delivery systems, as a result of sustained research and experimentation at universities and research laboratories and national centers worldwide. The research has focused on delivery systems in order to improve stability, bioavailability, and functionality of the entrapped bioactive. While questions still remain on the efficacy and safety of these delivery systems, and generalizations can not yet be inferred on the basis of the wide diversity of particles and bioactives, it is widely accepted that biodistribution, bioavailability, and other properties of the bioactive can be manipulated when delivered with polymeric nanoparticles. Pending proof of safety, polymeric nanoparticles are potential candidates for efficient delivery of bioactives in food, for improved quality and human health.

References

Abd El-Salam, M.H. and El-Shibiny, S. (2012) Formation and potential uses of milk proteins as nano delivery vehicles for nutraceuticals: A review. *International Journal of Dairy Technology,* 65(1), 13–21.

Ahlin, P., Kristl, J., Kristl, A. and Vrecer, F. (2002) Investigation of polymeric nanoparticles as carriers of enalaprilat for oral administration. *International Journal of Pharmaceutics,* 239(1–2), 113–20.

Alex, R. and Bodmeier, R. (1990) Encapsulation of water-soluble drugs by a modified solvent evaporation method. I. Effect of process and formulation variables on drug entrapment. *Journal of microencapsulation,* 7(3), 347–355.

Ameller, T., Marsaud, V., Legrand, P., Gref, R. and Renoir, J.M. (2004) Pure antiestrogen RU 58668-loaded nanospheres: morphology, cell activity and toxicity studies. *European Journal of Pharmaceutical Sciences,* 21(2–3), 361–70.

Anderson, J.M. and Shive, M.S. (1997) Biodegradation and biocompatibility of PLA and PLGA microspheres. *Advanced Drug Delivery Reviews,* 28(1), 5–24.

Anhorn, M.G., Mahler, H.C. and Langer, K. (2008) Freeze drying of human serum albumin (HSA) nanoparticles with different excipients. *International Journal of Pharmaceutics,* 363(1–2), 162–169.

Argin-Soysala, S., Kofinasb, P. and Martin-Lo, Y. (2009) Effect of complexation conditions on xanthan–chitosan polyelectrolyte complex gels. *Food Hydrocolloids,* 23, 202–209.

Arroyo-Maya, I.J., Rodiles-Lopez, J.O., Cornejo-Mazon, M., *et al.* (2012) Effect of different treatments on the ability of alpha-lactalbumin to form nanoparticles. *Journal of Dairy Science,* 95(11), 6204–6214.

Astete, C.E. and Sabliov, C.M. (2006) Synthesis and characterization of PLGA nanoparticles. *Journal of Biomaterials Science-Polymer Edition,* 17(3), 247–289.

Astete, C.E., Sabliov, C.M., Watanabe, F. and Biris, A. (2009) Ca^{2+} cross-linked alginic acid nanoparticles for solubilization of lipophilic natural colorants. *Journal of Agricultural and Food Chemistry,* 57(16), 7505–7512.

Astete, C.E., Dolliver, D., Whaley, M., Khachatryan, L. and Sabliov, C.M. (2011) Antioxidant poly(lactic-co-glycolic) acid nanoparticles made with alpha-tocopherol-ascorbic acid surfactant. *ACS Nano,* 5(12), 9313–9325.

Augst, A.D., Kong, H.J. and Mooney, D. J. (2006) Alginate hydrogels as biomaterials. *Macromolecular bioscience,* 6(8), 623–633.

Aukunuru, J.V., Ayalasomayajula, S.P. and Kompella, U.B. (2003) Nanoparticle formulation enhances the delivery and activity of a vascular endothelial growth factor antisense oligonucleotide in human retinal pigment epithelial cells. *The Journal of Pharmacy and Pharmacology,* 55(9), 1199–1206.

Avgoustakis, K. (2004) Pegylated poly(lactide) and poly(lactide-co-glycolide) nanoparticles: preparation, properties and possible applications in drug delivery. *Current drug delivery,* 1(4), 321–333.

Avgoustakis, K., Beletsi, A., Panagi, Z., Klepetsanis, P., Karydas, A.G. and Ithakissios, D.S. (2002) PLGA-mPEG nanoparticles of cisplatin: *in vitro* nanoparticle degradation, *in vitro* drug release and *in vivo* drug residence in blood properties. *Journal of Controlled Release,* 79(1–3), 123–135.

Bala, I., Hariharan, S. and Kumar, M.N.V.R. (2004) PLGA nanoparticles in drug delivery: The state of the art. *Critical Reviews in Therapeutic Drug Carrier Systems,* 21(5), 387–422.

Bansal, A., Kapoor, D.N., Kapil, R., Chhabra, N. and Dhawan, S. (2011) Design and development of paclitaxel-loaded bovine serum albumin nanoparticles for brain targeting. *Acta Pharmaceutica,* 61(2), 141–156.

Beletsi, A., Leontiadis, L., Klepetsanis, P., Ithakissios, D.S. and Avgoustakis, K. (1999) Effect of preparative variables on the properties of poly(DL-lactide-co-glycolide) methoxypoly(ethyleneglycol) copolymers related to their application in controlled drug delivery. *International Journal of Pharmaceutics,* 182(2), 187–197.

Berger, J., Reist, M., Mayer, J.M., Felt, O. and Gurny, R. (2004a) Structure and interactions in chitosan hydrogels formed by complexation or aggregation for biomedical applications. *European Journal of Pharmaceutics and Biopharmaceutics,* 57(1), 35–52.

Berger, J., Reist, M., Mayer, J.M., Felt, O., Peppas, N.A. and Gurny, R. (2004b) Structure and interactions in covalently and ionically crosslinked chitosan hydrogels for biomedical applications. *European Journal of Pharmaceutics and Biopharmaceutics,* 57(1), 19–34.

Borges, O., Cordeiro-Da-Silva, A., Romeijn, S.G., *et al.* (2006) Uptake studies in rat Peyer's patches, cytotoxicity and release studies of alginate coated chitosan nanoparticles for mucosal vaccination. *Journal of Controlled Release,* 114(3), 348–358.

Bouchemal, K., Briancon, S., Perrier, E. and Fessi, H. (2004) Nano-emulsion formulation using spontaneous emulsification: solvent, oil and surfactant optimisation. *International Journal of Pharmaceutics,* 280(1–2), 241–251.

Bouwmeester, H., Dekkers, S., Noordam, M.Y., *et al.* (2009) Review of health safety aspects of nanotechnologies in food production. *Regulatory Toxicology and Pharmacology,* 53(1), 52–62.

Brady, J.M., Cutright, D.E., Miller, R.A. and Barristone, G.C. (1973) Resorption rate, route, route of elimination, and ultrastructure of the implant site of polylactic acid in the abdominal wall of the rat. *Journal of Biomedical Materials Research,* 7(2), 155–66.

Britten, M. and Giroux, H.J. (2001) Acid-induced gelation of whey protein polymers: effects of pH and calcium concentration during polymerization. *Food Hydrocolloids,* 15(4–6), 609–617.

Bronich, T.K., Keifer, P.A., Shlyakhtenko, L.S. and Kabanov, A.V. (2005) Polymer micelle with cross-linked ionic core. *Journal of the American Chemical Society,* 127(23), 8236–8237.

Brownlow, S., Cabral, J.H.M., Cooper, R., *et al.* (1997) Bovine beta-lactoglobulin at 1.8 angstrom resolution – still an enigmatic lipocalin. *Structure,* 5(4), 481–495.

Cabra, V., Areguin, R., Galvez, A., Quirasco, M., Vasquez-Duhalt, R. and Farres, A. (2005) Characterization of a 19 kDa alpha-zein of high purity. *Journal of Agricultural and Food Chemistry,* 53(3), 725–729.

Chanasattru, W., Jones, O.G., Decker, E.A. and Mcclements, D.J. (2009) Impact of cosolvents on formation and properties of biopolymer nanoparticles formed by heat treatment of beta-lactoglobulin–pectin complexes. *Food Hydrocolloids,* 23(8), 2450–2457.

Chellat, F., Tabrizian, M., Dumitriu, S., *et al.* (2000) *In vitro* and *in vivo* biocompatibility of chitosan–xanthan polyionic complex. *Journal of Biomedical Materials Research,* 51(1), 107–116.

Chung, T.W., Huang, Y.Y., Tsai, Y.L. and Liu, Y.Z. (2002) Effects of solvent evaporation rate on the properties of protein-loaded PLLA and PDLLA microspheres fabricated by emulsion-solvent evaporation process. *Journal of Microencapsulation,* 19(4), 463–471.

Couvreur, P., Tulkens, P., Roland, M., Trouet, A. and Speiser, P. (1977) Nanocapsules – new type of lysosomotropic carrier. *Febs Letters,* 84(2), 323–326.

Couvreur, P., Kante, B., Roland, M., Guiot, P., Bauduin, P. and Speiser, P. (1979a) Polycyanoacrylate nanocapsules as potential lysosomotropic carriers – preparation, morphological and sorptive properties. *Journal of Pharmacy and Pharmacology,* 31(5), 331–332.

Couvreur, P., Kante, B., Roland, M. and Speiser, P. (1979b) Adsorption of anti-neoplastic drugs to polyalkylcyanoacrylate nanoparticles and their release in calf serum. *Journal of Pharmaceutical Sciences,* 68(12), 1521–1524.

Crisante, F., Francolini, I., Bellusci, M., Martinelli, A., D'ilario, L. and Piozzi, A. (2009) Antibiotic delivery polyurethanes containing albumin and polyallylamine nanoparticles. *European Journal of Pharmaceutical Sciences,* 36(4–5), 555–564.

Csaba, N., Gonzalez, L., Sanchez, A. and Alonso, M.J. (2004) Design and characterisation of new nanoparticulate polymer blends for drug delivery. *Journal of Biomaterials Science-Polymer Edition,* 15(9), 1137–1351.

Csaba, N., Caamano, P., Sanchez, A., Dominguez, F. and Alonso, M.J. (2005) PLGA:poloxamer and PLGA:poloxamine blend nanoparticles: new carriers for gene delivery. *Biomacromolecules,* 6(1), 271–278.

Dario, A.F., Hortencio, L.M.A., Sierakowski, M.R., Neto, J.C.Q. and Petri, D.F.S. (2011) The effect of calcium salts on the viscosity and adsorption behavior of xanthan. *Carbohydrate Polymers,* 84(1), 669–676.

De Jaeghere, F., Doelker, E. and Gurny, R. (1999) Nanoparticles. In: *Encyclopedia of Control Drug Delivery* (ed. E. Mathiowitz). John Wiley and Sons, Inc., New York, 641–664.

Dhandayuthapani, B., Poulose, A.C., Nagaoka, Y., *et al.* (2012) Biomimetic smart nanocomposite: *in vitro* biological evaluation of zein electrospun fluorescent nanofiber encapsulated CdS quantum dots. *Biofabrication,* 4(2), 025008.

Dillen, K., Vandervoort, J., Van Den Mooter, G., Verheyden, L. and Ludwig, A. (2004a) Factorial design, physicochemical characterisation and activity of ciprofloxacin-PLGA nanoparticles. *International Journal of Pharmaceutics,* 275(1–2), 171–187.

Dillen, K., Weyenberg, W., Vandervoort, J. and Ludwig, A. (2004b) The influence of the use of viscosifying agents as dispersion media on the drug release properties from PLGA nanoparticles. *European Journal of Pharmceutics and Biopharmaceutics,* 58(3), 539–549.

Dreis, S., Rothweller, F., Michaelis, A., Cinatl, J., Kreuter, J. and Langer, K. (2007) Preparation, characterisation and maintenance of drug efficacy of doxorubicin-loaded human serum albumin (HSA) nanoparticles. *International Journal of Pharmaceutics,* 341(1–2), 207–214.

Dumitriu, S. and Chornet, E. (1997) Immobilization of xylanase in chitosan–xanthan hydrogels. *Biotechnology Progress,* 13(5), 539–545.

Eley, J.G., Pujari, V.D. and Mclane, J. (2004) Poly(lactide-co-glycolide) nanoparticles containing coumarin-6 for suppository delivery: *in vitro* release profile and *in vivo* tissue distribution. *Drug Delivery,* 11(4), 255–261.

Elzoghby, A.O., El-Fotoh, W.S.A. and Elgindy, N.A. (2011) Casein-based formulations as promising controlled release drug delivery systems. *Journal of Controlled Release,* 153(3), 206–216.

Elzoghby, A.O., Samy, W.M. and Elgindy, N.A. (2012a) Albumin-based nanoparticles as potential controlled release drug delivery systems. *Journal of Controlled Release,* 157(2), 168–182.

Elzoghby, A.O., Samy, W.M. and Elgindy, N.A. (2012b) Protein-based nanocarriers as promising drug and gene delivery systems. *Journal of Controlled Release,* 161(1), 38–49.

Elzoghby, A.O., Samy, W.M. and Elgindy, N.A. (2013) Novel spray-dried genipin-crosslinked casein nanoparticles for prolonged release of alfuzosin hydrochloride. *Pharmaceutical Research,* 30(2), 512–522.

Esmaeilzadeh, P., Fakhroueian, Z. and Beigi, A.A.M. (2011) Synthesis of biopolymeric alpha-lactalbumin protein nanoparticles and nanospheres as green nanofluids using in drug delivery and food technology. *Journal of Nano Research,* 16, 89–96.

Fonseca, C., Simoes, S. and Gaspar, R. (2002) Paclitaxel-loaded PLGA nanoparticles: preparation, physicochemical characterization and *in vitro* anti-tumoral activity. *Journal of Controlled Release,* 83(2), 273–286.

Forster, S.P., Olveira, S. and Seeger, S. (2011) Nanotechnology in the market: promises and realities. *International Journal of Nanotechnology,* 8(6–7), 592–613.

Frassineti, G.L., Ibrahim, T., Zoli, W., *et al.* (2002) Docetaxel followed by gemcitabine in the treatment of advanced non-small cell lung cancer: a phase I study. *Tumori,* 88(2), 99–103.

Garcia, M., Forbe, T. and Gonzalez, E. (2010) Potential applications of nanotechnology in the agro-food sector. *Ciencia E Tecnologia De Alimentos,* 30(3), 573–581.

Giroux, H.J. and Britten, M. (2011) Encapsulation of hydrophobic aroma in whey protein nanoparticles. *Journal of Microencapsulation,* 28(5), 337–343.

Giroux, H.J., Houde, J. and Britten, M. (2010) Preparation of nanoparticles from denatured whey protein by pH-cycling treatment. *Food Hydrocolloids,* 24(4), 341–346.

Gomez-Estaca, J., Balaguer, M.P., Gavara, R. and Hernandez-Munoz, P. (2012) Formation of zein nanoparticles by electrohydrodynamic atomization: Effect of the main processing variables and suitability for encapsulating the food coloring and active ingredient curcumin. *Food Hydrocolloids,* 28(1), 82–91.

Govender, T., Stolnik, S., Garnett, M.C., Illum, L. and Davis, S.S. (1999) PLGA nanoparticles prepared by nano-precipitation: drug loading and release studies of a water soluble drug. *Journal of Controlled Release,* 57(2), 171–185.

Gulseren, I., Fang, Y. and Corredig, M. (2012a) Whey protein nanoparticles prepared with desolvation with ethanol: Characterization, thermal stability and interfacial behavior. *Food Hydrocolloids,* 29(2), 258–264.

Gulseren, I., Fang, Y. and Corredig, M. (2012b) Zinc incorporation capacity of whey protein nanoparticles prepared with desolvation with ethanol. *Food Chemistry,* 135(2), 770–774.

Gunasekaran, S., Ko, S. and Xiao, L. (2007) Use of whey proteins for encapsulation and controlled delivery applications. *Journal of Food Engineering,* 83(1), 31–40.

Haham, M., Ish-Shalom, S., Nodelman, M., *et al.* (2012) Stability and bioavailability of vitamin D nanoencapsulated in casein micelles. *Food and Function,* 3(7), 737–744.

Hasenhuettl, G.L. (2008) Overview of food emulsifiers. In: *Food Emulsifiers and Their Applications,* 2nd edn (eds G.L. Hasenhuettl and R.W. Hartel) Springer, Berlin, 1–9.

Honarkar, H. and Barikani, M. (2009) Applications of biopolymers I: chitosan. *Monatshefte Fur Chemie,* 140(12), 1403–1420.

Horne, D.S. (2002) Casein structure, self-assembly and gelation. *Current Opinion in Colloid and Interface Science,* 7(5–6), 451–461.

Horne, D.S. (2009) Casein micelles structure and stability. In: *Milk Proteins: From Expression to Food* (eds A. Thompson, M. Boland and H. Singh). Academic Press, San Diego, 133–162.

Hu, D.D., Lin, C.C., Liu, L., Li, S.N. and Zhao, Y.P. (2012) Preparation, characterization, and *in vitro* release investigation of lutein/zein nanoparticles via solution enhanced dispersion by supercritical fluids. *Journal of Food Engineering,* 109(3), 545–552.

Hu, Y.J., Liu, Y., Sun, T.Q., Bai, A.M., Lu, H.Q. and Pi, Z.B. (2006) Binding of anti-inflammatory drug cromolyn sodium to bovine serum albumin. *International Journal of Biological Macromolecules,* 39(4–5), 280–285.

Huang, Q.R., Yu, H.L. and Ru, Q.M. (2010) Bioavailability and Delivery of Nutraceuticals Using Nanotechnology. *Journal of Food Science,* 75(1), R50–R57.

Hurtado-Lopez, P. and Murdan, S. (2006) Zein microspheres as drug/antigen carriers: A study of their degradation and erosion, in the presence and absence of enzymes. *Journal of Microencapsulation,* 23(3), 303–314.

Ibrahim, N.K., Desai, N., Legha, S., *et al.* (2002a) Phase I and pharmacokinetic study of ABI-007, a cremophor-free, protein-stabilized, nanoparticle formulation of paclitaxel. *Clinical Cancer Research,* 8(5), 1038–1044.

Ibrahim, N.K., Samuels, B., Page, R., *et al.* (2002b) Efficacy and dose-dependent activity of ABI-007; a cremophor-free, nanoparticle paclitaxel, in first-line metastatic breast cancer: integrated results of 2 phase II trials. *Breast Cancer Research and Treatment,* 76, S131–S131.

Jahanshahi, M. and Babaei, Z. (2008) Protein nanoparticle: a unique system as drug delivery vehicles. *African Journal of Biotechnology,* 7(25), 4926–4934.

Jain, R.A. (2008) The manufacturing techniques of various drug loaded biodegradable poly(lactide-co-glycolide) (PLGA) devices. *Biomaterials,* 21(23), 2475–2490.

Jalil, R. and Nixon, J.R. (1990) Biodegradable poly(lactic acid) and poly(lactide-co-glycolide) microcapsules: problems associated with preparative techniques and release properties. *Journal of Microencapsulation,* 7(3), 297–325.

Jeon, H.J., Jeong, J.I., Jang, M.K., Park, Y.H. and Nah, J.W. (2000a) Effect of solvent on the preparation of surfactant-free poly(DL-lactide-co-glycolide) nanoparticles and norfloxacin release characteristics. *International Journal of Pharmaceutics,* 207(1–2), 99–108.

Jeon, H.J., Jeong, Y.I., Jang, M.K., Park, Y.H. and Nah, J.W. (2000b) Effect of solvent on the preparation of surfactant-free poly(DL-lactide-co-glycolide) nanoparticles and norfloxacin release characteristics. *International Journal of Pharmaceutics,* 207(1–2), 99–108.

Jones, O.G., Decker, E.A. and Mcclements, D.J. (2009) Formation of biopolymer particles by thermal treatment of beta-lactoglobulin-pectin complexes. *Food Hydrocolloids,* 23(5), 1312–1321.

Jones, O., Decker, E.A. and Mcclements, D.J. (2010a) Thermal analysis of beta-lactoglobulin complexes with pectins or carrageenan for production of stable biopolymer particles. *Food Hydrocolloids,* 24(2–3), 239–248.

Jones, O.G., Lesmes, U., Dubin, P. and Mcclements, D.J. (2010b) Effect of polysaccharide charge on formation and properties of biopolymer nanoparticles created by heat treatment of beta-lactoglobulin-pectin complexes. *Food Hydrocolloids,* 24(4), 374–383.

Julienne, M.C., Alonso, M.J., Amoza, J.L.G. and Benoit, J.P. (1992) Preparation of poly(D,L-lactide glycolide) nanoparticles of controlled particle-size distribution – application of experimental-designs. *Drug Development and Industrial Pharmacy,* 18(10), 1063–1077.

Jun, J.Y., Hoang, H.N., Paik, S.Y.R., Chun, H.S., Kang, B.C. and Ko, S. (2011) Preparation of size-controlled bovine serum albumin (BSA) nanoparticles by a modified desolvation method. *Food Chemistry,* 127(4), 1892–1898.

Kamau, S.M., Cheison, S.C., Chen, W., Liu, X.M. and Lu, R.R. (2010) Alpha-lactalbumin: its production technologies and bioactive peptides. *Comprehensive Reviews in Food Science and Food Safety,* 9(2), 197–212.

Ko, S. and Gunasekaran, S. (2006) Preparation of sub-100-nm beta-lactoglobulin (BLG) nanoparticles. *Journal of Microencapsulation,* 23(8), 887–898.

Konan, Y.N., Gurny, R. and Allemann, E. (2002) Preparation and characterization of sterile and freeze-dried sub-200 nm nanoparticles. *International Journal of Pharmaceutics,* 233(1–2), 239–252.

Konan, Y.N., Cerny, R., Favet, J., Berton, M., Gurny, R. and Allemann, E. (2003) Preparation and characterization of sterile sub-200 nm meso-tetra(4-hydroxylphenyl)porphyrin-loaded nanoparticles for photodynamic therapy. *European Journal of Pharmaceutical Sciences,* 55(1), 115–124.

Kontopidis, G., Holt, C. and Sawyer, L. (2002) The ligand-binding site of bovine beta-lactoglobulin: Evidence for a function? *Journal of Molecular Biology,* 318(4), 1043–1055.

Kontopidis, G., Holt, C. and Sawyer, L. (2004) Invited Review: beta-lactoglobulin: Binding properties, structure, and function. *Journal of Dairy Science,* 87(4), 785–796.

Kreuter, J. (1983a) Evaluation of Nanoparticles as Drug-Delivery Systems .1. Preparation Methods. *Pharmaceutica Acta Helvetiae,* 58(7), 196–209.

Kreuter, J. (1983b) Evaluation of nanoparticles as drug-delivery systems. 2. Comparison of the body distribution of nanoparticles with the body distribution of microspheres (diameter-greater-than-1-mu-m), liposomes, and emulsions. *Pharmaceutica Acta Helvetiae,* 58(8), 217–226.

Kreuter, J. (1983c) Evaluation of nanoparticles as drug-delivery systems. 3. Materials, stability, toxicity, possibilities of targeting, and use. *Pharmaceutica Acta Helvetiae,* 58(9–10), 242–250.

Kreuter, J. (1983d) Physicochemical characterization of polyacrylic nanoparticles. *International Journal of Pharmaceutics,* 14(1), 43–58.

Kreuter, J., Mills, S.N., Davis, S.S. and Wilson, C.G. (1983) Polybutylcyanoacrylate nanoparticles for the delivery of [norcholestenol-se-75. *International Journal of Pharmaceutics,* 16(1), 105–113.

Kuan, C.Y., Yee-Fung, W., Yuen, K.H. and Liong, M.T. (2012) Nanotech: propensity in foods and bioactives. *Critical Reviews in Food Science and Nutrition,* 52(1–3), 55–71.

Kumar, M.N., Mohapatra, S.S., Kong, X., Jena, P.K., Bakowsky, U. and Lehr, C.M. (2004) Cationic poly(lactide-co-glycolide) nanoparticles as efficient *in vivo* gene transfection agents. *Journal of Nanoscience and Nanotechnology,* 4(8), 990–994.

Kwon, H.Y., Lee, J.Y., Choi, S.W., Jang, Y.S. and Kim, J.H. (2001) Preparation of PLGA nanoparticles containing estrogen by emulsification–diffusion method. *Colloids and Surfaces a-Physicochemical and Engineering Aspects,* 182(1–3), 123–130.

Lamprecht, A., Ubrich, N., Yamamoto, H., *et al.* (2001) Biodegradable nanoparticles for targeted drug delivery in treatment of inflammatory bowel disease. *Journal of Pharmacology and Experimental Therapeutics,* 299(2), 775–781.

Langer, R. (1980) Polymeric delivery systems for controlled drug release. *Chemical Engineering Communications,* 6(1–3), 1–48.

Langer, R., Conn, H., Vacanti, J., Haudenschild, C. and Folkman, J. (1980a) Control of tumor-growth in animals by infusion of an angiogenesis inhibitor. *Proceedings of the National Academy of Sciences of the United States of America-Biological Sciences,* 77(7), 4331–4335.

Langer, R., Hsieh, D.S.T., Rhine, W. and Folkman, J. (1980b) Control of release kinetics of macromolecules from polymers. *Journal of Membrane Science,* 7(3), 333–350.

Lee, K.N., Ye, Y., Carr, J.H., Karem, K. and D'souza, M.J. (2011) Formulation, pharmacokinetics and biodistribution of Ofloxacin-loaded albumin microparticles and nanoparticles. *Journal of Microencapsulation,* 28(5), 363–369.

Lee, S.H., Heng, D., Ng, W.K., Chan, H.K. and Tan, R.B.H. (2011) Nano spray drying: A novel method for preparing protein nanoparticles for protein therapy. *International Journal of Pharmaceutics,* 403(1–2), 192–200.

Lee, S.J., Jeong, J.R., Shin, S.C., *et al.* (2005) Magnetic enhancement of iron oxide nanoparticles encapsulated with poly(D,L-latide-co-glycolide) *Colloids and Surfaces A – Physicochemical and Engineering Aspects,* 255(1–3), 19–25.

Lee, W.K., Park, J.Y., Yang, E.H., *et al.* (2002) Investigation of the factors influencing the release rates of cyclosporin A-loaded micro- and nanoparticles prepared by high-pressure homogenizer. *Journal of Controlled Release,* 84(3), 115–123.

Lertsutthiwong, P., Noomun, K., Jongaroonngamsang, N., Rojsitthisak, P. and Nimmannit, U. (2008) Preparation of alginate nanocapsules containing turmeric oil. *Carbohydrate Polymers,* 74(2), 209–214.

Livney, Y. D. (2010) Milk proteins as vehicles for bioactives. *Current Opinion in Colloid and Interface Science,* 15(1–2), 73–83.

Lopez-Quintela, M. A. (2003) Synthesis of nanomaterials in microemulsions: formation mechanisms and growth control. *Current Opinion in Colloid and Interface Science,* 8(2), 137–144.

Luo, Y.C., Zhang, B.C., Cheng, W.H. and Wang, Q. (2010) Preparation, characterization and evaluation of selenite-loaded chitosan/TPP nanoparticles with or without zein coating. *Carbohydrate Polymers,* 82(3), 942–951.

Luo, Y.C., Zhang, B.C., Whent, M., Yu, L.L. and Wang, Q. (2011) Preparation and characterization of zein/chitosan complex for encapsulation of alpha-tocopherol, and its *in vitro* controlled release study. *Colloids and Surfaces B-Biointerfaces,* 85(2), 145–152.

Luo, Y.C., Teng, Z. and Wang, Q. (2012) Development of zein nanoparticles coated with carboxymethyl chitosn for encapsulation and controlled release of vitamin D_3. *Journal of Agricultural and Food Chemistry,* 60(3), 836–843.

Mainardes, R.M. and Evangelista, R.C. (2005) PLGA nanoparticles containing praziquantel: effect of formulation variables on size distribution. *International Journal of Pharmaceutics,* 290(1–2), 137–144.

Martins, S., Sarmento, B., Souto, E.B. and Ferreira, D.C. (2007) Insulin-loaded alginate microspheres for oral delivery – Effect of polysaccharide reinforcement on physicochemical properties and release profile. *Carbohydrate Polymers,* 69(4), 725–731.

Mehravar, R., Jahanshahi, M. and Saghatoleslami, N. (2009) Production of biological nanoparticles from alpha-lactalbumin for drug delivery and food science application. *African Journal of Biotechnology,* 8(24), 6822–6827.

Mehta, S.K., Kaur, G. and Verma, A. (2011) Fabrication of plant protein microspheres for encapsulation, stabilization and *in vitro* release of multiple anti-tuberculosis drugs. *Colloids and Surfaces A – Physicochemical and Engineering Aspects,* 375(1–3), 219–230.

Mohammed, F.A. and Hassan, M.A. (2003) Formulation and evaluation of ciprofloxacin hydrochloride and norfloxacin microspheres prepared by an enhanced emulsion-solvent evaporation process. *STP Pharma Sciences,* 13(5), 319–327.

Mu, L. and Feng, S.S. (2003) A novel controlled release formulation for the anticancer drug paclitaxel (Taxol): PLGA nanoparticles containing vitamin E TPGS. *Journal of Controlled Release,* 86(1), 33–48.

Murugeshu, A., Astete, C., Leonardi, C., Morgan, T. and Sabliov, C.M. (2011) Chitosan/PLGA particles for controlled release of alpha-tocopherol in the GI tract via oral administration. *Nanomedicine,* 6(9), 1513–1528.

Nagpal, K., Singh, S.K. and Mishra, D.N. (2010) Chitosan nanoparticles: a promising system in novel drug delivery. *Chemical and Pharmaceutical Bulletin,* 58(11), 1423–1430.

Nakache, E., Poulain, N., Candau, F., Orecchioni, A.M. and Irache, J.M. (2000) Biopolymer and polymer nanoparticles and their biomedical applications. In: *Handbook of Nanostructured Materials and Nanotechnology* (ed. H.S. Nalwa). Academic Press, New York, 5(11), 577–635.

Neethirajan, S. and Jayas, D.S. (2011) Nanotechnology for the food and bioprocessing industries. *Food and Bioprocess Technology,* 4(1), 39–47.

Nesamony, J., Singh, P.R., Nada, S.E., Shah, Z.A. and Kolling, W.M. (2012) Calcium alginate nanoparticles synthesized through a novel interfacial cross-linking method as a potential protein drug delivery system. *Journal of Pharmaceutical Sciences,* 101(6), 2177–2184.

Nickols-Richardson, S.M. and Piehowski, K.E. (2008) Nanotechnology in nutritional sciences. *Minerva Biotecnologica,* 20(3), 117–126.

Niwa, T., Takeuchi, H., Hino, T., Kunou, N. and Kawashima, Y. (1993) Preparations of biodegradable nanospheres of water-soluble and insoluble drugs with D,L-lactide glycolide copolymer by a novel spontaneous emulsification solvent diffusion method, and the drug release behavior. *Journal of Controlled Release,* 25(1–2), 89–98.

Oakenfull, D.G. (1991) The chemistry of high-methoxyl pectins. In: *The Chemistry and Technology of Pectin* (ed. R.H. Walter). Academic Press, San Diego.

Obeidat, W.M. and Price, J.C. (2003) Viscosity of polymer solution phase and other factors controlling the dissolution of theophylline microspheres prepared by the emulsion solvent evaporation method. *Journal of Microencapsulation,* 20(1), 57–65.

Oster, C.G., Wittmar, M., Unger, F., Barbu-Tudoran, L., Schaper, A.K. and Kissel, T. (2004) Design of amine-modified graft polyesters for effective gene delivery using DNA-loaded nanoparticles. *Pharmaceutical Research,* 21(6), 927–931.

Pahuja, P., Arora, S. and Pawar, P. (2012) Ocular drug delivery system: a reference to natural polymers. *Expert Opinion on Drug Delivery,* 9(7), 837–861.

Panagi, Z., Beletsi, A., Evangelatos, G., Livaniou, E., Ithakissios, D.S. and Avgoustakis, K. (2001) Effect of dose on the biodistribution and pharmacokinetics of PLGA and PLGA-mPEG nanoparticles. *International Journal of Pharmaceutics,* 221(1–2), 143–152.

Panyam, J., Williams, D., Dash, A., Leslie-Pelecky, D. and Labhasetwar, V. (2004) Solid-state solubility influences encapsulation and release of hydrophobic drugs from PLGA/PLA nanoparticles. *Journal of Pharmaceutical Sciences,* 93(7), 1804–1814.

Patel, A., Hu, Y.C., Tiwari, J.K. and Velikov, K.P. (2010) Synthesis and characterisation of zein-curcumin colloidal particles. *Soft Matter,* 6(24), 6192–6199.

Patel, A.R., Heussen, P.C.M., Hazekamp, J., Drost, E. and Velikov, K.P. (2012) Quercetin loaded biopolymeric colloidal particles prepared by simultaneous precipitation of quercetin with hydrophobic protein in aqueous medium. *Food Chemistry,* 133(2), 423–429.

Paul, W. and Sharma, C.P. (2000) Chitosan, a drug carrier for the 21st century: a review. *Stp Pharma Sciences,* 10(1), 5–22.

Pietzonka, P., Walter, E., Duda-Johner, S., Langguth, P. and Merkle, H.P. (2002) Compromised integrity of excised porcine intestinal epithelium obtained from the abattoir affects the outcome of *in vitro* particle uptake studies. *European Journal of Pharmaceutical Sciences,* 15(1), 39–47.

Podaralla, S. and Perumal, O. (2010) Preparation of Zein Nanoparticles by pH Controlled Nanoprecipitation. *Journal of Biomedical Nanotechnology,* 6(4), 312–317.

Portnaya, I., Cogan, U., Livney, Y.D., Ramon, O., Shimon, K., Rosenberg, M. and Danino, D. (2006) Micellization of bovine beta-casein studied by isothermal titration microcalorimetry and cryogenic transmission electron microscopy. *Journal of Agricultural and Food Chemistry,* 54, 5555–5561.

Prabha, S. and Labhasetwar, V. (2004) Critical determinants in PLGA/PLA nanoparticle-mediated gene expression. *Pharmaceutical Research,* 21(2), 354–364.

Prakobvaitayakit, M. and Nimmannit, U. (2003) Optimization of polylactic-co-glycolic acid nanoparticles containing itraco\nazole using 2(3) factorial design. *AAPS PharmSciTech,* 4(4), 1–9.

Qi, J.N., Yao, P., He, F., Yu, C.L. and Huang, C.O. (2010) Nanoparticles with dextran/chitosan shell and BSA/chitosan core-Doxorubicin loading and delivery. *International Journal of Pharmaceutics,* 393(1–2), 176–184.

Rashidi, L. and Khosravi-Darani, K. (2011) The applications of nanotechnology in food industry. *Critical Reviews in Food Science and Nutrition,* 51(8), 723–730.

Rastogi, R., Sultana, Y., Aqil, M., Ali, A., Kumar, S., Chuttani, K. and Mishra, A.K. (2007) Alginate microspheres of isoniazid for oral sustained drug delivery. *International Journal of Pharmaceutics,* 334(1–2), 71–77.

Ravi Kumar, M.N., Bakowsky, U. and Lehr, C.M. (2004) Preparation and characterization of cationic PLGA nanospheres as DNA carriers. *Biomaterials,* 25(10), 1771–1777.

Reddy, I.M. and Kinsella, J.E. (1990) Interaction of beta-lactoglobulin with kappa-casein in micelles as assessed by chymosin hydrolysis – effect of temperature, heating time, beta-lactoglobulin concentration, and pH. *Journal of Agricultural and Food Chemistry,* 38(1), 50–58.

Reddy, I.M., Kella, N.K.D. and Kinsella, J.E. (1988) Structural and conformational basis of the resistance of beta-lactoglobulin to peptic and chymotryptic digestion. *Journal of Agricultural and Food Chemistry,* 36(4), 737–741.

Regier, M.C., Taylor, J.D., Borcyk, T., Yang, Y. and Pannier, A.K. (2012) Fabrication and characterization of DNA-loaded zein nanospheres. *Journal of Nanobiotechnology,* 1044.

Rodiles-Lopez, J.O., Arroyo-Maya, I.J., Jaramillo-Flores, *et al.* (2010) Effects of high hydrostatic pressure on the structure of bovine alpha-lactalbumin. *Journal of Dairy Science,* 93(4), 1420–1428.

Ron, N., Zimet, P., Bargarum, J. and Livney, Y.D. (2010) Beta-lactoglobulin-polysaccharide complexes as nanovehicles for hydrophobic nutraceuticals in non-fat foods and clear beverages. *International Dairy Journal,* 20(10), 686–693.

Sahoo, S.K., Panyam, J., Prabha, S. and Labhasetwar, V. (2002) Residual polyvinyl alcohol associated with poly (D,L-lactide-co-glycolide) nanoparticles affects their physical properties and cellular uptake. *Journal of Controlled Release,* 82(1), 105–114.

Sarmento, B., Ferreira, D., Veiga, F. and Ribeiro, A. (2006) Characterization of insulin-loaded alginate nanoparticles produced by ionotropic pre-gelation through DSC and FTIR studies. *Carbohydrate Polymers,* 66(1), 1–7.

Sarmento, B., Ribeiro, A.J., Veiga, F., Ferreira, D.C. and Neufeld, R.J. (2007) Insulin-loaded nanoparticles are prepared by alginate ionotropic pre-gelation followed by chitosan polyelectrolyte complexation. *Journal of Nanoscience and Nanotechnology,* 7(8), 2833–2841.

Saxena, V., Sadoqi, M. and Shao, J. (2004) Indocyanine green-loaded biodegradable nanoparticles: preparation, physicochemical characterization and *in vitro* release. *International Journal of Pharmaceutics,* 278(2), 293–301.

Schmitt, V., Leal-Calderon, F. and Bibette, J. (2003) Preparation of monodisperse particles and emulsions by controlled shear. *Topics in Current Chemistry,* 227, 195–215.

Semo, E., Kesselman, E., Danino, D. and Livney, Y.D. (2007) Casein micelle as a natural nano-capsular vehicle for nutraceuticals. *Food Hydrocolloids,* 21(5–6), 936–942.

Shapira, A., Assaraf, Y.G. and Livney, Y.D. (2010a) Beta-casein nanovehicles for oral delivery of chemotherapeutic drugs'. *Nanomedicine-Nanotechnology Biology and Medicine,* 6(1), 119–126.

Shapira, A., Markman, G., Assaraf, Y.G. and Livney, Y.D. (2010b) Beta-casein-based nanovehicles for oral delivery of chemotherapeutic drugs: drug-protein interactions and mitoxantrone loading capacity. *Nanomedicine-Nanotechnology Biology and Medicine,* 6(4), 547–555.

Shen, Z.Y., Li, Y., Kohama, K., Oneill, B. and Bi, J.X. (2011) Improved drug targeting of cancer cells by utilizing actively targetable folic acid-conjugated albumin nanospheres. *Pharmacological Research,* 63(1), 51–58.

Shi, L. and Gunasekaran, S. (2008) Preparation of Pectin-ZnO Nanocomposite. *Nanoscale Research Letters,* 3(12), 491–495.

Shpigelman, A., Israeli, G. and Livney, Y.D. (2010) Thermally-induced protein-polyphenol co-assemblies: beta lactoglobulin-based nanocomplexes as protective nanovehicles for EGCG. *Food Hydrocolloids,* 24(8), 735–743.

Shpigelman, A., Cohen, Y. and Livney, Y.D. (2012) Thermally-induced beta-lactoglobulin-EGCG nanovehicles: Loading, stability, sensory and digestive-release study. *Food Hydrocolloids,* 29(1), 57–67.

Singh, P., Kumari, K., Tomar, V., *et al.* (2011) A novel method to chemically bind thiazolidine-2,4-dione through cross-linked chitosan nanoparticles using malanodialdehyde as a cross-linker. *Canadian Journal of Chemistry – Revue Canadienne De Chimie,* 89(11), 1332–1342.

Somchue, W., Sermsri, W., Shiowatana, J. and Siripinyanond, A. (2009) Encapsulation of alpha-tocopherol in protein-based delivery particles. *Food Research International,* 42, 909–914.

Song, C.X., Labhasetwar, V., Murphy, H., *et al.* (1997) Formulation and characterization of biodegradable nanoparticles for intravascular local drug delivery. *Journal of Controlled Release,* 43(2–3), 197–212.

Sun, Q.S., Dong, J., Lin, Z.X., Yang, B. and Wang, J.Y. (2005) Comparison of cytocompatibility of zein film with other biomaterials and its degradability *in vitro*. *Biopolymers,* 78(5), 268–274.

Takada, S., Yamagata, Y., Misaki, M., Taira, K. and Kurokawa, T. (2003) Sustained release of human growth hormone from microcapsules prepared by a solvent evaporation technique. *Journal of Controlled Release,* 88(2), 229–242.

Tantra, R., Tompkins, J. and Quincey, P. (2010) Characterisation of the de-agglomeration effects of bovine serum albumin on nanoparticles in aqueous suspension. *Colloids and Surfaces B – Biointerfaces,* 75(1), 275–281.

Tice, T.R. and Cowsar, D.R. (1984) Biodegradable controlled-release parenteral systems. *Pharmaceutical Technology,* 1126–34.

Tonnesen, H.H. and Karlsen, J. (2002) Alginate in drug delivery systems. *Drug Development and Industrial Pharmacy,* 28(6), 621–630.

Vandervoort, J. and Ludwig, A. (2002) Biocompatible stabilizers in the preparation of PLGA nanoparticles: a factorial design study. *International Journal of Pharmaceutics,* 238(1–2), 77–92.

Vandervoort, J., Yoncheva, K. and Ludwig, A. (2004) Influence of the homogenisation procedure on the physicochemical properties of PLGA nanoparticles. *Chemical and Pharmaceutical Bulletin,* 52(11), 1273–1279.

Wang, A.Z., Langer, R. and Farokhzad, O.C. (2012) Nanoparticle delivery of cancer drugs. *Annual Review of Medicine,* 63, 185–198.

Wang, Q. and Padua, G.W. (2007) AGFD 124-Controlled self-organization of zein nanostructures for encapsulation of active food ingredients. *Abstracts of Papers of the American Chemical Society,* 233, 46–46.

Wang, X. H., Du, Y.M., Li, Y., Li, D. and Sun, R.C. (2011) Fluorescent identification and detection of *Staphylococcus aureus* with carboxymethyl chitosan/CdS quantum dots bioconjugates. *Journal of Biomaterials Science – Polymer Edition,* 22(14), 1881–1893.

Weber, C., Coester, C., Kreuter, J. and Langer, K. (2000a) Desolvation process and surface characterisation of protein nanoparticles. *International Journal of Pharmaceutics,* 194(1), 91–102.

Weber, C., Kreuter, J. and Langer, K. (2000b) Desolvation process and surface characteristics of HSA-nanoparticles. *International Journal of Pharmaceutics,* 196(2), 197–200.

Weber, C., Reiss, S. and Langer, K. (2000c) Preparation of surface modified protein nanoparticles by introduction of sulfhydryl groups. *International Journal of Pharmaceutics,* 211(1–2), 67–78.

Win, K.Y. and Feng, S.S. (2005) Effects of particle size and surface coating on cellular uptake of polymeric nanoparticles for oral delivery of anticancer drugs. *Biomaterials,* 26(15), 2713–2722.

Wu, Y.P., Luo, Y.G. and Wang, Q. (2012) Antioxidant and antimicrobial properties of essential oils encapsulated in zein nanoparticles prepared by liquid-liquid dispersion method. *Lwt-Food Science and Technology,* 48(2), 283–290.

Xu, H.L., Jiang, Q.R., Reddy, N. and Yang, Y.Q. (2011) Hollow nanoparticles from zein for potential medical applications. *Journal of Materials Chemistry,* 21(45), 18227–18235.

Xue, D.Q. and Sethi, R. (2012) Viscoelastic gels of guar and xanthan gum mixtures provide long-term stabilization of iron micro- and nanoparticles. *Journal of Nanoparticle Research,* 14(11), 1239.

Yang, L., Cui, F., Cun, D.M., Tao, A., Shi, K. and Lin, W.H. (2007) Preparation, characterization and biodistribution of the lactone form of 10-hydroxycamptothecin (HCPT)-loaded bovine serum albumin (BSA) nanoparticles. *International Journal of Pharmaceutics,* 340(1–2), 163–172.

Yu, S.Y., Yao, P., Jiang, M. and Zhang, G.Z. (2006) Nanogels prepared by self-assembly of oppositely charged globular proteins. *Biopolymers,* 83(2), 148–158.

Zhang, Y. and Zhuo, R.X. (2005) Synthesis and *in vitro* drug release behavior of amphiphilic triblock copolymer nanoparticles based on poly(ethylene glycol) and polycaprolactone. *Biomaterials,* 26(33), 6736–6742.

Zhang, Y.Y., Yang, Y., Tang, K., Hu, X. and Zou, G.L. (2008) Physicochemical characterization and antioxidant activity of quercetin-loaded chitosan nanoparticles. *Journal of Applied Polymer Science,* 107(2), 891–897.

Zhao, L. M., Shi, L.E., Zhang, Z.L., *et al.* (2011) Preparation and application of chitosan nanoparticles and nanofibers. *Brazilian Journal of Chemical Engineering,* 28(3), 353–362.

Zhao, X. and Goveas, J.L. (2001) Size selection in viscoelastic emulsions under shear. *Langmuir,* 17(13), 3788–3791.

Zigoneanu, I.G., Astete, C.E. and Sabliov, C. M. (2008) Nanoparticles with entrapped alpha-tocopherol: synthesis, characterization, and controlled release. *Nanotechnology,* 19(10), 105606.

Zimet, P., Rosenberg, D. and Livney, Y.D. (2011) Re-assembled casein micelles and casein nanoparticles as nano-vehicles for omega-3 polyunsaturated fatty acids. *Food Hydrocolloids,* 25(5), 1270–1276.

Zimmer, A.K., Zerbe, H. and Kreuter, J. (1994) Evaluation of pilocarpine-loaded albumin particles as drug-delivery systems for controlled delivery in the eye. 1. *In vitro* and *in vivo* characterization. *Journal of Controlled Release,* 32(1), 57–70.

Zimmer, A.K., Chetoni, P., Saettone, M.F., Zerbe, H. and Kreuter, J. (1995) Evaluation of pilocarpine-loaded albumin particles as controlled drug-delivery systems for the eye. 2. Coadministration with bioadhesive and viscous polymers. *Journal of Controlled Release,* 33(1), 31–46.

Zohri, M., Nomani, A.R., Haririan, I., Ehsani, M. and Mirdamadi, S. (2010) Chitosan/alginate nanoparticles for bactericidal protein delivery in food. *Journal of Biotechnology,* 150, S294–S295.

Zou, T., Li, Z., Percival, S.S., Bonard, S. and Gu, L.W. (2012) Fabrication, characterization, and cytotoxicity evaluation of cranberry procyanidins–zein nanoparticles. *Food Hydrocolloids,* 27(2), 293–300.

Zweers, M.L., Engbers, G.H., Grijpma, D.W. and Feijen, J. (2004) *In vitro* degradation of nanoparticles prepared from polymers based on DL-lactide, glycolide and poly(ethylene oxide). *Journal of Controlled Release,* 100(3), 347–356.

18 Encapsulation of bioactive compounds using electrospinning and electrospraying technologies

Loong-Tak Lim

Department of Food Science, University of Guelph, Guelph, Ontario, Canada

18.1 INTRODUCTION

Encapsulation is a versatile technique to enclose bioactives within polymeric solid, liquid or semi-solid shells. The objective of encapsulation is to protect active food components from deleterious environmental factors (e.g., light, oxygen, water). Another motive of encapsulation is to facilitate the delivery of active components in food, such as converting those that are in liquid state into dry particulates to simplify the handling process. Recent trends in the field of encapsulation have been focused on the development of methods to encapsulate various bioactives at the submicron level for various reasons, such as to enhance reactivity of the material with the surrounding medium, increased bioavailability through facilitating the absorption through the intestine and skin, and improved organoleptic properties of fortified food products.

Nanoparticles can be formed using "top-down" and "bottom-up" approaches. In the former, materials are broken down physically and/or chemically into smaller sizes, while in the latter, materials are assembled from molecular level into a nanosized complex system. Encapsulation typically exploits the bottom-up approach by which nanocapsules are typically produced via polymerization of monomers, self-assembly of polymers, salting out, precipitation, solvent evaporation, emulsification, and so on (Sanguansri and Augustin, 2006; Weiss *et al.*, 2006; Lim, 2011; Rao and Geckeler, 2011). Some of these methods are reviewed in detail elsewhere in this book, and therefore will not be further elaborated here. This chapter will focus on summarizing the advances of encapsulation of active compounds using electrospinning and electrospraying. While there are many variants of equipment set-up for these techniques, they are all based on the same basic principle of utilizing electrostatic force to spray polymer solutions/melts into small droplets (electrospraying) or spin them into ultrafine fibers (electrospinning). Through the manipulation of solution properties (e.g., concentration, viscosity, surface tension, electrical conductivity) and optimization of electrostatic processing conditions, fibers and particles with diameters ranging from hundreds of nanometers to tens of microns are achievable. This chapter reviews the basic principles of electrostatic encapsulation and discusses the key processing and solution parameters involved. A literature review is presented on recent advances on the encapsulation of food and related bioactives using electrospinning and electrospraying technologies.

Nanotechnology and Functional Foods: Effective Delivery of Bioactive Ingredients, First Edition.
Edited by Cristina M. Sabliov, Hongda Chen, and Rickey Y. Yada.

18.2 BASIC PRINCIPLES OF ELECTROSPINNING AND ELECTROSPRAYING

18.2.1 Electrospinning

The fundamentals of electrospinning are not new; literature related to electrospinning appeared as early as the 1930s and 1940s in a series of US patents assigned to Formhals and Gladding (Baumgarten, 1971). However, electrostatic forming methods started to receive heightened attention only within the past decade, due to the surge in interest on nanosized materials. In a typical process, the polymer fluid to be electrospun is delivered to a capillary spinneret by a pump (Figure 18.1). The application of electrical potential causes the pendant droplet at the spinneret to elongate towards the grounded collector. The resulting conical geometry is known as the "Taylor cone": named after Geoffrey Taylor, who studied in great detail the effects of an electric field on the shape and jetting phenomena of liquid droplets (Taylor, 1966; Taylor, 1969; Yarin *et al.*, 2001). When the applied voltage exceeds a certain critical level, the build-up of electrostatic charge repulsion on the surface of the droplet will eventually overcome the surface tension of the solution, causing a jet to erupt from the apex of the Taylor cone. Several complex phenomena occur as the jet takes flight in the air, including shearing of the visco-elastic polymer solution (extensional flow), chaotic whipping of the polymer jet, which is critical for fiber size reduction, evaporation of solvent, and solidification/deposition of the fibers on the grounded collector. To increase electrospinning production throughput, equipment that incorporates multiple

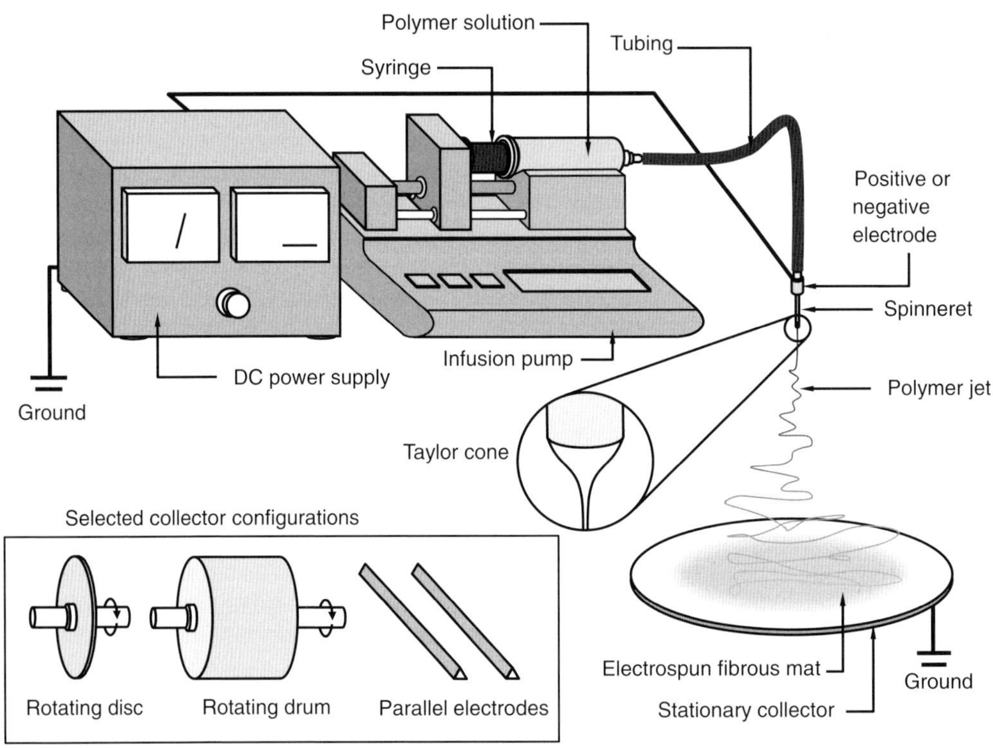

Figure 18.1 Typical experimental set-up for electrospinning, showing the major components, including DC power supply, infusion pump (or other pump), syringe to hold the polymer solution (or other reservoir), tubing to transfer the solution spinneret, and collector (stationary plate and other configurations).

spinnerets and free-surface electrospinning have been invented (Cengiz *et al.*, 2009; Zhou *et al.*, 2009). Depending on the rheological, molecular, and electrical characteristics of the polymer solution, fibers with submicron diameter can be produced (Grafe, 2003; Huang *et al.*, 2003).

Several studies have attempted to electrospin polymer melts instead of solutions. This is achieved by using a heated enclosure, extruder or laser to melt the polymer and electrically charge the melt with high voltage as described above (Lyons *et al.*, 2004; Zhou *et al.*, 2006; Ogata *et al.*, 2007). The absence of solvent in this process is attractive since many organic solvents are nonfood grade. Nevertheless, due to the elevated temperature involved, the polymer and the bioactive to be encapsulated may be susceptible to thermal degradation. Moreover, the resulting fibers tend to be larger in size as compared to those produced from polymer solutions, due to premature solidification of the fiber during spinning, which can hamper the chaotic whipping and stretching of the polymer jet. Because of these limitations, the majority of electrospinning studies have been focused on using polymer solutions.

18.2.2 Electrospraying

Similar to electrospinning, electrospraying involves charging a liquid located at the tip of a capillary nozzle with electric potential of several thousand volts (Cloupeau and Prunet-Foch, 1994). Instead of producing fibers seen in electrospinning processes, electrospraying produces aerosol droplets due to fission of the charged polymer solution. This is prevalent in polymer solutions that exhibit low viscoelasticity, such as those prepared from low-molecular-weight polymer, low solution concentration, and polymers that adopt compact morphology when dispersed in solution (Yu *et al.*, 2006; Regev *et al.*, 2010).

In electrospraying, since the aerosol particles are charged, they are self-dispersing, which is advantageous to prevent the aggregation of droplets, allowing the production of smaller droplets and a narrower size distribution as compared to a conventional spray-drying process (Jaworek, 2008; Quispe-Condori *et al.*, 2011). Another benefit of electrospraying is that the deposition efficiency is higher as compared to spray drying, since the charged droplets can be attracted towards a grounded or oppositely charged collector. The technique is very versatile and can be adapted for spraying a colloidal suspension of core material using different experimental configurations (Jaworek and Sobczyk, 2008).

18.2.3 Important polymer solution properties that affect electrospinning and electrospraying

Electrospinning and electrospraying behaviors of polymer solutions are governed by a number of parameters. First, the polymer should be dissolved in a solvent that has properties conducive for polymer jet formation. The charge repulsion established on the surface of the pendant droplet must be high enough to overcome the surface tension of the solution. Thus, polymer solutions that exhibit high surface tension will require greater applied voltage to initiate jetting than those with low surface tension. Polymer solutions with high surface tension tend to produce beads due to the tendency for the solution to form spherical droplets to reduce the surface area per unit mass. Similarly, polymer solutions of lower viscosity often produce beads due to the more dominant effect of surface tension than polymer–polymer chain entanglement. Here, solvent molecules tend to congregate into spherical droplets, rather than interacting with the polymer chains (Ramakrishna *et al.*, 2005).

In order to form continuous fiber, the polymer solutions must have optimal viscoelastic properties, which are determined mainly by the molecular weight of the polymer, polymer–solvent interaction, and concentration. Polymers with high molecular weight tend to promote chain entanglement in compatible solvents – an effect important for stabilizing the polymer jet during electrospinning, as well as preventing the formation of beaded fibers and droplets. On the other extreme, polymer solutions that are very viscous can be problematic, since jetting of the solution could be impeded (Alborzi *et al.*, 2010). The morphology of many proteins, especially globular protein, exhibits strong pH-dependent behaviors. When the pH of the polymer solution is below and above the isoelectric point, the net positive and negative electrostatic charges on the protein surface will result in repulsion, causing increased viscosity and increased protein–solvent interaction (Vega Lugo and Lim, 2008, 2012). The addition of a reducing agent (e.g., β-mercaptonol) capable of reducing the disulphide bonds in globular proteins and the use of a solvent (e.g., trifluoroethanol) to induce denaturation will improve the electrospinnability of proteins due to increased polymer chain entanglement (Dror *et al.*, 2008).

Another factor to be considered is the vapor pressure of the solvent. The solvent should be volatile enough to produce solidified fibers or particles as they deposit on the collector. For electrospinning fibers, solvents that are highly volatile may cause premature solidification of the polymer jet during spinning, thereby preventing the stretching of fiber essential for fiber size reduction. Electrospinning or electrospraying of polymer solutions containing highly volatile solvents tend to be unstable due to the formation of a skin layer on the surface of the Taylor cone as the solvent evaporates. To overcome this challenge, a second miscible but less volatile liquid, which may or may not be the solvent for the polymer, could be added to depress the vapor pressure of the main solvent. For instance, dimethylformamide has been added to chloroform to depress the vapor pressure of chloroform to enable continuous electrospinning of poly(lactic acid) (Zhou and Lim, 2009). The proportion of the binary solvent will need to be optimized to balance the polymer solubility and solvent evaporation rate.

The formation of a jet requires the migration of charge to the surface in order to establish the charge repulsion to overcome the surface tension of the polymer solution. Thus, electrical conductivity is another factor that governs the electrospinning process. Solutions of high conductivity will cause the surface charge to transfer more readily downstream, and therefore reduces the surface charge density and electrostatic force required to initiate jetting (Feng, 2002). In highly conductive solutions, electrical charges can also slip through the fluid during spinning, reducing the effective repulsion force needed for stretching the fiber. By comparison, solutions of lower electrical conductivity tend to allow the charges to "stick" on the polymer jet. As these charges are enacted by the electric field, the electrostatic force established is transmitted into the fluid, effectively stretching and thinning the jet (Reneker *et al.*, 2000; Feng, 2002; Carroll and Joo, 2006;). By contrast, such requirements are not as critical for electrospraying, since fission of a polymer solution to produce fine aerosol droplets is the main goal, instead of stretching of a stable polymer jet. On another extreme, a solution that is a non-conductor will be difficult to spin or spray due to the lack of surface charge build-up.

In addition to the solution properties discussed above, process parameters such as voltage, polarity of electrode, spinneret–collector distance, solution flow rate, and spinneret diameter, are also important variables that will affect the production throughput, process stability, and morphology of the resulting electrospun materials (Gaskell, 1997; Huang *et al.*, 2003; Ramakrishna *et al.*, 2005; Teo and Ramakrishna, 2006; Xie *et al.*, 2006).

18.3 ENCAPSULATION OF BIOACTIVES USING ELECTROSPINNING AND ELECTROSPRAYING

18.3.1 Biopolymer encapsulants

Encapsulation of food bioactives is often achieved using edible biopolymers derived from protein, polysaccharides and their derivatives. While many of these biopolymers are readily soluble in water, electrospinning of these aqueous biopolymer solutions tends to be challenging. Instead of forming a cone-jet structure (essential for achieving a stable electrospinning process), charging of these aqueous biopolymer solutions causes the pendant droplet at the spinneret tip to elongate, wiggle, and break away as large droplets of irregular in size (Vega-Lugo and Lim, 2008). The difficulty of electrospinning aqueous biopolymer solutions has been attributed to the polyelectrolytic nature of the polymer, which weakens the electrostatic charge build-up (Huang *et al.*, 2004; Zhang *et al.*, 2005). By using less conductive solvents (e.g., formic acid, acetic acids, trifluoroethanol, hexafluoroisopropanol, trifluroacetic acid:dichloromethane blend, ethanol), proteins and polysaccharides that are otherwise not spinnable in aqueous solution could be electrospun readily (Buchko *et al.*, 2001; Matthews *et al.*, 2002; Huang *et al.*, 2004; Min *et al.*, 2004; Li *et al.*, 2005; Stephens *et al.*, 2005; Woerdeman *et al.*, 2005; Zhang *et al.*, 2005; Torres-Giner *et al.*, 2008; Li *et al.*, 2009; Wongsasulak *et al.*, 2010). Another strategy to overcome the difficulty of electrospinning aqueous biopolymer solutions is by the addition of polymer aids, such as poly(ethylene oxide) (PEO), hydroxypropylmethylcellulose (HPMC), and poly(vinyl alcohol) (PVA) (Jin *et al.*, 2002; Xie and Hsieh, 2003; Yi *et al.*, 2004; Buttafoco *et al.*, 2006; Frenot *et al.*, 2006; Lu *et al.*, 2006; Bhattarai and Zhang, 2007; Safi *et al.*, 2007; Schiffman and Schauer 2008). The reasons for the positive effects of PEO and PVA are unclear, although it has been suggested that increased interfacial viscoelasticity and reduced conductivity are the main contributors (Yu *et al.*, 2006; Vega-Lugo and Lim, 2008, 2012; Regev *et al.*, 2010). In their reviews, Schiffman and Schauer (2008) and Ramakrishna *et al.* (2005) exhaustively summarized various solvents suitable for electrospinning of biopolymers.

18.3.2 Encapsulation of bioactives in Electrospun fibers

Fortification and enrichment are common strategies to increase the nutrition values of food. However, these approaches are not without challenges. First, the added nutrients and bioactives may react with other components present in the food matrix, reducing bioavailability. Second, these nutrients and bioactive compounds may be inherently unstable and/or possess undesirable organoleptic properties, which may impact consumers' acceptance of the products. Therefore, the added compounds must be protected. Several studies have investigated the use of electrospun fiber for this purpose. For instance, Li *et al.* (2009) encapsulated (−)-epigallocatechin gallate (EGCG), a green tea polyphenol that has been implicated for a number of health benefits (e.g., prevention of cancer and heart disease), in electrospun zein fibers. By co-dissolving 20% w/w EGCG in 20% w/w zein in aqueous ethanol solution (70:30 ethanol:water), followed by electrospinning at 12 kV at 10 cm spinneret–collector distance, fibers ranging from 150 to 600 nm were obtained. Freshly electrospun fibers were less effective at immobilizing the EGCG upon immersion in water (82% recovery) as compared to fibers that were aged under dry conditions for at least 1 day (>98% recovery) before exposing to water. Fourier transform infrared spectroscopy (FTIR) demonstrated that changes in secondary structure of the protein due to hydrogen bonding and hydrophobic interactions might have contributed to the stabilization of EGCG in zein fibers in water (Li *et al.*, 2009). This method not only is promising for stabilizing the polyphenol, but also for masking the astringency of polyphenols in fortified foods products. A similar approach was adopted by Fernandez *et al.* (2009) to encapsulate β-carotene in electrospun zein fibers to increase the photostability of the colorant/antioxidant. These researchers dissolved zein:β-carotene (31.35:1.65 by weight) in aqueous ethanol (80:20 ethanol:water by volume), followed by electrospinning at 12 kV at 10 cm spinneret–collector distance. The resulting fibers had diameters that ranged from 340 to 3580 nm. One limitation of encapsulating β-carotene in the zein fiber is the limited solubility of the compound in aqueous ethanol solvent. At the concentration investigated by Fernandez *et al.* (2009), a fraction of β-carotene was not dissolved and existed as aggregates dispersed in the zein fiber. Nevertheless, Raman imaging spectroscopy showed that an increased light stability of β-carotene was observed due to the physical entrapment.

Electrospun fibers have also been studied for the protection of folic acid, which is a coenzyme in the metabolism of amino acids and nucleotides in the human body. It is an essential micronutrient, the deficiency of which has been implicated in the occurrence of heart diseases, cancers, and neural tube defect in baby at birth (Caudill, 2004). Fortification of food products with the micronutrient has been one of the strategies to increase the dietary intake of folic acid. However, the folic acid molecule is susceptible to hydrolysis degradation below pH 5 to form *p*-aminobenzoylglutamic acid and 6-methylpterin. To stabilize the micronutrient in acidic products, various polysaccharide encapsulants have been investigated. For example, folic acid has been encapsulated in sodium alginate–pectin fibers by electrospinning (Alborzi *et al.*, 2010, 2013). In this approach, folic acid was first dissolved in alkaline water (pH 11.5) at 0.05% w/w to form a homogeneous solution, followed by the addition of the alginate and pectin mixture (70:30 sodium alginate:pectin). To enable the electrospinning process, PEO (900 kDa) was added at a 30:70 PEO:alginate–pectin level to produce solutions with 3–5% w/w polymer concentrations. The folic-acid containing polymer solutions were electrospun at 20 kV at 18 cm spinneret–collector distance to produce fibers that were further cross-linked in ethanolic $CaCl_2$ and aqueous $CaCl_2$ solutions. Using the electrospinning process, fibers of less than 300 nm in diameter were produced when low-viscosity sodium alginate was used (250 cP; 2% solution at 25°C). By contrast, beads (2–3 μm diameter) and fibers (<100 nm diameter) coexisted when high-viscosity (2500 cP; 2% solution at 25°C) sodium alginate was used. A significant increase in stability was observed for folic acid encapsulated in the alginate–pectin–PEO blend, with 97% retention of the initial quantity after 41 days storage in the dark in acidified water (pH 3), but only 10% retention was reported when exposed to fluorescence light. In contrast, the nonencapsulated counterparts were completely degraded within the first day of exposure to the acidic condition, when exposed to light, and in the dark only 8% was retained. The FTIR and nuclear magnetic resonance (NMR) data suggested that the increased stability of folic acid was likely due to physical entrapment, rather than specific chemical interaction between folic acid and the polymer carrier (Alborzi *et al.*, 2013). Potentially the encapsulation method may be used for the protection of micronutrients in acidic products, such as fruit juices and beverages.

Besides chemical compounds, viable bacteria cells have also been encapsulated in electrospun fibers by researchers. Lopez-Rubio *et al.* (2009) encapsulated a probiotic, *Bifidobacterium animalis* subsp. *lactis* Bb12, in PVA fibers, by electrospinning 10% w/w aqueous PVA solution concurrently with a suspension of the bacteria in skimmed milk. In their coaxial spinneret set-up, the core material (cell suspension) flowed through the inner tube while the shell material (PVA solution) was pumped through the annular space between the inner and outer tubes. Electrospinning was conducted at 11 kV and the spinneret–collector distance was 13 cm. Despite the high voltage and mechanical shear applied during the electrospinning process, the *B. animalis* Bb12 cells encapsulated within the PVA fibers were viable for 40 and 130 days at room and refrigeration temperatures, respectively. By contrast, a significant decrease in viability was observed when the bacteria were not encapsulated. Townsend-Nicholson and Jayasingh (2006) also encapsulated living cells in electrospun fibers by using a similar coaxial spinneret set-up, by pumping a suspension of living cells and polydimethylsiloxane solution through the inner and outer tubes, respectively. They reported that the electrospinning spinning process did not affect the viability of living cells. These methods are potentially useful for the development of novel probiotic dairy products, such as in conjunction with the use of probiotic soluble dietary as the encapsulant polymer. In fthe ermentation process, this encapsulation technique may be useful for immobilizing viable bacteria and yeasts to facilitate the retrieval and reuse of the microorganisms.

18.3.3 Electrosprayed particles as encapsulants

Nanoencapsulation by electrostatic spraying can be achieved by spraying colloidal suspension of core materials that could be liquid (may be stabilized by emulsifier) or solid nanoparticles (Figure 18.2). One of the examples here is the encapsulation of omega-3 fatty acid to protect it from oxidative degradation. Torres-Giner *et al.* (2010) encapsulated docosahexaenoic acid (DHA) in zein capsules by electrospraying a solution of a DHA–zein mixture in 85% w/w aqueous ethanol. The applied voltage, nozzle–collector distance, and flow rate of 12 kV, 15 cm, and 0.20 mL/h, were used respectively. The diameters of the resulting capsules were approximately 490 to 530 nm in diameter, interconnected with nanosized fibers. These researchers found that the amount of secondary oxidation products (propanal, 2,4-heptadienal and 2,4,7-decatrienal) formed in encapsulated fatty acid were much lower than the nonencapsulated counterpart (Torres-Giner *et al.*, 2010).

Lopez-Rubio and Lagaron (2012) encapsulated β-carotene in whey protein concentrate (WPC) micro- and nanocapsules using an electrospraying method. The WPC encapsulant was dispersed in water at 10–20% w/w concentrations and adjusted to pH 4.5–8 using NaOH or citric acid.

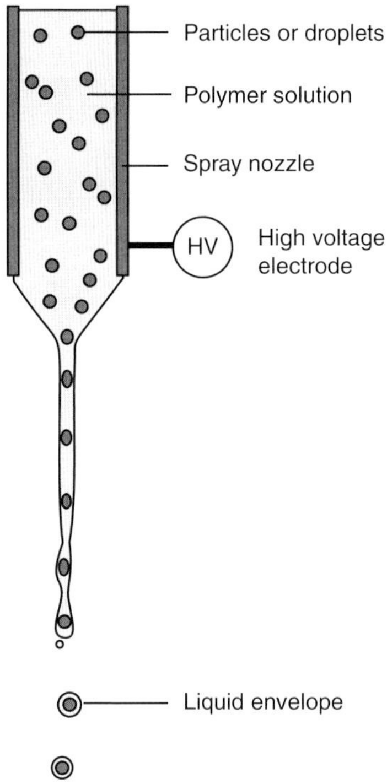

Figure 18.2 Nozzle configuration for electrospray encapsulation using simple nozzle. (Data from: Cloupeau and Prunet-Foch, 1990; Loscertales *et al.*, 2002; Jaworek, 2008.)

Below pH 4.5 and above pH 8.5, the solutions were not suitable for electrospraying due to the formation of gel and irregular electrosprayed mass, respectively. The optimal pH and WPC concentrations for electrospraying were found to be 5.5 and 40%, respectively. These authors incorporated β-carotene into the fiber by first preparing a concentrated solution of β-carotene in glycerol (5 mg/mL), followed by dispersing an aliquot of this concentrate into the aqueous WPC solution to attain a final 20% w/w solution for electrospraying. The stability of encapsulated β-carotene against UV light was evaluated using infrared spectroscopy. No detectable changes in FTIR spectra was reported, indicating that the electrosprayed capsules were capable of protecting the carotenoid (López-Rubio and Lagaron, 2012). Using a similar set-up (15 cm spinneret–collector distance; 15 μL/min flow rate, 25–30 kV applied voltage; 0.6 mm spinneret ID), Trotta *et al.* (2009) electrosprayed stearic acid dissolved in a series of alcohols (ethanol, *n*-propanol, *n*-butanol, *n*-pentanol, and *n*-hexanol) at 5–7.5% w/w lipid concentrations. Stearic acid dissolved in the alcohols alone produced particles that were irregular in size and shape. On the other hand, a stearic acid:ethylcellulose blend (4.5:0.5 w/w) prepared using the same alcohols produced more uniform particles. Ethanol produced sprayed materials with a platelet-like morphology of submicron thicknesses, while *n*-propanol solvent produced wrinkly particles of approximately 1 μm in size and *n*-butanol resulted in sprayed platelets and particles that were irregular in size. By contrast, *n*-pentanol produced more uniform beads that were about 2 μm in size. Finally, *n*-hexanol, which is the least volatile among the solvents, produced irregular lipid films due to coalescence of sprayed particles (Trotta *et al.*, 2009).

The encapsulation methods discussed above are all based on spraying the bioactive-containing polymer solution onto a solid collector. In order to prevent the fusion of particles, adequate evaporation of solvent must occur before they arrived on the collector. Alternatively, the solution may be sprayed onto a collector containing a coagulation bath that induces cross-linking reaction in the encapsulant polymer, thereby strengthening the shell and/or rendering it insoluble when exposed to water. One of the biopolymers that has been studied the most for this purpose is sodium alginate, which is well known for its ability to form firm gel quickly when its aqueous solution is exposed to Ca^{2+} cations. Recently, electrostatic atomization techniques have been used to produce small alginate beads to encapsulate sugar invertase enzyme (Watanabe *et al.*, 2001). In this study, sodium alginate solutions (10–30 g/L polymer concentration) containing 1.0 g/L of the enzyme was electrosprayed using a set-up as depicted in Figure 18.3. With this method, these researchers reported that wet calcium cross-linked beads of about 100 μm in size could be formed. In a similar approach, Fukui *et al.* (2010) electrosprayed aqueous alginate solution (1.5 % w/w; pH 7.5; positively charged) into aqueous chitosan solution (0.5% w/w; pH 3.6; negatively charged) containing 20 mM acetic acid to form complex shell structures. With this set-up, because the negatively charged sodium alginate droplets come in contact with the positively charged chitosan polymers, a polyelectrolyte complex was formed due to electrostatic interaction at the alginate–chitosan interface. The key experimental conditions used were: nozzle–collector distance, 5 cm; applied voltage, 22.5 kV; and inner/outer nozzle diameters 130/310 μm. The resulting microcapsules, ranging from 115 to 145 μm in diameter, have a relatively narrow size variation (8% coefficient of variation). These researchers were successful in encapsulating protein, dextran, polymeric microspheres, and yeast cells within the alginate–chitosan capsules (Fukui *et al.*, 2010). In another similar study, Chang *et al.* (2012) prepared electrospun alginate fibers with a chitosan sheath. They noted that

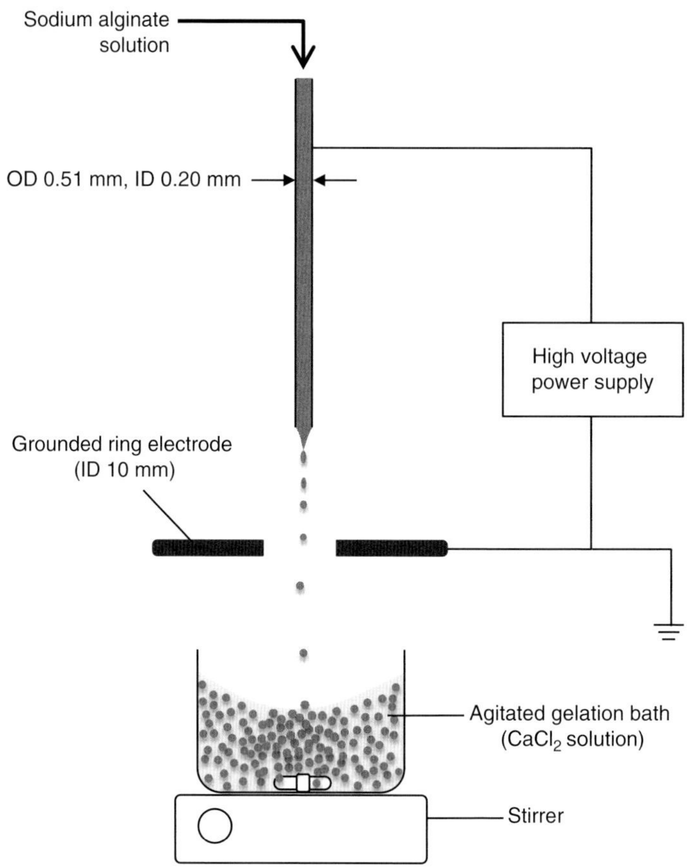

Figure 18.3 Schematic experimental set-up for the production of cross-linked sodium alginate particles in aqueous $CaCl_2$ solution with the aid of electrostatic spraying. (Adapted from Watanabe *et al.*, 2001.)

the addition of ethanol to the chitosan coagulation bath, at a 50% ethanol (w/w) level, facilitated the formation of alginate fiber due to the reduction of surface tension and specific gravity of the coagulation solution, which has the effect of allowing the fiber to submerge rapidly into the bath and to coagulate. When exposed to the saline solution, which is known to disintegrate calcium alginate due to the exchange of Ca^{2+} with Na^{+}, 25% loss of alginate was reported for fibers with a chitosan sheath, while the alginate fiber cross-linked in the $CaCl_2$ bath encountered 65% loss in alginate (Chang *et al.*, 2012). This result shows that the chitosan bath treatment could be advantageous to stabilize electrospun alginate fibers and particles, especially if these carriers are to be exposed to high NaCl food matrices.

Jung *et al.* (2011) reported a similar approach to the aforementioned studies but with a subtle but important equipment modification in order to electrospray ethanolic plant extract from *Sophora flavescens*, a medicinal perennial herb that is commonly used in northeast Asia. Their electrospraying set-up was arranged in a point-to-office plate configuration, which positioned a stainless steel capillary (0.1 and 0.32 mm outer and inner diameters, respectively) just 7 mm from an orifice plate (1 mm diameter orifice; 10 mm thickness). The capillary was charged to several thousand volts while the orifice plate was grounded. The particles produced were spherical and nonagglomerated due to mutual electrostatic repulsion between the charged particles. Scanning and transmission electron microscopy (TEM) analyses revealed that the resulting nanoparticles were of bimodal distribution, with averaged sizes of < 50 nm and 120–250 nm. Although this set-up allows the formation of small particles that are promising for nanotechnology applications, the researchers reported an extremely low flow rate of 30 μL/h at an applied voltage of 3.8 kV. Further investigation using a scale-up of the production method will be needed in order to make this process commercially viable. One of the possible configurations to increase the electrospray throughput could be by using multiple electrospray atomizers emitted from an array of holes or spinnerets, similar to that depicted in Figure 18.4. This approach can elevate the spraying throughput up to several hundred milliliters per hour (Bocanegra *et al.*, 2005a; Deng *et al.*, 2006).

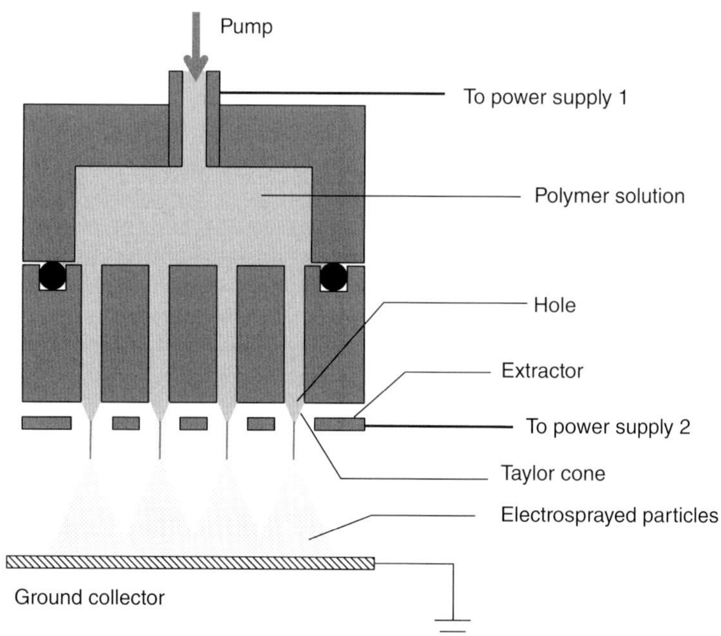

Figure 18.4 Typical atomizer set-up for producing multiple jets, using extractor concept to prevent multiplexing charge repulsion. The use of an extractor is essential for high compactness array electrospray atomizers to prevent the space charge due to cloud of charged droplets, resulting in electric field shielding of neighboring Taylor cone menisci. (Adapted from Bocanegra *et al.*, 2005a.)

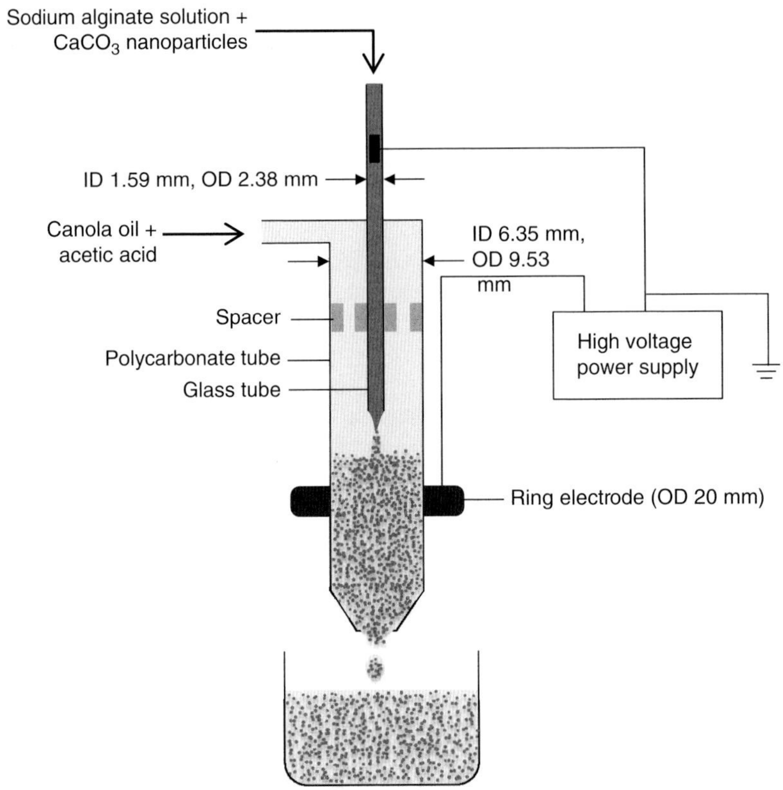

Figure 18.5 Electrodispersion of alginate into a liquid phase for controlled gelation of alginate particles using $CaCO_3$ particles. (Adapted from Zhao *et al.*, 2007.)

Electrospraying is typically carried out in the air, although it is also possible to electrospray particles in a liquid medium. Zhao *et al.* (2007) developed an innovative electrodispersion process to control the gelation of alginate beads that were atomized into an acidified vegetable oil containing acetic acid (Figure 18.5). In their approach, $CaCO_3$ nanoparticles of about 40 nm were dispersed into an aqueous alginate solution with the aid of a surfactant (dioctyl sulfosuccinate sodium salt). The strategic use of $CaCO_3$ here is to prevent the uncontrolled gelation of alginate that prevents the optimal formation of uniform beads. As the particles were sprayed into the acidic oil, the acid reacted with $CaCO_3$, liberating Ca^{2+} ions that induced the gelation of the alginate beads in a controlled manner (Zhao *et al.*, 2007). This method resulted in alginate beads ranging from 10 to 100 μm, depending on the voltage applied. The dried alginate particles had a much smaller size (~5 μm) as compared to those produced by a conventional gelation-bath approach (55 μm to 1.2 mm). It is noteworthy that in order to effectively disperse the polymer solution into the continuous phase, the former should be sufficiently low in viscosity and density. Moreover, the electrical conductivity of the dispersed phase should be higher than the continuous phase (Jaworek, 2008).

18.3.4 Composite capsule structures

Direct dispersion of bioactives in polymer solution followed by electrospinning and electrospraying is the simplest approach to encapsulate these compounds. However, this approach has a drawback in that a certain fraction of the dispersed compound is distributed near or on the surface of the capsules. These unprotected species are vulnerable to degradation due to exposure of undesirable external environmental factors. To overcome this shortcoming, capsules of well-defined core and shell structures have been developed (Figure 18.6). Here, the core–shell structures are prepared from two liquids, separate by an interface, by co-spraying them in a coaxial nozzle. The two liquids are injected separately into two capillary tubes positioned in a concentric manner. Careful selection of the outer and inner liquid pair and optimization of flow rates are important to produce proper encapsulation of the active compound. For instance, unfilled outer liquid droplets can form if the outer liquid jet breaks up too fast as compared to the inner jet. When the velocity of the inner jet is much faster than the outer jet, full encapsulation of the inner liquid will be hampered. Mei and Chen (2007) used the ratios of relaxation length

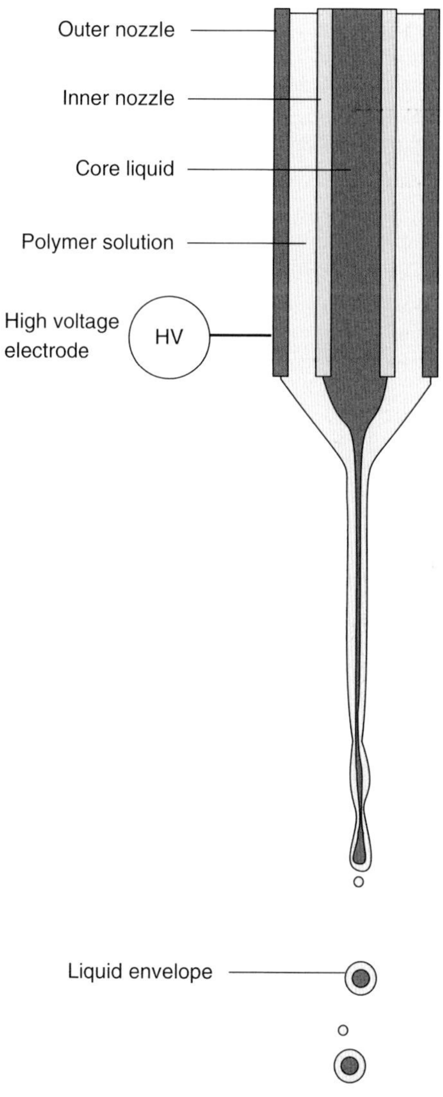

Figure 18.6 Nozzle configuration for electrospray encapsulation using a coaxial nozzle. (Data from: Cloupeau and Prunet-Foch, 1990; Loscertales *et al.*, 2002; Jaworek, 2008.)

(r_0^*/r_i^*) and inertial break-up length (R_0^*/R_i^*) for the inner and outer jets to predict the feasibility of the formation of a core and shell for a given pair of liquids:

$$r^* = \left(Q\varepsilon\varepsilon_o / K\right)^{1/3} \tag{18.1}$$

$$R^* = \left(\rho Q^2 / \gamma\right)^{1/3} \tag{18.2}$$

where Q, K, ε, ρ, and γ are flow rate, electrical conductivity, dielectric constant, density, and surface tension of the solution, respectively. The formation of core–shell droplets is feasible when $r_0^*/r_i^* < 500$ and $R_0^*/R_i^* < 0.015$ (Mei and Chen, 2007).

Several studies have explored the use of coaxial electrospray techniques for producing core–shell structures for protecting various bioactive core materials. For example, a coaxial spinneret has been used for the electrospraying of cocoa butter microcapsules with a single aqueous core surrounded by the cocoa butter shell (Bocanegra *et al.*, 2005b). In this study, the cocoa butter was heated to 60°C, which is well above its melting point (~37°C), in order to keep the cocoa butter in a liquid state and to provide a suitable viscosity for electrospraying. Because the outer cocoa butter has a relatively lower electrical conductivity than the inner aqueous liquid, the latter became the driving liquid in which most of the electrical current was injected. This method is promising for the encapsulation of aqueous ingredients (e.g., flavour compounds, salts, minerals, vitamins, enzyme), in lipid-base shells. The disadvantage of this melt electrospraying is that the particles were relatively larger (~10–20 μm) as compared to polymer solution spraying. Electrospraying of polymer solutions resulted in droplets that decreased in size as the solvent evaporated, but solidification of the molten droplets did not result in appreciable size reduction. A similar coaxial set-up was adopted by Juan *et al.* (2006) to produce core–sheath nanofibers, but with a reversed hydrophobic core and hydrophilic sheath layer (Díaz *et al.*, 2006). The outer biocompatible polymer used was 20% w/w poly(vinyl pyrrolidone) dissolved in *N,N'*-dimethylformamide, while the inner fluid was an industrial oil. The hydrophilic shell and hydrophobic core structures are promising in drug delivery and cosmetic applications, where the hydrophobic actives need to be encapsulated in biocompatible hydrophilic materials to penetrate through the skin.

Jiang *et al.* (2005) adopted an innovative approach to spin coaxial nanofibers with a poly(ε-caprolactone) shell and a poly(ethylene glycol) (PEG) core, wherein bovine serum albumin or lysozyme were dispersed. In this approach, two solutions were dosed into a spinneret at controlled rates and spun simultaneously to produce a coaxial hybrid fiber. The researchers reported that the composite fibers provided a steady release of protein for ~10 to 30 days. A similar coaxial electrospray process was used to encapsulate proteins (bovine serum albumin and lysozyme) in biodegradable poly(DL-lactide-co-glycolide) (PLGA) microparticles (Xie *et al.*, 2008). Particles size varied from 5–10 μm, as the PLGA shell solution (acetonitrile as solvent) varied from 5 to 15% w/v polymer concentration. The inner aqueous protein solutions, which contained 10–20% w/v bovine serum albumin or lysozyme, were prepared in 40% w/v aqueous PEG solvent. An *in vitro* study showed that protein release was sustained for more than 30 days and the activity of lysozyme released was more than 90% (Xie *et al.*, 2008). These studies showed that electrospinning with the aid of a coaxial spinneret could be promising for encapsulating bioactives to isolate them from the deleterious factors in the surrounding environment.

18.4 OTHER NONEDIBLE APPLICATIONS

18.4.1 Active and intelligent food packaging

Packaging is an integral component of preservation systems, and is critical to the protection of food products during their sales and distribution. In response to the changing industry's needs (e.g., cost-saving, reduction of waste, increased shelf-life) and consumer's expectations (e.g., end-use convenience, high quality, safe), many packaging innovations have emerged. One example here is active packaging that can "sense" the changes in the internal/external conditions of the package, and respond dynamically by activating some physicochemical mechanisms to protect the product. Some of these mechanisms include moisture sorption, ethylene removal, and oxygen scavenging (Labuza and Breene, 1989; Ozdemir and Floros, 2004; Day, 2008; Scully, 2009).

Antimicrobial active packages are especially attractive for extending the shelf life of food products that are susceptible to microbial spoilage. This variant of active packaging system is capable of inhibiting the growth of spoilage microorganisms through the controlled release of antimicrobial compounds directly onto the product, or indirectly to the product surface by emitting it into the package headspace in the case where the antimicrobial compound is volatile. In general, the release of antimicrobial agents is diffusion controlled (Appendini and Hotchkiss, 2002). Furthermore, release-triggering mechanism may be present to ensure that the antimicrobial agent is released only under conditions susceptible to microbial proliferation, such as elevated water activity and temperature. Vega-Lugo and Lim (2009) exploited the moisture sensitivity of electrospun fibers prepared from a soy protein isolate (SPI)/PEO blend electrospun fibers for controlled release of allyl isothiocyanate (AITC). Briefly, SPI (15% w/w), PEO (0.6%) and Triton X-100 (1%) were dissolved in 1% NaOH solution at 60°C for 2 h to denature the protein. β-Cyclodextrin was then dissolved in the polymer solution, cooled to 24 ± 2°C, followed by the addition of AITC at 0.4:2.4 AITC:β-cyclodextrin molar ratio. The solution was then electrospun into fibrous mats using a set-up similar to Figure 18.1, producing fibers that were less than 300 nm in diameter. The use of cyclodextrin here is beneficial to reduce AITC evaporative loss during spinning. At 0% relative humidity (RH), the release of AITC from the electrospun fibers was negligible. However, exposing the fibers to greater than 50% RH resulted in the release of AITC, the rate and amount of which were proportional to the exposed RH. The interactive behavior of these antimicrobial nanofibers may be promising in active packaging applications for foods in order to ensure that the release of AITC takes place only under elevated humidity conditions, such as within a food package. Conceivably, a similar concept can be applied for encapsulation of other antimicrobial agents for shelf-life extension of food.

In another active packaging example, electrospun fibers have been used as a carrier to indirectly produce antimicrobial species to inhibit the growth of microorganism in milk by leveraging the lactoperoxidase (LP) system. The LP system is a naturally occurring antimicrobial mechanism in bovine milk. In the presence of a trace amount of hydrogen peroxide (H_2O_2), the LP system catalyzes the oxidation of thiocyanate (SCN^-, another component present in milk), generating hypothiocyanate ions ($OSCN^-$) and other intermediate products that possess antimicrobial properties. The oxidation of the sulphydryl (SH) group in microbial enzymes and proteins by these products is believed to be responsible for the antimicrobial actions of the LP system (Aune and Thomas, 1977; Barrett *et al.*, 1999; Seifu *et al.*, 2005). In one study, Zhou and Lim (2009) immobilized glucose oxidase (GOX) in nonwoven poly(lactic acid) (PLA) membranes in order to activate the LP system. By oxidizing glucose present naturally in milk using GOX, the resulting H_2O_2 can trigger the formation of $OSCN^-$ and other intermediate antimicrobial compounds. To prepare the bioactive fibers, aqueous GOX was dispersed in PLA–chloroform solution with the aid of sorbitan monopalmitate (SMP) and sonication. The emulsion was then electrospun into a GOX-containing fibrous mat. The exposure of fresh milk to the GOX-containing electrospun membrane resulted in an increase in $OSCN^-$ concentration for up to about 9 h, showing that the naturally occurring SCN^- and glucose in milk were adequate to activate the LP-catalyzed reactions. The GOX encapsulated in the electrospun PLA fibers exhibited about 19 times greater enzymatic activity as compared to a continuous film cast from the same emulsion. The increased enzymatic activity could be attributed to the larger surface area of the electrospun fibrous membrane than the cast film carrier. The bioactive electrospun mat may be applied as a small label or patch on the product contact surface of milk packaging to increase the stability of milk. The Food and Agriculture Organization (FAO) of the United Nations and the International Dairy Federation have developed protocols to add thiocyanate and H_2O_2 in milk, in the form of sodium thiocyanate and sodium carbonate peroxyhydrate, respectively, during transportation and storage of milk in regions of the world where refrigeration infrastructures are lacking (FAO, 1999). The GOX electrospun mat may be used as an alternative approach to deliver a controlled amount of H_2O_2 to activate the LP system.

Oxygen-sensitive food products are often packaged in modified atmosphere packaging (MAP) wherein the package headspace is modified with zero or reduced oxygen atmosphere. To ensure that the modified atmosphere is maintained, it is desirable for the modified atmosphere packages to be equipped with an indicator that detects the headspace oxygen concentration to reveal if the package is compromised. One version of oxygen indicators involves the use of TiO_2 to activate a redox dye for oxygen detection. The TiO_2 particles, in rutile and anatase crystalline forms, are a white pigment commonly used in packaging. In the nanometer range, these particles exhibit photocatalytic properties useful for various redox reactions. Lee *et al.* (2005) developed an oxygen indicator ink by encapsulating TiO_2 nanoparticles (80:20 anatase:rutile; about 30 nm in diameter), a mild reducing agent (triethanolamine), and methylene blue in hydroxyethyl cellulose film. By exposing the oxygen film with UV radiation, the methylene blue is activated to a colorless oxygen-sensitive form. Upon exposure to oxygen, the active indicator is oxidized to produce blue coloration, the intensity of which is proportional to the amount of oxygen reacted. Using the same concept, recently Mihindukulasuriya (2012) encapsulated TiO_2 nanoparticles, glycerol, and methylene blue in electrospun and PEO fibers. Briefly, PEO powder was dispersed in 70% aqueous ethanol to produce a 5% w/w polymer solution. Methylene blue, glycerol, and TiO_2 nanoparticles were then added to prepare the fiber-forming solutions containing different component weight ratios. The solutions were electrospun at 6–8 kV at 20 cm spinneret–collector distance to produce an indicator membrane. For comparison, the fiber-forming solution was cast as a continuous film, instead of electrospinning into fiber, in order to evaluate the effect of surface area on the PEO carrier polymer. Oxygen indicators prepared by the electrospinning method required shorter UVA exposure time (5 s) to activate, as compared to those produced by casting (at least 30 s). The different indicator behaviors can be attributed to the different morphologies of the indicators produced using electrospinning and casting methods. The cast film was more compact in morphology, whereas the electrospun membrane was structurally more open. Another possible reason for the enhanced sensitivity of the electrospun indicator is that the electrospinning process involved high shear that might have facilitated the dispersion of the TiO_2 nanoparticles in the direction of flight trajectory. The rapid evaporation of solvent rapidly immobilized the TiO_2 within the PEO fiber matrix, preventing the nanoparticles from agglomerating. By contrast, during the film casting process, the evaporation of solvent was relatively slow. The lack of shear and surface-tension effect of the solvent might have induced the aggregation of TiO_2 particles during drying. The clumpy features on the micrograph for the cast film indicator may be indicative of the formation of the TiO_2 aggregates. This study shows that the electrospun TiO_2-based indicator could be promising for oxygen detection in modified-atmosphere-packaging applications.

18.4.2 Immobilization of enzymes

Enzymes are used in many industrial and food processes to catalyze specific reactions under mild conditions. In many applications, the use of free enzymes is undesirable due to the their instability. Various enzyme immobilization techniques are therefore developed to address this issue. By virtue of their porous and nonwoven morphology, electrospun fibers are ideal carrier to stabilize the enzymes, allowing their easy recovery and facilitating their reuse. Immobilizing of enzymes in electrospun fibers is largely based on either encapsulating the enzyme within the fiber matrix or attaching the enzyme on the surface of the fibers. In the former, the enzyme could be dispersed into an aqueous polymer solution and electrospun. This approach results in high enzyme loading (as high as 50% of the fiber), good enzymatic activity, and fast response, making the enzyme-containing electrospun membrane ideal electrodes for biosensing applications (Wang *et al.*, 2009). However, since the polymers used here are water-soluble (e.g., proteins, polysaccharides, PEO, PVA), post-spinning cross-linking of fibers may be needed. For example, Xie and Hsieh (2003) investigated the use of electrospun casein fibers for the encapsulation of lipase enzymes in PEO or PVA blended with casein. To render the fiber insoluble in water, cross-linking of amine groups from casein and hydroxyl groups from PVA and PEO was achieved with 4,4'-methylenebis(phenyl diisocyanate). The activity of the encapsulated lipase was 100-fold lower than that of the free counterpart, but six times more active than that encapsulated in cast membrane from the same polymer–enzyme solution (Xie and Hsieh, 2003).

To encapsulate enzymes within water-insoluble polymer, a water-in-oil emulsion is required. To this end, Yang *et al.* (2008) electrospun an emulsion containing a lysozyme dispersed phase and a continuous phase of poly(D,L-lactic acid) (PDLLA) solution prepared using chloroform as a solvent. The lysozyme was first dissolved in a phosphate buffer solution (PBS; 50 mM, pH 7.4) along with methylcellulose. The aqueous phase was then added to the organic continuous phase of PDLLA and emulsified using ultrasonication in an ice bath, and the resulting emulsion was electrospun using the conventional simple spinneret. With this approach, Yang *et al.* (2008) reported the formation of core–sheath fibers where the lysozyme was encapsulated within the polymeric shell material. Since chloroform evaporates faster than water, the viscosity of the outer PDLLA layer increased faster than the aqueous lysozyme–methyl-cellulose phase. This caused the elongated aqueous phase to move inwards and coalesce to form the core, rather than being dispersed on the surfaces. The core–sheath structures produced using the emulsion method are similar to those prepared using the coaxial spinneret described earlier, with an obvious advantage that the experimental set-up is much simpler than the former approach. The formation of a core–sheath structure will be affected by the concentration of the dispersed phase: a high proportion of the emulsion droplets will disrupt the formation of core–sheath fibers. Yang *et al.* (2008) reported rapid initial release of enzyme from the fibers during the first 12 h in PBS at pH 7.4, resulting in 27–47% release of the enzyme for fibers loaded with 1% and 3% enzyme, respectively. This was followed by a slower phase of release that lasted for about 5–15 days at the two enzyme loadings, respectively. These results show that the enzyme was not entirely encapsulated within the core. The release profile of the active compound can be adjusted by coating the surface of the ultrafine fibers with another polymer with a different barrier property as compared to the base carrier. As reported by Zeng *et al.* (2005), the release of bovine serum albumin and luciferase from PVA electrospun fibers can be manipulated by depositing poly(*p*-xylylene) coating using chemical vapor deposition polymerization. The hydrophobic poly(*p*-xylylene) coating protected the PVA from water and allowed for controlled releases of protein and enzyme (Zeng *et al.*, 2005).

Although direct encapsulation of enzyme within the electrospun fibers is convenient, it is inevitable that this approach will limit the enzyme activity. Furthermore, there will be some enzymes distributed to the surface that can cause leakage loss. To effectively anchor the enzyme to the surface, researchers have covalently attached enzyme molecules to the surface of various electrospun fiber carriers. For instance, Jia *et al.* (Jia *et al.*, 2002) developed a bioactive nonwoven polystyrene-based fibrous mat immobilized with α-chymotrypsin. The functionalized polystyrene was produced first by attaching nitrophenyl end-groups to the polystyrene chain by reacting the polymer with 4-(dimethylamino)-pyridine (DMAP) in toluene. The functionalized polystyrene was then dissolved in methyl ethyl ketone:*N*,*N*-dimethylformamide (1:1) containing 0.5% w/w LiCl, and electrospun using a typical set-up. The addition of LiCl is important to produce uniform electrospun fibers, with average diameter of about 120 nm, by preventing the formation of beads. The resulting electrospun fibers were attached with α-chymotrypsin in a borate buffer (pH 8.2) at room temperature for 36 h. This method allows an enzyme loading of 1.4% (w/w), corresponding to over 27.4% monolayer surface coverage of the nanofibers, and provided 65% enzymatic activity as compared to the native enzyme in aqueous solutions. By contrast, in organic solvents (hexane and isooctane), the activity was more than three orders of magnitude as compared to the nonimmobilized counterpart (Jia *et al.*, 2002). While attaching immobilizing enzymes to electrospun nonofibers is an attractive strategy to increase loading efficiency and reduce mass transfer limitation, potentially this approach may impact enzymatic activity due to the reduced freedom of the immobilized enzyme, nonspecific interaction between the enzyme and the supporting fiber surface resulting in undesirable conformational changes, and hindrance of electron transfer of the supporting enzyme that catalyzes redox reactions (Wang *et al.*, 2009). Approaches that may be taken to overcome these limitations include: (i) modification of synthetic electrospun fiber surfaces to increase their biocompatibility towards the enzyme to be immobilized; (ii) introduction of spacer arms between the supporting surface and enzyme to reduce the steric hindrance; (iii) increase electrical conductivity of fibers through blending with conductive filler, in particular in enzyme-based sensor applications; (iv) combinations of the above (Wang *et al.*, 2009).

18.4.3 Biomedical applications

Encapsulation of medicinal compounds in biodegradable carriers is one of most commonly used methods for controlling the release of drugs and for protecting the bioactives from undesirable gastrointestinal conditions. Kim *et al.* (2004) attempted to use electrospun PLGA fibers as a carrier for cefoxitin sodium, a hydrophilic antibiotic. A large burst of drug release was observed within the first 6 h. The authors attributed the burst release to the concentration of drug on the fiber surface. Probably the sorption of moisture may have also caused the PLGA to plasticize, triggering an enhanced release of the antibiotic. Kenawy *et al.* (2002) compared the controlled release of tetracycline hydrochloride from fibers and cast films of poly(ethylene-co-vinylacetate), PLA and their 50:50 blend carriers. Similar to the observation reported by Kim *et al.* (2004), a burst release of the active compound was observed with the PLA. In contrast, poly(ethylene-co-vinylacetate) and the 50:50 blend fiber carriers provided a steady release of tetracycline hydrochloride over 5 days. As expected, the release of the drug in cast films was lower than the electrospun mats due to their smaller surface area. Xu *et al.* (2005) electrospun a water-in-oil emulsion containing a water-soluble anticancer agent, doxorubicin hydrochloride emulsified in a PEG-poly(L-lactic acid) (PLLA) copolymer, using sodium dodecyl sulfate as an emulsifying agent. The emulsified drug provided a sustained release, while the drug added without the emulsifier exhibited an uncontrolled burst release. By exposing the fiber to proteinase K enzyme, the rate of drug release was accelerated due to the combined drug diffusion and enzymatic degradation of the carrier. This result is different from the study by Zeng *et al.* (2003), who electrospun a lipophilic drug, rifampin (for tuberculosis) in PLLA ultrafine fibers. Due to the strong compatibility of the drug and the polymer, no release was observed for up to 7 days, unless proteinase K was added. The drug release in this case was controlled primarily by the degradation of the polymer carrier. Thus, it is important to take into account the compatibility between the active compound and polymer carrier to achieve a desirable release behavior.

Son *et al.* (2006) developed a method to prepare antimicrobial cellulose acetate fibers containing silver nanoparticles. The method is based on electrospinning acetone:water (80:20 w/w) cellulose acetate solution (10% w/w) dispersed with 0.5% w/w $AgNO_3$, using a typical spinneret–plate electrospinner set-up (17 kV applied voltage, 10 cm spinneret–collector distance, 0.84 mm spinneret ID, and 3 mL/h flow rate). After electrospinning, the fibers were exposed to UV radiation at 245 nm, causing photoreduction of silver ions to silver nanoparticles that had an average size of 21 nm. Their TEM analysis of the cross-section of the fibers showed that the silver ions were mainly deposited on the surface of the electrospun cellulose acetate fibers. The nonwoven membranes electrospun from the cellulose acetate fibers exhibited strong antimicrobial activity against Gram-negative microorganisms (*Escherichia coli*, *Klebsiella pneumonia*, *Pseudomonas aeruginosa*) (Son *et al.*, 2006). Other drugs have also been encapsulated in electrospun fibers. Kim *et al.* (2004) encapsulated antibiotic cefoxin sodium in electrospun PLGA nonwoven membranes, which effectively inhibited growth of *Staphylococcus aureus* in both agar and liquid media. The same research group also encapsulated DNA in an electrospun PLGA random copolymer and a poly(D,L-lactide)–PGA block copolymer. These types of fibers are promising for therapeutic application, gene delivery for tissue engineering, antimicrobial filters, and wound or post-surgical dressing applications to prevent infections.

Recently, an electrospraying method based on a coaxial spinneret was adopted for one-step preparation of PLGA-coated drug particles, without using the emulsion process (Lee *et al.*, 2010). The model lipophilic drug used was Budesonide, which is a glucocorticoid steroid for the treatment of asthma and noninfectious rhinitis. The drug was dissolved in acetonitrile as the solution for the inner spinneret, while the PLGA polymer solution was pumped through the outer spinneret. Particles as small as 165 nm were obtained when 0.5% w/w polymer concentration was used. As polymer concentration increased to 5% w/w, particles of larger size (~1200 nm) were obtained. Expectedly, the drug release rates of smaller particles were more rapid than the large ones due to the larger surface area of the former. Additionally, the researchers also reported that the penetration of water into the particles increased with decreasing particle size, inducing particle swelling and eventual rupture (Lee *et al.*, 2010).

18.5 PROSPECTS OF ELECTROSTATIC ENCAPSULATION TECHNOLOGY AND CONCLUSIONS

In summary, electrospinning and electrospraying are promising techniques for micro and nanoencapsulation of bioactives to enhance their stability, release characteristics, and payload in capsules. Moreover, these techniques allow the engineering of a wide spectrum of carrier systems through exploiting different composite and multilayered structures. Going forward, research on the development of innovative capsules, such as using electrospraying and electrospinning methods, will continue to emerge, not only to stabilize the bioactives, but also to enhance their bioavailability by using techniques that produce submicron capsules. For nanodelivery systems, study of carriers using separation, imaging, and analytical techniques will be important to understand their risk and benefit (Luykx *et al.*, 2008). Optimization of encapsulation technology is expected to take a more holistic approach of not only studying the release behaviors of bioactives alone, but also considering interactions between the bioactives with other food components, microbial flora in the gut, and complex gastrointestinal conditions (Parada and Aguilera 2007). Along with this trend, development of new methods to assess the efficacy of encapsulated bioactives, both *in vitro* and *in vivo*, are expected to make inroads in encapsulation research. Encapsulation throughput of spinneret–based encapsulation electrospraying and electrospinning processes, in general, is lower than other well-established techniques. Further research on using this innovative approach, such as on free-surface electrospinning and hybridization of spray drying using an electrohydrodynamic approach, is likely to find more applications for micro-/nanoencapsulation.

References

Alborzi, S., Lim, L.-T. and Kakuda, Y. (2010) Electrospinning of sodium alginate–pectin ultrafine fibers. *Journal of Food Science*, 75(1), C100–C107.

Alborzi, S., Lim, L.-T. and Kakuda, Y. (2013) Encapsulation of folic acid and its stability in sodium alginate-pectin-poly(ethylene oxide) electrospun fibres. *Journal of Microencapsulation. Journal of Microencapsulation*, 30(1), 64–71.

Appendini, P. and Hotchkiss, J.H. (2002) Review of antimicrobial food packaging. *Innovative Food Science and Emerging Technologies*, 2, 113–126.

Aune, T.M. and Thomas, E.L. (1977) Accumulation of hypothiocyanate ion during peroxidase-catalyzed oxidation of thiocyanate ion. *European Journal of Biochemistry*, 80, 209–214.

Barrett, N., Grandison, A.S. and Lewis, M. (1999) Contribution of the lactoperoxidase system to the keeping quality of pasteurized milk. *Journal of Dairy Science*, 66, 73–80.

Baumgarten, P.K. (1971) Electrostatic spinning of acrylic microfibers. *Journal of Colloid and Interface Science*, 36(1), 71–79.

Bhattarai, N. and Zhang, M. (2007) Controlled synthesis and structural stability of alginate-based nanofibers. *Nanotechnology*, 18, 1–10.

Bocanegra, R., Galán, D., Márquez, M., Loscertales, I.G. and Barrero, A. (2005a) Multiple electrosprays emitted from an array of holes. *Aerosol Science*, 36, 1387–1399.

Bocanegra, R., Gaonkar, A.G., Barrero, A., Loscertales, I.G., Pechack, D. and Marquez, M. (2005b) Production of cocoa butter microcapsulesusing an electrospray process. *Journal of Food Science*, 70, E492–E497.

Buchko, C.J., Kozloff, K.M. and Martin, D.C. (2001) Surface characterization of porous, biocompatible protein polymer thin films. *Biomaterials*, 22, 1289–1300.

Buttafoco, L., Kolkman, N.G., Engbers-Buijtenhuijs, P., Poot, A.A., Dijkstra, P.J., Vermes, I. and Feijen, J. (2006) Electrospinning of collagen and elastin for tissue engineering applications. *Biomaterials*, 27, 724–734.

Carroll, C.P. and Joo, Y.L. (2006) Electrospinning of viscoelastic Boger fluids: modeling and experments. *Physics of Fluids*, 18(053102), 1–14.

Caudill, M. (2004) The role of folate in reducing chronic and developmental disese risk – an overview. *Journal of Food Science*, 69, SNQ55–67.

Cengiz, F., Krucińska, I., Gliścińsk, E., Chrzanowski, M. and Göktepe, F. (2009) Comparative analysis of aarious alectrospinning methods of nanofibre formation. *Fibres and Textiles in Eastern Europe*, 17(1), 13–19.

Chang, J.-J., Lee, Y.-H., Wu, M.-H., Yang, M.-C. and Chien, C.-T. (2012) Preparation of electrospun alginate fibers with chitosan sheath. *Carbohydrate Polymers*, 87, 2357–2361.

Cloupeau, M. and Prunet-Foch, B. (1990) Electrostatic spraying of liquids: Main functioning modes. *Journal of Electrostatics*, 25, 165–184.

Cloupeau, M. and Prunet-Foch, B. (1994) Electrohydronamics spraying functioning modes – a critical review. *Journal of Aerosol Science*, 25(6), 1021–1036.

Day, B.P.F. (2008) Chapter 1: Active ackaging of food. In: *Smart Packaging Technologies for Fast Moving Consumer Goods* (ed. J. Kerry and B. Paul). John Wiley & Sons, Ltd, Chichester, 1–18.

Deng, W., Klemic, J.F., Li, X., Reed, M.A. and Gomez, A. (2006) Increase of electrospray throughput using multiplexed microfabricated sources for the scalable generation of monodisperse droplets. *Aerosol Science*, 37, 696–714.

Díaz, J.E., Barrero, A., Márquez, M. and Loscertales, I.G. (2006) Controlled encapsulation of hydrophobic liquids in hydrophilic polymer nanofibers by co-electrospinning. *Advanced Functional Materials*, 16, 2110–2116.

Dror, Y., Ziv, T., Makarov, V., Wolf, H., Admon, A. and Zussman, E. (2008) Nanofibers Made of Globular Proteins. *Biomacromolecules*, 9, 2749–2754.

FAO. (1999) *Manual on the Use of the LP-system in Milk Handling and Preservation*. Food and Agriculture Organization of the United Nations, Rome.

Feng, J.J. (2002) The stretching of an electrified non-Newtonian jet – a model for electrospinning. *Physics of Fluids*, 14(11), 3912–3926.

Fernandez, A., Torres-Giner, S. and Lagaron, J.M. (2009) Novel route to stabilization of bioactive antioxidants by encapsulation in electrospun fibers of zein prolamine. Food Hydrocolloids 23, 1427–1432.

Frenot, A., Henriksson, M.W. and Walkenstrom, P. (2006) Electrospinning of Cellulose-Based Nanofibers. *Journal of Applied Polymer Science*, 103, 1473–1482.

Fukui, Y., Maruyama, T., Iwamatsu, Y., Fujii, A., Tanaka, T., Ohmukai, Y. and Matsuyama, H. (2010) Preparation of monodispersed polyelectrolyte microcapsules with high encapsulation efficiency by an electrospray technique. *Colloids and Surfaces A: Physicochemical and Engineering Aspects*, 370, 28–34.

Gaskell, S.J. (1997) Electrospray – principles and practice. *Journal of Mass Spectrometry*, 32, 677–688.

Grafe, T. (2003) Polymeric nanofibers and nanofiber webs: A new class of nonwovens. *International Nonwovens Journal*, Spring, 51–55.

Huang, Z.-M., Zhang, Y.-Z., Kotaki, M. and Ramakrishna, S. (2003) A review on polymer nanofibers by electrospinning and their applications in nanocomposites. *Composites Science and Technology*, 63, 2223–2253.

Huang, Z.-M., Zhang, Y.Z., Ramakrishna, S. and Lim, C.T. (2004) Electrospinning and mechanical characterization of gelatin nanofibers. *Polymer*, 45, 5361–5368.

Jaworek, A. (2008) Electrostatic micro- and nanoencapsulation and electroemulsification: a brief review. *Journal of Microencapsulation*, 25(7), 443–468.

Jaworek, A. and Sobczyk, A.T. (2008) Electrospraying route to nanotechnology – an overview. *Journal of Electrostatics*, 66, 197–219.

Jia, H., Zhu, G., Vugrinovich, B., Kataphinan, W., Reneker, D.H. and Wang, P. (2002) Enzyme-carrying polymeric nanofibers prepared via electrospinning for use as unique biocatalysts. *Biotechnology Progress*, 18, 1027–1032.

Jiang, H., Hu, Y., Li, Y., Zhao, P., Zhu, K. and Chen, W. (2005) A facile technique to prepare biodegradable coaxial electrospun nanofibers for controlled release of bioactive agents. *Journal of Controlled Release*, 108, 237–243.

Jin, H.-J., Fridrikh, S.V., Rutledge, G.C. and Kaplan, D.L. (2002) Electrospinning *Bombyx mori* silk with poly(ethylene oxide). *Biomacromolecules*, 3, 1233–1239.

Jung, J.H., Park, S.Y., Lee, J.E., Nho, C.W., Lee, B.U. and Bae, G.N. (2011) Electrohydrodynamic nano-spraying of ethanolic natural plant extracts. J. *Aerosol Science*, 42, 725–736.

Kenawy, E.-R., Bowlin, G.L., Mansfield, K., Layman, J., Simpson, D.G., Sanders, E.H. and Wnek, G.E. (2002) Release of tetracycline hydrochloride from electrospun poly(ethylene-co-vinylacetate), poly(lactic acid), and a blend. *Journal of Controlled Release*, 81, 57–64.

Kim, K., Luu, Y.K., Chang, C., Fang, D., Hsiao, B.S., Chu, B. and Hadjiargrou, M. (2004) Incorporation and controlled release of a hydrophilic antibiotic using poly(lactide-co-glycolide)-based electrospun nanofibrous scaffolds. *Journal of Controlled Release*, 98, 47–56.

Labuza, T.P. and Breene, W. (1989) Application of "active packaging" technologies for the improvement of shelf-life and nutritional quality of fresh and extended shelf-life foods. *Bibliotheca Nutritio et Dieta (Basel)*, 43, 252–259.

Lee, S.-K., Sheridan, M. and Mills, A. (2005) Novel UV-activated colorimetric oxygen indicator. *Chemistry of Materials*, 17, 2744–2751.

Lee, Y.-H., Mei, F., Bai, M.-Y., Zhao, S. and Chen, D.-R. (2010) Release profile characteristics of biodegradable-polymer-coated drug particles fabricated by dual-capillary electrospray. *Journal of Controlled Release*, 145, 58–65.

Li, M., Mondrinos, M.J., Gandhi, M.R., Ko, F.K., Weiss, A.S. and Lelkes, P.I. (2005) Electrospun protein fibers as matrices for tissue engineering. *Biomaterials*, 26, 5999–6008.

Li, Y., Lim, L.T. and Kakuda, Y. (2009) Electrospun zein fibers as carriers to stabilize (−)-epigallocatechin gallate. *Journal of Food Science*, 74(3), C233–C240.

Lim, L.-T. (2011) Chapter 13: Nanostructure packaging technologies. In: *Modified Atmosphere Packaging for Fresh-Cut Fruits and Vegetables* (ed. A.L. Brody, H. Zhuang, and J.H. Han). John Wiley &Sons, Chichester, 241–266.

López-Rubio, A. and Lagaron, J.M. (2012) Whey protein capsules obtained through electrospraying for the encapsulation of bioactives. *Innovative Food Science and Emerging Technologies*, 13, 200–206.

Lopez-Rubio, A., Sanchez, E., Sanz, Y. and Lagaron, J.M. (2009) Encapsulation of living bifidobacteria in ultrathin PVOH electrospun fibers. *Biomacromolecules*, 10, 2823–2829.

Loscertales, I.G., Barrero, A., Guerrero, I., Cortijo, R., Marquez, M. and Ganan-Calvo, AM. (2002) Micro/nano encapsulation via electrified coaxial liquid jets. *Science*, 295, 1695–1698.

Lu, J.-W., Zhu, Y.-L., Guo, Z.-X., Hu, P. and Yu, J. (2006) Electrospinning of sodium alginate with poly(ethylene oxide). *Polymer*, 47, 8026–8031.

Luykx, D.M.A., Peters, R.J.B., van Ruth, S.M. and Bouwmeester, H. (2008) A review of analytical methods for the identification and characterization of nano delivery systems in food. *Journal of Agricultural and Food Chemistry*, 56, 8231–8247.

Lyons, J., Li, C. and Ko, F. (2004) Melt-electrospinning part I: processing parameters and geometric properties. *Polymer*, 45, 7597–7603.

Matthews, J.A., Wnek, G.E., Simpson, D.G. and Bowlin, G.L. (2002) Electrospinning of collagen nanofibers. *Biomacromolecules*, 3, 232–238.

Mei, F. and Chen, D.-R. (2007) Investigation of compound jet electrospray – particle encapsulation. *Physics of Fluids*, 19, 103303.

Mihindukulasuriya, S. (2012) *Investigations of heat seal parameters and oxygen detection in flexible packages*. PhD Thesis, Department of Food Science, University of Guelph.

Min, B.-M., Lee, S.W., Lim, J.N., You, Y., Lee, T.S., Kang, P.H. and Park, W.H. (2004) Chitin and chitosan nanofibers: electrospinning of chitin and deacetylation of chitin nanofibers. *Polymer*, 45, 7137–7142.

Ogata, N., Yamaguchi, S., Shimada, N., Lu, G., Iwata, T., Nakane, K. and Ogihara, T. (2007) Poly(lactide) nanofibers produced by a melt-electrospinning system with a laser melting device. *Journal of Applied Polymer Science*, 104, 1640–1645.

Ozdemir, M. and Floros, J.D. (2004) Active food packaging technologies. *Critical Reviews in Food Science and Nutrition*, 44(3), 185–193.

Parada, J. and Aguilera, J.M. (2007) Food microstructure affecs the bioavailability of several nutrients. *Journal of Food Science*, 72(2), R21–R32.

Quispe-Condori, S., Saldaña, M.D.A. and Temelli, F. (2011) Microencapsulation of flax oil with zein using spray and freeze drying. *LWT – Food Science and Technology*, 44, 1880–1887.

Ramakrishna, S., Fujihara, K., Teo, W.-E., Lim, T.-C. and Ma, Z. (2005) *An Introduction to Electrospinning and Nanofibers*. World Scientific.

Rao, J.P. and Geckeler, K.E. (2011) Polymer nanoparticles: Preparation techniques and size-control parameters. *Progress in Polymer Science*, 36, 887–913.

Regev, O., Vandebril, S., Zussman, E. and Clasen, C. (2010) The role of interfacial viscoelasticity in the stabilization of an electrospun jet. *Polymer*, 51, 2611–2620.

Reneker, D.H., Yarin, A.L., Fong, H. and Koombhongse, S. (2000) Bending instability of electrically charged liquid jets of polymer solutions in electrospinning. *Journal of Applied Physics*, 87, 4531–4547.

Safi, S., Morshed, M., Hosseini Ravandi, S.A. and Ghiaci, M. (2007) Study of electrospinning of sodium alginate, blended solutions of sodium alginate/poly(vinyl alcohol) and sodium alginate/poly(ethylene oxide) *Journal of Applied Polymer Science*, 104, 3245–3255.

Sanguansri, P. and Augustin, M.A. (2006) Nanoscale materials development – a food industry perspective. *Trends in Food Science and Technology*, 17, 547–556.

Schiffman, J.D. and Schauer, C. (2008) A review: Electrospinning of biopolymer nanofibers and their applications. *Polymers Reviews*, 48, 317–352.

Scully, A. (2009) Active Packaging. In: *The Wiley Encyclopedia of Packaging Technology*, 3rd edn (ed. K.L. Yam). John Wiley & Sons, Inc., Hoboken, NJ, 2–9.

Seifu, E., Buys, E.M. and Donkin, Ef. (2005) Significance of teh lactoperoxidase system in the dairy industry and its potential applications – a review. *Trends in Food Science and Technology*, 16, 137–154.

Son, W.K., Youk, J.H. and Park, W.H. (2006) Antimicrobial cellulose acetate nanofibers containing silver nanoparticles. *Carbohydrate Polymers*, 65, 430–434.

Stephens, J.S., Fahnestock, S.R., Farmer, R.S., Kiick, K.L., Chase, D.B. and Rabolt, J.F. (2005) Effects of electrospinning and solution casting protocols on the secondary structure of a genetically engineered dragline spider silk analogue investigated via Fourier transform raman spectroscopy. *Biomacromolecules*, 6, 1405–1413.

Taylor, G. (1966) The force exerted by an electric field on a long cylinder conductor. *Proceedings of the Royal Society of London, Series A: Mathematical and Physical Sciences*, 291, 145–158.

Taylor, G. (1969) Electrically driven jets. Proc. Roy. Soc. Lond. A, 313, 453–475.

Teo, W.E. and Ramakrishna, S. (2006) A review on electrospinning design and nanofiber assemblies. Nanotechnology 17: R89–R106.

Torres-Giner, S., Ocio, M.J. and Lagaron, J.M. (2008) Development of active antimicrobial fiber based chitosan polysaccharide nanostructures using electrospinning. *Engineering in Life Sciences*, 8(3), 303–314.

Torres-Giner, S., Martinez-Abad, A., Ocio, M.J. and Lagaron, J.M. (2010) Stabilization of a nutraceutical omega-3 fatty acid by encapsulation in ultrathin electrosprayed zein prolamine. *Journal of Food Science*, 75(6), N69–N79.

Townsend-Nicholson, A. and Jayasingh, S.N. (2006) Cell electrospinning: a unique biotechnique for encapsulating living organisms for generating active biological microthreads/scaffolds. *Biomacromolecules*, 7, 3364–3369.

Trotta, M., Cavalli, R., Trotta, C.R.B. and L.C. (2009) Electrospray technique for solid lipid-based particle production. *Drug Development and Industrial Pharmacy*, 36, 431–438.

Vega-Lugo, A.-C. and Lim, L.-T. (2008) Electrospinning of soy protein isolate nanofibers. *Journal of Biobased Materials and Bioenergy*, 2, 223–230.

Vega-Lugo, A.C. and Lim, L.T. (2009) Controlled release of allyl isothiocyanate using soy protein and poly(lactic acid) electrospun fibers. *Food Research International*, 42(8), 933–940.

Vega-Lugo, A.-C. and Lim, L.-T. (2012) Effects of poly(ethylene oxide) and pH on the electrospinning of whey protein isolate. *Journal of Polymer Science Part B Polymer Physics*, 50, 1188–1197.

Wang, Z.-G., Wan, L.-S., Liu, Z.-M., Huang, X.-J. and Xu, Z.-K. (2009) Enzyme immobilization on electrospun polymer nanofibers: An overview. *Journal of Molecular Catalysis B: Enzymatic*, 56, 189–195.

Watanabe, H., Matsuyama, T. and Yamamoto, H. (2001) Preparation of immobilized enzyme gel particles using an electrostatic atomization technique. *Biochemical Engineering Journal*, 8, 171–174.

Weiss, J., Takhistiv, P. and Julian, M.D. (2006) Functional materials in food nanotechnology. *Journal of Food Science*, 71(9), R107–R116.

Woerdeman, D.L., Ye, P., Shenoy, S., Parnas, R.S., Wnek, G.E. and Trofimova, O. (2005) Electrospun fibers from wheat protein: investigation of the interplay between molecular structure and the fluid dynamics of the electrospinning process. *Biomacromolecules*, 6, 707–712.

Wongsasulak, S., Patapeejumruswong, M., Weiss, J., Supaphol, P. and Yoovidhya, T. (2010) Electrospinning of food-grade nanofibers from cellulose acetate and egg albumen blends. *Journal of Food Engineering*, 98, 370–376.

Xie, J. and Hsieh, Y.-l. (2003) Ultra-high surface fibrous membranes from electrospinning of natural proteins: casein and lipase enzyme. *Journal of Materials Science*, 38, 2125–2133.
Xie, J., Lim, L.K., Phua, Y., Hua, J. and Wang, C.-H. (2006) Electrohydrodynamic atomization for biodegradable polymeric particle production. *Journal of Colloid and Interface Science*, 302, 103–112.
Xie, J., Ng, W.J., Lee, L.Y. and Wang, C.-H. (2008) Encapsulation of protein drugs in biodegradable microparticles by co-axial electrospray. *Journal of Colloid and Interface Science*, 317, 469–476.
Xu, X., Yang, L., Xu, X., Wang, X., Chen, X., Liang, Q., Zeng, J. and Jing, X. (2005) Ultrafine medicated fibers electrospun from W/O emulsions. *Journal of Controlled Release*, 108, 33–42.
Yang, Y., Li, X., Qi, M., Zhou, S. and Weng, J. (2008) Release pattern and strictural integrity of lysozyme encapsulated in core-sheath structured poly(DL-lactide) ultrafine fibers prepared by emulsion electrospinning. *European Journal of Pharmaceutics and Biopharmaceutics*, 69, 106–116.
Yarin, A.L., Koobhongse, S. and Reneker, DH. (2001) Taylor cone and jetting from liquid droplets in electrospinning of nanofibers. *Journal of Applied Physics*, 90, 4836–4846.
Yi, F., Guo, Z.-X., Hu, P., Fang, Z.-X., Yu, J. and Li, Q. (2004) Mimetics of eggshell membrance protein fibers by electrospinning. *Macromolecular Rapid Communications*, 25, 1038–1043.
Yu, J.H., Fridrikh, S. and Rutledge, GC. (2006) The role of elasticity in the formation of electrospun fibers. *Polymer*, 47, 4789–4797.
Zeng, J., Xu, X., Chen, X., Liang, Q., Bian, X., Yang, L. and Jing, X. (2003) Biodegradable electrospun fibers for drug delivery. *Journal of Controlled Release*, 92, 227–231.
Zeng, J., Aigner, A., Czubayko, F., Kissei, T., Wendorff, J.H. and Greiner, A. (2005) Poly(vinyl alcohol) nanofibers by electrospinning as a protein delivery system and the retardation fo enzyme release by additional polymer coatings. *Biomacromolecules*, 6, 1484–1488.
Zhang, Y., Ouyang, H., Lim, C.T., Ramakrishna, S. and Huang, Z.-M. (2005) Electrospinning of gelatin fibers and gelatin/PCL composite fibrous scaffolds. *Journal of Biomedical Materials Research Part B: Applied Biomaterials*, 72, 156–165.
Zhao, Y., Carvajal, M.T., Won, Y.-Y. and Harris, M.T. (2007) Preparation of calcium alginate microgel beads in an electrodispersion reactor using an internal source of calcium carbonate nanoparticles. *Langmuir*, 23, 12489–12496.
Zhou, F.L., Gong, R.H. and Porat, I. (2009) Mass production of nanofibre assemblies by electrostatic spinning. *Polymer International*, 58(4), 331–342.
Zhou, H., Green, T.B. and Joo, Y.L. (2006) The thermal effects on electrospinning of polylactic acid melts. *Polymer*, 47, 7497–7505.
Zhou, Y. and Lim, L.-T. (2009) Activation of lactoperoxidase system in milk by PLA-immobilized glucose oxidase. *Journal of Food Science*, 74(2), C170–C176.

19 Risks and ethics in the context of food nanotechnology and the delivery of bioactive ingredients

Paul B. Thompson

Departments of Philosophy, Community Sustainability and Agricultural, Food and Resource Economics, Michigan State University, East Lansing, Michigan, USA

19.1 INTRODUCTION

Risk is, by its very nature, inherently normative or value-laden. Actions or circumstances are risky when there is some unwanted, averse, or "bad" result that accompanies or follows them. Deploying the idea or concept of risk in connection with food nanotechnology highlights the potential for such unwanted or averse outcomes. Furthermore, there is widespread agreement among food science professionals that novel food chemistries *should be* evaluated for the potential of deleterious effects. In this, food scientists have reached a consensus on a norm: a principle that articulates actions that individuals and organizations are strongly advised if not obligated to undertake in connection with developing new food products or interventions in the food system (Jensen and Sandøe, 2002). It is both the implicitly value-laden nature of risk and the agreement on norms of responsible conduct that provides the basis for ethics in the context of food nanotechnology and the delivery of bioactive ingredients.

Although the scope of ethics for risk assessment is potentially quite extensive, this chapter will emphasize value judgments that result in pervasive and significant differences in the way that risk assessments for nanotechnology and bioactive ingredients might be conducted. First will be a general discussion of ethical issues in food safety. Next it will be helpful to contrast two significantly different paradigms for understanding what would count as responsible or adequate to potential hazards arising in connection with virtually any new technology. A *rational choice* approach will be contrasted with one that stresses risk imposition and *informed consent*. Following this theoretical discussion of risk ethics, the chapter will discuss how risk management and risk communication for nanotechnologies and foods with bioactive ingredients face specific challenges given the politicized nature of food safety in the wake of debates over genetically engineered crops. The chapter concludes with some "take home messages" for scientists and policy makers working on risk assessment and policy formulation for food nanotechnology.

Nanotechnology and Functional Foods: Effective Delivery of Bioactive Ingredients, First Edition.
Edited by Cristina M. Sabliov, Hongda Chen, and Rickey Y. Yada.

19.2 FOOD RISK EVALUATION AND MANAGEMENT

Food consumption has been increasingly subsumed under institutions of commercial exchange ever since the dawn of the industrial revolution. Whereas many people might have once been involved in the primary production of much of their food, today that is the exception, rather than the rule. People purchase food from vendors operating in a plethora of modes, and these vendors themselves are but one element in a complex food supply chain. While in some domains of commercial exchange a rule of caveat emptor or simple merchantability may still prevail, the safety and integrity of the food supply came to be recognized as a public good early in the twentieth century. Protecting the safety and integrity of food is thus a fiduciary responsibility – a duty to look out for the welfare of parties who are unable to control their own fate – that is jointly shared by the food industry and by state and local governments.

However, executing this fiduciary responsibility has been fraught with unexpected complications. Effective governance for public goods presupposes at least a modicum of public trust in the individuals and institutions charged with protecting the public interest. But public trust can be eroded either by malfeasance or by misunderstanding. Food safety policy for nanotechnologies should thus be made with cognizance of two ways in which the underlying logic of the policy is subject to misinterpretation. First, experts commonly make a distinction between novel and normal risk in their practice, yet seldom recognize this distinction explicitly in representing the logic of their practice to the public. Second, the discussion of safety for genetically engineered foods has left us with a process versus product distinction that is apt to cause confusion in an era of food nanotechnology.

19.2.1 Food safety as a fiduciary responsibility

The acceptance of a clear governmental role in assuring food safety occurred in tandem with the recognition that private firms have incentives to engage in unsafe practices, at least in so far as such practices remain undetected by their customers. Upton Sinclair's *The Jungle* (1906) portrayed a meat industry equally insensitive to the interests of their workers and the pallets of the customers. A muckraking book by Arthur Kallet and F.J. Schlink from the 1930s also raised questions about the food industry's commitment to safe and wholesome products (Kallet and Schlink, 1935). Similar exposes have continued to appear right to the present day. While the food industry as a whole has an interest in maintaining the safety of the food supply, individual firms have incentives to control costs and to conceal mistakes that pose hazards to the consumers of their products (Early, 2002).

Informal risk assessments have always played a role in exercising responsibility for the safety of the food system, both within private industry and within governmental agencies. During the early decades of the twentieth century, the presumptive framework for risk assessment was contamination. In the wake of Louis Pasteur's work on the germ theory of disease, a public official might run through a mental checklist, looking to ascertain where key opportunities for contamination might arise. Of course, mental checklists have not disappeared, but today they are more likely to be specified formally, as in guidance on Hazard Analysis and Critical Control Point (HACCP) procedures for food production, processing and distribution (Pierson, 1992). Whether executed as an informal checklist backed up by a general understanding of scientific principles or explicitly specified as a detailed procedure for hazard identification and exposure assessment, risk assessment is fundamental to both industry and governmental roles in executing fiduciary responsibilities to the broader public for the overall safety of the food supply.

19.2.2 Food safety and public confidence

Effective execution of fiduciary responsibility presupposes not only competence by the executor, but confidence on the part of the party to whom the responsibility is owed. In the case of food safety, this means that the complex of industry and governmental agencies must be able to deliver two discrete goods. First, the food supply must be *safe*, given some reasonable standard of safety that is widely accepted by the general public. Second, the general public must have *confidence* in the safety of the food supply and in the ability and willingness of food safety authorities to assure the agreed upon standard of safety. Arguably, food safety officials in industrialized countries have done a better job in making the food system comply with a reasonable standard of safety than they have in generating public confidence. In general, the sources and measures of public trust or confidence are not well understood. There are a number of competing models for defining trust, and there is no consensus among social psychologists about the dimensions or indicators of public confidence in key social institutions (Lobb, 2005).

The discussion that follows throughout the chapter will highlight features of risk analysis and risk communication that involve value orientations in the characterization, management, and communication of risk. Disagreement or lack of clarity about values may be associated with deficits in public trust or confidence (Brom, 2000). However, simply explaining values or ethical commitments should not be regarded as an adequate approach to developing trust and confidence in novel food technologies. As will be discussed below, the present-day context for evaluation of nanotechnologies and bioactive ingredients for food is shaped by recent events, including both highly publicized failures in food safety and poorly understood policy decisions for other emerging technologies. Arguably, confidence has been shaken in part by a failure to adequately communicate *what* is uncertain and what is being debated when experts first develop methods for ascertaining the safety of novel technologies. And the evolution of the process versus product distinction in food safety policy is a second source of confusion.

19.2.3 Novel and normal risk

Other chapters in this volume discuss some of the ways in which particles at the nanoscale may pose challenges to classic methods in toxicology and food-safety risk assessment. In this respect, nanomaterials and the introduction of novel bioactive ingredients into foods pose a conceptual – I would say "philosophical" – problem for established procedures for risk assessment and risk management. Risk assessment is a procedure that uses quantitative and empirically tested methods to attain reasonable confidence about the predicted human health impact of allowing a substance into the food supply, but of necessity, outcomes cannot be known with absolute certainty. The phrases "reasonable confidence" and "absolute certainty" reflect the fact that risk assessments can be either more or less definitive in the results they achieve. In many – arguably most – cases, materials or substances under review are amenable to one or more toxicological methods in use by risk assessors. The risks identified by these methods can be referred to as *normal risks*. One cannot exclude the possibility that new hazards or modes of exposure will be discovered, nor are standard risk assessment methods infallible in all cases. To assert that risks are normal is simply to say that existing methods and the current state of knowledge are viewed as adequate for the purpose of attaining reasonable confidence.

Emerging technologies frequently raise doubt about the adequacy of existing risk assessment methods. There may be widespread consensus about the challenge that a new substance or product poses to existing methods, or doubts may be unevenly shared both among food safety experts and by individuals lacking expertise in the methods of hazard identification and exposure quantification. In either case, these *novel risks* will be subjected to a period of discussion, debate, and experimentation with methods for risk assessment. Such debate and disagreement is typical for scientific progress, and is usually followed by growing consensus among members of the expert community (though sometimes only after an extended period of heated dispute) (Thompson and Hannah, 2012). However, in the context of a fiduciary responsibility and suspicion about the motives and goodwill of key actors, differences of opinion about the novelty of a risk have secondary effects on public confidence. The fact that experts are debating has the potential for two interrelated effects that will be discussed below. First, it brings attention to the materials, substances, or products being debated, where they might not otherwise be even noticed outside a small group of risk assessment professionals. Second, the fact that experts disagree is interpreted as evidence for risk. Note that among experts, debate about whether risk assessment methods are adequate does not necessarily imply anything about the actual potential for hazards. Debate implies only that some experts doubt that current methods provide a basis for judgment. Yet beyond the expert community, this nuanced point is highly likely to be missed. The fact that experts do not agree will be taken as evidence that an emerging technology is risky.

19.2.4 Process versus product

The introduction of food crops with germplasm altered through tools of recombinant DNA has affected thinking on food safety in ways that have yet to be fully appreciated. One key policy result was a focus on the distinction between product and process. Within the context of genetically engineered foods, the process was understood to mean the exploitation of rDNA to alter plant genomes. The product was the specific plant variety or animal breed being assessed for risk. United States regulators adopted the policy that it is appropriate to regulate the product and not the process. This has widely been interpreted to mean that the processes used in development or manufacture of a product should not increase the burdens of proof for risk assessment. Other food safety agencies have taken the stance that the use of rDNA techniques justifies a higher level of regulatory scrutiny, in part simply because public constituencies demand it. This difference in risk management strategy reflects a key instance of the way that considerations relating to public confidence can penetrate into risk assessment methods (Dunlop, 2000).

At the same time, it is worth noting that many aspects of food safety regulation are explicitly process oriented, and justifiably so. The HACCP. a methodology for assessing and managing risks of microbial contamination in products that reach food consumers by explicit attention to key junctures in the production process that are most vulnerable to contamination (Simonsen *et al.*, 1987), is a prominent case in point. Food safety policy thus involves an implicit distinction that is being drawn between processes used in product development and processes utilized in product manufacture that has not been explicitly recognized, and is not mentioned in communications with nonexpert communities. This distinction may have been fully justified in the case of genetic engineering, where the process of plant transformation occurs in the development of a crop variety and plays no role in the production processes (such as Good Agricultural Practice (GAP) rules) that would normally be covered by HACCP. It is not clear how it could be applied in the case of nanoscale ingredients. Like the distinction between normal and novel risk, the distinction between process risk and product risk involves value judgments that, while quite plausibly justifiable, remain unarticulated and unevaluated in the absence of an exercise designed explicitly to do identify and examine underlying values in risk assessment practice.

19.3 RISK AND ETHICS: AN OVERVIEW

Although there is little disagreement about the existence of some normative basis for risk assessment in the context of food safety and new product development, the association of the word "ethics" with risk assessment is still not well understood. Stated simply, ethics is an activity that consists in deliberately articulating the normative basis for risk assessment and then subjecting the key normative principles or assumptions in risk assessment to a critical review. In this sense, ethics is not a particularly unusual or mysterious component of risk assessment activities. The process or activity of specifying and critically evaluating the normative elements in risk assessment is amenable to theorization and methodological development. Within the discipline of philosophy and applied ethics programs (such as bioethics), the word 'ethics' is used much like other terms that indicate domains of research and theory development, such as 'toxicology' or 'chemistry': the word refers alternately to the theory developed, the activity of theory development, and to the phenomena being theorized.

Nevertheless, development of theory for specifying or evaluating norms and values is too frequently attenuated and downplayed in the development or conduct of risk analysis and risk management, especially within food safety and toxicological contexts. As a legally mandated component of the regulatory process, risk assessments become routinized. The normative commitments that guide a risk assessment may be specified by administrative procedures, obviating the need or opportunity for scientifically trained participants in the risk assessment process to participate in their specification or critical review. As a result, normative commitments become standardized, leaving no role for critical evaluation. To the extent that norms and values become implicit in specified procedures for risk assessment, they may be taken as given elements of risk assessment methodologies, and risk assessment practitioners may neither see any need for explicit articulation of them nor be fully aware of what underlying value assumptions may be operative.

The distinction between risk assessment and risk management suggests that many key ethical judgments will be deferred to decision making that occurs once a scientific assessment of risks has been completed. This distinction is meaningful, but it is important to recognize that normatively is implied even at the stage of hazard identification, where some value orientation must be in place in order to classify an outcome or state of affairs as adverse, unwanted, potentially harmful, or hazardous. Exposure assessment also involves key value judgments, such as partitioning the set of all possible outcomes into discrete classes for the purposes or the identification of reference populations so that elevated levels of risk can be characterized (Cranor 1998; Thompson, 2003). Throughout this paper, ethics is the activity of identifying the various junctures at which value judgments play key roles in facilitating risk assessment *and* risk management, as well as the critical evaluation of judgments with an eye toward more effectively executing the overall goals and purposes of risk assessment.

19.3.1 The ethics of rational choice under uncertainty or risk

The standard normative model of decision making for evaluating the risk of novel technology was in fact articulated as a general approach to ethics by Jeremy Bentham in 1789. Bentham argued that *all* decision making should conform to a model in which the decision maker carefully anticipates both benefits and harms that would accrue as a result of selecting each of the several potential options available. Bentham, who characterized benefit and harm broadly in terms of pleasure and pain, recognized that the estimation of harm would need to account for the intensity or extent of an adverse outcome (both with respect to the degree of pain as well as the number of parties affected) as well as the propinquity or likelihood that painful outcomes would actually materialize. Knowing this, Bentham indicated that an ethically responsible decision maker will choose the course of action with the greatest expectation of net benefit, once costs or expected harms had been deducted. Bentham's rule for ethics has been characterized as the utilitarian maxim: choose so as to produce the greatest good for the greatest number (Bentham, 1789).

By emphasizing outcomes and characterizing the potential for adverse outcomes in terms of the degree of harm (or pain) and its likelihood, Bentham anticipated the overall theoretical approach to decision making under uncertainty that is most frequently associated with risk assessment and risk management in the present day. Bentham's overall approach has undergone more than two centuries of refinement by economists and decision theorists, but several key ethical implications have remained unchanged. The first is a feature that philosophers have characterized as *consequentialism*: ethics is above all a process of evaluating and comparing the outcome that a possible course of action can be expected to have. In the case of technological risk assessment, the new technology (or class of technologies) is treated as a possible course of action that will have both beneficial and potentially harmful outcomes. Ethical evaluation of new technology involves identification and comparison of those outcomes. Second, the approach to comparing options involves *quantitative ranking*: some method for ordering outcomes is necessary. Third is *welfarism*: the ethical significance of consequences is defined in terms of impact on the health, wealth or well-being of affected parties. Finally, the approach emphasizes *optimization*: ethics requires a decision maker to choose the course of action that will produce the best outcome, all things considered (Sen, 1987).

Decision making for new food technologies has departed from some elements of Bentham's framework while continuing to embrace others. Specifically, the ethical acceptability of food technologies employs quantitative ranking methods in a manner that does not support an optimizing approach. Worldwide regulatory procedures for food safety, for example, utilize a *de minimus* approach in which *any* significant level of risk from a new food technology is deemed unacceptable. Potential benefits, either in the form of healthful diets or economic benefits, are not taken into consideration. Thus food-safety decision making makes no attempt to achieve "the greatest good for the greatest number." Interestingly, however, decision making for evaluating the residue of agricultural pesticides *does* weigh risks from incidental exposure to residues in food consumption against the benefits of using the pesticide from the perspective of farm profitability and environmental impact. There are thus significant variations in which a rational choice approach to risks might be developed and applied.

What remains standard among the various approaches deriving from Bentham's initial approach is that risk is understood to have ethical significance exclusively in terms of outcomes, and that outcomes are understood in terms of adverse impact on some aspect of the welfare or well-being of persons (or other beings) affected by the action in question. Given these standard elements, rational choice theory understands ethics primarily in terms of developing and defending a decision rule that utilizes the quantitative characterization of adverse impact as a key input. The decision rule may or may not involve comparison with other options, and it may or may not utilize optimization (or in Bentham's terms maximizing net benefit). In addition, owing the logical impossibility of eliminating any possibility of an adverse outcome, rational choice procedures require some way of configuring their decision rule so that *some* nonzero quantitative estimate of risk can be found acceptable. If risk/benefit weighing is not used, so that benefits can be seen as compensating for potential adverse outcomes, a maxim such as "least possible" may be developed as a substitute.

A full discussion of the different approaches to risk management that are available is beyond the scope of this chapter. As noted above, the particular approach may be specified by organizational policies or, in the case of regulatory risk assessment, by authorizing legislation. However, when people do articulate contrasting approaches to risk management, and especially when they discuss which approach is most appropriate in a given context, this *is* ethics, as ethics has been defined above. Furthermore, to the extent that the *relevance* of risk is understood in terms of the expected value associated with a given practice, policy, or course of action, the ethical justification of a particular decision rule is limited to a particular tradition in ethics that descends from Bentham's utilitarianism, translated through two centuries of economic analysis and rational choice theory. Yet this is not the only way to conceptualize the ethical relevance of risk.

19.3.2 Ethics and the imposition of risk

An extensive review of the use of human subjects in harmful and potentially harmful biomedical research occurred in the 1970s. Key episodes included experimentation on people held in concentration camps by Nazi doctors, the "Tuskegee study" in which syphilis was allowed to run its course in a population of African-Americans who were never offered treatment, and the Willowbrook study in which severely retarded children were exposed to live hepatitis virus in order to test the effectiveness of gamma globulin. In each case, researchers rationalized the harm and exposure to a risk of harm to these human research subjects by convincing themselves that benefits accruing from the studies in question would more than offset the risks borne by research subjects. *The Belmont Report* published in 1979 articulated ethical principles for research involving human subjects that drew upon a significantly different approach to the ethics of risk (NIH, 1979).

Rational choice approaches the ethics of risk by assigning a value to the potential for unwanted outcomes that reflects both the degree or intensity of harm and the likelihood that harm will materialize, and then moves on to consider whether risks are acceptable or should be mitigated. The approach recommended in *The Belmont Report* begins with the assumption that only those who will bear the risk can adequately evaluate its moral significance. While they are entitled to full information about the nature and likelihood of harm, as well as potentially offsetting benefits, they are under no obligation to weigh risks rationally (though they are certainly free to do so), or even to conceptualize risk using the risk assessment notions of hazard and exposure. The *informed consent* approach to risk focuses first on whether people are truly able to accept or reject a risk, and secondarily on the information that must be provided to them in order to ensure that coercion or deception does not compromise their freedom (Faden and Beauchamp, 1986.).

The primary interface between informed consent and the biological sciences occurs in the human subjects approval process that virtually all researchers in developed countries have been obligated to follow for over three decades. It is, however, entirely possible to apply the informed consent approach to many situations in which people will be exposed to risk. Something very much like informed consent has been the model for understanding the ethics of liability for risk in employment contracts for a very long time (Bohlen, 1906a, b). Without necessarily implying that a standard of informed consent could be applied in the case of food nanotechnologies, it is nonetheless useful to be clear on how informed consent suggests a somewhat different conceptualization of the task for ethics.

A rational choice approach might recommend exactly the same result as an informed consent approach in a given case; that is, both might find the same practices to be acceptable or unacceptable in ethical terms. However, while the rational choice approach centers on the value of outcomes, the informed consent approach focuses on the integrity of the choice situation for the person who will bear risks. The fact that one party is undertaking an activity that exposes a second party to risk generates the need for ethical consideration, and it is the first party's responsibility to respect the autonomy or dignity of the risk bearer that generates the basis for an ethical reflection. Thus the literature on informed consent is populated by studies that examine questions such as whether inducements in the form of compensation could be considered coercive, or that question how informed consent can be effected for risk-bearing human subjects that are illiterate (Raich *et al.*, 2001), socially disempowered (Torczyner, 1991), or, in the extreme cases, mentally incompetent (Redding, 1993). As a result, the informed consent approach is relatively unengaged with risks that are not associated with some human action that initiates the conditions of exposure: there is no literature on informed consent for bearing the risks of an earthquake or tsunami. As will be discussed at more length below, the key point to take from the informed consent tradition is that intentional actions that expose people to risk may trigger a set of ethical considerations that differ from the valuation and trade-off approaches to ethics that are typical of rational choice.

19.3.3 Ethical responsibilities in risk communication

Classic instances of risk communication arise when experts become aware of an underappreciated or poorly understood hazard, and then need to communicate this information to members of the nonexpert public. Health risks associated with smoking or sexual behavior fit this pattern, as do hazards that are associated with natural hazards (such as tsunamis and earthquakes) or anthropogenic climate change. In such instances, the point of the communication effort to apprise people of the fact that there are actions that they are advised to consider, even in the case of natural hazards. It would be pointless, and perhaps even unethical, to divert people from their business if there is nothing that they could do to alter the situation. Risk communications should include both information about the potential hazard, as well as advice or instruction as to what efforts will be or should be undertaken to ameliorate harm or reduce the probability that harm will materialize (Witte and Allen, 2000).

A somewhat different type of communication ensues in the wake of panic, especially in cases where no cause for alarm can be justified. Such cases of false alarm are not infrequent. They might be described as communications that are intended to *counter* a previous signal, such as a literally false alarm, that has already placed people on alert, leading them to expect that further messages will clarify what steps need to be taken (Otway and Wynne, 1989). Such messages might use expressions like "Nothing to be concerned about," "All clear," or "Everything is fine; go on about your business." In such circumstances, communication is effectively telling people that the exceptional, risky circumstance is either over or never occurred. They are telling people that there is (now, at least) "no risk".

Professionals in risk assessment and risk management have been exceedingly uncomfortable with "no risk" messaging, perhaps presuming that people will take this to mean zero probability of harm. Indeed, in a litigious society such a presumption is quite reasonable, at least from a legal perspective. The alternative, however, may be to apprise people of facts about hazards and exposure mechanisms that are of utterly no practical use in the conduct of ordinary life, and this alternative has ethical implications, as well. First, it may appear to endorse the claim that everything a person does should be subjected to a deliberative weighing of risk and benefit, whether the potential for harm is significant or not, and even in cases where people have no meaningful alternatives. Such a message is arguably quite irresponsible from an ethical perspective. Second, and more pertinently, apprising people of even insignificant potential harms has the effect of bringing this potential to their attention. As a result, risk communications that may have originated in the effort to forestall or avoid a panic can have exactly the opposite result (Katz, 2001). This will especially tend to be the case when people suspect the motives of risk communicators who may have ulterior motives. Striking the balance between shilling for industry, on the one hand, and needlessly scaring the pants off people proves to be one of the most difficult ethical problems in risk communication (Thompson, 2012).

19.4 THE POLITICIZED CONTEXT OF FOOD NANOTECHNOLOGY AND FUNCTIONAL FOODS

Nanotechnology and foods with bioactive ingredients are moving through regulatory review at a time when the political context for risk assessment, scrutiny by civil society groups, and eventual market acceptability have been shaped by a number of key events. Food scares – generally unassociated with novel technology – have commanded significant attention from national media (Caswell, 2006). Many have noted that events such as the Chernobyl incident and the incidence of mad cow disease created a hypersensitivity to risk among the European public (Frewer *et al.*, 2002). At the same time, social movement theorists have noted the makings of an alternative food movement among people who see large-scale agriculture and corporate control over the food supply chain as both an environmental threat and a form of social injustice (Tovey, 1997; Hassanein, 2003; Woods, 2003). Difficulties encountered by genetically engineered foods provide a key instance, and these foods have been proposed as a model for risks associated with nanotechnologies of all kinds (Berubi, 2006; Sandler, 2006; Sandler and Kay, 2006). This context provides a basis for more specific ethical factors to bear in mind when evaluating risks from nanotechnologies and functional foods.

19.4.1 Food, culture, and dietary choice

First and foremost, one must not lose sight of the fact that nonexperts will form their initial opinions on nanotechnologies and bioactive ingredients against the background of a dense network of cultural values. People choose what they eat based on ethnic background, their personal social identity, their friends and family, cost, and convenience. Within this framework of values, safety will typically be taken for granted. Consumers in industrial societies presume that any marketable food will be, in principle, safe, though they are at the same time mindful of expiration dates and sanitary conditions (Grunert, 2002). Given an act-classifying orientation to risk, foods available in industrial retail outlets are "not risky" for the majority of developed country consumers. To the extent that this is true, a given individual will be especially sensitive to any message or indication that leads to a reclassification. A reference to risks from nanotechnologies is likely to move the food item in question from the "not risky" mental basket into to the "needs attention and deliberation" basket. Given the plethora of food choices that most consumers face, this may be enough reason to simply avoid any food product in the "risky" basket, irrespective of objective facts about the likelihood of harm actually being realized. This should *not* be regarded as irrational. Consumers simply have other things on their mind.

The dense cultural thicket of food-related values has a further consequence for the way that consumers experience their food choices. For most people most of the time, a food choice is *either* a relatively trivial affair in which price or convenience may dominate, *or* it is a choice that is closely tied to their cultural identity, to their experience of solidarity with ethnic, cultural, generational, social, and even political values. People who are choosing to buy organic or a particular brand may be experiencing a social or political connection as much as they are basing a choice on perceived health or quality benefits (Sparks and Shepherd, 1992; Lindeman and Stark, 1999). Given this background it is not unreasonable for them to evaluate the extent to which a potential item of food choice is or is not consistent with a wide array of identity-preserving values, and given *this* one should expect that anything they interpret as a barrier to their ability to make such an evaluation will be perceived as a threat to their autonomy. Autonomy here simply indicates the extent to which a person is able to choose in a manner consistent with their own values. Elements of labeling that either omit or conceal key information will challenge autonomy.

The challenge to autonomy is ethically significant for two distinct reasons. First, consumer autonomy is itself viewed by many as important and as intrinsically valuable. Being "free to choose" is a precondition for an ethically defensible market economy, and may be of particular importance when it comes to food choice (Rubel and Streiffer, 2005; Streiffer and Hedemann, 2005). Secondarily, threats to autonomy or to an individual's ability to maintain key affiliations or elements of a cultural identity may not directly pose a health hazard, but they are experienced as threats, nonetheless. They are thus readily interpreted within an ethical framework of risk. What is more, to the extent that these threats are taken to reveal something about the persons or organizations that are either offering the product or are involved in the threat to autonomy, it is quite reasonable to translate this threat into a more generalized set of concerns about the product, concerns that may quite reasonably extend to health. The reasoning goes something like this: if this company has so little respect for my social or political values as to deceive me, I should also worry that they have little regard for my health.

19.4.2 The social amplification of risk in the context of food politics

Social amplification of risk is a social-science conceptual framework for research on a variety of non-biophysical factors that influence the perception of risk. The framework presumes that biophysical hazards and conditions of exposure are interpreted by human beings through cognitive, social, and linguistic categories that can both amplify and attenuate a given level of risk that can be explained strictly through the analysis of biophysical factors. Cognitive or social amplifiers thus have the potential to explain why people perceive risks from genetic engineering as much more serious or threatening than risk assessments focused exclusively on biological hazards would imply. Cognitive or social factors that amplify or attenuate perception of risk could also potentially explain why people tend to express greater concern over food safety risks from pesticide residues as opposed to microbial contamination (Bruhn and Schutz, 1999; Redmond and Griffith, 2004).

A full development of the social amplification of risk literature is well beyond the scope of this chapter. However, it is worth noting that some cognitive factors or amplifiers clearly serve as potential sources of bias or faulty risk perception, while others represent factors that truly affect the level of risk borne by certain individuals or groups. In the first category, social psychologists have identified a large number of heuristic devices that people use to cope with probabilistically complex problems, and many can result in systematic over- or underestimates of risk. Framing biases, for example, are in evidence when quantitatively identical information is presented in terms of potential gains as opposed to potential loses. Subjects reliably overestimate a risk framed in terms of loss, and will readily accept quantitatively identical risk conditions when the probabilistic dimension is framed in terms of potential benefits (Tversky and Kahneman, 1981; Kahneman *et al.*, 1991). Such cognitive factors "bias" lay judgments of risk. In such cases, nonexperts are likely to misperceive the level of risk that they face in a given situation.

However, there are cases where social factors quite legitimately affect the level of risk that subjects actually face. Transactions in which parties have unequal access to information are an obvious case in point. The proverbial "used car salesman" has more detailed information about the quality of the particular item for sale than any prospective buyer. A seller might know whether the car is a "lemon," for example a car with a significant history of problems. In contrast, a new car would be presumed to represent average quality, something that both buyer and seller could know. The unequal information of parties to the used-car transaction is what makes buying a used car risky (Akerlof, 1970). Technically complex information may create a situation of information inequality that is analogous. Nonexperts simply cannot know what the experts know, and must trust the expert in much the same way that a buyer must trust a used-car dealer. If there are social factors that lead nonexperts to distrust expert advice, these factors amplify the risks of an emerging technology in much the same way that a used-car salesman's incentives and privileged knowledge amplify the risks of buying a used car (Thompson, 2007).

Given the politicized context of emerging food technologies, experts are not in a position to simply assume that members of the lay public will trust their judgment. What is more, civil society groups who oppose new technologies or the existing power of key actors in the food industry can exploit both legitimate and illegitimate social risk amplifiers to create an impression that those offering new technologies are not to be trusted. In this respect, ethical concerns about consumer autonomy and freedom of choice join the ranks of other factors that have been studied in the social amplification of risk paradigm. Scientists from neutral organizations (such as government laboratories, regulatory agencies, or universities) may be very nearly as vulnerable to strategic campaigns that amplify perceptions of risk as are employees of the food industry or technology firms. In the latter case, the economic incentive to benefit from sales of an emerging technology reinforce the analogy to a used car: these individuals both know more about the risks and are in a position to benefit from their knowledge. But consumers are increasingly wary of the incentive structure of putatively neutral scientists, and have noted that staff from universities or regulatory agencies may have been recruited from industry ranks (Meghani and Kuzma, 2011).

The upshot is that a structured and publicly visible activity of considering the full range of ethical questions associated with nanotechnologies and functional foods may be more than simply a fiduciary responsibility of food safety scientists. It may be a strategic necessity for achieving the level of trust that is needed for all actors to operate successfully in the politicized context of emerging food technologies. If so, then an ethics effort may need to go well beyond the amount of deliberation, debate, and critical discussion that would be needed for individual risk scientists or risk managers to assure themselves of the integrity and validity of their assumptions and framework. More extensive communication and public participation efforts may be needed to ensure an adequate level of public trust. Social scientists are actively experimenting with methodologies for undertaking the kind of activities that could both satisfy public concerns and reliably steer risk assessment activities (Barben *et al.*, 2007; Kuzma, 2007). At this juncture, it may be appropriate to close with a few simple guidelines for scientists working to develop food nanotechnologies or bioactive ingredients, as well as for scientists working to understand the risks.

19.5 CONCLUSION: ETHICAL PITFALLS TO AVOID IN ADDRESSING POSSIBLE RISKS FROM FOOD NANOTECHNOLOGY

19.5.1 Do not assume that a science-based view of risk is the only defensible perspective

Although scientific expertise is indispensable, people trained in disciplines such as toxicology, biochemistry, measurement methods, or physiology are in no position to have special insight into the cultural values that people bring to food consumption, or to the way that policies and actions of food industry firms can threaten or challenge those values. It is not unreasonable to articulate concern about such threats and challenges using the language of risk. As such, nanotechnologies and functional foods can generate entirely legitimate concerns that have nothing whatsoever to do with biophysical mechanisms. Lay and social science perspectives may provide important insight into these factors.

19.5.2 Do not overstate OR understate what science can accomplish

Although scientists are often cautious about implying more certainty than results warrant, it is equally important to adopt a stance that reflects the degree of confidence that one actually has. This generally means going beyond what one would say in a journal article, and even beyond a simple translation of what one would say in a journal article into ordinary concepts. It may be important to help people understand the context in which scientific information is relevant both to policy decision and to their personal estimate of the risks and benefits of a new technology (Pielke Jr., 2007).

19.5.3 Do not ignore criticism

Professional norms often lead scientists to think that if it was not published in a peer reviewed journal, differing opinions can safely be ignored. Criticisms can take the form of generalized comments on science or technology, but can also involve direct challenges to findings or results. Although some critics are so convinced of their views that engagement with them is pointless, it is better to presume that criticism is an invitation to better understand the perspective or information that a can be offered from an alternative perspective, so that it may be reflected in a more adequate understanding of risks and how they should be managed.

19.5.4 Do not exaggerate your commitment to public involvement and stakeholder participation

The overall thrust of this chapter supports involvement between scientists and risk assessors, on the one hand, and the broader public, including critics from civil society, on the other. However, undertaking such an engagement should not be done in a manner that leads participants to think that they can have an impact on decisions that have already been made. It is important to understand where there is leeway in risk management policies or in the design or marketing strategy for a product, and to both structure activities of public engagement accordingly and to ensure that participants understand limitations from the outset. Failed efforts at public participation can be worse than no effort at all.

References

Akerlof, G.A. (1970) The market for "lemons": quality uncertainty and the market mechanism. *The Quarterly Journal of Economics*, 84(3), 488–500.

Barben, D., Fisher E., Selin, C. and Guston, D.H. (2007) Anticipatory governance of nanotechnology: foresight, engagement, and integration. In *The Handbook of Science and Technology Studies* (ed. E.J. Hackett, O. Amsterdamska, M. Lynch and J. Wajcman). The MIT Press, Cambridge, MA; 979–1000.

Bentham, J. (1789) *An Introduction to the Principles of Morals and Legislation*. The Clarendon Press, Oxford, UK.

Berubi, D. (2006) *Nanohype: The Truth behind the Nanotechnology Buzz*. Prometheus Books, Amherst, NY.

Bohlen, F.H. (1906a) Voluntary assumption of risk. I. *Harvard Law Review*, 20(1), 14–34.

Bohlen, F.H. (1906b) Voluntary assumption of risk. II. *Harvard Law Review*, 20(2), 91–115.

Brom, F. (2000) Food, consumer concerns, and trust: food ethics for a globalizing market. *Journal of Agricultural and Environmental Ethics*, 12(2), 127–139.

Bruhn, C.M. and Schutz, H.G. (1999) Consumer food safety knowledge and practices. *Journal of Food Safety*, 19(1), 73–87.

Caswell, J.A. (2006) A food scare a day: Why aren't we better at managing dietary risk? *Human and Ecological Risk Assessment*, 12(1), 9–17.

Cranor, C. (1998) The Normative Nature of Risk Assessment: Features and Possibilities, *Risk: Health, Safety and Environment*, 8, 123–136.

Dunlop, C. (2000) GMOs and regulatory styles. *Environmental Politics*, 9(2), 149–55.

Early, R. (2002) Food ethics: a decision making tool for the food industry? *International Journal of Food Science and Technology*, 37(4), 339–349.

Faden, R.R. and Beauchamp, T.L. (1986) *A History and Theory of Informed Consent.* Oxford University Press, Oxford, UK.

Frewer, L.J., Miles, S. and Marsh. R. (2002) The media and genetically modified foods: evidence in support of social amplification of risk. *Risk Analysis*, 22(4), 701–711.

Grunert, K.G. (2002) Current issues in the understanding of consumer food choice. *Trends in Food Science and Technology*, 13(8), 275–285.

Hassanein, N. (2003) Practicing food democracy: a pragmatic politics of transformation. *Journal of Rural Studies*, 19(1), 77–86.

Jensen, K.K. and Sandøe, P. (2002) Food safety and ethics: the interplay between science and values. *Journal of Agricultural and Environmental Ethics*, 15(3), 245–253.

Kahneman, D., Knetsch, J.L. and Thaler, R.H. (1991) Anomalies: the endowment effect, loss aversion, and status quo bias. *The Journal of Economic Perspectives*, 5(1), 193–206.

Kallet, A. and Schlink, F.J. (1935) *100,000,000 Guinea Pigs*. Grosset & Dunlap, New York.

Katz, S.B. (2001) Language and persuasion in biotechnology communication with the public: how to not say what you're not going to not say and not say it. *AgBioForum*, 4(2), 93–97.

Kuzma, J. (2007) Moving forward responsibly: Oversight for the nanotechnology–biology interface. *Journal of Nanoparticle Research*, 9(1), 165–182.

Lindeman, M. and Stark, K. (1999) Pleasure, pursuit of health or negotiation of identity? Personality correlates of food choice motives among young and middle-aged women. *Appetite*, 33(1), 141–161.

Lobb, A. (2005) Consumer trust, risk and food safety: A review. *Food Economics – Acta Agriculturae Scandinavica, Section C*, 2(1), 3–12.

Meghani, Z. and Kuzma, J. (2011) The "revolving door" between regulatory agencies and industry: a problem that requires reconceptualizing objectivity. *Journal of Agricultural and Environmental Ethics*, 24(6), 575–599.

NIH. (1979) *The Belmont Report: Ethical Principles and Guidelines for the Protection of Human Subjects of Research*. The National Commission for the Protection of Human Subjects of Biomedical and Behavioral Research, National Institutes of Health, Washington, DC.

Otway, H. and Wynne, B. (1989) Risk communication: paradigm and paradox. *Risk Analysis*, 9(2), 141–145.

Pielke Jr, R. (2007) *The Honest Broker: Making Sense of Science in Policy and Politics*. Cambridge Univesity Press, Cambridge, UK.

Pierson, M.D. and Corlett, D.A. (1992) *HACCP: Principles and Applications*. Van Nostrand Reinhold, New York.

Raich, P.C., Plomer, K.D. and Coyne, C.A. (2001) Literacy, comprehension, and informed consent in clinical research. *Cancer Investigation*, 19(4), 437–445.

Redding, R.E. (1993) Children's competence to provide informed consent for mental health treatment. *Washington and Lee Law Review*, 50, 695.

Redmond, E.C. and Griffith, C.J. (2004) Consumer perceptions of food safety risk, control and responsibility. *Appetite*, 43(3), 309–313.

Rubel, A. and Streiffer, R. (2005) Respecting the automony of European and American consumers: defending positive labels on GM foods. *Journal of Agriculture and Environmental Ethics*, 18, 75–84.

Sandler, R. and Kay, W.D. (2006) The GMO–nanotech (dis)analogy. *Bulletin of Science. Technology and Society*, 26(1), 57–62.

Sen, A.K. (1987) *On Ethics and Economics*. Basil Blackwell, Oxford, UK.

Simonsen, B., Bryan, F.L., Christian, J.H.B., Roberts, T.A., Tompkin, R.B. and Silliker, J.H. (1987) Prevention and control of food-borne salmonellosis through application of hazard analysis critical control point (HACCP). *International Journal of Food Microbiology*, 4(3), 227–247.

Sparks, P. and Shepherd, R. (1992) Self-identity and the theory of planned behavior: assessing the role of identification with "green consumerism". *Social Psychology Quarterly*, 55(4), 388–399.

Streiffer, R. and Hedemann, T. (2005) The political import of interinsic objections to genetically engineered food. *Journal of Agricultural and Environmental Ethics*, 18, 191–210.

Thompson, P.B. (2007) *Food Biotechnology in Ethical Perspective,* 2nd edn. Springer, Dordrecht, NL.

Thompson, P.B. (2012) Ethics and risk communication. *Science Communication*, 34(5), 618–641.

Thompson, P.B. and Hannah, W. (2012) Novel and normal risk: where does nanotechnology fit in? In *Enviromental Ethics: What Really Matters? What Really Works?* (ed. D. Schmitz and E. Willett). Oxford University Press, New York; 609–622

Torczyner, J. (1991) Discretion, judgment, and informed consent: ethical and practice issues in social action. *Social Work*, 36(2), 122–128.

Tovey, H. (1997) Food, environmentalism and rural sociology: on the organic farming movement in Ireland. *Sociologia Ruralis*, 37(1), 21–37.

Tversky, A. and Kahneman, D. (1981) The framing of decisions and the psychology of choice. *Science*, 211(4481), 453–458.

Witte, K. and Allen, M. (2000) A meta-analysis of fear appeals: implications for effective public health campaigns. *Health Education and Behavior*, 27(5), 591–615.

Woods, M. (2003) Deconstructing rural protest: the emergence of a new social movement. *Journal of Rural Studies*, 19(3), 309–325.

20 Consumer perceptions of nanomaterials in functional foods

William K. Hallman and Mary L. Nucci

Department of Human Ecology, Rutgers University, New Brunswick, New Jersey, USA

20.1 FUNCTIONAL FOODS

The concept of functional foods was first established in Japan in the late 1980s when the Japanese Ministry of Health and Welfare created a regulatory framework to delineate a category of foods specifically consumed to provide particular health benefits (Food for Specified Health Uses), but which are not considered drugs (Bailey, 2005). The US Food and Drug Administration (FDA) has no similar regulatory category (Ross, 2000), and there is no universally accepted definition of functional foods (American Dietetic Association, 2009). However, most descriptions of functional foods have at their core the idea that they are "dietary components that may provide a health benefit beyond basic nutrition" (International Food Information Council, 2011).

The market for functional foods is extraordinarily large, with current worldwide sales estimated at more than $168 billion; 2.5 times as large as that for vitamins and dietary supplements (Euromonitor, 2010). Sales of functional foods continue to increase in the United States (NPD Group, 2012), Europe (Khan *et al.*, 2013), and Japan (Leatherhead Food Research, 2011), and are expanding in other regions, including China, Brazil, India, Mexico, Poland, Russia, and South Korea (NutraIngredients, 2012).

In the United States, consumption of functional foods has become mainstream (Kapsak *et al.*, 2011). A survey by the Academy of Nutrition and Dietetics (2011) found that American consumers say they have recently heard more about foods with health-related benefits, and nearly half say that they actively seek information about "nutrition and healthy eating." A 2011 survey of Americans by the International Food Information Council found that nearly three-quarters believe that functional foods play "a great role" in maintaining and improving overall health, and 90% could name at least one functional food and its associated benefit. The survey also found that the majority of Americans believe that certain functional foods can improve heart (79%), bone (81%), eye (66%), circulatory (74%), immune (79%), and digestive health (78%). In addition, 80% of Americans believe that certain foods and beverages can increase their physical energy or stamina, improve their mental performance (69%), and enhance their overall appearance (68%).

The most frequent reasons consumers purchase functional foods include: maintenance of overall well-being and health, weight loss, improvement of physical health, and to prevent, or to ameliorate a specific health condition (Betoret *et al.*, 2011; Kapsak *et al.*, 2011). Roberfroid (2007) suggests

Nanotechnology and Functional Foods: Effective Delivery of Bioactive Ingredients, First Edition.
Edited by Cristina M. Sabliov, Hongda Chen, and Rickey Y. Yada.

that using functional foods for these reasons makes sense in light of the increasing cost of health care, the steady increase in life expectancy, and the desire of older people for improved quality of their later years.

Khan *et al.* (2013) further suggests that consumers are particularly motivated to purchase functional foods for two main reasons: (i) to help prevent chronic illnesses such as cardiovascular disease, Alzheimer's disease, and osteoporosis; (ii) to optimize their health and well-being, by increasing energy and stamina, and boosting the immune system. In fact, nearly 30% of Americans say that they purchase products with labels that claim to improve a specific health condition (Sloan, 2012).

Goetzke and Spiller (2014) show evidence that consumption of functional foods is particularly linked to passive strategies to improve or maintain health. This is consistent with Bech-Larsen and Grunert's (2003) view that "Functional foods presumably enable the consumer to lead a healthier life without changing eating habits." Similarly, the Academy of Nutrition and Dietetics (2011) concluded that, "Consumers prefer eating their way to health, eating inherently healthy foods rather than reducing their intake of foods they like that may not be as healthy." Thus, consumption of functional foods may be seen by consumers as a way to improve health without denying themselves the pleasures of foods that they would prefer not to exclude from their diets. In some cases, consumers may also consider including functional foods in their diets as an effective counterbalance to their intake of unhealthy foods.

Because consumers are so interested in purchasing functional foods, there are clear advantages to manufacturers who are able to advertise that their products have specific health benefits. Labeling foods with information about their dietary components and associated health benefits has been shown to improve the perceived value of specific products (Koponen *et al.*, 2012), resulting in increased sales (Levy and Stokes, 1987; Freimuth *et al.*, 1988). In addition, consumers are willing to pay a premium of 50% to 200% more for products with health claims when compared to similar foods without those advertised benefits (Munene, 2006). Accordingly, some functional products in the dairy category cost 30–50% more than similar conventional foods (Menrad, 2003).

With a primary focus on its potential profitability, the functional foods market is attractive to food manufacturers and marketers. However, despite consumers' increased interest in purchasing products with health benefits, and willingness to pay a premium for the privilege, developers of functional foods need to proceed with caution. As Khan *et al.* (2013) note, the failure rate for the development of even conventional foods is exceptionally high (Hoban, 1998); and the product development process for new functional food products is "complex, expensive and risky" (van Kleef *et al.*, 2002; Siro *et al.*, 2008).

To reduce the risk of failure, improving product development of functional foods depends on understanding both the potential market for particular products, and the technologies that can be used to create them. As Betoret *et al.* (2011) put it, "Product development requires detailed knowledge of the products and the customers." They argue that high reported failure rates for new functional foods in the international market place is the result of a failure to link what is known about customer attitudes about potential products with the scientific disciplines involved in the new product development process. Moreover, van Kleef *et al.* (2002) showed that there is a considerable mismatch between the predictions of experts (nutritionists, marketers, food technologists) concerning the kinds of new functional foods that would be appealing to consumers and those that consumers say they would like to try.

Nanotechnology holds the potential to create an array of functional foods that appeal to consumers while addressing nutritional and health needs. Emerging applications of nanoparticles, nanoemulsions, nanocomposites, and nanostructured materials (Chen *et al.*, 2006; Weiss *et al.*, 2006) detailed in the previous chapters, offer the science necessary to fundamentally transform functional foods. But like other food technologies, the acceptance of nanotechnology-based functional foods is dependent on factors that lie outside the control of the food industry. Recognizing the drivers of public perceptions of nanotechnology, and tailoring products that are consonant with those perceptions will be key to its acceptance by the consuming public. Therefore, this chapter is designed to address what is known about consumer perceptions of nanotechnology and its application to food products that may be useful in informing the development of new functional foods made possible through the use of nanomaterials.

20.2 MEDIA COVERAGE OF NANOTECHNOLOGY

For most adults, media communications are the primary source for learning about science and technology once an individual has left formal education (Jarman *et al.*, 2012). Indeed, the media are a critical source of information on new technologies. The way that the media frames the discussions around science and technology is important, because these set the boundaries for debates around scientific issues, including what topics are worth considering and what information is relevant (Nelkin, 1995; Gibbons, 1999; Ten Eyck *et al.*, 2001). In doing so, the media plays an important role in establishing public perceptions of risk and benefit related to new technologies (Bauer *et al.*, 1998).

For nanotechnology, the role of the media is no different, and is key to shaping public perceptions and acceptance (Dudo *et al.*, 2011). Therefore, it is important to understand how the media has portrayed nanotechnology over the past decade and a half.

Since 2000, the level of media coverage of nanotechnology has tended to increase over time (Friedman and Egolf, 2005; Weaver and Bimber, 2008). As is common with early developments in science, the first media stories about nanotechnology tended towards the positive, with frames that discussed the developing field in terms of scientific discovery, progress or business (Gaskell *et al.*, 2003; Lewenstein *et al.*, 2004; Anderson *et al.*, 2005; Friedman and Egolf, 2005; Gorss and Lewenstein, 2005; Stephens, 2005; Wilkinson *et al.*, 2007). Ethical, legal, or social implications (ELSI) frames were less common in the early years of media coverage (Friedman and Egolf, 2005; Stephens, 2005). Even in articles exploring risk, the number of positive assessments was about equal with the number of negative assessments (Lewenstein *et al.*, 2004).

As the technology moved from research and development into consumer goods, the media coverage shifted. Consumer-related themes such as quality, safety, and packaging increased in 2007 and 2009 (Dudo *et al.*, 2011), and concerns related to potential toxicity, environmental damage, regulations, and other uncertainties relevant to nanoscale materials began to appear more frequently after 2005 (Borm, 2002; Anderson *et al.*, 2005; Friedman and Egolf, 2005; Oberdörster *et al.*, 2005; Handy and Shaw, 2007; Wilkinson *et al.*, 2007; Weaver *et al.*, 2009). Regardless, throughout the years, the dominant media frame for nanotechnology has been clear optimism related to the potential benefits associated with nanotechnology innovation, mixed with expressions of uncertainty regarding both the unknown risks (and benefits) related to the technology (Anderson *et al.*, 2005).

20.3 NANOTECHNOLOGY INVESTMENTS AND PRODUCTS

Because of the useful properties of nanoscale materials, the use of nanotechnology has influenced nearly every major research area within the physical, biological, materials, and engineering sciences (Roco, 2003). As a result, nanotechnology is broadly viewed as a transformative technology, with applications across broad sectors of the economy, including functional foods (Sargent, 2013).

To exploit the potential of nanotechnology, the public and private sectors have invested billions of dollars on research and development. Those investments are also growing significantly. A National Science Foundation sponsored study (Lux Research, 2014) found that between 2010 and 2013, global funding for emerging nanotechnology increased by as much as 45% per year. The study also revealed a dramatic growth in revenue resulting from nano-enabled products worldwide. Revenues grew from \$339 billion in 2010, to \$731 billion in 2012, and then to more than \$1 trillion in 2013. These numbers include a near tripling of revenue from nanotechnology products in the United States, which grew from \$110 billion in 2010 to \$318 billion in 2013 (Lux Research, 2014).

Given the investments in nanotechnology research and development and the revenues generated by existing nanotechnology products, it is not surprising to find that the use of nanoscale materials is increasing across a broad variety of product categories. An inventory maintained by the Pew Project on Emerging Nanotechnologies (2014) shows that nanomaterials are already used in more than 1600 consumer products; a 24% increase since 2010 (Pew Project on Emerging Nanotechnologies, 2013). These products include paints, adhesives, ultrathin polymer coatings, filtration systems, building materials, home furnishings, appliances, automotive goods, clothing and other textiles, cookware, sports equipment, sunscreens, cosmetics, personal care products, cleaning and laundry products, toys, games, electronics, computers, packaging, food, dietary supplements, and nutrition products.

20.4 NANOTECHNOLOGY AND FOOD AND DIETARY SUPPLEMENTS

The Pew Project on Emerging Nanotechnologies (2013) lists a number of food and dietary supplements thought to contain nanomaterials. Many of these are comprised of materials that consumers may not typically associate with food ingredients. For example, one of the most widely used nanomaterials known to be used in food applications is titanium dioxide, which is often used as a pigment to provide whiteness and opacity to foods and beverages. Other nanomaterials reported to be in use within food and dietary supplement products include nanoscale calcium, magnesium, zeolite, micelles, cellulose, colloidal silver, cobalt, copper, silicon, silicon oxide, zinc oxide, liposome, gold, and platinum.

Other nanoscale materials may be in use, however; in the United States, the FDA does not enforce specific requirements for labeling food or dietary supplement products containing nanoscale ingredients. In addition, some manufacturers label their products in ways that suggest that they make use of nanoscale ingredients or materials that they do not actually contain. In contrast, other manufacturers have removed any reference to nanotechnology in product descriptions to avoid potential negative public responses to their products (Scheufele, 2006). Therefore, it is difficult to create a complete catalog of products containing nanoscale ingredients or a listing of nanomaterials that might be included in them.

20.5 PUBLIC AWARENESS OF NANOTECHNOLOGY

For the consumer, nanotechnology is both literally and metaphorically invisible. By definition, nanomaterials (between 1 and 100 nm) are unable to be seen without the use of powerful microscopes. Unless a manufacturer or marketer of a product enabled by nanotechnology specifically mentions its use on a label or in an advertising campaign, it would be extraordinarily unlikely that a consumer would know that the product contained nanomaterials.

Therefore, notwithstanding its use in thousands of consumer products, and more than 20 years of media coverage, studies demonstrate that the public knows very little about nanotechnology or its core concepts; with the majority unable to even name a product that makes use of nanotechnology (Bainbridge, 2002; Cobb and Macoubrie, 2004; Dowling, 2004; Waldron *et al.*, 2006; Nucci and Hallman, in press).

Despite efforts by the National Nanotechnology Initiative, the Nanoscale Informal Science Education Network, and other associations to promote knowledge of nanotechnology through formal and informal (i.e., museums and science centers) educational programming (Nanoscale Informal Science Education Network, 2014; National Nanotechnology Initiative, undated a), research has shown that most Americans currently know little about, and many lack familiarity even with the term, nanotechnology (Bainbridge, 2002; Cobb and Macoubrie, 2004; Gaskell *et al.*, 2005; Lee *et al.*, 2005; Vandermoere *et al.*, 2010; Nucci and Hallman, 2012, in press).

In 2004, the majority of Americans (80%) indicated that they had heard little to nothing about nanotechnology and had difficulty in correctly answering questions about the technology (Cobb and Macoubrie, 2004). In a follow-up survey a year later, there was little change in self-reporting of knowledge of nanotechnology. Fifty four percent of those surveyed reported that they knew nothing about nanotechnology, 17% felt they knew something about nanotechnology, and 26% said they knew a little (Macoubrie, 2005).

Other studies performed in 2005, and again in 2007, showed no increase in the awareness of nanotechnology (Pew Project on Emerging Nanotechnologies, 2005, 2007). Only 6% said they had heard a lot about nanotechnology. The super majority of 92% indicated that they had heard some to nothing at all (Pew Project on Emerging Technologies, 2007). Through 2009, knowledge of nanotechnology remained unchanged (Scheufele *et al.*, 2009; Priest *et al.*, 2010) and in a Harris poll conducted in 2012, Americans still reported having heard only a little about nanotechnology (Harris, 2012).

20.6 PUBLIC PERCEPTIONS OF NANOTECHNOLOGY

Most studies indicate that despite knowing little about nanotechnology, the public tends to have a positive or neutral perception of the technology (Gaskell *et al.*, 2005; Cobb, 2005; Scheufele and Lewenstein, 2005; Priest *et al.*, 2011). There is some evidence that the balance of benefits and risks are shifting more to risk concerns (Priest *et al.*, 2010). Yet, even without having a clear mental model or representation of the technology, opinions about it appear well developed (Priest *et al.*, 2011).

However, complicating the potential acceptance of the application of nanotechnology and nanoscale materials is the fact that, like biotechnology, nanotechnology does not represent a single technological innovation, or a particular product, or even a tangible "thing," but rather encompasses a domain of multiple innovations and technical possibilities. Yet, there is some tendency for consumers to think of nanotechnology as a substance, rather than as the ability to manipulate matter at the nanoscale (Nucci and Hallman, in press).

Critically, the public's acceptance of new technologies, especially those with wide-ranging applications can diverge sharply depending on their use. For example, when the public perceives clear benefits from an application, such as biotechnology-based treatments for Parkinson's disease or diabetes, acceptance is higher than for applications where benefits are not as clearly apparent, such as with genetically modified (GM) food (Gaskell *et al.*, 2003), or with irradiated food where the perceived risks are more determinant of behavior than the clear health benefits (Cardello *et al.*, 2007; Huang *et al.*, 2007). Moreover, while health, environmental, and production related benefits (Onyango *et al.*, 2004) are known to be critical to consumer acceptance of new technologies, issues of morality, democracy, and uncertainty are also major factors in determining the consumers' level of support for products that use those technologies (Siegrist and Cvetkovich, 2000; Hallman *et al.*, 2002; Brown and Ping, 2003).

Like biotechnology, consumers are likely to judge nanotechnology both in terms of what it is (scientifically), and what it does. Indeed, much of the research regarding biotechnology indicates that there is a "hierarchy of approval" (Hallman *et al.*, 2002, 2003, 2004) regarding the characteristics, uses, and benefits of food biotechnology, that influences consumer acceptance. For example, Gaskell *et al.* (2000) found that consumers are more positive about "red" biotechnologies (medical or health applications) than they are about "green" biotechnologies (food and agriculture applications).

This hierarchy of approval appears to apply to the use of nanotechnology. Similar to perceptions of genetic modification technologies, the specific benefit and perceived use of nanotechnology is critical to acceptance, with medical applications ranking highest (31%) among consumers, followed by better consumer products at 27%, general progress (12%), environmental protection (8%), safer and better food (6%), energy, economy, electronics (4%), and benefits to soldiers and security (3%) (Macoubrie, 2005).

When asked about specific applications of nanotechnology, American consumers appear willing to use certain products even with health or safety risks. For example, nanotechnology was perceived to be less risky than herbicides, chemical disinfectants, handguns, and food preservatives (Currall *et al.*, 2006).

Unfortunately, measures of perceptions and intentions have been shown to be unreliable predictors of behavior (Fishbein and Ajzen, 1975; Townsend and Cambell, 2004); what people say they will do, and what they actually do are often very different. Even when perceived benefits are high, such as with nanotechnology packaging, acceptance is low (Currall *et al.*, 2006), indicating that benefit alone does not guarantee acceptance. This is similar to findings from a study of consumer acceptance of GM foods, where even clearly communicated benefits still resulted in low willingness to consume (Cox *et al.*, 2004).

20.7 ASSESSING CONSUMER RESPONSES TO FOOD NANOTECHNOLOGY

In studies examining consumer acceptance of new foods and new food technologies, a complex equation of distal (technological characteristics, consumer, social system) and proximal factors (cost/benefit, risk perceptions, control) (Ronteltap *et al.*, 2007), along with optimistic bias, information source, price, health status, trust in food companies (Frewer *et al.*, 2003), along with the technology used (Lähteenmaki *et al.*, 2006), and the manufacturing process (Caporale and Monteleone, 2004) have all been shown to influence consumers' risk/benefit decisions about food. It is important to retain the perspective that unlike other technologies, food is physically incorporated into the body (Rozin, 1999), and in consequence food technologies are characterized by risk factors of greatest concern to consumers (e.g., involuntary, invisible, uncontrolled, latent effects; Slovic, 1987). As a result, food applications of nanotechnology are judged as more risky than applications such as computers, building materials, and car paints (Siegrist, 2010). Within food applications, the use of nanotechnology is seen as more acceptable in packaging than in the food products themselves (Stampfli *et al.*, 2010).

When specific applications of food-related nanotechnology were presented to Swiss consumers, barcodes for food security, UV-protection packaging, and stronger packaging film were ranked highest for benefits and lowest for risks, while nanotechnology-modified foods, health-promoting food and forage, and nutritional supplement capsules were ranked lowest for benefits and highest for risks (Siegrist *et al.*, 2008). In the same study repeated in Mexico, barcodes for guaranteed food security, health promoting green tea, bread with nutritional value, UV-protection packaging, and cancer-preventing supplements were found highest for benefits and lowest for risks, while individually modifiable foods, bacteria detection spray, oxygen-blocking plastic bottle, stronger packaging film, and antibacterial food containers were found lowest for benefits and highest for risks (Lopez-Vasquez *et al.*, 2012).

Several studies have investigated consumer perceptions of and intentions to purchase food-related nanotechnology products, but they have done so in the abstract, without presenting consumers with "actual" products. Consumers in New Zealand presented with the idea of low fat lamb and beef modified using nanotechnology indicated support for modified meat and a willingness to purchase it (Cook and Fairweather, 2007). When presented with a series of hypothetical examples of nanotechnology-based foods or food packaging, consumers in Switzerland expressed little willingness to purchase any of the products, although they were more willing to purchase foods packaged using nanotechnology than foods prepared using nanotechnology (Siegrist *et al.*, 2007a).

In a survey in 2008, consumers in Switzerland ranked 19 likely nanotechnology food products on feelings toward the product, worries about product risk, control over product risks, voluntariness of exposure, health effects, and contact with the application. Nanotechnology food packaging was seen as less risky and more beneficial than any other nanotechnology food application (Siegrist *et al.*, 2008).

In a 2009 survey, Swiss consumers asked about willingness to purchase either ice cream or yogurt enhanced with natural or nanotechnology-derived additives for cancer prevention were hesitant about purchasing foods produced using nanotechnology. The authors concluded that the added health benefits did not provide enough incentive to consumers to influence their willingness to purchase a nanotechnology food product (Siegrist *et al.*, 2009).

20.8 MULTIMODAL RESEARCH

In research performed by our research group at Rutgers, we examined consumer willingness to taste food products ostensibly created using nanotechnology, and their likelihood to purchase the sample food product and other products as a function of benefit characteristics. Our multimodal research project funded by the US Department of Agriculture used interviews, a national survey, and a food tasting experiment to assess consumers' perceptions and acceptance/rejection of food nanotechnology (Nucci and Hallman, 2012, in press).

In a series of interviews to assess consumers' perceptions and acceptance/rejection of nanotechnology, we endeavored to mimic the path towards decision making about an unknown field: exposure to the subject, information gain about the subject, and access to the subject (through descriptions of hypothetical nanotechnology-enabled food products). Qualitative interviews were used to evaluate latent knowledge about nanotechnology in general and food nanotechnology specifically before and after being given the opportunity to read the National Nanotechnology Initiative (NNI) brochure, *Nanotechnology: Big Things from a Tiny World* (National Nanotechnology Initiative, undated b).

Consistent with previous studies of consumer's knowledge of nanotechnology (Cobb and Macoubrie, 2004; Gaskell *et al.*, 2005; Lee *et al.*, 2005; Macoubrie, 2005; Vandermoere *et al.*, 2010) we found that very few of the participants reported knowing anything about nanotechnology prior to reading the NNI brochure, noting comments such as they "honestly can't think of anything," "don't know technology," "it doesn't ring a bell," "I've heard of it but I really don't know anything. I'm sure someone will go 'nano nano,'" or "this would be something I couldn't figure out and understand. I'm not technologically savvy."

When asked what came to mind when they heard the word "nanotechnology," participants mentioned technologies such as computers, iPods, chips, jump drives, or lasers, medical therapies such as miniature cameras for *in vitro* examination and monitoring or repairing human cells), and popular culture, including *Mork and Mindy*, James Bond, and *Stargate*. Only three of the 30 participants linked nanotechnology to food and food preparation. One participant felt that nanotechnology was used for "Instant fractions of a second, computer-generated devices to help make decisions about food quality or food analysis," while another commented that it had something to do with food composition. Referencing their mention of *Mork and Mindy*, the third participant commented that "I think of a planet like ours with fields of vegetables really, and something flat like a satellite flying over and just, like, getting images of these elements that, whatever, vegetables or what have you." Notably, later in the interview, two of the three participants acknowledged that their mention of food was influenced by the recruitment materials for the study, which said that it concerned food and new food technologies.

Many knew that the term nanotechnology was somehow linked to concepts of small size ("Because I just think of something little when you say 'nano.' I think my kids have something that's 'nano' and it's small."; "Well, I don't know much about it, but 'nano' usually means something small."; "The smaller it is, the better."; "Very small but intelligent."), often associating it with the benefits of miniaturization. Participants commented that "I guess it's compressing more and more and smaller and smaller and smaller"; "it can do all the same stuff, but it comes in a much smaller package, like almost like a fifth of the size or less"; "Nanotechnology is just making things smaller"; and "Assuming I'm correct about the miniaturization, then I would say that it's – you know, an evolution of an existing technology that has, you know, somehow become miniaturized".

When asked specifically about familiar products that currently use nanotechnology, very few were able to provide any examples: the majority could not name *any* products. Categories of products thought to incorporate nanotechnology included clothing (shirts), and electronics (microwaves, mp3 players, computer chips, cell phones, video games, hard drives, computers, iPods). Medical (miniature cameras, pharmaceuticals) or industrial applications (robotics, carbon fiber tubes) were mentioned by several participants.

After reviewing the NNI brochure, slightly less than half the participants used terms about size (small, minimize, shrink, or miniaturization). Instead, participants mentioned "green concepts" related to cleaning up the environment or nature, or constructs such as cost-effectiveness, improving products, greater efficiency, better solutions, better materials, higher standards, stronger, lighter, or safer.

One participant noted that there was "not much about food" in the NNI brochure. In fact, there were only three references to food/water applications of nanotechnology in the pamphlet: "Cosmetics and food producers are 'nano-sizing' some ingredients, claiming that improves their effectiveness" (National Nanotechnology Initiative, undated b, p. 4); "which has been shown to neutralize bacteria, including *E. Coli*, in water" (National Nanotechnology Initiative, undated b, p. 6); and "low cost technology for cleaning arsenic from drinking water" (National Nanotechnology Initiative, undated b, p. 8).

Participants had more difficulty when asked about the connection between nanotechnology and food, often struggling to come up with a link between the two. Several mentioned processed, engineered, or efficient food as an outcome of using nanotechnology, while others felt that the use of nanotechnology was associated with healthy foods, making foods better, downsizing meals, nutraceuticals, or low calorie foods. Focusing on the technology aspect, a number of participants felt that nanotechnology would be connected to food through new or novel cooking technologies, microwaves or cooking equipment; or improvements in agricultural production through increased yield, hybridization of plants or animals, food safety/food quality monitoring, prevention of illness in animals, or replacement of pesticides. Quick meals, futuristic foods, unhealthy/processed foods, food production, or food safety were each mentioned by a single participant. One participant commented: "I really doubt there would be too much nanotechnology in the food we actually eat. That I think would be pushing the boundaries a little. . . If people don't know what's inside their food – you know, that's technically, I think, breaking the law."

Building on this qualitative research, we conducted a nationally representative survey of 1210 American adults using Knowledge Networks© Internet panel members. Using both a set of general survey questions and a set of quantitative experiments, the Internet-based study was designed to better understand the acceptable characteristics and uses of nanomaterials in food, establish the baseline parameters of acceptability of nanotechnologies in food, understand who may be most open to nanotechnology foods, and understand factors that influence acceptance.

The study began by using an open-ended question, asking participants to describe the first thought or image that came to mind when they read or heard the word "nanotechnology." These thoughts/images were then categorized by the researchers. Consistent with the expected general lack of public awareness of nanotechnology, 29% could produce *no* thought or image related to nanotechnology. For the majority, the first thought was related to scale: 17% "small," 9% "small robots or machines," 8% "small computer components," and for 7% "small technology." For 4%, the first thought concerned new or advanced technology. Other thoughts and images involved (each 3% or less) health/medicine, popular culture, space, cell phones, cameras, and other electronic devices, and other ideas. Significantly, fewer than 2% of the respondents produced *positive affective* responses (e.g., "awesome"), and fewer than 2% of the respondents produced *negative affective* responses (e.g., "yuk").

The respondents were then told that "Nanotechnology allows scientists to manipulate materials at the level of tiny molecules." They were then asked (in an open-ended question) about what their first thought or image was with regard to "food nanotechnology," and their responses were again captured and categorized by the researchers using the same categories as above. In response, 18% reported that they had no first thought or image. However, more than half (53%) had clearly negative first thoughts or images; 31% simply produced negative affective responses, others reported particular concerns about the technology or what they imagined to be its applications (e.g., additives, preservatives, robots in foods). In contrast, 9% produced first thoughts or images that were positive (e.g., improving foods, healthier foods, progress). Only 5% reported thoughts or images related to ideas of small scale (e.g., small particles, small portions).

The results help to demonstrate why people's imagined starting points regarding nanotechnology appear to be so important. Respondents whose first thoughts or images were anchored within the realm of nonedible applications of nanotechnology had negative reactions to the idea of nanotechnology as applied to food. This was consistent with the results of our qualitative research, which indicated that once anchored by nonedible applications of nanotechnology, participants simply could not extend their conceptions of the meaning of nanotechnology to an application involving food.

The results showed that because most people do not know much (or anything) about nanotechnology, how nanotechnology is described can affect levels of acceptance. In an experiment, participants were exposed to a description of nanotechnology that focused only on the idea that nanotechnology materials are extremely small, or a description that indicated that nanotechnology materials were both small and behave differently than their larger counterparts, or a description that indicated that nanotechnology materials were small, behave differently than their larger counterparts, and that the ways nanotechnology materials behave are often difficult to predict. Compared to those exposed to the definition that focused only on the idea that nanomaterials are extremely small, those who read the description indicating that the way nanotechnology materials behave are often difficult to predict were significantly less likely to indicate that they would purchase foods labeled as containing "nanotechnology" in the grocery store, eat nanotechnology foods, serve nanotechnology foods to family and friends, and to recommend nanotechnology food to a friend.

The results also demonstrated that describing the benefits of specific applications of food nanotechnology increases acceptance somewhat. Yet, as predicted, some benefits of nanotechnology products seem more desirable to the public than others. For example, health benefits derived through nanotechnology appear to be more acceptable to consumers than simply improving sensory qualities, such as color, sweetness, or other taste parameters.

However, the study using a quantitative experiment demonstrated that any health benefit achieved using nanotechnology also needed to be consistent with consumer perceptions of the existing health qualities of the products. For example, consumers indicated that they would be significantly more willing to purchase nanotechnology-enabled heart-healthy omega 3 fatty acids paired with a product already perceived as healthy (such as peanut butter) rather than if it is paired with a product perceived as unhealthy (such as hot dogs). The study found that the application of nanotechnology to certain products, such as nutraceuticals, is seen as more desirable than their use in products such as milk or baby food.

The study also found that, as predicted, because consumers do not understand that nanomaterials have properties that are quite different from their bulk counterparts, consumer perceptions of the bulk materials also appear to extend to their use as nanomaterials. Thus, perceptions of the compatibility of a bulk material as a food ingredient appear to influence the acceptability of the use of that material in foods at the nanoscale. For example, nanomaterials derived from plant-based materials are seen as more acceptable than those derived from metals.

This research also reaffirms previous research by Siegrist *et al.* (2007a) that food packaging containing nanomaterials is seen as more acceptable than foods that contain nanomaterials themselves. However, our research also found that nano-enabled coatings for food products (even those that may be washed off products) have a similar level of perceived acceptability as nanomaterials that are incorporated within food products.

It was apparent, regardless of whether the participants were surveyed or personally interviewed, that not only was there a low level of understanding of nanotechnology, no participant was able to describe the key attributes of nanotechnology identified in most authoritative definitions; that is, the ability to engineer, understand, and manipulate matter at the nanoscale, and the capability to generate and make use of the novel properties of materials at the molecular level (Center for Responsible Nanotechnology, 2008; Food and Drug Administration, 2012; National Nanotechnology Initiative, undated c).

In our food tasting study, we used a revealed preferences approach, conducting an experiment in a sensory laboratory designed to capture participant judgments about the perceived hedonic qualities of food products they were led to believe had a particular benefit achieved through the use of nanotechnology, either through incorporation in the product itself, or in the product packaging. A total of 161 participants rated the qualities of cherry tomatoes or chocolate ice cream. The control samples they tasted were presented as normal products purchased from the supermarket. The participants also tasted cherry tomatoes that were either presented as having packaging that used nanotechnology to preserve freshness or that were presented as having packaging that used nanotechnology with antimicrobial properties to enhance food safety. In the chocolate ice cream conditions, some of the participants tasted ice cream that they were told contained probiotics made possible by nanotechnology, and some tasted ice cream that incorporated nanotechnology to prevent the formation of ice crystals, improving its texture and keeping it fresher longer. In fact, all of the samples were identical, all were commercial products purchased at a local supermarket, and none made use of nanotechnology. The only difference among the experimental conditions was the information provided to the participants.

The results of the food tasting experiment illustrate the power of suggestion, and reinforces the maxim that "perception is reality." Most of the participants in each of the experimental conditions told the researchers that they would be willing to purchase the nanotechnology food products they tasted. Three-quarters (75%) said that they would buy the tomatoes with the nanotechnology packaging that kept them fresher longer, and 82% reported that they would buy the tomatoes with the nanotechnology packaging that enhanced food safety. Eighty percent said that they would buy the ice cream with nanotechnology designed to prevent icing, and 86% said that they would buy the chocolate ice cream with nanotechnology probiotics.

This rate of agreement to purchase was not surprising given that all of the participants were willing to participate in a food tasting experiment advertised as involving "new food technology" in the first place. Not surprisingly, all of the participants scored very low on the Food Technology Neophobia Scale (Cox and Evans, 2008), indicating that they are inclined to try new foods; so the task of trying novel foods is likely to have been appealing to them. It was also not surprising given the demand characteristics of the experiment, which had participants tasting familiar products in a university-sponsored setting.

However, the most interesting results pertain to the approximately 20% of the participants who said that they would not be willing to purchase the nanotechnology food products. In the case of each nanotechnology product, those who said that they were unwilling to purchase it indicated that those products had less sensory appeal (color, sweetness, firmness, creaminess, flavor, etc.) than the control products. This, despite the fact that the product purported to contain nanotechnology was *identical* to the control product. In other words, although the participants tasted two samples of the same product, the thought that one of them had nanotechnology in it made it seem to taste worse. Those unwilling to purchase those products literally expressed a distaste for them.

What this integrated research strongly suggests is that most people have very little conception of what nanotechnology is, why it is important, or in what products they are likely to find it. At issue for the acceptance of nanotechology-enabled products is the fact that lack of knowledge of nanotechnology may require consumers to draw upon analogies to past technologies, such as asbestos, dioxin, Agent Orange, or nuclear power, which confounds acceptance (Macoubrie, 2005), personal perspectives on naturalness, religion, science, technology, and nature (Lee *et al.*, 2005; Brossard *et al.*, 2008;

Scheufele *et al.,* 2008; Siegrist, 2008; Vandermoere *et al.,* 2010), trust in science (Lee and Scheufele, 2006; Brossard and Nisbet, 2007), business (Cobb and Macoubrie, 2004), or the food industry (Siegrist *et al.*, 2007b; Siegrist, 2008).

Our research demonstrated that there was extremely low acceptance of food nanotechnology, and that the public's acceptance of food nanotechnology was not homogeneous, but rather diverged sharply depending on several factors.

It was clear that images were key to the perceptions about nanotechnology. Qualitative interview results showed that the NNI images of nanotechnology in the brochure, *Nanotechnology: Big Things from a Tiny World* (National Nanotechnology Initiative, undated b), were essential to participants' developing (*de novo* in most cases) any perceptions of the technology. After viewing images of nanotechnology applications in the brochure, it was apparent that the participants came to understand nanotechnology as a substance or product rather than as a scaling technology that alters the behavior of materials (Center for Responsible Nanotechnology, 2008; Food and Drug Administration, 2012; National Nanotechnology Initiative, undated c). Survey results showed that the existing mental image of nanotechnology impacted the acceptance of food nanotechnology. Respondents whose first thoughts or images were anchored within the realm of nonedible applications of nanotechnology had negative reactions to the idea of nanotechnology as applied to food. For food nanotechnology, this meant that participants sometimes understood nanotechnology as a foreign substance that is put into a food; a critical challenge to a food industry increasingly questioned by consumers about the healthfulness of food components and ingredients (Damewood, 2014; Kindy, 2014).

20.9 CONCLUSIONS

By definition, materials at the nanoscale are literally invisible. Critically, with regard to the potential acceptance of nanotechnology applications, nanotechnology is also *figuratively* invisible to the consumer. The public's dearth of knowledge about the technology itself, and their lack of awareness of the applications of nanotechnology in the products they are already using makes its use imperceptible to most consumers (e.g., Scheufele, 2006). At this point in its history, the rapidly expanding use of nanotechnology appears to have been helped by its invisibility. While consumer and environmental advocates have campaigned for greater regulatory control over nanotechnology, a sustained lack of consumer awareness has resulted in little public opposition to its use.

However, lack of public opposition is not the same as public acceptance. Studies suggest that when consumers are made aware of nanotechnology, their initial response is distrust and unease (Gaskell *et al.*, 2005; Siegrist *et al.*, 2007b). In fact, initial efforts to advertise the use of nanotechnology as a desirable attribute of consumer products (such as cosmetics) appeared to increase consumer perceptions of risk related to the product (Siegrist and Keller, 2011). As a result, while the use of nanotechnology in consumer products has dramatically expanded, most manufacturers have removed any mention of nanotechnology from their product labels (Scheufele, 2006).

Even as the word "nanotechnology" is disappearing from product labels, it remains clear that public perceptions of nanotechnology are critical to its future success. The promise of nanotechnology is as much dependent on consumers' openness to and acceptance of nanotechnology products as it is on the ability to create them. This is especially true of its application to food and agriculture.

What our work, and that of others suggests, is the importance of comprehending *how* people come to understand the meaning of nanotechnology. Critical to this is understanding people's cognitive starting points when it comes to nanotechnology. Unfortunately, because people know so little about nanotechnology and nanomaterials, their starting points seem firmly rooted in what they think they know about materials at the bulk scale.

While a great deal of formal and informal science communication focuses on just how small nanomaterials are, in essence, many people appear to believe that nanomaterials are just tiny versions of the materials with which they are already familiar. What much of the public does not appear to understand is that the real utility of nanotechnology lies in the fact that nanoscale materials have unique physical, chemical, and biological properties that are different from those that they exhibit at either the level of the single atom or at the macro "bulk" scale. As a result, people have difficulty making the leap from what they know to be true about materials in the visible world to what they must imagine is true at the invisible nanoscale.

People's starting points concerning the uses and utility of nanotechnology are also important. If consumers come to associate nanomaterials with electronics, tennis racquets, and wrinkle-resistant pants, it may be difficult for them to conceive of how, or why, nanotechnology may safely enhance the foods they eat.

Their starting points concerning what they believe nanotechnology represents is also important. In reality, nanotechnology may be best thought of as a collection of innovations at the nanoscale. Yet, for many, nanotechnology appears to be perceived as a *substance*. Because they appear to generalize what they know from the bulk scale to the nanoscale, and tend to associate nanomaterials with distinctly nonfood items, many people tend to approach the thought of nanomaterials in their foods as contamination, and not an enhancement. As a result, they are much more sanguine about the use of nanotechnology as applied to food packaging than they are about its incorporation into actual food products.

Because most people do not really understand the nature of nanotechnology or how it works, there is a tendency for the public to try to use what they think they understand about other technologies (with similar characteristics) as a starting point for their perceptions of nanotechnology (Frewer *et al.*, 2011). For many, the closest analogy to food nanotechnology of which they can conceive, is genetic modification. In fact, Stampfli *et al.* (2010) showed that people's attitudes toward gene technology was one of the strongest predictors of their perceptions of both the benefits and risks of food applications of nanotechnology, and ultimately, their willingness to purchase foods containing nanomaterials.

Understanding the likely parameters of consumer acceptance of nanotechnology is particularly important in better directing the future development of nanotechnologies intended to be applied to food products. Yet, only a small number of studies have examined public perceptions of food nanotechnology; therefore, our understanding is quite limited.

It is clear that perceived risks and benefits of the applications of nanotechnology to food products matters in overall acceptance, and that some benefits are perceived as more desirable than others. For example, our work shows that health benefits derived through nanotechnology were rated by consumers as being more acceptable than simply improving sensory qualities, such as color, sweetness, or other hedonic factors. But, those health benefits also need to be consistent with consumer perceptions of the baseline health qualities of the products enhanced through nanotechnology. Consumers appear to be more willing to accept additional health benefits to products that they already perceive to be healthy rather than adding them to products that they view as unhealthy. It is also clear that because certain kinds of food products carry cultural meanings beyond simple nutrition, the incorporation of nanomaterials in products such as milk and baby food may be seen as particularly unacceptable. However, understanding these and other parameters of acceptance in much greater detail is necessary to guide the development of functional foods that make use of nanomaterials.

Acknowledgments

This material is based upon work supported by the US Department of Agriculture (USDA), Cooperative State Research, Education, and Extension Service (CSREES), National Research Initiative Grant #: 2008-01415. Any opinions, findings, and conclusions or recommendations expressed in this material are those of the authors and do not necessarily reflect the views of the USDA, or of Rutgers, The State University of New Jersey.

References

Academy of Nutrition and Dietetics. (2011) *Nutrition and you: Trends 2011 public opinion on food and nutrition: 20 years of insights.* [Online] Available: http://www.eatright.org/media/content.aspx?id=7639 (accessed 4 April 2014).

American Dietetic Association (2009) Position of the American Dietetic Association: Functional foods. *Journal of the American Dietetic Association*, 109, 735–746.

Anderson, A., Allan, S., Petersen, A. and Wilkinson, C. (2005) The framing of nanotechnologies in the British newspaper press. *Science Communication*, 27, 200–220.

Bailey, R. (2005) Functional foods in Japan: FOSHU ("Foods for Specified Health Uses") and "Foods with Nutrient Function Claims". In: *Regulation of Functional Foods and Nutraceuticals: A Global Perspective* (ed. C.M. Hasler), Blackwell Publishing, Ames, IO; 247–262.

Bainbridge, W.S. (2002) Public attitudes towards nanotechnology. *Journal of Nanoparticle Research*, 4, 561–570.

Bauer, M.W., Durant, J. and Gaskell, G. (1998) *Biotechnology in the Public Sphere*. Science Museum, London.

Bech-Larsen, T. and Grunert, K.G. (2003) The perceived healthiness of functional foods: A conjoint study of Danish, Finnish and American consumers' perception of functional foods. *Appetite*, 40, 9–14.

Betoret, E., Betoret, N., Vidal, D. and Fito, P. (2011) Functional foods development: Trends and technologies. *Trends in Food Science and Technology*, 22, 498–508.

Borm, P.J.A. (2002) Particle toxicology: From coal mining to nanotechnology. *Inhalation Toxicology*, 14, 311–324.

Brossard, D. and Nisbet, M.C. (2007) Deference to scientific authority among a low information public: Understanding U.S. opinion on agricultural biotechnology. *International Journal of Public Opinion Research*, 19, 24–52.

Brossard, D., Scheufele, D.A., Kim, E. and Lewenstein, B.V. (2009) Religiosity as a perceptual filter: Examining processes of opinion formation about nanotechnology. *Public Understanding of Science*, 18, 546–558.

Brown, J.L. and Ping, Y. (2003) Consumer perception of risk associated with eating genetically engineered (GE) soybeans is less in the presence of a perceived consumer benefit. *Journal of the American Dietetic Association*, 103, 208–214.

Caporale, G. and Monteleone, E. (2004) Influence of information about manufacturing process on beer acceptability. *Food Quality and Preference*, 15, 271–278.

Cardello, A.V., Schutz, H.G. and Lesher, L.L. (2007) Consumer perceptions of foods processed by innovative and emerging technologies: A conjoint analytic study. *Innovative Food Science and Emerging Technologies*, 8, 73–83.

Center for Responsible Nanotechnology. (2008) *What is Nanotechnology?* [Online] Available: http://crnano.org/whatis.htm (accessed 4 April 2014).

Chen, H., Weiss, J. and Shahidi, F. (2006) Nanotechnology in nutraceuticals and functional foods. *Food Technology*, 60(3), 30–32, 34–36.

Cobb, M.D. (2005) Framing effects on public opinion about nanotechnology. *Science Communication*, 27, 221–239.

Cobb, M.D. and Macoubrie, J. (2004) Public perceptions about nanotechnology: Risks, benefits and trust. *Journal of Nanoparticle Research*, 6, 395–405.

Cook, A.J. and Fairweather, J.R. (2007) Intentions of New Zealanders to purchase lamb or beef made using nanotechnology. *British Food Journal*, 109, 675–688.

Cox, D.N. and Evans, G. (2008) Construction and validation of a psychometric scale to measure consumers' fears of novel food technologies: The food technology neophobia scale. *Food Quality and Preference*, 19, 704–710.

Cox, D.N., Koster, A. and Russell, C.G. (2004) Predicting intentions to consume functional foods and supplements to offset memory loss using an adaptation of protection motivation theory. *Appetite*, 43, 55–64.

Currall, S.C., King, E.B., Lane, N., Madera, J. and Turner, S. (2006) What drives public acceptance of nanotechnology? *Nature Nanotechnology*, 1, 153–155.

Damewood, K. (2014) Food ingredients: Trade secrets vs. public disclosure. *Food Safety News*. [Online] Available: http://www.foodsafetynews.com/2014/03/food-ingredients-trade-secrets-versus-public-disclosure/#.U0V0afldW-g (accessed 19 April 2014).

Dowling, A. (2004) *Nanoscience and nanotechnologies: Opportunities and uncertainties.* [Online] Available: http://www.nanotec.org.uk/finalReport.htm (accessed 18 April 2014).

Dudo, A., Choi, D.H. and Scheufele, D.A. (2011) Food nanotechnology in the news. Coverage patterns and thematic emphases during the last decade. *Appetite*, 56, 78–89.

Euromonitor (2010) *Cardiovascular Health: A Key Area of Functional Food and Drinks Development.* Euromonitor International, London.

Fishbein, M. and Ajzen, I. (1975) *Belief, Attitude, Intention and Behavior: An Introduction to Theory and Research.* Addison-Wesley, Massachusetts.

Food and Drug Administration. (2012) *Nanotechnology.* [Online] Available: http://www.fda.gov/scienceresearch/specialtopics/nanotechnology/default.htm (accessed 19 April 2014).

Freimuth, V.S., Hammond, S.L. and Stein, J.A. (1988) Health advertising: Prevention for profit. *American Journal of Public Health,* 78, 557–561.

Frewer, L., Scholderer, J. and Lambert, N. (2003) Consumer acceptance of functional foods: issues for the future. *British Food Journal*, 105, 714–731.

Frewer, L.J., Bergmann, K., Brennan, M., *et al.* (2011) Consumer response to novel agri-food technologies: Implications for predicting consumer acceptance of emerging food technologies, *Trends in Food Science and Technology,* 22, 442–456.

Friedman, S.M. and Egolf, B.P. (2005) Nanotechnology: Risks and the media. *IEEE Technology and Society Magazine*, 24(4), 5–11.

Gaskell, G., Allum, N., Bauer, M.W., Jackson, J., Howard, S. and Lindsey, N. (2003) *Ambivalent GM Nation? Public Attitudes to Biotechnology in the UK, 1991–2002.* Life Sciences in European Society Report, London School of Economics and Political Science.

Gaskell, G., Ten Eyck, T., Jackson, J. and Veltri, G. (2005) Imagining nanotechnology: Cultural support for technological innovation in Europe and the United States. *Public Understanding of Science*, 14, 81–90.

Gibbons, M. (1999) Science's new social contract with society. *Nature*, 402, C81–C84.

Goetzke, B.I. and Spiller, A. (2014) Health-improving lifestyles of organic and functional food consumers. *British Food Journal,* 116, 510–526.

Gorss, J.B. and Lewenstein, B.V. (2005) *The Salience of Small: Nanotechnology Coverage in the American Press, 1986–2004.* International Communication Association.

Hallman, W.K., Adelaja, A.O., Schilling, B.J. and Lang, J.T. (2002*) Public Perceptions of Genetically Modified Food: Americans Know Not What They Eat.* Food Policy Institute Report RR-0302-001. [Online] Available: http://foodpolicy.rutgers.edu/docs/pubs/2002_Public_Perceptions_of_Genetically_Modified_Food.pdf (accessed 19 April 2014).

Hallman, W.K., Hebden, W., Aquino, H., Cuite, C. and Lang, J. (2003) *Public Perceptions of Genetically Modified Foods: A National Study of American Knowledge and Opinion.* Food Policy Institute Report RR-1003-004. [Online] Available: http://foodpolicy.rutgers.edu/docs/pubs/2003_Public_Perceptions_of_Genetically_Modified_Foods.pdf (accessed 19 April 2014).

Hallman, W.K., Hebden, W.C., Cuite, C.L., Aquino, H.L. and Lang, J.T. (2004) *Americans and GM Food: Knowledge, Opinion and Interest in 2004.* Food Policy Institute report RR-1104-007 [Online] Available: http://foodpolicy.rutgers.edu/docs/pubs/2004_Americans%20and%20GM%20Food_Knowledge%20Opinion%20&%20Interest%20in%202004.pdf (accessed 19 April 2014).

Handy, R.D. and Shaw, B.J. (2007) Toxic effects of nanoparticles and nanomaterials: Implications for public health, risk assessment and the public perception of nanotechnology. *Health, Risk and Society*, 9(2), 125–144.

Harris. (2012) *Nanotechnology Awareness May Be Low, But Opinions Are Strong*. [Online] Available: http://www.harrisinteractive.com/NewsRoom/HarrisPolls/tabid/447/ctl/ReadCustom%20Default/mid/1508/ArticleId/1073/Default.aspx (accessed 19 April 2014).

Hoban, T.J. (1998) Improving the success of new product development. *Food Technology,* 52, 46–49.

Huang, C. L., Wolfe, K. and McKissick, J. (2007) Consumers' willingness to pay for irradiated poultry products. *Journal of International Food and Agribusiness Marketing*, 19(2–3), 77–95.

International Food Information Council. (2011) *Functional Foods/Foods for Health Consumer Trending Survey.* [Online] Available: http://www.foodinsight.org/Resources/Detail.aspx?topic=2011_Functional_Foods_Foods_For_Health_Consumer_Trending_Survey (accessed 20 November 2011).

Jarman, R., McClune, B., Pyle, E. and Braband, G. (2012) The critical reading of the images associated with science-related news reports: Establishing a knowledge, skills, and attitudes framework. *International Journal of Science Education Part B*, 2(2), 103–129.

Kapsak, W.R., Rahavi, E.B., Childs, N.M. and White, C. (2011) Functional foods: Consumer attitudes, perceptions, and behaviors in a growing market. *Journal of the American Dietetic Association*, 111, 804–810.

Khan, R.S., Grigor, J., Winger, R. and Win, A. (2013) Functional food product development – opportunities and challenges for food manufacturers. *Trends in Food Science and Technology,* 30, 27–37.

Kindy, K. (2014) Are secret, dangerous ingredients in your food? Federal eye: keeping tabs on your Government. *Washington Post*. [Online]. Available: http://www.washingtonpost.com/blogs/federal-eye/wp/2014/04/07/are-secret-possibly-dangerous-ingredients-in-your-food/ (accessed 4 April 2014).

Koponen, A., Sandell, M., Salminen, S. and Lenoir-Wijnkoop, I. (2012) Nutrition economics: towards comprehensive understanding of the benefits of nutrition. *Microbial Ecology in Health and Disease,* 23, 46–50.

Lähteenmäki, L., Lyly, M. and Urala, N. (2006) Consumer attitudes towards functional foods. In: *Understanding Consumers of Food Products*, 412–427.

Leatherhead Food Research (2011) *Future Directions for the Global Functional Foods Market*, [Online]. Available: http://www.leatherheadfood.com/UserFiles/pdfs/publicationspromopdf/Functional-Directions-for-the-Global-Functional-Foods-Market-2011.pdf (accessed 18 April 2014).

Lee, C.J. and Scheufele, D.A. (2006) The influence of knowledge and deference toward scientific authority: A media effects model for public attitudes towards nanotechnology. *Journalism and Mass Communication*, 83, 819–834.

Lee, C.J., Scheufele, D.A. and Lewenstein, B.A. (2005) Public attitudes toward emerging technologies: Examining the interactive effects of cognitions and affect on public support for nanotechnology. *Science Communication*, 27, 240–267.

Levy, A.S. and Stokes, R.C. (1987) Effects of a health promotion advertising campaign on sales of ready-to-eat cereals. *Public Health Reports*, 102, 398–403.

Lewenstein, B.V., Radin, J. and Diels, J. (2004) Nanotechnology in the media: A preliminary analysis. In: *Nanotechnology: Societal Implications II: Individual Perspectives* (ed. C. Roco and W.S. Bainbridge). Springer, Dordrecht; 258–265.

López-Vázquez, E., Brunner, T.A. and Siegrist, M. (2012) Perceived risks and benefits of nanotechnology applied to the food and packaging sector in Mexico. *British Food Journal*, 114, 197–205.

Lux Research. (2014) *Nanotechnology update: corporations up their spending as revenues for nano-enabled products increase*. Available: http://nano.gov/sites/default/files/energy_-_flynn.pdf (accessed 25 February 2015).

Macoubrie, J. (2005) *Informed Public Perceptions of Nanotechnology and Trust in Government.* Woodrow Wilson International Center Project on Emerging Nanotechnologies, Washington, DC. [Online] Available: http://www.pewtrusts.org/uploadedFiles/wwwpewtrustsorg/Reports/Nanotechnologies/Nanotech_0905.pdf (accessed 18 April 2014).

Menrad, K. (2003) Market and marketing of functional food in Europe. *Journal of Food Engineering*, 56(2–3), 181–188.

Munene, C.N. (2006) *Analysis of consumer attitudes and their willingness to pay for functional foods*. Unpublished Masters Thesis, Department of Agricultural Economics and Agribusiness, Louisiana State University. [Online] Available: http://etd.lsu.edu/docs/available/etd-01052006-121418/unrestricted/Munene_thesis.pdf (accessed 18 April 2014).

Nanoscale Informal Science Education Network (2014). *Nanoscale Informal Science Education Network*. [Online] Available: http://www.nisenet.org/

National Nanotechnology Initiative. (undated a) *Education*. [Online] Available: http://www.nano.gov/education-training.

National Nanotechnology Initiative. (undated b) *Nanotechnology: Big Things from a Tiny World*. [Online] Available: http://www.nano.gov/node/240 (accessed 18 April 2014).

National Nanotechnology Initiative. (undated c) *What it is and How it Works*. [Online] Available: http://www.nano.gov/nanotech-101/what (accessed 18 April 2014).

Nelkin, D. (1995) *Selling Science: How the Press Covers Science and Technology*. Freeman, New York.

NPD Group (2012) *The majority of U.S. Adults consider themselves to be healthy, but eating behaviors show only 20 percent classified as "most health eaters," reports NPD.* [Press release] Available from: https://www.npd.com/wps/portal/npd/us/news/press-releases/pr_120904/ (accessed 18 April 2014).

Nucci, M. and Hallman, W. (2012) Food and scientific illiteracy. *Museums and Social Issues*, 7(1), 59–70.

Nucci, M. and Hallman, W.(In press) The role of public (mis)perceptions in the acceptance of new food technologies: Implications for food nanotechnology applications. In *Communication Practices in Engineering, Manufacturing, and Research for Food, Drug, and Water Safety.* (ed. D. Wright). Wiley-IEEE Press, 000–000.

NutraIngredients (2012), *Global nutraceutical ingredient market worth €18.5bn in 2015*. [Online] Available: http://www.nutraingredients.com/Consumer-Trends/Global-nutraceutical-ingredient-market-worth-18.5bn-in-2015 (accessed 6 January 2012).

Oberdörster, G., Oberdörster, E. and Oberdörster, J. (2005) Nanotoxicology: An emerging discipline evolving from studies of ultrafine particles. *Environmental Health Perspectives*, 113, 823–839.

Onyango, B., Govindasamy, R. and Nayga, R. (2004) *Measuring U.S. consumer preferences for genetically modified foods using choice modeling experiments: The role of price, product benefits and technology.* Food Policy Institute, Working Paper No. WP1104-017. [Online] Available: http://dafre.rutgers.edu/documents/ramu/aaeachoicemodeling.pdf (accessed 18 April 2014).

Pew Project on Emerging Technologies. (2005*) Informed Public Perceptions of Nanotechnology and Trust in Government.* [Online] Available: www.pewtrusts.org/our_work.aspx?category=338 (accessed 18 April 2014).

Pew Project on Emerging Technologies. (2007) *Awareness of and attitudes toward nanotechnology and federal regulatory agencies.* [Online] Available: www.pewtrusts.org/our_work.aspx?category=338 (accessed 18 April 2014).

Pew Project on Emerging Nanotechnologies (2013) *Inventory Finds Increase in Consumer Products Containing Nanoscale Materials.* [Online] Available:http://www.nanotechproject.org/news/archive/9242/ (accessed 18 April 2014).

Pew Project on Emerging Nanotechnologies. (2014) *Nanotechnology Consumer Products Inventory.* [Online] Available:http://www.nanotechproject.org/inventories/consumer/ (accessed 18 April 2014).

Priest, S., Greenhalgh, T. and Kramer, V. (2010) Risk perceptions starting to shift? US citizens are forming opinions about nanotechnology. *Journal of Nanoparticle Research*, 12, 11–20.

Priest, S., Lane, T., Greenhalgh, T., Hand, L.J. and Kramer, V. (2011) Envisioning emerging nanotechnologies: A three-year panel study of South Carolina citizens. *Risk Analysis*, 31, 1718–1733.

Roberfroid, M.B. (2000) Concepts and strategy of functional food science: The European perspective. *American Journal of Clinical Nutrition,* 71, 1660S–1664S.

Ronteltap, A., van Trijp, J.C.M., Renes, R.J. and Frewer, L.J. (2007) Consumer acceptance of technology-based food innovations: Lessons for the future of nutrigenomics. *Appetite*, 49, 1–17.

Roco, M.C. (2003) Nanotechnology: convergence with modern biology and medicine. *Current opinion in biotechnology,* 14(3), 337–346.

Ross, S. (2000) Functional foods: the Food and Drug Administration perspective. *The American Journal of Clinical Nutrition*, 71, 1735S.

Rozin, P. (1999) Food is fundamental, fun, frightening, and far-reaching. *Social Research,* 9–30.

Sargent, J.F. Jr. (2013) *The National Nanotechnology Initiative: Overview, Reauthorization and Appropriations Issues*. Congressional Report No. RL34401, Library of Congress Congressional Research Service, Washington DC. [Online] Available: https://www.fas.org/sgp/crs/misc/RL34401.pdf (accessed 18 April 2014).

Scheufele, D.A. (2006) Manufacturers removing nano labels from product descriptions. *Nanopublic: Public Opinion, Media, and Technologies.* [Online] Available: http://www.nanopublic.com/2006/11/manufacturers-removing-nano-labels.html (accessed 18 April 2014).

Scheufele, D.A. and Lewenstein, B.V. (2005) The public and nanotechnology: How citizens make sense of emerging technologies. *Journal of Nanoparticle Research*, 7, 659–667.

Scheufele, D.A., Corley, E.A., Shih, T.J., Dalrymple, K.E. and Ho, S.S. (2009) Religious beliefs and public attitudes to nanotechnology in Europe and the US. *Nature Nanotechnology*, 4, 91–94.

Siegrist, M. (2008) Factors influencing public acceptance of innovative food technologies and products. *Trends in Food Science and Technology*, 19, 603–608.

Siegrist, M. (2010) Predicting the future: Review of public perception studies of nanotechnology. *Human and Ecological Risk Assessment*, 16, 837–846.

Siegrist, M. and Cvetkovich, G. (2000) Perception of hazards: The role of social trust and knowledge. *Risk Analysis*, 20, 713–720.

Siegrist, M and Keller, C. (2011) Labeling of nanotechnology consumer products can influence risk and benefit perceptions. *Risk Analysis*, 31, 1762–1769.

Siegrist, M., Cousin, M., Kastenholz, H. and Wiek, A. (2007a) Public acceptance of nanotechnology foods and food packaging: The influence of affect and trust. *Appetite*, 49, 459–466.

Siegrist, M., Wiek, A., Helland, A. and Kastenholz, H. (2007b) Risks and nanotechnology: The public is more concerned than experts and industry. *Nature Nanotechnology*, 2, 67.

Siegrist, M., Stampfli, N., Kastenholz, H. and Keller, C. (2008) Perceived risks and perceived benefits of different nanotechnology foods and nanotechnology food packaging. *Appetite*, 51, 283–290.

Siegrist, M., Stampfli, N. and Kastenholz, H. (2009) Acceptance of nanotechnology foods: a conjoint study examining consumers' willingness to buy. *British Food Journal*, 111, 660–668.

Siró, I., Kápolna, E., Kápolna, B. and Lugasi, A. (2008) Functional food. Product development, marketing and consumer acceptance – a review. *Appetite,* 51, 456–467.

Sloan, E.A. (2012) *Top 10 Functional Food Trends.* Institute of Food Technologists. [Online] Available: http://www.ift.org/food-technology/past-issues/2012/april/features/top-10-functional-food-trends.aspx?page=viewall (accessed 18 April 2014).

Slovic, P. (1987) Perception of risk. *Science*, 236, 280–285.

Stampfli, N., Siegrist, M. and Kastenholz, H. (2010) Acceptance of nanotechnology in food and food packaging: a path model analysis. *Journal of Risk Research*, 13(3), 335–347.

Stephens, L.F. (2005) News narratives about nano S&T in major U. S. and non-U.S. newspapers. *Science Communication*, 27, 175–199.

Ten Eyck, T.A., Thompson, P.B. and Priest, S.H. (2001) Biotechnology in the United States of America: Mad or moral science? In *Biotechnology 1996–2000: The Years of Controversy* (eds G. Gaskell and M.W. Bauer). NMSI Trading Ltd, Science Museum, London; 307–318.

Townsend, E. and Campbell, S. (2004) Psychological determinants of willingness to taste and purchase genetically modified food. *Risk Analysis*, 24, 1385–1393.

Van Kleef, E., van Trijp, H.C.M., Luning, P. and Jongen, W.M.F. (2002) Consumer-oriented functional food development: how well do functional disciplines reflect the 'voice of the consumer'? *Trends in Food Science and Technology*, 13, 93–101.

Vandermoere, F., Blanchemanche, S., Bieberstein, A., Marette, S. and Roosen, J. (2010) The morality of attitudes toward nanotechnology: about God, techno-scientific progress, and interfering with nature. *Journal of Nanoparticle Research*, 12, 373–381.

Waldron, A.M., Spencer, D. and Batt. C. (2006) The current state of public understanding of nanotechnology. *Journal of Nanoparticle Research*, 8, 569–575.

Weaver, D.A. and Bimber, B. (2008) Finding news stories: A comparison of searches using LexisNexis and Google News. *Journalism and Mass Communication Quarterly*, 85, 515–530.

Weaver, D.A., Lively, E. and Bimber, B. (2009) Searching for a frame. News media tell the story of technological progress, risk, and regulation. *Science Communication*, 31(2), 139–166.

Weiss, J., Takhistov, P. and McClements, D.J. (2006) Functional materials in food nanotechnology. *Journal of Food Science*, 71(9), R107–R116.

Wilkinson, C., Allan, S., Anderson, A. and Petersen, A. (2007) From uncertainty to risk? Scientific and news media portrayals of nanoparticle safety. *Health Risk and Society*, 9(2), 145–157.

21 Safety assessment of nano- and microscale delivery vehicles for bioactive ingredients

Qasim Chaudhry and Laurence Castle

The Food and Environment Research Agency, Sand Hutton, York, UK

21.1 BACKGROUND

Although food applications of nanosized ingredients are currently only few and far between, one of the developing research and development themes relates to nano- and microsized vehicles designed to deliver useful bioactive substances in supplements, nutraceuticals, cosmeceuticals, and other health/food products. These very small sized carriers provide a new way of addressing the quest of the modern-day consumer for more choices in food products and at the same time consciousness towards health and well-being. Examples of the typical products and applications of the micro-/nanocarrier technology have been discussed in the preceding chapters. The technology involves formulation of bioactive substances at micro- or nanoscales in the form of micelles, liposomes, or encapsulates from natural or synthetic polymeric materials.

The basic concepts behind the delivery systems seem to have emerged from the medicines sector, where there has been a longstanding research and development on micro-/nanoscale formulation to enhance the uptake and targeted delivery of pharmacologically active compounds and certain therapeutic macromolecules (e.g., Nishiyamaa and Kataoka, 2006; Peer *et al.*, 2007; Mundargi *et al.*, 2008). Microencapsulation has also been used by the health/food sector for many years for better dispersion of nutrients and supplements, protection from degradation during processing, controlled release, taste/odor masking, etc. In this context, the methods used for development of micro-/nanodelivery systems are not new as such, and nanoencapsulation appears to be the next logical technological advancement of microencapsulation. However, what is distinctive for nanoscale formulation is the potential for enhancement in the uptake of supplements through the oral route (Carr *et al.*, 1996; Florence, 2005; des Rieux *et al.*, 2006), as well as improvements in taste, consistency, and texture of the formulated food products (Chaudhry *et al.*, 2010; McClements, 2011). Examples discussed in Chapter 10 of this book include improvement in the low bioavailability of polyphenol antioxidants through diet using a nanoscale formulation; enhancement in bioavailability, cellular uptake, and efficacy of curcumin–PLGA (polylactic polyglycolic acid co-polymer) nanoparticles; epigallocatechin-3-gallate encapsulated in PLA–PEG (poly(lactic acid–poly(ethylene glycol)); and quercetin encapsulated in solid lipid particles. A significant enhancement in the uptake of health-promoting but poorly absorbed supplements through diet, or their

Nanotechnology and Functional Foods: Effective Delivery of Bioactive Ingredients, First Edition.
Edited by Cristina M. Sabliov, Hongda Chen, and Rickey Y. Yada.

targeted release in the body, should be a welcome benefit for the consumer in general, and some population groups in particular (e.g. the elderly, patients, athletes).

Although the use of very small sized substances and materials has opened up a new avenue for consumer benefits, it may also have safety implications for consumer's health. Such concerns over safety have arisen from a number of studies, which show that nanoparticles may penetrate biological membranes that protect the cells and organs from entry of (larger) particles (e.g., see reviews in Chaudhry *et al.*, 2008; EFSA, 2009; FAO/WHO, 2010). Also, it is currently not fully understood how different nanoparticles, if available internally, would interact with various biological entities and processes. In view of this, the use of very small sized (especially nanoscale) vehicles needs to consider not only the potential benefits, but also any possible risk to the consumer's health. This chapter is aimed at discussing some basic concepts that could provide pointers to safety evaluation of the micro-/nanovehicles carrying bioactive substances intended for use in health/food products.

21.2 THEORETICAL CONSIDERATIONS FOR SAFETY ASSESSMENT

A number of studies have shown that properties, behavior, and toxicological effects of some submicron (especially nanosized) particles may differ from conventional bulk equivalents. However, most of these studies have been carried out on insoluble/biopersistent particles (e.g., metals/metal oxides). Micro-/nanoformulations (e.g., emulsions) and carriers (e.g., encapsulates) composed of "soft" degradable materials have not received the same level of evaluation for safety. According to the European Food Safety Authority's (ESFA) guidance on the risk assessment of nanoscience and nanotechnologies in food/feed (EFSA, 2011), "Food and feed may contain components that have internal structures that individually could be present at the nanoscale, e.g. naturally occurring liposomes, micelles or crystals. However, 'natural' components are considered within the context of this ENM [engineered nanomaterial] Guidance, only if they have been deliberately produced to have nanoproperties, or used e.g. to encapsulate bioactive compounds."

In regard to micro-/nanoformulation and encapsulation of bioactive substances, it needs to be remembered that they are generally claimed for:

- enhanced uptake/bioavailability of the substances formulated/encapsulated at micro-/nanoscales;
- altered physicochemical properties of particulate substances (especially at nano-scale).

For example, micro-/nanoscale formulations have been claimed to render a water-soluble substance fat-dispersible, and vice versa (e.g. US Patent US5668183-A "Water dispersible compositions"; Garti, 2006; NovaSOL®, undated).

The enhanced absorption and bioavailability of a bioactive substance through micro-/nanoformulation and encapsulation may give rise to atypically high plasma levels. Similarly, a significant change in properties may make it possible for the particle to have a different behavior and fate in the body than the conventional equivalent. A combination of increased uptake/bioavailability, altered biokinetics/biodistribution, and potential for the nano-sized particles to interact with biological systems close to the molecular level, may bring about harmful consequences that are otherwise not foreseen for conventional forms of the same substances.

It is, however, also important to highlight that any size-related departure in properties, behavior, or biological effects of a vehicle would intrinsically hinge upon the integrity of the physical form at micro-/nanoscales. Where a vehicle loses its physical form at micro- or nanoscales, for exmple as a result of digestion or metabolic breakdown, the properties and/or effects of the resulting chemical constituents would not be expected to be any different from the bulk equivalents.

In this context, it is equally important to note that, unlike other routes of exposure (e.g., inhalation, dermal), any micro-/nanoparticles entering the body through the oral route would have to go through a number of destructive processes in the digestive system. For example, the particles will be exposed to strong acidic conditions, digestive enzymes, bile salts, etc, in the gastrointestinal (GI) tract. This could lead to transformation, or partial or complete breakdown, resulting in the loss of physical form at the micro-/nanoscale. The vehicles may also bind with other (undigested) components of food, which may reduce their absorption and translocation across the GI tract. Knowing the fate and behavior of the micro-/nanoparticles in food and in the body is therefore of utmost importance for safety assessment. A key determinant in the safety assessment of a micro-/nanocarrier, therefore, is whether it can retain the micro-/nanostructure in the GI tract, and whether it can translocate from the GI tract to other parts of the body. A conceptual scheme highlighting safety and testing considerations for micro-/nanovehicles carrying bioactive substances is presented in Figure 21.1.

In regard to health/food applications of the micro-/nanocarriers, it is also important to emphasize that our foodstuffs are largely made of natural materials (proteins, carbohydrates, lipids) that are composed of structures at the micro- or nanoscale. Any larger sized food materials are also broken down to smaller and smaller sizes during the digestive process, and they in theory go through the nanoscale before being completely assimilated. Expanding this analogy to micro- or nanosized vehicles, it is logical to think that the micro-/nanocarriers that are composed of natural food materials should not raise any special safety concerns as they are likely to be digested in the GI tract. Thus the focus of consumer safety concerns needs to be on those micro-/nanovehicles that are composed of stable, recalcitrant to digestion, and biopersistent materials, which after oral intake could translocate intact, or in fragmented forms, to other parts of the body. Keeping these aspects in view, micro-/nanovehicles derived from "soft" degradable natural materials, for example emulsions and liposomes, etc., could be considered a low-risk category as they should not pose any additional risk to that of the constituent chemicals. Other vehicles based on polymeric encapsulates may need additional safety considerations depending on the nature of the vehicle composition, fate in the GI tract, and uptake and biodistribution profiles of the bioactives and/or any vehicle–bioactive conjugates (Figure 21.1).

21.2.1 Nature of the vehicle composition

For safety assessment of micro-/nanovehicles carrying bioactive compound(s), it is not only important to consider the nature of the encapsulating material and the encapsulated bioactive(s), but also the two together. As discussed in detail in the preceding chapters, such carrier systems can be composed of different matrices. In relatively simple forms, these can be oil-in-water emulsions, micelles, liposomes, nanocarriers made of long-chain saturated fatty acids, solid lipid nanoparticles, or lipid- or protein-based vesicles. They can also be in the form of encapsulates made of both natural and synthetic polymers. The natural polymers generally used for this purpose include those based on polysaccharide polymers – such as alginate, carrageenan, xanthan gum, chitosan, modified starch, starch dendrimers, and nanocellulose – as well as protein-based, such as chicken egg albumin, milk protein (casein), whey proteins (α-lactalbumin, β-lactoglobulin), collagen, gelatine, etc. Coacervates between proteins and negatively charged polysaccharides have also been used for micro-/nanoencapsulation. The synthetic polymeric materials used for this purpose include polylactic acid; poly(ε-caprolactone), poly(methyl methacrylates), and poly(alkyl cyanoacrylates), etc. Other synthetic but biocompatible and biodegradable polymers, such as poly(glycolic acid) and poly(D,L-lactide-co-glycolide), have also been used in biomedical applications.

As discussed above, the main consideration for safety assessment hinges upon whether or not a micro-/nanovehicle is broken down and the contents are released in the GI tract. As highlighted in Figure 21.1, bioactive substance(s) released in the GI tract should not be considered as being any different from the conventional (bulk) equivalents in a safety assessment. However, where an intact or

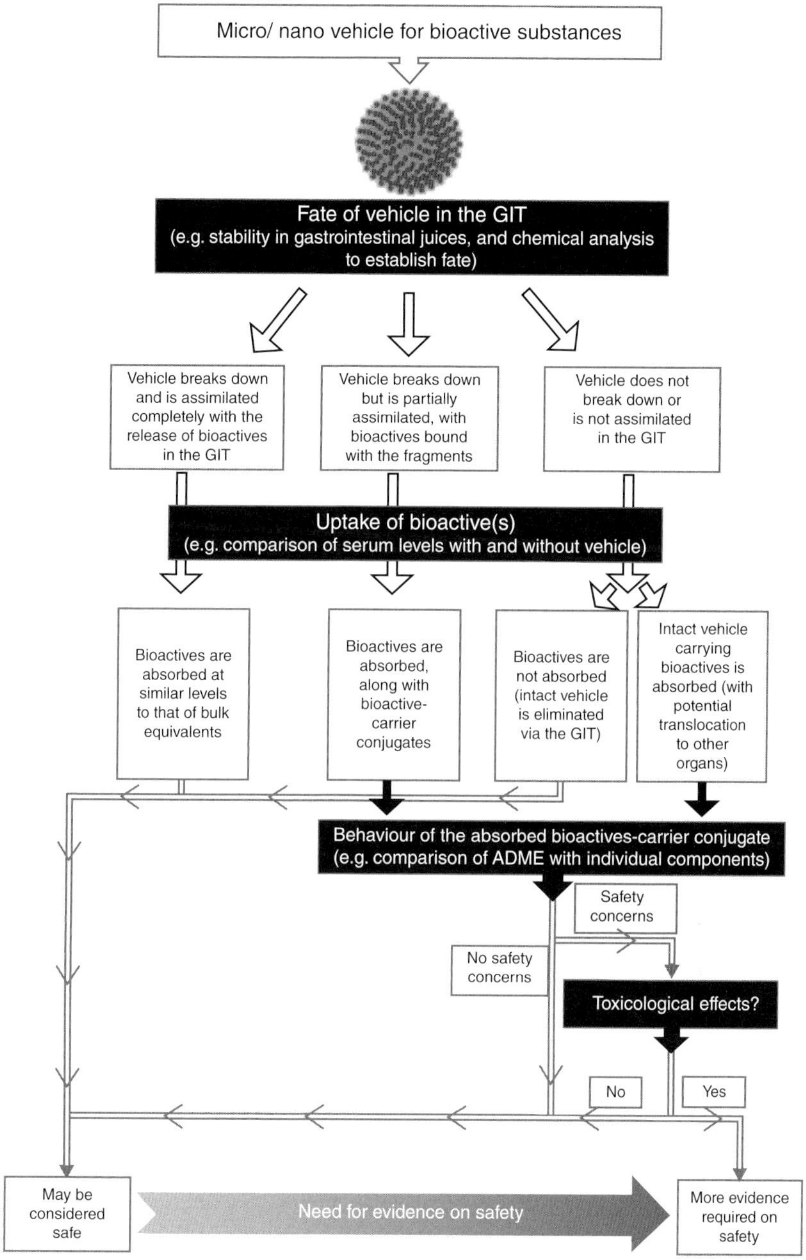

Figure 21.1 Considerations for safety assessment of micro-/nanovehicles in heath/food products (assuming that bioactive substances are not harmful to consumer health). GIT, gastrointestinal tract; ADME, absorption, distribution, metabolism, and excretion.

partially degraded vehicle carrying bioactive substance(s) can translocate from the GI tract to the circulatory system, its subsequent absorption, tissue distribution, and bioavailability may be different from those expected from conventional equivalents. Knowing the fate of micro-/nanovehicles is therefore crucial for safety assessment.

21.2.2 Fate of micro-/nanovehicles in the GI tract

In theory, a micro-/nanoformulation or encapsulation of bioactives made with natural or synthetic polymer matrix may present the following three scenarios, which can guide the assessment of their safety in food products.

1. A micro-/nanovehicle carrying bioactive substance may release bioactives in the GI tract following complete digestion of the vehicle. Thereafter conventional safety assessment should apply to the resulting chemical entities (i.e., the bioactives and the vehicle components). However, it should be considered that any formulation/encapsulation at micro- and nanoscales would also involve the use of a relatively much greater proportion of surfactants or vehicle components to the encapsulated bioactive substance(s). Consideration may therefore be also needed regarding the safety of the surfactants/vehicle components at the levels used in a micro-/nanoformulation.
2. A micro-/nanovehicle carrying bioactive substance may release bioactives only partially in the GI tract due to incomplete break up of the vehicle, and/or due to chemical binding between bioactives and the vehicle components that may have resulted, for example, from cross-linking during an encapsulation process. The resulting conjugates of vehicle fragments and bioactives may have different properties and biokinetic behavior from either of them alone. Also, being of very small size, some vehicle-bioactive conjugates may translocate from the GI tract to other organs. Such a scenario would require further assessment of absorption, distribution, metabolism, and excretion (ADME) profiles of the bioactive–vehicle conjugates. An analogy may be the case of gelatine, which is used in the encapsulation of health/food ingredients. Native gelatine is known to be nonimmunogenic. However, the encapsulation process involves cross-linking within and between gelatine molecules. This results in cross-linked gelatine, which is immunogenic. Antibodies to cross-linked gelatine have recently been raised for use in the detection of gelatine nanoparticles in beverage products under the Nanloyse project (Dehalu *et al.*, 2012). This points to the possibility that some micro-/nanodelivery vehicles translocating from the GI tract may pose a risk of immunological and other adverse effects. As discussed in Chapter 10, an irony of the supposedly safe use of "natural" substances for micro-/nanoscale formulation is that the immunological concerns are more likely to arise from a multicomponent "natural" material than from a well-characterized "synthetic" polymeric material of a known uniform composition.
3. A micro-/nanovehicle carrying bioactive substance may be recalcitrant to digestion and therefore may not release bioactives in the GI tract. This can lead to the following two scenarios.

    i. Where the intact vehicle carrying bioactives is eliminated from the GI tract without absorption of the bioactives or the vehicle. However, this is an unlikely scenario because such formulations, albeit probably safe, would be of little commercial use in health/food applications.
    ii. Where the intact vehicle carrying bioactives can translocate from the GI tract, and through the circulatory system reach the liver and other internal organs. This is where a detailed investigation into toxicokinetics may be necessary. Due to size constraints posed by intestinal mucosa (the mucus lining), epithelial membrane, and other biological barriers, this scenario is more likely to happen in the case of nanoscale vehicles than those at the microscale. Under such a scenario, knowing the ADME profile and the potential effects of the internally available nanovehicle carrying bioactive substances would become important. As the micro-/nanocarrier systems are likely to be designed to deliver useful nutrients/supplements, the effects may not necessarily be adverse and could in fact be beneficial. However, if such a vehicle is to be used in health/food applications, it is essential that it undergoes more in-depth investigation into any unexpected changes in properties, biodistribution behavior, and toxicological effects. It is also important that safety assessment of micro-/nanovehicles is carried out considering any gross changes in

solubility/dispersion, and other nano-related aspects, such as agglomeration/aggregation behavior, etc. These aspects have been highlighted in more detail in the guidance from EFSA (2011) on risk assessment of nanomaterials in food/feed, and from SCCS (2012) on risk assessment of nanomaterials in cosmetics.

It is evident from a large body of published literature on delivery carriers developed for drug delivery, and limited information relating to health/food applications, that the submicron-sized vehicles (especially in the nanoscale) can translocate from the GI tract (e.g., Florence, 1997; Palapied *et al.*, 2011). For polymeric micelles, the mechanism of absorption has been considered to occur through passive diffusion of the drug, or through endocytosis of the drug-loaded micelles. The polymeric carrier itself may be partly absorbed both as a monomer and in the form of a micelle. For polymeric nanoparticles, the cellular absorption is considered to mainly take the endocytosis route, and fate of the vehicle depends on the stability of the polymer and the nanoparticle form (Plapied *et al.*, 2011).

For translocation of particles across the GI tract, they have to first cross the barrier of intestinal mucosa, which is thought to restrict the passage of substances over 55 nm in diameter. Certain surface modifications, such as with poly(ethylene glycol), a substance commonly used to coat drug particles to prevent uptake by phagocytic immune cells, can increase translocation of particles as large as 200 nm (Samuel *et al.*, 2007). Any particles getting through the mucus layer then have to cross enterocyte cells in the intestinal epithelium, or M-cells in Peyer's patches, mainly via a paracellular route and to a limited extent via a transcellular pathway (see review by Chaudhry *et al.*, 2008). Using engineered nanoparticles, it has been shown that translocation of insoluble/ biopersistent particles from the GI tract is greater for nanosized particles than larger ones (Desai *et al.*, 1996; des Rieux *et al.*, 2006), and that particles of a lower nanometer scale can cross the mucus layer faster than the larger ones (Hillyer and Albrecht, 2001; Hoet *et al.*, 2004). In a study by Desai *et al.* (1996) biodegradable nanoparticles (100 nm) and microparticles (0.5, 1.0 and 10 μm) of a poly(lactic acid)–poly(glycolic acid) (50:50) co-polymer along with bovine serum albumin as a model protein, were formulated by a water-in-oil-in-water emulsion solvent evaporation technique. The uptake of the particles was studied in rat *in situ* using the intestinal loop model. The uptake of 100 nm nanoparticles by the intestinal tissue was 15–250-fold higher compared to the larger size microparticles. The diffusion of nanoparticles was observed throughout the submucosal layers, whereas microparticles were predominantly localized in the epithelial lining. Interestingly, the uptake of the particles was up to 200-fold higher in the Peyer's patches than other intestinal parts, even though Peyer's patches only comprise around 1% of the total intestinal surface area. A later *in vitro* study by Desai *et al.* (1997) in Caco-2 cells showed that the uptake of PLGA nanoparticles (100 nm), containing bovine serum albumin as a model protein and 6-coumarin as a fluorescent marker, was much greater than the larger microparticles. Apart from size and surface coating, a number of other factors also influence the overall systemic uptake of particles via the GI route. For example, charged particles, especially those made of positively charged polymers, may not translocate so efficiently because of electrostatic repulsion and entrapment in the mucus layer (Hoet *et al.*, 2004).

It therefore appears that although translocation of any micro-/nanosized bioactive-carrying vehicle across the GI tract is theoretically possible, it is more likely in the case of vehicles that are nanosized, are not composed of positively charged polymers, and are recalcitrant to breakdown in the digestive system.

21.3 CONSIDERATION FOR SAFETY ASSESSMENT

In general, safety assessment of materials composed of nanosized particles is a new and developing field. Due to current knowledge gaps in relation to properties, behavior, and effects, safety assessment of nanosized substances is carried out on a case-by-case basis (EFSA, 2011; SCCS, 2012). Chaudhry *et al.* (2008, 2010), FAO/WHO (2010), and EFSA (2009, 2011) have highlighted safety considerations relating to the use of nanomaterials in food products. In general, micronized materials, especially with particle sizes in the nanoscale, have a greater surface area and may show distinctively different behavior in relation to particle agglomeration/aggregation, biokinetics, biological interactions, and effects, compared to conventional bulk equivalents. At present there is little understanding of the potential interaction of materials and substances with living systems at the micro-/nanoscale. However, due to the very small size, biological interactions of nanosized substances may take place close to the molecular level, which may lead to toxicological effects of some nanovehicles different to those from bulk equivalents.

However, as mentioned before such a departure in properties, behavior, and effects of nanomaterials is dependent on the physical integrity of the nanostructure. Where a nanomaterial loses its nanostructure, the resulting chemical forms will also lose any specific properties/behavior relating to micro-/nanosize. Thus safety assessment should look beyond the chemical nature of a micro-/nanovehicle, and also consider its physical form, fate, and behavior, the bioactive substances contained therein, as well as the combined properties, behavior, and effect of the bioactive and vehicle material together.

Although these knowledge gaps make it difficult to assess the risk of bioactive-carrying micro-/nanovehicles to a consumer, a careful consideration of the materials and applications can provide a conceptual basis for a risk assessment (as highlighted in Figure 21.1). It is, nevertheless, difficult to be prescriptive and so formulations and their applications should be considered on a case-by-case basis. Points that should be considered in a safety assessment are considered below.

The likely starting premise for safety assessment in most cases is that both the bioactive substance(s) and the materials used for the micro-/nanodelivery vehicle are either food ingredients, are derived from food ingredients, or have been evaluated separately as food additives, supplements, etc. Thus to a first approximation they can be considered either as having a safe history of use, or they have been judged to be safe, individually. That being so, a safety assessment can then focus on what, if any, are the consequences of bringing together the bioactive and the delivery material.

21.3.1 Specifications and purity

In the first instance it can be recommended to consider the specifications of the separate ingredients, including their composition and impurities, and consider what role if any these specifications may play in ensuring their safety. What additional factors may come into play when they are combined for the intended application? Does the new formulation meet the specifications for its individual components? Are these specifications complete and adequate for the intended use?

21.3.2 Reaction and breakdown products

The possible formation of any undesirable chemical reaction or breakdown products should be addressed both theoretically and by chemical analysis. The encapsulation/formulation process may be quite gentle or it could involve chemical reaction(s), and/or high-energy processes such as high mechanical sheer, pressure, temperature, and ultrasonics, which could promote the formation of reaction products. The stability of the formulation should also be considered, both "on the shelf" and once formulated into a food or a supplement. Some delivery vehicles may have a poor stability in food products, and their contents may leak or diffuse out over time. The high surface area could, for example, promote a more extensive hydrolysis, oxidation, or photolysis than a conventional formulation. These tests for undesirable reaction products should involve storage tests and food processing and cooking if relevant.

21.3.3 Fate and behavior of the bioactive

The fate and behavior of the formulation should be studied following ingestion. As shown in Figure 21.1, this could in the first instance be done in terms of *in vitro* digestion using gastrointestinal juices, followed by chemical analysis to establish the fate of the bioactive(s) and the polymeric materials. Where needed (e.g., for recalcitrant vehicles), this may be followed through using *in vivo* studies. This will ensure that the release characteristics of the bioactive are understood. In practice, the release of bioactives could be accelerated or retarded by the vehicle design (giving a higher or lower peak plasma concentration, for example), and/or the release may be at a higher or a lower overall percentage bioavailability (giving a higher or lower area under the curve). The pharmacology/toxicology of the bioactive should then be considered in the light of these experimental findings.

21.3.4 Fate and behavior of the delivery material

A material balance should be performed for the delivery material to understand its fate and behavior as far as possible. In case of doubt it may be necessary to resort to an ADME study *in vivo* using labeled ingredients to help track the nano-/micromaterials, both associated in the delivery vehicles and as individual "solubilized" substances.

21.3.5 Unwanted side-effects of the delivery material

Once the bioactive is released *in vivo*, the fate of the vehicle material should be considered carefully. Is it simply dispersed (no longer nano- or microstructured) and excreted, possibly with metabolism? Or is it metabolically inert – in which case is there the potential for it absorbing or adsorbing ("mopping-up") essential vitamins or minerals and reducing their bioavailability as an unwanted side-effect?

21.3.6 Considerations of allergenicity and immunotoxicity

If the nature of the delivery material suggests (e.g., is derived from a milk protein or similar), and if alterations in this feature cannot be discounted, then allergenicity and immunotoxicity effects should be investigated.

Potential allergenicity can be assessed by comparison of the amino acid sequence with those of known allergens. The EFSA guidance (2012) describes the type of testing that can be used to establish immunotoxicity or immunomodulatory effects of a food additive. For example, a 90-day study in rats (OECD TG 408) under Tier-1 provided data on parameters that may be indicative of immunotoxic or immunomodulatory effect – such as changes in spleen and thymus weights relative to body weight in the absence of overt toxicity, histopathological changes in these and other organs of the immune system (e.g., bone marrow, lymph nodes, Peyer's patches), as well as changes in total serum protein, albumin:globulin ratio, and in the hematological profile of the animals, notably in lymphocyte numbers and in the total and differential blood-cell counts. Further studies under Tier-2 (e.g., OECD TG 443) and Tier-3 (e.g., mitogen stimulation assays for B and T cells; or functional analysis on natural killer cell functional analysis, etc., or alternative methods using human cells from the umbilical cord, such as hematopoietic progenitor clonogenic assays) may be necessary if the results indicate immunotoxic potential.

In the final analysis, if the studies outlined above do not give reassurance on the safety of use, then a guideline chronic toxicity study may be needed.

21.4 CONCLUSIONS

In brief, data from the general assumptions over safety of micro-/nanovehicles, such as those based on the history of safe use of the individual bulk constituents, should not be used as a surrogate to show that the application is safe to the consumer. That said, it is likely that, by virtue of good design, most micro-/nanodelivery vehicle systems will fall into the least problematic category, and their safety can be evaluated in a rather straightforward way. Only in unusual cases – for example where a vehicle is recalcitrant to breakdown and can translocate to various organs, and/or can bring about gross changes in the uptake and biodistribution of bioactives – might it prove necessary to conduct extensive chemical and toxicity testing. However, being aware of the full spectrum of safety concerns and in turn being aware of the tools and considerations available to address these concerns, will help to promote the careful design, development, and commercialization of good micro-/nanodelivery vehicles for bioactives.

References

Carr, K.E., Hazzard, R.A., Reid S. and Hodges, G.M. (1996) The effect of size on uptake of orally administered latex microparticles in the small intestine and transport to mesenteric lymph nodes. *Pharmaceutical Research*, 13, 1205–1209.

Chaudhry, Q., Scotter, M., Blackburn, J., *et al.* (2008) Applications and implications of nanotechnologies for the food sector. *Food Additives and Contaminants*, 25(3), 241–258.

Chaudhry, Q., Castle, L. and Watkins, R. (eds) (2010) *Nanotechnologies in Food*. Royal Society of Chemistry Publishers, London.

Dehalu, V., Weigel, S., Rebe, S., Grombe, R., Löbenberg, R. and Delahaut P. (2012) Production and characterization of antibodies against crosslinked gelatin nanoparticles and first steps toward developing an ELISA screening kit. *Analytical and Bioanalytical Chemistry*, 403(10), 2851–2857.

Des Rieux, A., Fievez, V., Garinot, M., Schneider, Y.J. and Preat, V. (2006) Nanoparticles as potential oral delivery systems of proteins and vaccines: a mechanistic approach. *Journal of Controlled Release*, 116, 1–27.

Desai, M.P., Labhasetwar, V., Amidon, G.L. and Levy, R.J. (1996) Gastrointestinal uptake of biodegradeable microparticles: effect of particle size. *Pharmaceutical Research*, 13, 1838–1845.

Desai, M.P., Labhasetwar, V., Walter, E., Levy, R.J. and Amidon, G.L. (1997) The mechanism of uptake of biodegradable microparticles in Caco-2 cells is size dependent. *Pharmaceutical Research*, 14(11), 1568–1573.

EFSA. (2009) Scientific Opinion of the Scientific Committee on a request from the European Commission on the potential risks arising from nanoscience and nanotechnologies on food and feed safety. *The European Food Safety Authority Journal*, 958, 1–39.

EFSA. (2011) Guidance on the risk assessment of the application of nanoscience and nanotechnologies in the food and feed chain, *The European Food Safety Authority Journal*, 2011;9(5):2140 [36 pp.]. doi:10.2903/j.efsa.2011.2140.

EFSA. (2012) Panel on Food Additives and Nutrient Sources added to Food (ANS); Guidance for submission for food additive evaluations. *The European Food Safety Authority Journal*, 10(7), 2760. doi:10.2903/j.efsa.2012.2760.

FAO/WHO. (2010) *Meeting report of Expert meeting on the application of nanotechnologies in the food and agriculture sectors: potential food safety implications*. Food and Agriculture Organization of the United Nations and World Health Organization, Rome; 129 pp. Available: www.fao.org/docrep/012/i1434e/i1434e00.pdf (accessed 4 January 2015).

Florence, A.T. (1997) The oral absorption of micro- and nanoparticulates: neither exceptional nor unusual. *Pharmaceutical Research*, 14, 259–266.

Florence A.T. (2005) Nanoparticle uptake by the oral route: Fulfilling its potential? *Drug Discovery Today: Technologies*, 2, 75–81.

Garti, N. (2006) NutraLease – NSSL Technology, Ltd. Food goes nano – nano architectures as delivery platforms for nutraceuticals. *Second EC Workshop on Food-Health-Well Being*, Brussels, November 21–22. Available: ftp://ftp.cordis.europa.eu/pub/food/docs/garti-n_en.pdf (accessed 4 January 2015).

Hillyer, J.F. and Albrecht, R.M. (2001) Gastrointestinal persorption and tissue distribution of differently sized colloidal gold nanoparticles. *Journal of Pharmaceutical Sciences*, 90, 1927–1936.

Hoet, P.H.M., Bruske-Hohlfeld, I. and Salata O.V. (2004) Nanoparticles: known and unknown health risks. *Journal of Nanobiotechnology*, 2, 1–15.

McClements, D.J. (2011) Edible nanoemulsions: fabrication, properties, and functional performance. *Soft Matter*, 7, 2297–2316. doi: 10.1039/c0sm00549e.

Mundargi, R.C., Babu, V.R., Rangaswamy, V., Patel, P. and Aminabhavi, T.M. (2008) Nano/micro technologies for delivering macromolecular therapeutics using poly(D,L-lactide-co-glycolide) and its derivatives. *Journal of Controlled Release*, 125(3), 193–209.

Nishiyama, N. and Kataoka, K. (2006) Current state, achievements, and future prospects of polymeric micelles as nanocarriers for drug and gene delivery, *Pharmacology and Therapeutics*, 112(3), 630–648.

NovaSOL®. (Undated) *Crystal Clear Solutions – Overview*. Available: www.aquanova.de/media/public/pdf_produkte%20unkosher/NovaSOL_OVERVIEW.pdf (accessed 4 January 2015).

Peer, D., Karp, J.M., Hong, S., Farokhzad, O.C., Margalit R. and Langer, R. (2007) Nanocarriers as an emerging platform for cancer therapy. *Nature nanotechnology*, 2, 751–760.

Plapied, L., Duhem, N., des Rieux A. and Préat, V. (2011) Fate of polymeric nanocarriers for oral drug delivery. *Current Opinion in Colloid and Interface Science*, 16, 228–237.

Samuel, K., Lai, D., O'Hanlon, E., *et al*. (2007) Rapid transport of large polymeric nanoparticles in fresh undiluted human mucus. *Proceedings of the National Academy of Sciences*, 104, 1482–1487.

SCCS. (2012) *Guidance on the Safety Assessment of Nanomaterials in Cosmetics,* SCCS/1484/12, Scientific Committee on Consumer Safety. Available: http://ec.europa.eu/health/scientific_committees/consumer_safety/docs/sccs_s_005.pdf (accessed 5 January 2015).

22 Evidence-based regulation of food nanotechnologies: a perspective from the European Union and United States

Diana Bowman,[1] Qasim Chaudhry,[2] and Anna Gergely[3]

[1]*Department of Health Management and Policy and Risk Science Center, University of Michigan, Ann Arbor, Michigan, USA*
[2]*The Food and Environment Research Agency, Sand Hutton, York, UK*
[3]*Steptoe & Johnson LLP, Brussels, Belgium*

22.1 INTRODUCTION

The global food landscape is ever evolving, and so are the legislative and regulatory instruments that govern its safety from farm to fork. As this chapter seeks to highlight, the emergence of new technological processes and products within the agri-food sector produces a moving target, and poses new challenges to the available regulatory instruments. In response, the availability of new knowledge and scientific evidence for risk requires the regulatory frameworks to evolve, and/or to be recast where necessary.

Like any technological novelty, nanotechnology-derived materials, processes, products, and applications have brought a promise of new benefits, but at the same time have also highlighted a number of health, environmental, societal, and regulatory challenges. The current regulatory frameworks have also come under a close scrutiny over whether they would be capable of "capturing" and controlling the potential risk of engineered nanomaterials (or nanomaterials). In this context, this chapter aims to provide an overview of the food-related applications of nanotechnologies, and the potential risks that they might pose to the consumer. It also discusses the regulatory frameworks that are relevant to food applications in the European Union (EU) and the United States – the world's two largest markets for foods – and provides an up-to-date account of the developments that are taking place in these two jurisdictions.

22.1.1 Why nanotechnologies in food?

A particular feature of the recent developments in nanosciences and nanotechnologies has been the rapid pace of innovation and the breadth of applications across a wide range of sectors, including the agri-food sector that itself has a multitude of subsectors and branches which span from farm to fork. The competitive nature of the agri-food sector, consumer demands for innovative, yet safe, and affordable food products, and regulatory pressures to safeguard consumer health and environmental sustainability, mean that the sector is continually seeking new technologies to improve efficiency, safety, and cost-effectiveness of the feedstocks,

Nanotechnology and Functional Foods: Effective Delivery of Bioactive Ingredients, First Edition.
Edited by Cristina M. Sabliov, Hongda Chen, and Rickey Y. Yada.

processes, and products. The sheer size of the global food-retailing sector alone – estimated to be worth around $US4 trillion annually (US Department of Agriculture, 2009) – means that a successful new product or application can potentially attain a very large market value worldwide. With the value of the sector predicted to only increase with time, there can be little surprise that the agri-food sector has been an attractive target of new technologies that can bring in innovation improving the existing processes and products. Pressures from consumers and regulators are also driving the industry further to look for new ways to, for example, reduce the use of agrochemicals on the farm, lower the levels of fat, salt, artificial colors, flavors, and preservatives, and improve the nutritional value, hygiene, safety, and security of the food products.

In this context, recent advances in nanosciences and nanotechnologies have opened up a whole new array of possibilities for addressing many of the industry's current and future needs (Cientifica, 2006; Chaudhry *et al.*, 2008; Chaudhry *et al.*, 2010b).

The use of nanomaterials in preference to bulk equivalents, which may have been used by the industry for decades, has arisen due to certain advantages. Nanosizing can enable manipulation of material properties to suit a particular application, and may also generate novel functionalities. For example, nanosizing of agrochemicals may improve efficacy and thus enable a reduction in the use of biocides, pesticides, and veterinary medicines in food production (Watson *et al.*, 2011). It may also enable a better control over field applications (e.g. through the use of slow release formulations), and fortification of animal feeds with nano-sized supplements, antimicrobials, and detoxifying substances.

A number of inorganic nanomaterials are known to be used in food, health-food, and food-packaging applications. These include nanoforms of transition metals such as silver, titanium dioxide and iron, of alkaline earth metals such as calcium and magnesium, and of nonmetals such as selenium and silicates (Chaudhry *et al.*, 2008; Woodrow Wilson International Centre for Scholars, 2011).

As this book has discussed, the use of nanotechnologies has also enabled the development of nano- and microsized organic carrier systems for delivery of nutrients, supplements, and other bioactive ingredients in food products. They generally comprise bioactive substances that have been formulated or encapsulated at the nanoscale in liposomes, micelles, or protein-based carriers. In addition to delivery of bioactive substances, these nanocarriers can also be used for protecting them from degradation during processing. Nanoformulated substances have also been claimed for enhanced bioavailability, improved antimicrobial activity, and other health benefits (Chaudhry and Groves, 2010). Nanoscale processing of food materials has also been used to develop improved or new tastes, textures, and mouth sensations (e.g., see Clegg *et al.*, 2009).

Another advantage of sizing or formulating food ingredients at the nanoscale is the improved dispersibility of water-insoluble additives, such as colors, flavors, preservatives, and supplements, without the use of additional fat or surfactants.

Also under research and development are the next-generation nanomaterials, which comprise nanoscale materials that are functional(ized), such that they can add a certain functionality to the matrix: for example, a preservative action.

Despite the numerous anticipated and projected benefits, when compared to other sectors such as cosmetics and personal-care products, food-related applications of nanotechnologies are still emergent and marginal (Cientifica, 2006; Chaudhry *et al.*, 2010a; Chaudhry and Castle, 2011). They are, nevertheless, widely anticipated to grow in the future, and are seen to have the potential to transform the whole of the food chain (e.g., see Huang, 2012). Amongst other factors that might shape the future of nanotechnology application for the food sector, clarity on the likely regulatory hurdles would certainly facilitate the transformation of many new concepts into commercial reality.

22.1.2 What's new to regulate?

For the regulatory frameworks to adequately address the safety issues, they have to take account of the diverse types and distinctive properties of nanomaterials and their likely applications, as well as the associated uncertainties over the potential (new) risks that they may pose. The nanomaterials that are, or may be, used in food related applications, comprise a diverse range in terms of chemical nature and physical forms. These may differ in terms of the different morphological forms that can be produced in (e.g., different sizes, shapes, coatings, surface modifications, etc.), and the possible different or new properties they may possess compared to the corresponding bulk forms. It is therefore not surprising that the first, and major, regulatory challenge has so far been to develop a definition that can provide an overarching description of the materials, and that can be used as a basis for providing clear boundaries and thresholds to enable regulation of nanomaterial additives in food.

Other major challenges include establishing the availability and adequacy of analytical tools to detect and measure any nanosized additives in (complex) food matrices, and providing guidance for risk assessment in the face of uncertainty. This is because insoluble particulate materials may behave differently from conventional chemical forms (which can be solubilized). Therefore, existing analytical methods that are valid for conventional chemicals are still undergoing validation for detection and characterization of nanoparticles (e.g., see the EU project Nanolyse). Similarly, the use of existing toxicity testing methods for nanomaterials requires a number of nanospecific considerations (Scientific Committee on Consumer Safety, 2012) in order to obtain valid results. The current lack of uniform measurement tools, and uncertainty over the behavior, fate, and effects of nanoparticles, have so far made it difficult to set any specific permissible limits for nanosized additives in food in any jurisdiction. For these reasons, the main emphasis has been to use the existing regulatory instruments to control the risk, and adapt and/or recast where necessary when dictated by the scientific evidence for risk.

22.1.3 Deconstructing the definition debate and implications for regulation

The basis of many of the current uncertainties over materials and products of nanotechnologies seems to be due to the lack of a clear definition that adequately covers the distinctive features and properties of a nanosized material. A number of definitions in this regard have been proposed (Scientific Committee on Emerging and Newly Identified Health Risks, 2010; NICNAS, 2010; Health Canada, 2011; Environmental Risk Management Authority, 2012), as has the need for improving the description of nanomaterial through the use of criteria other than the size parameter alone, including, for example, volume specific surface area (VSSA) (Kreyling *et al.*, 2010). In the United Kingdom, for instance, the House of Lords' Science and Technology Committee (2010) published a detailed report on the findings of their inquiry into nanotechnologies and food. The report recommended "…any regulatory definition of nanomaterials proposed at a European level, in particular in the Novel Foods Regulation, should not include a size limit of 100 nm but instead refer to "the nanoscale" to ensure that all materials with a dimension under 1000 nm are considered" (House of Lords, 2010, p 112). Further recommendations contained in the House of Lords' Report on Nanotechnologies and Food favored an approach based on a change in functionality (how the substance interacts with the body) rather than an approach relying on size range alone (see also Holdren *et al.*, 2011; FDA, 2012).

Another difficulty in this regard has been that particle size distribution of a material is usually quoted as a mean (± standard deviation), which does not provide a full representation of the particle sizes in the nanoscale. A report by the Scientific Committee on Emerging and Newly Identified Health Risks adopted on 6 July 2010 proposed the use of mean particle size ±3× the standard deviation (to cover the size range adequately), and where appropriate, the use of VSSA (i.e. where VSSA $\geq 60\ m^2/cm^3$ the material will be regarded a nanomaterial), and limiting the definition to "purposely designed nanomaterials" (Scientific Committee on Emerging and Newly Identified Health Risks, 2010, p. 27).

Although a material could be defined as nanomaterial on the basis of these criteria, uncertainties remained over those materials that were not engineered as nanomaterials (e.g., micronized materials) but contained a fraction in the nanoscale. There are several examples of such materials, but two known examples are titanium dioxide and fumed silica – both of which are approved food additives (E551 and

E171, respectively). This needed setting a threshold for the nanofraction in a particulate material in order to regard it as a nanomaterial for regulatory purposes. In view of this, and following public consultation in 2010, the European Commission adopted an overarching regulatory definition of nanomaterial in October 2011. (Commission Recommendation 2011/696/EU on the definition of nanomaterial (October 18, 2011). Pursuant to the Commission Recommendation 2011/696, a

> "nanomaterial" means a natural incidental or manufactured material containing particles, in an unbound state or as an aggregate or as an agglomerate and where, for 50% or more of the particles in the number size distribution, one or more external dimensions is in the size range 1nm–100nm.

The recommendation is meant to provide guidance for tailoring more specific definitions under the different policies and legislation applied within the EU and the European Economic Area (European Commission, 2010, p. 4). While the adoption should, in principle, have clarified a number of regulatory uncertainties due to the current lack of a clear definition of nanomaterial, the definition has been subject to a criticism and debate (e.g., see Maynard, 2011; Stamm, 2011). A major element of this debate is centered on the current lack of validated analytical methods for measurement of nanomaterial levels in food in terms of particle number concentration for implementation of the proposed 50% threshold.

Currently, the only EU definition of nanomaterial, relevant to food applications, is provided under point (t) of Article 2(2) of the Food Information Regulation (European Commission Regulation No. 1169/2011). For the purposes of the Food Labelling Regulation, Article 2 defines an engineered nanomaterial to mean:

> any intentionally produced material that has one or more dimensions of the order of 100 nm or less or that is composed of discrete functional parts, either internally or at the surface, many of which have one or more dimensions of the order of 100 nm or less, including structures, agglomerates or aggregates, which may have a size above the order of 100 nm but retain properties that are characteristic of the nanoscale. Properties that are characteristic of the nanoscale include:
>
> i. those related to the large specific surface area of the materials considered;
>
> And/or
>
> ii. specific physico-chemical properties that are different from those of the non-nanoform of the same material.

Regulation 1169/2011 expressly acknowledges that the definition of a nanomaterial adopted within its text may need to be adjusted in line with technical and scientific progress (see Article 2(2)). Specifically, paragraph 51 of the Preamble states that, "food information rules should be able to adapt to a rapidly changing social, economic, and technological environment."

In due course, other relevant regulatory frameworks in the EU are likely to adopt/adapt the nanomaterial definition in view of the criteria provided in the European Commission Recommendation 2011/969.

Compared to the EU, the US Food and Drug Agency (FDA) has not established a regulatory definition of nanotechnology or related terms. In June 2011, FDA issued draft guidance on considering whether an FDA-regulated product contains nanomaterials or otherwise involves the use of nanotechnologies. Despite the absence of a formal definition, in the case of nanotechnology applications, the

> FDA will ask: (1) whether an engineered material or end product has at least one dimension in the nanoscale range (approximately 1 nm to 100 nm); or (2) whether an engineered material or end product exhibits properties or phenomena, including physical or chemical properties or biological effects, that are attributable to its dimension(s), even if these dimensions fall outside the nanoscale range, up to one micrometer (FDA, 2011).

While applicable to new products, these considerations may also apply when manufacturing changes alter the dimensions, properties, or effects of an FDA-regulated product or any of its components.

Looking further afield, Australia's industrial chemical regulatory scheme, the National Industrial Chemicals Notification and Assessment Scheme (NICNAS), has adopted new administrative procedures for the notification and assessment of "new" industrial nanomaterials, as defined by their own working definition (NICNAS, 2010). Under this definition, an "industrial nanomaterial" means,

> … industrial materials intentionally produced, manufactured or engineered to have unique properties or specific composition at the nanoscale, that is a size range typically between 1 nm and 100 nm, and is either a nano-object (i.e. that is confined in one, two, or three dimensions at the nanoscale) or is nanostructured (i.e. having an internal or surface structure at the nanoscale) (NICNAS, 2010: 1).

It is not clear, at this time, how many chemicals have been captured under these new procedures.

22.1.4 Potential risk

Like the early stages of other new technologies, nanotechnologies are currently associated with a high level of uncertainty over the safety of nanomaterials (Maynard *et al.*, 2006; Scientific Committee on Consumer Products, 2007; Chaudhry *et al.*, 2008). In relation to food applications, the concerns range from potential health effects emanating from consumption of nanomaterials via food and drinks, to environmental impacts from the use of nanotechnology-derived agrochemicals and packaging materials. However, for evidence-based risk regulation, it is important to segregate a real, or potential risk, from a hypothetical or "perceived" one.

A number of studies have indicated that the current risk-assessment paradigm, in use for conventional substances (including hazard identification, exposure assessment, hazard characterization, followed by risk characterization), is also applicable to nanomaterials (e.g., see Rocks *et al.*, 2008; European Food and Safety Authority, 2009; Scientific Committee on Emerging and Newly Identified Health Risks, 2009; OECD, 2009; Chaudhry *et al.*, 2010a). However, at the same time, the need for nanorelated considerations and possible adaptations in the current testing methods have also been highlighted to take account of the special features of nanomaterials (Balbus *et al.*, 2007; Rocks *et al.*, 2008).

It is well known that, when compared to larger sized materials, nanomaterials may behave differently. A particular departure from other (larger) particulate materials is the ability of nanoparticles to penetrate biological membrane barriers that normally prevent the entry of insoluble particles into cells and tissues. This has added a new dimension to the toxicology of particulate materials, as insoluble particles with potential to cause harmful effects may reach new targets in the body (Hillyer and Albrecht, 2001; Hoet *et al.*, 2004; Florence, 2005; des Rieux *et al.*, 2006). In regard to food applications, any risk emanating from the use of nanosized materials would be intrinsically dependent on the possibility of systemic exposure. This in turn would be dependent on the chemical nature and physicochemical properties of the nanomaterials that may be consumed via food and drink.

It is important to note that nanomaterials are – at the same time – chemical(s) as well as particles. Any nanospecific property and/or effect of a nanomaterial would intrinsically depend on the physical integrity of its nanostructure. Where a nanomaterial loses nanostructure due to digestion, metabolic breakdown, solubilization, degradation, etc., the resulting chemical constituents will not be expected to behave any differently from the nonnano-equivalent. Thus nano- and microscale vehicles designed for delivery of bioactive ingredients based on "soft" natural materials, including, for example, emulsions, liposomes, and lipid/protein encapsulates, need to be differentiated as a potentially low risk category from the insoluble and biopersistent nanomaterials. This is because such delivery systems are most likely to be degraded, digested, and assimilated within the GI tract, and the resulting chemical constituents dealt with in the body in a similar way as their conventional equivalents. They should, therefore, not pose any additional health risk to that of the bulk equivalents. There may, however, be some exceptions; for example, where nanoscale formulation can make an additive or ingredient more bioavailable, as this may give rise to atypically high target fluid and tissue levels of the substance. Another situation may be where formulating a substance at nano- or microscale makes it harmful due to a change in biokinetic behavior. However, it is unlikely that such a substance/formulation will be used deliberately in food applications. The consumer safety concerns therefore relate mainly to those nanomaterials that are insoluble, indigestible, and biopersistent. If such "hard" nanomaterials are used in food applications, this will provide a direct route of entry of the nanoparticles to other parts of the body where they might cause some harmful effects. Such concerns are, however, largely due to the current knowledge gaps. For example, little is known about the possible behavior, interaction, fate, and effects of most nanoparticles within and outside the human GI tract. It is possible that, like "soft" nanomaterials, most "hard" nanomaterials will not remain in a free particulate form in food products, or in the GI tract, due to agglomeration, aggregation, and binding with other food components. They may also be degraded or transformed in the GI tract, and hence not remain available for translocation in particle form to other parts of the body.

Thus those natural food materials that are digested/assimilated in the GI tract, and are not biopersistent – for example, processed natural food materials in the form of nanostructures – should not pose any additional nanospecific risk to the consumer. The main concerns relate to the use of insoluble, indigestible, and biopersistent nanoadditives (e.g., some metals/oxides), and functionalized nanomaterials in food and drinks – the toxicological properties of which are not fully known.

Other applications, such as food packaging, would pose a risk to consumer health only if nanoparticles were able to migrate from packaging materials into the packaged foodstuffs.

The use of nanosized or nanoscale formulated agrochemicals for food production may pose a risk to a farm worker of exposure to hazardous nanomaterials, and to consumers if nanomaterial residues are carried over to food products.

Unlike other exposure routes, such as dermal or inhalation, the orally ingested materials are subjected to an active digestive system. Throughout the passage of food through the GI tract, foodstuffs undergo various degrading conditions, from a harsh acid environment in the stomach, to enzymic breakdown, and reactions with bile acids, etc. This leads to the breakdown of even larger food particles. The nutrients and other substances are able to cross the GI barrier only when they have been broken down to smaller components. The passage of particles from the mucus lined GI epithelium is generally limited, although there is emerging evidence that, within the nanoscale (1–100 nm), smaller particles are more likely to translocate out of the GI tract than the larger particles (Hillyer and Albrecht, 2001; Hoet *et al.*, 2004; des Rieux *et al.*, 2006). For example, metals and metal oxides nanoparticles are typically taken up by M cells of the Peyer's patches and passed to the macrophages, where they may accumulate and appear as pigmentation in cells at the base of human intestinal lymphoid aggregates (Powel *et al.*, 2000, 2010).

The most important determinant of a nano-related risk therefore is whether nanoparticles added to food can remain, translocate from the GI tract, and reach other parts of the body, in insoluble particle form.

As highlighted in Chapters 11–18 of this book, most nano- and microscale delivery systems for bioactive compounds comprise natural encapsulating materials that are likely to be degraded in the GI tract. Such a delivery system may thus only be of concern if the encapsulating materials are not digested or degraded in the GI tract and are able to carry the encapsulated substances to the circulatory system from where they can reach to other unintended systemic areas. In such a case, the biokinetics and biodistribution of the substances contained in the nanocarrier may be different from that of conventional forms. As increased bioavailability of some substances may lead to health risks, it would be important to know whether a nanoscale delivery system would increase the bioavailability of certain bioactive substances when consumed via food.

An approach used in risk assessment, where uncertainties and knowledge gaps are high, is the so-called "precautionary principle". This approach is meant to provide a reasoned and structured framework for action in the face of scientific uncertainties or absence of scientific consensus. The recourse to the precautionary principle presupposes that potentially harmful effects deriving from a product or process have been identified, and that existing scientific evaluation does not allow the risk to be determined with sufficient certainty, and provides grounds for assigning responsibility for producing the scientific evidence necessary for a comprehensive risk assessment. The precautionary principle is incorporated into EU food law in Article 7 of Regulation 178/2002, which states that where, following an assessment of available information, the possibility of harmful effects on health is identified but scientific uncertainty persists, provisional risk management measures necessary to ensure the high level of health protection chosen in the Community may be adopted, pending further scientific information for a more comprehensive risk assessment. Despite the current uncertainties over the potential risk of some nanomaterials, there does not seem to be sufficient scientific basis to invoke the precautionary principle in the various food and related applications of nanomaterials.

In view of the diverse range of food-related applications of nanomaterials, and the current uncertainties over the risks they might pose to consumer health and/or the environment, it would be challenging to regulate them as different risk categories. However, it is also important that regulatory frameworks are able to differentiate the low-risk applications, such as "soft" delivery systems, from the other high-risk applications, such as metals and metal oxides, on the basis of the available scientific evidence. The European Food and Safety Authority (EFSA) has published a Scientific Opinion, which provides structured guidance on risk assessment of nanoscience and nanotechnology applications to food and feed (European Food and Safety Authority, 2011). The document provides a systematic decision route to risk assessment, with guidance on the types of tests and methods that may be required for physicochemical characterization and toxicological evaluation of nanomaterials in food/feed products. This is where a specific definition that sets down clear boundaries for regulated nanomaterials in food-related applications would bring clarity for many nanoscale applications that fall into the no- or low-risk category in order to facilitate their adoption and acceptance by the food sector.

22.2 REGULATORY DEVELOPMENTS IN NANOTECHNOLOGIES

Most countries around the world already have regulatory frameworks that require pre-market safety evaluation of certain food and related products. These frameworks are the most likely routes to regulatory oversight of nanotechnology-derived food products. These include, for example, laws relating to general food safety, food additives, novel foods, specific health claims, chemical safety, food-contact materials, and water quality. These are further underpinned in many jurisdictions through private law, such as tort law. Other specific regulations relevant to nanotechnology applications relate to the use of certain types of chemicals in food production/ protection, such as biocides, pesticides, veterinary medicines, etc. Environmental regulations are also likely to capture the use of nanotechnologies in food packaging and agri-food production applications (Chaudhry *et al.*, 2008; Gergely *et al.*, 2010). A few studies have assessed the adequacy of existing regulatory frameworks to control the potential risks of nanotechnologies to the consumer and the environment (e.g. see Chaudhry *et al.*, 2006; Davies, 2006; Taylor, 2006; Ludlow *et al.*, 2007; Hodge *et al.*, 2010). A consensus from these studies is that the current developments in this field are not taking place in any regulatory vacuum, and the risks of nanomaterials will be controlled under the existing frameworks (FSA, 2006; Chaudhry *et al.*, 2008; Gergely *et al.*, 2010). The European Commission Communication of 2008 on regulatory aspects of nanomaterials also concluded that, "current legislation covers in principle the potential health, safety and environmental risks in relation to nanomaterials." The European Commission Communication of 2012, the second regulatory review on nanomaterials, was not so definitive in its language. Instead, greater emphasis was placed on the technical challenges, the need for case-by-case risk assessment requirements, and transparency (European Commission, 2012a, b).

The regulatory gap studies so far have regarded the existing frameworks relating to food and food-contact materials in different jurisdictions, such as the EU, the United States, and Australia, broad enough to "capture" nanotechnology applications in the food sector (Ludlow *et al.*, 2007; Chaudhry *et al.*, 2008; Gergely *et al.*, 2010). This position has been reinforced with the release of a joint memorandum from the US Executive Office of the President to all Department and Agency Heads (Holdren *et al.*, 2011). Specifically, the memorandum sought to remind the intended audience that the "existing regulatory statutes provide a firm foundation for the regulation and oversight of nanomaterials" and that "building consumer trust and confidence" in the framework was integral to the success of the technology (Holdren *et al.*, 2011). While recognizing that the frameworks may be required to "evolve" over time in response to scientific developments, the Executive Office stressed the importance of employing a "science-based" approach to the regulation of nanomaterials. In their words, agencies must continue to employ an "evidence-based and application-specific" approach in order to protect human and environmental health and safety (Holdren *et al.*, 2011). The draft food guidance released by the FDA similarly supports this position (FDA, 2012).

There are, nevertheless, a few current uncertainties that seem to have arisen from the lack of understanding of, for example, a regulatory definition of nanosized food additives, permissible limits for nanoadditives, and an exclusive premarket approval system for nanoenabled food products. Recent regulatory developments have therefore been aimed at addressing such uncertainties. In this regard, the EU has played a particularly leading role in paving the way for evidence-based regulation of the materials and products of nanotechnologies.

22.2.1 European Union

Among the most relevant horizontal (cross-cutting) measures in the EU are general food laws such as European Commission Regulation No. 178/2002, which sets out the general principles and requirements for food within the EU and provides for the establishment of the EFSA to ensure that the food placed on the market is safe, and that the producer placing it on the market is liable for its safety. Foods containing nanomaterials and/or foods manufactured using nanotechnologies are covered under Regulation 178/2002, due to the Regulation's general provisions and all-inclusive scope. Examples of general food laws in other countries include the Federal Food, Drug and Cosmetic Act (FDC Act) in the United States, which is administered by the FDA, and the Australian and New Zealand Food Standards Code.

European Commission Regulation No. 1169/2011 on the provision of food information to consumers (known as the Food Labeling Regulation) contains new food-labeling rules for the EU market generally, and includes a number of nanospecific provisions. Of particular note to this chapter is that Regulation No. 1169/2011 establishes a new EU labeling requirement aimed at "informing consumers of the presence of engineered nanomaterials in food" (Preamble, 25).

The Food Labeling Regulation sets down specific labeling requirements for all food ingredients containing nanomaterials. Pursuant to Article 18, ingredients in food that are present in a nanomaterial form are to be listed clearly on the label and followed by the word "nano" in brackets. This is analogous to the approach adopted by the Parliament for cosmetic products. This labeling requirement is additional to the other mandatory particulars set out by the Regulation. Responsibility for labeling requirements and the relevant safety testing of nanomaterials is placed on the "food business operator under whose name or business name the food is marketed, or, if the operator is not established in the Union, the importer into the Union market" (Article 8(1)). In doing so, it builds on the requirements set down in Regulation 178/2002.

Another relevant horizontal legislation in the EU is Directive 2001/95/EC of 3 December 2001 on General Product Safety (in force since 15 January 2004, replacing Directive 92/59/EC), which embodies the main principle that only safe products can be placed on the market. Due to its broad and horizontal scope, the General Product Safety Directive applies to risks that are not covered by other specific EU provisions on products. Thus it also applies to products containing nanomaterials, with the onus of ensuring safety of such products resting with the person who places them on the market. Other notable horizontal regulation relevant to nanotechnology applications is the EU's chemicals regulation REACH (Registration, Evaluation, Authorisation of Chemicals – European Commission Regulation No. 1907/2006, in effect from 1 June 2007), which requires registration of all substances that are produced and/or marketed in the European Union in quantities above 1 tonne/year – as such, in preparations, or in articles. While important provisions of REACH do not apply to substances used in food (registration, evaluation, authorisation), REACH covers the environmental impacts of nanomaterials contained in food-packaging materials.

Within the context of the EU's vertical (sector/product specific) measures regulating certain end uses, the Cosmetics Regulation (European Commission Regulation No. 1223/2009: OJ L 342, of 22.12.2009, p. 59), for example, specifically covers the use of nanomaterials, through the provision of a definition of nanomaterial, and a mechanism for reporting, labeling, and safety evaluation of the products containing nanomaterials (Gergely and Coroyannakis, 2009). Thus, the Cosmetics Regulation provides the first regulatory definition of a nanomaterial. In accordance with Article 2(1)(k), "nanomaterial" means an insoluble or biopersistant and intentionally manufactured material with one or more external dimensions, or an internal structure, on the scale from 1 to 100 nm. Thus the definition covers only insoluble or biopersistent materials (e.g., nanosized metal, metal oxide, carbon materials), and not soluble or nonbiopersistent materials (e.g., liposomes, nanoemulsions). There is also a provision in the Regulation, under Article 2 (3), which provides for a possible future adjustment of the definition if there is an internationally agreed definition, as follows:

> in view of the various definitions of nanomaterials published by different bodies and the constant technical and scientific developments in the field of nanotechnologies, the Commission shall adjust and adapt point (k) of paragraph 1 to technical and scientific progress and to definitions subsequently agreed at international level.

Discussions are currently underway to revise the definition under Cosmetics Regulation to align it with the European Commission Recommendation 2011/969. As, and when, it is revised, it is likely to provide more clarity on whether or not a material falls under the definition of a nanomaterial. This is especially important for materials that are mainly composed of larger sized particles but also contain a fraction of particles in the nanoscale. This fraction might be insignificant on a mass basis but could easily fall into the > 50% number particle range. As the Regulation requires cosmetic products containing nanomaterials to be notified to the Commission 6 months prior to being placed on the market, and the nanoscale ingredients to be labeled (name of ingredient, followed by "nano" in brackets) it is crucial that the definition provides an unambiguous basis for compliance. If there are concerns over the safety of a nanomaterial, the Commission will refer it to the Scientific Committee on Consumer Safety for opinion (Gergely and Coroyannakis, 2009; Bowman *et al.*, 2010).

Other vertical legislation, such as European Commission Regulation 258/97 relating to Novel Foods and Novel Food Ingredients, establishes a mandatory premarket approval system for all novel foods. Under Article 1 of Regulation 258/97, a novel food is defined as a food or food ingredient not having a significant history of human consumption within the Community prior to May 1997 and which is covered by one of the several defined categories. Categories most relevant to nanotechnologies are contained in Article 1(c)

> Foods and food ingredients with a new or intentionally modified primary molecular structure.

and 1(f)

> Foods and food ingredients to which has been applied a production process not currently used, where that process gives rise to significant changes in the composition or structure of the foods or food ingredients which affect their nutritional value, metabolism or level of undesirable substances.

of Regulation 258/97.

Taking into consideration the current and projected applications of nanotechnologies in food, it is more likely that nanotechnology-derived food products would fall under the second category. In line with the provisions defining novel food, however, the Regulation would apply only if its nutritional value, metabolism, or level of undesirable substances was substantially altered in relation to its macroscale counterpart.

This Regulation went through a revision in order to "streamline the authorisation procedure, develop a more adjusted safety assessment system for traditional food from third countries...and clarify the definition of novel food, including new technologies with an impact on food" (European Commission, 2008). The draft new Regulation had an elaborated definition of a nanomaterial. The definition in Article 3(2)(c) of the draft Regulation on novel foods as adopted by the Council in its Common Position at first reading provided that:

> intentionally produced material that has one or more dimensions of the order of 100 nm or less or that is composed of discrete functional parts, either internally or at the surface, many of which have one or more dimensions of the order of 100 nm or less, including structures, agglomerates or aggregates, which may have a size above the order of 100 nm but retain properties that are characteristic of the nanoscale. Properties that are characteristic of the nanoscale include:
>
> i. those related to the large specific surface area of the materials considered;
>
> And/or
>
> ii. specific physico-chemical properties that are different from those of the non-nanoform of the same material.

However, on 29 March 2011, the conciliation on novel foods failed after more than 3 years of negotiations. This failure was not related to the debate concerning nanomaterials, or the definition of a nanomaterial when used in food as introduced by the draft Novel Food Regulation, as demonstrated by the fact that the above definition of nanomaterial was finally adopted in the Food Information Regulation (FIR). In July 2011, during the plenary vote of the EU Parliament in the second reading of FIR an agreement was reached, which as well as the proposed definition also covered labeling requirements for nanomaterials in food.

Another vertical regulation relates to food additives. In 2010, European Commission Regulation No. 1333/2008 on food additives entered into force in the EU, which, in conjunction with the Regulations on food enzymes (European Commission Regulation No. 1334/2008) and on food flavorings (European Commission Regulation No. 1334/2008), has established a common authorization system. Importantly, Article 12 of the new food additive Regulation 1333/2008 provides for the periodic reevaluation of additives that are produced by a process, or from starting substances, which significantly differ from the ones approved. Specifically:

> When a food additive is already included in a Community list and there is a significant change in its production methods or in the starting materials used, or there is a change in particle size, for example through nanotechnology, the food additive prepared by those new methods or materials shall be considered as a different additive and a new entry in the Community lists or a change in the specifications shall be required before it can be placed on the market (Article 12).

A new risk assessment would therefore be required not only for all new additives, but also for already authorized additives if there has been a significant change as outlined in Article 12. A "significant difference" could mean, *inter alia*, "a change of the production method from extraction from a plant to production by fermentation using a micro-organism or a genetic modification of the original microorganism, a change in starting materials, or a change in particle size, including the use of nanotechnology" (Preamble 12 of European Commission Regulation No. 1332/2008).

This provision would therefore "capture" a new, or an already approved, additive if it was used in the nanoscale. The principle of continuous observation and reevaluation is also fostered by Article 26, which obliges producers or users of a food additive to inform the Commission of any new scientific or technical information that may affect the safety assessment of the food additive. In this regard, the EFSA panel on food additives and nutrient sources added to food has in its draft "guidance for submission for food additive evaluations" addressed the data and information requirements for nanomaterials. (For the purposes of this chapter, other relevant legislative instruments include the *Dietary Supplement Health and Education Act of 1994*, the *Food Additive Amendment Act of 1958*, and the Code of Federal Regulations.)

It is also of note that, in general, safety aspects of feed are treated on similar lines to that of food in the EU. The EFSA Guidance on the potential risks of nanotechnologies in food/feed (European Food and Safety Authority, 2009), and on risk assessment of nanomaterials in food/feed applications (European Food and Safety Authority, 2011) have both highlighted that the guidance applies equally to food and feed applications of nanomaterials. The Guidance (European Food and Safety Authority, 2011) notes "[t]his ENM [engineered nanomaterial] Guidance is aimed at all interested parties, e.g. applicants and risk assessors. For the purpose of this ENM Guidance, ENM in feed will in general be treated in a similar way as those in food, since the impact on animals is likely to be similar to that on humans".

In this regard, the EFSA panel on additives and products of substances used in animal feed (FEEDAP), that assesses the safety and efficacy of feed additives, has included the concept of "size" of substances in its technical guidance documents for the preparation of dossiers for feed additives. Thus any application for a new feed additive will need to provide a chapter on particle size.

Specific Health Claims legislation in the EU (European Commission Regulation No. 1924/2006) requires information on the role of the nutrient or an accepted diet–health relationship, as well as composition of the product. Essentially, this will require health-food manufacturers claiming health benefit of a nanoproduct to back up their claims with supporting scientific data.

The regulatory framework relating to food-contact materials (FCM) in the EU is European Commission Regulation No. 1935/2004 on materials and articles intended to come into contact with food (the "Framework Regulation"). Article 3 of the Framework Regulation requires that all FCMs be manufactured in accordance with good manufacturing practice to avoid under normal or foreseeable conditions of use a transfer of their constituents to food in quantities that could endanger human health, bring about an unacceptable change in the composition of food, or a deterioration of the food's organoleptic characteristics. Thus, substances not covered by other specific obligatory requirements may be used in FCMs in compliance with applicable EU legislation, provided they are safe for intended use and that the substances and finished products in which they are used comply with the Framework

Regulation and national legislation of the Member States in which they are marketed, subject to the principle of mutual recognition. Demonstrating that a substance used in the manufacture of FCMs complies with the general safety requirements of Article 3 of the Regulation would need establishing that it is covered by an applicable listing and that it is used in accordance with the potential restrictions in that listing. In other words, all FCMs have to be safe for the intended application in all conditions. The same requirements apply to nanomaterials when used in food-contact applications. Thus, even if clearance for the bulk form material exists, the intended use of the nanoform as a FCM has to be demonstrated as safe.

As from May 1, 2011, under Regulation (EU) No. 10/2011 on measures for plastic materials and articles intended for food-contact applications, nanomaterials can be used only if specifically listed in Annex I of the Regulation. Article 9(2) of the Regulation states that "[s]ubstances in nanoform shall only be used if explicitly authorised and mentioned in the specifications in Annex I". Similarly, Article 6 of European Commission Regulation No. 450/2009 on active and intelligent materials and articles intended to come into contact with food identifies "substances deliberately engineered to particle size which exhibit functional physical and chemical properties that significantly differ from those at a larger scale" and restrict their application unless specifically authorized.

Only a few nanomaterials are currently authorized for use in food-contact materials in the EU. These include silicon dioxide and carbon black, based on the listing of the originally petitioned equivalents, which turned out to be meeting the definition of substances in nanoforms and titanium nitride. The latter was petitioned as a specific nanomaterial and is authorized subject to the restrictions listed in Annex I.

The legislation relating specifically to quality of potable (or drinking) water is also likely to capture the use of nanomaterials in water treatment. In this regard, relevant guidelines are provided by the World Health Organization (WHO) guidelines for drinking-water quality, and WHO/FAO/UNEP guidelines for the safe use of wastewater for agriculture. These are further supplemented by WHO guidelines on groundwater recharge with reclaimed (waste, desalinated, etc.) water.

22.2.2 United States

The US legislative instrument of most relevance to this chapter is the FDC Act, administered by the FDA. The FDA also has jurisdiction over other product categories, including drugs, devices, and cosmetic products. In relation to foods, the FDA – through its Center for Food Safety and Applied Nutrition – is required by law to ensure that foods that fall within the regulatory scope of the FDC Act are safe. Pursuant to the definition of a food as set out in s.201(f) of the FDC Act. A food is defined to mean: (1) articles used for food or drink for man or other animals, (2) chewing gum, and (3) articles used for components of any such article. This includes whole foods, food additives, dietary supplements. A dietary supplement is defined under s.201(ff) of the FDC Act to include: "(1) means a product (other than tobacco) intended to supplement the diet that bears or contains one or more of the following dietary ingredients: (A) a vitamin; (B) a mineral; (C) a herb or other botanical; (D) an amino acid; (E) a dietary substance for use by man to supplement the diet by increasing the total dietary intake; or (F) a concentrate, metabolite, constituent, extract, or combination of any ingredient described in clause (A), (B), (C), (D), or (E);...", "generally recognised as safe" (GRAS) food ingredients, and FCMs. This is achieved through pre-market authorization requirements for food and color additives, with the safety of dietary supplements and GRAS food ingredients regulated through post-market oversight mechanisms. Unsurprisingly, the FDC Act does not specifically refer either to the use of nanotechnologies as a process or the inclusion of nanoscale ingredients in foods or FCM. However, as noted by the FDA Taskforce itself, the general nature of the FDC Act and most of the subsidiary Act enforced by the FDA (2007, p. 4), "are general in nature by design... offering flexibility to accommodate products made with new technologies or containing new kinds of materials".

The adequacy of the FDA's existing regulatory framework with regards to nanotechnologies across each of the product categories it regulates, including foods, has been assessed by Taylor (2006, 2008) and the FDA Taskforce (2007). Taylor's (2006, p. 51) analysis of the FDA's framework and resourcing led him to conclude that "while the FDA has most of the legal tools it needs to regulate most of the products of nanotechnology, significant gaps in authority remain." In relation to food this included, for example, a lack of power to acquire information pertaining to new nano-foods". In their opinion, the FDA (2007, p. iii) believed that for product categories subject to pre-market approval, including food and color additives,

> the agency's authorities are generally comprehensive; the adequacy of the framework was not, in their view, altered by the use of nanotechnologies in processing or the presence of nanoscale ingredients. However, for products not subject to pre-marketing authorization processes, including whole foods, dietary supplements, and GRAS food ingredients, the Agency noted that its oversight capacity [was] less comprehensive.

In recognition of the current lack of knowledge and regulatory challenges potentially posed by some categories and their incorporation of nanotechnologies, the FDA Taskforce made a number of recommendations pertaining to data requirements and regulation perspectives. Given the breadth of these two reports, this chapter does not provide an in-depth analysis of the framework. (The adequacy, or in their view, the inadequacies of the FDA's regulatory framework and nanofoods has also been highlighted by Friends of the Earth Australia. See Miller and Senjen (2008, pp. 21–24).)

The regulatory status of FCMs under the FDC Act is codified in Title 21 (Food and drugs), Chapter 9. It is based on the general safety requirements for food, as described above. Based on these requirements, as implemented notably by the pre-market approval process, FCMs are also placed under the responsibility of the FDA. More specifically, the FDC Act prohibits the adulteration of food, and deems all substances that are intended, or may reasonably be expected, directly or indirectly, to become components of food or affect its characteristics, including FCMs, to be food additives. Food additives are automatically deemed to be unsafe and to cause food to be adulterated, unless the intended use of these substances is permitted under FDA's food additive regulations, found in Title 21 of the Code of Federal Regulations (21 CFR). Various exemptions from FDA's pre-market approval requirement are included in the FDC Act as implemented by FDA's regulations and case law, so that components that are not specifically listed as permitted for food-contact applications in 21 CFR may still be used in compliance with the FDC Act and FDA's food additive regulations within the scope of these exemptions.

The use of any substance in a food-contact application remains first and foremost subject to the general safety requirement of the FDC Act, reinforced by FDA's GMP regulation, 21 CFR,§ 174.5(a)(2), which requires any food contact substance to be of a "purity suitable for its intended use," and demands that the potential level of migration into food of any component of a FCM to be safe. Safety is defined in the food additive regulations as "reasonable certainty in the minds of competent scientists that the substance is not harmful under the intended conditions of use" (see 21 CFR, § 170.3(i)). We note that while section 409(c)(3)(A) of the FDC Act establishes that no food additive shall be deemed by FDA to be "safe" if the additive is found "to induce cancer when ingested by man or animal, or if it is found, after tests which are appropriate for the evaluation of safety of [the additive], to induce cancer in man or animal," this general prohibition on the use of carcinogenic additives only applies to the "substance that is actually intended for use in food or for food contact," while all "non-functional chemicals present in that substance would be called the 'constituents' of the additive" (see 21 CFR, § 170.3(i)). Such constituents include impurities, residual reactants, intermediates, manufacturing aids, and the products of side reactions and chemical degradation. Based on this distinction between the "food additive" and its "constituents," FDA may permit the use of an additive that is noncarcinogenic if it contains "safe" levels of carcinogenic constituents.

Against this backdrop, in April 2012, the FDA released draft guidance material for industry in relation to the use of nanotechnologies in food ingredients and food-contact substances. The focus of the draft food guidance is on the manufacturing processes. It does not articulate regulatory definition for the term nanotechnology or nanomaterials. The intent of the document is to "describe the factors [industry] should consider when determining whether a significant change in manufacturing process for a food substance already in the market" (FDA, 2012, p. 4). Through the draft food guidance, the FDA addressed their concern about changes in food manufacturing processes that involve nanotechnologies. The FDA notes, for example, that

> although traditionally manufactured food substances can sometimes include particles with size distributions that extend in to the nanometer range, recent advances in technology can allow the intentional production of food substances with particle size distributions more fully in the nanometer range, resulting in new properties not seen in traditionally manufactured food substances (2012, p. 13).

Accordingly, the FDA sets out its position of denying nanoengineered chemicals the status of GRAS; they shall, instead, be reviewed as food additives. A caveat is, however, provided by the FDA in regards to this decision. Their focus is not nanotechnologies *per se*, but rather manufacturing changes that may affect the safety of the final product. In their own words,

> FDA does not categorically judge all products containing nanomaterials or otherwise involving application of nanotechnology as intrinsically benign or harmful. Rather for nanotechnology-derived and conventionally manufactured food products alike, FDA considers the characteristics of the finished product and the safety of its intended use (FDA, 2012, p. 13).

A similarly important component of the draft guidance is the distinction that the FDA makes between food ingredients containing nanomaterials and identical ingredients when used at a larger particle size. The draft food guidance states that, "safety studies conducted with a food substance manufactured as relatively large particles may have little relevance to the safety of that food substance when manufactured in a significantly smaller particle size" (FDA, 2012, p. 12). This arguably suggests a movement by the Agency towards stricter testing standards of nanomaterials prior to market entry. The guidance is, however, nonbinding and as of March 2013, had yet to be finalized.

22.3 CONCLUSIONS

As this chapter has sought to illustrate, the scientific uncertainties over nanomaterial properties and effects have also had an impact on the regulatory and broader policy developments. In general, various regulatory assessment studies have indicated that most regulations already in place are adequate to cover potential risks of nanomaterials. More recently, regulatory guidance on potential risks and risk assessment issues has become available both in the EU and the United States. First examples of regulatory opinions on risk assessment of cosmetic ingredients are also becoming available in EU, where discussions are ongoing over shaping up the relevant definitions and thresholds. Other regulatory tools, such as the notification portal for nanosized cosmetic ingredients, and labeling requirements in the EU, are all adding up to bring a clearer regulatory landscape for the use of nanomaterials in consumer products.

These developments have occurred over a relatively short period of time. As recently as 2008 it appeared that the legislative *status quo* was generally likely to be maintained by governments until the state of the scientific art had evolved significantly. However, it is clear from the discussion presented in this chapter that significant activities have been undertaken in the past 5 years in relation to nomenclature, risk assessment, and evidence-based regulation of nanotechnologies. We are, however, still witnessing an increasing diversity in how jurisdictions will approach the regulatory challenges.

The European Parliament has shown its willingness to only pass relevant legislative instruments that include nanospecific provisions. The Cosmetic Regulation and the Food Labeling Regulation appear to be the first of many such relevant frameworks that will incorporate a definition of a nanomaterial for regulatory purposes, and additional regulatory requirements – including labeling – for end-products containing nanomaterials. In contrast, the current US Administration has shown unwillingness to adopt such an approach, focusing instead on the development and use of guidance material for industry. The regulatory disparities between the two jurisdictions appear only to be enhanced over the short to medium term. Importantly, a lack of harmonization between the two trading partners is not unique to nanotechnologies, but a broader reflection of the scientific, cultural, and philosophical values towards regulation.

Looking more broadly, the development and publication of regulatory and scientific opinions, legislation, and guidance material by expert bodies within the EU and the US may assist other jurisdictions with ensuring the robustness of their regulatory regimes. For countries that lack the resources, infrastructure, and/or expertize to effectively grapple these challenges at this time, there is a need to ensure that only safe products and applications of nanotechnologies are permitted on the market. Such documents and instruments should provide decision makers in these countries with cohesive guidance and help ensure that stringent regulatory controls remain effective.

References

Balbus, J.M., Maynard, A.D., Colvin, V.L., *et al.* (2007) Hazard assessment for nanoparticles–report from an interdisciplinary workshop. *Environmental Health Perspective*, 115, 1654–1659.

Bowman, D.M., Van Calster, G. and Friedrichs, S. (2010) Nanomaterials and the regulation of cosmetics. *Nature Nanotechnology*, 5(2), 92.

Bowman, D.M. and Ludlow, K. (2013) Assessing the impact of a 'for government' review on the nanotechnology regulatory landscape. *Monash University Law Review*, 38(3), 168–212.

Chaudhry, Q. and Castle, L. (2011) Food applications of nanotechnologies – An overview of opportunities and challenges for developing countries. *Trends in Food Science and Technology*, 22, 595–603.

Chaudhry, Q. and Groves, K. (2010) Nanotechnology Applications for Food Ingredients, Additives and Supplements. In: *Nanotechnologies in Food* (ed. Q. Chaudhry, L. Castle and R. Watkins). Royal Society of Chemistry, London; 69–85.

Chaudhry, Q., Blackburn, J., Floyd, P., *et al.* (2006) *Final Report: A Scoping Study to Identify Gaps in Environmental Regulation for the Products and Applications of Nanotechnologies*. Department for Environment, Food and Rural Affairs, London.

Chaudhry, Q., Scotter, M., Blackburn, J., *et al.* (2008) Applications and implications of nanotechnologies for the food sector. *Food Additives and Contaminants*, 25(3), 241–258.

Chaudhry, Q., Bouwmeester, H. and Hertel, R.F. (2010a) The current risk assessment paradigm in relation to regulation of nanotechnologies. In: *International Handbook on Regulating Nanotechnologies* (ed. G.A. Hodge, D.M. Bowman and A.D. Maynard). Edward Elgar, Cheltenham; 124–143.

Chaudhry, Q., Castle, L. and Watkins R. (eds). (2010b) *Nanotechnologies in Food*. Royal Society of Chemistry, London.

Cientifica (2006) *Nanotechnologies in the Food Industry*. [Online] Available: www.cientifica.com/www/details.php?id=47 (accessed 9 January 2015).

Clegg, S.M., Knight, A.I., Beeren, C.J.M. and Wilde, P.J. (2009) Fat reduction whilst maintaining the sensory characteristics of fat using multiple emulsions. *Fifth International Symposium on Food Rheology and Structure*, Zurich, June15–18; 238–241.

Davies, J.C. (2006) *Managing the Effects of Nanotechnology*. Project on Emerging Nanotechnologies, Washington, DC.

Des Rieux, A., Fievez, V., Garinot, M., Schneider, Y.-J. and Preat V. (2006) Nanoparticles potential oral delivery systems of proteins vaccines: A mechanistic approach. *Journal of Controlled Release*, 116, 1–27.

Environmental Risk Management Authority. (2012) *Cosmetic Products Group Standard (Amended July 2012)*. Environmental Protection Authority, New Zealand Government, Wellingon.

European Commission (2008) *Proposal for a Regulation of the European Parliament and of the Council on novel foods and amending Regulation (EC) No. XXX/XXX [common procedure]*. COM(2007) 872 Final, 14 January, European Commission, Brussels.

European Commission. (2010) *Proposal for a Definition of the Term "Nanomaterial" that the European Commission Intends to Use as an Overarching, Broadly Applicable Reference Term for any European Union Communication or Legislation Addressing Nanomaterials*. Directorate-General Environment, Brussels.

European Commission. (2012a) *Second Regulatory Review on Nanomaterials*. COM(2012) 572 Final, 3 October. European Commission, Brussels.

European Commission. (2012b) *Commission Staff Working Paper – Types and Uses of Nanomaterials, Including Safety Aspects*. SWD(2012) 288 Final, 3 October. European Commission, Brussels.

European Food and Safety Authority. (2009) *The Potential Risks Arising from Nanoscience and Nanotechnologies on Food and Feed Safety – Scientific Opinion of the Scientific Committee (Question No EFSA-Q-2007-124a)*. EFSA, Brussels.

European Food and Safety Authority. (2011) Guidance on the risk assessment of the application of nanoscience and nanotechnologies in the food and feed chain. *EFSA Journal*, 9(5), 2140.

Florence, AT. (2005) Nanoparticle update by the oral route: fulfilling its potential? *Drug Discovery Today: Technologies*, 2, 75–81.

FDA. (2007) *Nanotechnology – A Report of the U.S. Food and Drug Administration Nanotechnology Task Force*. Food and Drug Administration, Washington, DC.

FDA. (2011) *Considering Whether an FDA-Regulated Product Involves the Application of Nanotechnology – Guidance for Industry: Draft Guidance*, June. Food and Drug Administration, Washington, DC. [Online] Available: http://www.fda.gov/regulatoryinformation/guidances/ucm257698.htm (accessed 20 March 2013).

FDA. (2012) *Guidance for Industry: Assessing the Effects of Significant Manufacturing Process Changes, Including Emerging Technologies, on the Safety and Regulatory Status of Food Ingredients and Food Contact Substances, Including Food Ingredients that are Color Additives*. Food and Drug Administration, Washington, DC.

FSA. (2006) *A Review of Potential Implications of Nanotechnologies for Regulations and Risk Assessment in Relation to Food. Draft Report of FSA Regulatory Review*. Food Standards Agency, London.

Gergely, A. and Coroyannakis, L. (2009) Nanotechnology in the EU cosmetics regulation. *Household and Personal Care Today*, 3, 28–30.

Gergely, A., Chaudhry, Q. and Bowman, D.M. (2010) Regulatory perspectives on nanotechnologies in food and food contact materials. In: *International Handbook on Regulating Nanotechnologies*, (ed. G.A. Hodge, D.M. Bowman and A.D. Maynard). Edward Elgar, Cheltenham, 321–341.

Health Canada. (2011) *Policy Statement on Health Canada's Working Definition for Nanomaterial*, 26 May. [Online] Available: http://www.hc-sc.gc.ca/sr-sr/pubs/nano/pol-eng.php (accessed 18 March 2013).

Hillyer, J.F. and Albrecht, R.M. (2001) Gastrointestinal persorption and tissue distribution of differently sized colloidal gold nanoparticles. *Journal of Pharmaceutical Sciences*, 90, 1927–1936.

Hoet, P.H., Bruske-Hohlfeld, I. and Salata, O. (2004) Review: nanoparticles – known and unknown health effects. *Journal of Nanobiotechnology*, 2(12), 1–15.

Hodge, G.A., Bowman, D.M. and Maynard, A.D. (eds) (2010) *International Handbook on Regulating Nanotechnologies*. Edward Elgar, Cheltenham.

Holdren, J.P., Sunstein, C.R. and Siddiqui, I.A. (2011) *Memorandum for the Heads of Executive Departments and Agencies: Policy Principles for the U.S. Decision-Making Concerning Regulation and Oversight of Applications of Nanotechnology and Nanomaterials*, 9 June. Executive Office of the President, Washington, DC.

House of Lords Science and Technology Committee. (2010) *1st Report of Session 2009–10: Nanotechnologies and Food, Volume I*. The Stationery Office Limited, London.

Huang, Q. (ed.). (2012) *Nanotechnology in the Food, Beverage and Nutraceutical Industries*. Woodhead Publishing, Oxford.

Kreyling, W.G., Behnke, M., and Chaudhry, Q. (2010) A complementary definition of nanomaterial. *Nano Today*, 5, 165–168.

Ludlow, K., Bowman D.M. and Hodge G.A. (2007) *A Review of Possible Impacts of Nanotechnology on Australia's Regulatory Framework*. Monash Centre for Regulatory Studies, Melbourne.

Maynard, A.D. (2011) Don't define nanomaterials. *Nature*, 475, 31.

Maynard, A.M., Aitken, R., Butz, T., *et al.* (2006) Safe handling of nanotechnology. *Nature*, 444, 267–269.

Miller, G. and Senjen, R. (2008) *Out of the Laboratory and on to Our Plates: Nanotechnology in Food and Agriculture*. FoEA Australia, Europe and US, Melbourne.

Nanolyse. (2010) *Nanoparticles in Food: Analytical Methods for Detection and Characterization*. [Online] Available: www.nanolyse.eu (accessed 22 March 2013).

NICNAS (2010) *Guidance on New Chemical Requirements for Notification of Industrial Nanomaterials*. National Industrial Chemicals Notification and Assessment Scheme. [Online] Available: http://www.nicnas.gov.au/current_issues/Nanotechnology/Guidance%20on%20New%20Chemical%20Requirements%20for%20Notification%20of%20Industrial%20Nanomaterials.pdf (accessed 18 March 2013).

OECD. (2009) *Preliminary Review of OECD Test Guidelines for their Applicability to Manufactured Nanomaterials*, ENV/CHEM/NANO(2009)6/REV1. Working Party on Manufactured Nanomaterials, Environment Directorate Organisation For Economic Co-operation and Development, Paris.

Powell, J.J., Harvey, R.S., Ashwood, P., Wolstencroft, R., Gershwin, M.E., and Thompson, R.P. (2000) Immune potentiation of ultrafine dietary particles in normal subjects and patients with inflammatory bowel disease. *Journal of Autoimmunity*, 14(1), 99–105.

Powell, J.J., Faria, N., Thomas-McKay, E. and Pele, L.C. (2010) Origin and fate of dietary nanoparticles and microparticles in the gastrointestinal tract. *Journal of Autoimmunity*, 34(3), J226–J233.

Rocks, S., Pollard, S., Dorey, R., Levy, L., Harrison, P. and Handy, R (2008) *Comparison of Risk Assessment Approaches for Manufactured Nanomaterials*. Department for Food, Environment and Rural Affairs, London.

Scientific Committee on Emerging and Newly Identified Health Risks. (2009) *Risk Assessment of Products of Nanotechnologies*, 19 January 2009. Directorate-General for Health and Consumers, Brussels.

Scientific Committee on Consumer Products. (2007) *Opinion on Safety of Nanomaterials in Cosmetic Products*. Health and Consumer Protection Directorate-General, Brussels.

Scientific Committee on Consumer Safety. (2012) *Guidance on the Safety Assessment of Nanomaterials in Cosmetics*, Health and Consumer Protection Directorate-General, Brussels.

Scientific Committee on Emerging and Newly Identified Health Risks. (2010) *Scientific Basis for the Definition of the Term "Nanomaterial,"* Directorate-General for Health and Consumers, Brussels.

Stamm, H. (2011) Risk factors: nanomaterials should be defined. *Nature*, 476, 399.

Taylor, M.R. (2006) *Regulating the Products of Nanotechnology: Does FDA Have the Tools It Needs?* Project on Emerging Nanotechnologies, Washington, DC.

Taylor, M.R. (2008) *Assuring the Safety of Nanomaterials in Food Packaging: The Regulatory Process and Key Issues*. Project on Emerging Nanotechnologies, Washington, DC.

US Department of Agriculture. (2009) *Global Food Markets: Global Food Industry Structure*. [Online] Available: www.ers.usda.gov/Briefing/GlobalFoodMarkets/Industry.htm (accessed 20 July 2011).

Watson, S.B., Gergely, A. and Janus, E.R. (2011) Where is "Agronanotechnology" heading in the United States and European Union? *American Bar Association Natural Resources and Environment*, Summer, 8–12.

Woodrow Wilson International Centre for Scholars. (2011) *The Project on Emerging Nanotechnologies – Consumer Product Inventory*. [Online] Available at: http://www.nanotechproject.org/inventories/consumer/ (accessed 22 March 2013).

Index

Nanotechnology and Functional Foods: Effective Delivery of Bioactive Ingredients, First Edition.
Edited by Cristina M. Sabliov, Hongda Chen, and Rickey Y. Yada.

Food Science and Technology from Wiley Blackwell